Flat Water:
A History of Nebraska and Its Water

Robert D. Kuzelka, Project Manager

Charles A. Flowerday, Editor

Robert N. Manley, Contributing Editor/Consulting Historian

Bradley C. Rundquist, Editorial Associate

Sally J. Herrin and Charles A. Flowerday, Photography Editors

Resource Report No. 12

March 1993

Conservation and Survey Division
Institute of Agriculture and Natural Resources
University of Nebraska-Lincoln

Cover: Whitetail Creek, Keith County, by Anne Burkholder. This painting illustrates the intimate connection between groundwater and surface water as groundwater seeps onto the floodplain and feeds into the creek. Courtesy of Donald F. McGinley.

The Conservation and Survey Division of the University of Nebraska is the agency designated by statute to investigate and interpret the geologically related natural resources of the state, to make available to the public the results of these investigations, and to assist in the development and conservation of these resources.

The division is authorized to enter into agreements with federal agencies to engage in cooperative surveys and investigations in the state. Publications of the division and the cooperating agencies are available from the Conservation and Survey Division, University of Nebraska, Lincoln, Nebraska 68588-0517.

It is the policy of the Conservation and Survey Division, as it is of the University of Nebraska-Lincoln, not to discriminate on the basis of and to provide information and educational programs to all regardless of sex, age, handicap, race, color, religion, marital status, veteran's status, national or ethnic origin or sexual orientation.

Publication and price lists are furnished upon request.

Flat water: a history of Nebraska and its water/ Robert D. Kuzelka, project manager: Charles A. Flowerday, editor: Robert N. Manley, contributing editor/consulting historian: Bradley C. Rundquist, editorial associate.

291 p. 21.55 cm. — (Resource report/Conservation and Survey Division, Institute of Agriculture and Natural Resources, University of Nebraska-Lincoln; no. 12)

Includes bibliographic references and index.

ISBN 1-56161-001-1 (pbk.)

1. Water-supply—Nebraska—History. 2. Hydrology—Nebraska. 3. Water resources development—Nebraska—History. 4. Water supply-Government policy—Nebraska—History. 5. Nebraska—History. I. Kuzelka, Robert D. II. Flowerday, Charles A. III. Series: Resource report (University of Nebraska—Lincoln. Conservation and Survey Division); no. 12.

TD224.N18F58 1993

333.91'15'09782—dc20 92-72544 CIP

Contents

Introduction

by Robert D. Kuzelka

You Know You're a Nebraskan . . .

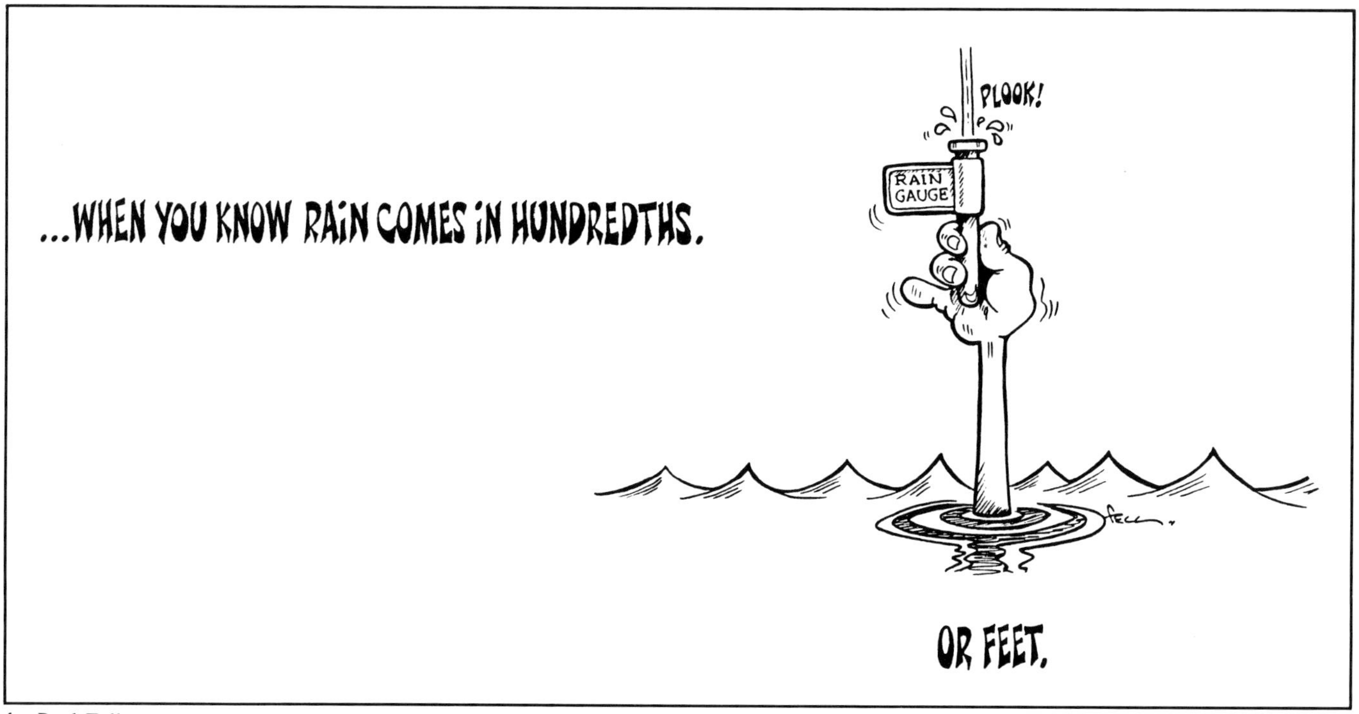

by Paul Fell

Courtesy of Roger Welsch and J and L Lee Company, Lincoln, Nebraska

Nebraskans are preoccupied with water! To them rainfall is always a newsworthy topic, as Paul Fell, political cartoonist and co-author of *You Know You're A Nebraskan . . .*, so accurately has captured in his cartoon about the variations in precipitation in the state. These annual, seasonal and geographical variations of precipitation were well known to both the state's natives and immigrants. Nebraska's rivers and their valleys provided the routes for travel and plains for settlements. Furthermore, the state is named after a derivation of the word for the major trans-state river, the Platte—in the Oto and Omaha languages approximately "Nebraska," loosely translated as "broad, flat water." The French gave the river its present name, meaning "flat."

More recently, the state has laid claim to and put to use its vast supply of underground water for irrigation, drinking and livestock supplies. It is our "Rocky Mountains," as touted by the Nebraska Groundwater Foundation's Sue Seacrest.

This preoccupation comes not only from the presence and scope of Nebraska's water, but from its potential for development. It is this potential, both realized and imagined, that caused this anthology to be planned, written and assembled.

Initially, the book was to be a modest history marking the centennial of the Nebraska State Irrigation Association, which was formed in November of 1893. The concept grew and matured with wise counsel from Nebraska historian Robert Manley, who advised that it include as many aspects and constituencies of the history of the state's water as possible. Finally, it is now presented here as a history of the state and its water, a history that is a collection of people, events and tales, a record of past events and an examination of what the future might bring.

The book opens with views of Nebraska and its water prior to 1900. By that year, some would argue, most the state's water potential and problems had been realized or foreseen. Then the reader has the opportunity to evaluate that argument. Eight chapters explore varied facets of Nebraska's water history between 1900 and 1993. Finally, three perspectives and three "commentaries" are offered on the future of Nebraska and its water.

The entire work is laced with vignettes (sidebars), endpieces (appendices), biographies and curiosities about water-related persons, places, events and things. In addition, it is a pictorial of the state and this empowering natural resource. As a result we offer you a book to be read as a whole or in parts with equal interest and enjoyment.

To help the reader appreciate the character of the book and Nebraska's water, the editors, authors and project management team offer a few maps, two charts and a table to help profile important water-related conditions.

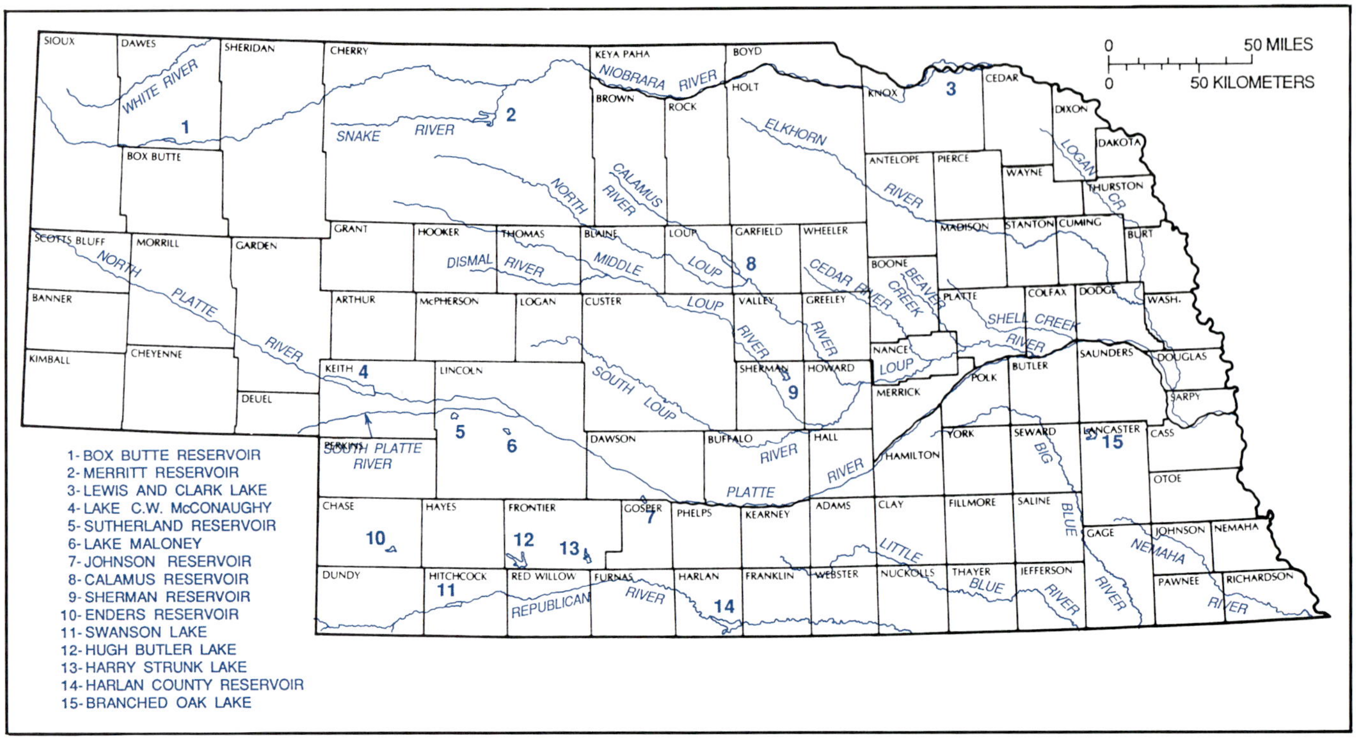

The base map shows the major streams and current major reservoirs. These reservoirs are the result of a tremendous amount of organizing, campaigning, lobbying, wrangling and fighting, most of which are chronicled in this water history.

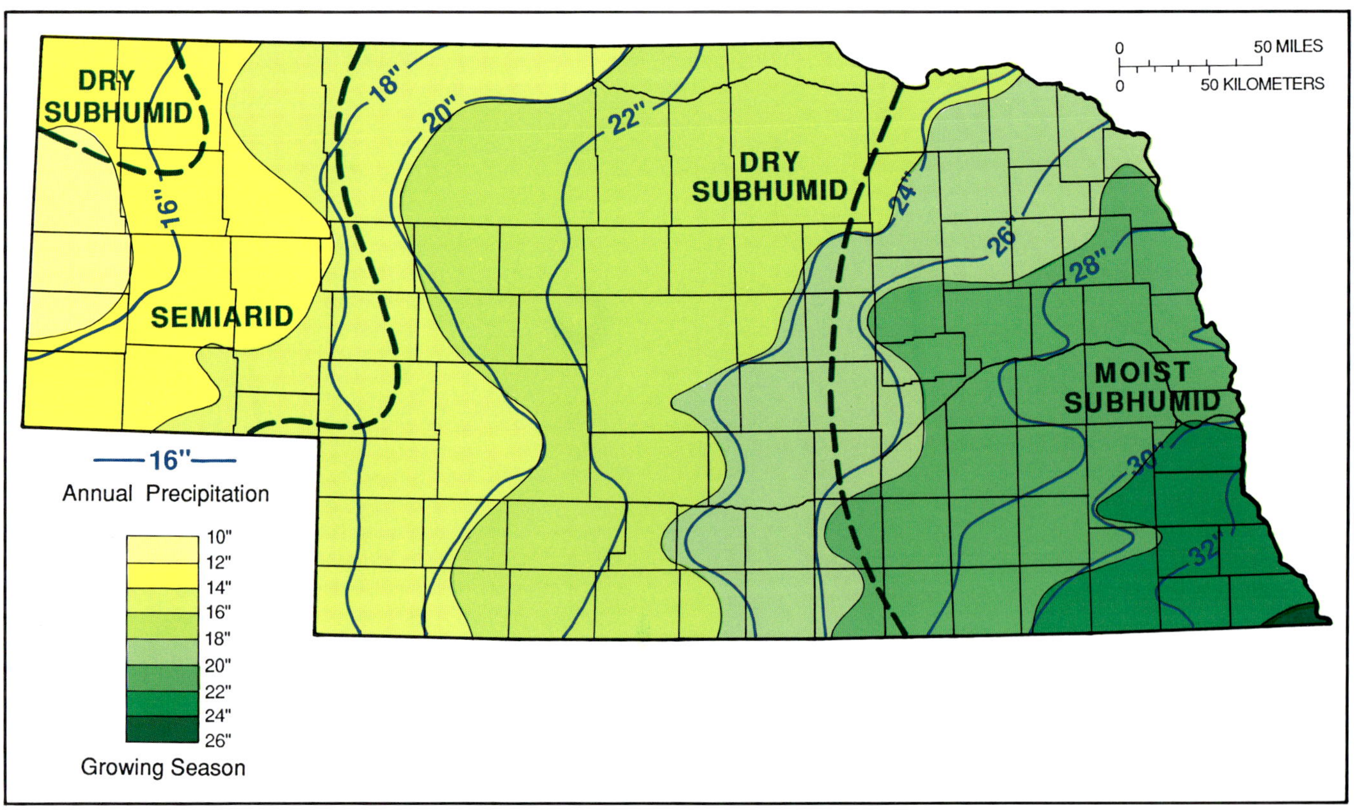

Average annual (normal) and average growing season precipitation

The combination of rainfall distribution patterns and characteristic growing season gave the explorers of Nebraska little cause for rejoicing about the economic future of the state. Early agriculturalists at times found the annual and growing-season precipitation quite discouraging. And often they found the variations from the average rainfall even more discouraging. The maps of normal annual and growing-season precipitation show these conditions as present averages. The chart on the next page of average annual departures from normal precipitation shows them in historical perspective.

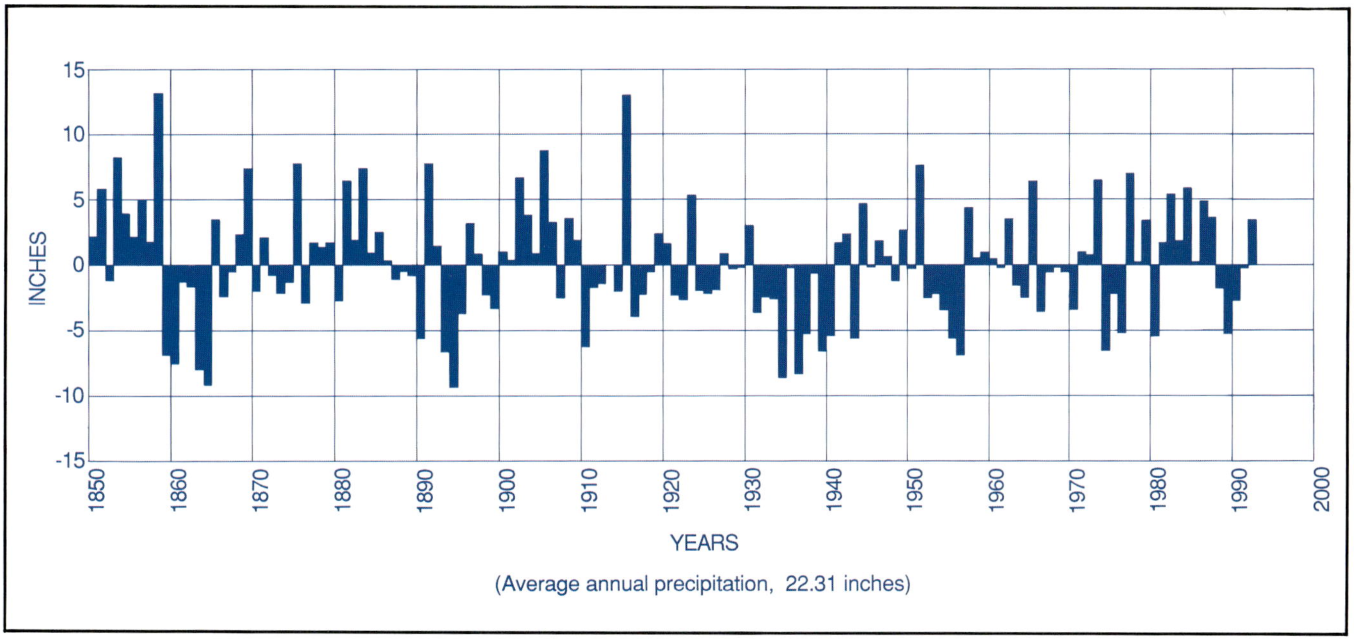

Average annual departures from normal precipitation

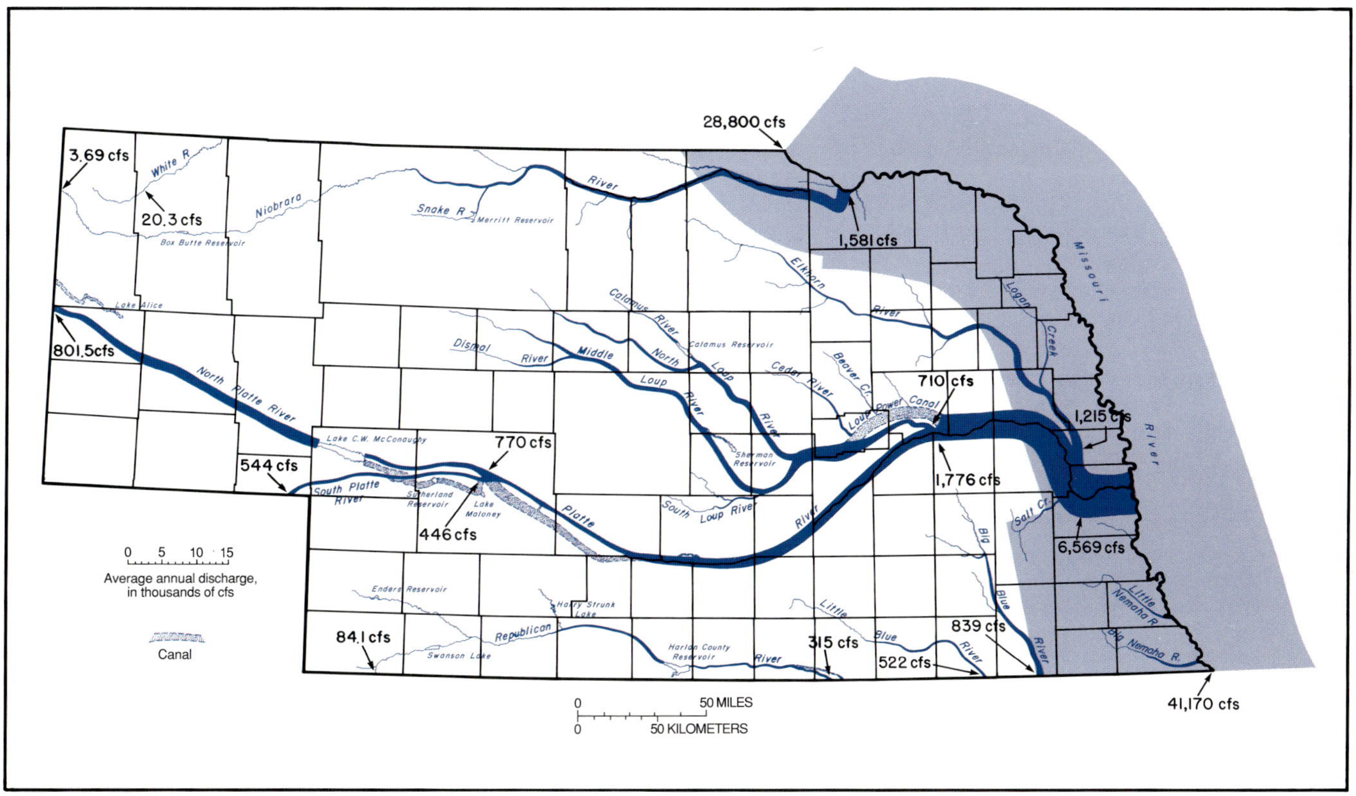

Streamflow of Nebraska's major rivers

On the other hand, the river courses and their annual flow formed an early cause for optimism about potential energy and crop development. Although the map of streamflow shows current flows after considerable river development and management, the flow quantities are of historical significance. Of importance are the two differing characters of Nebraska rivers. The Niobrara, Elkhorn and Loups are fed primarily by groundwater and flow with great annual and seasonal consistency. The remaining rivers depend on runoff from within and outside of the state. They have significant annual and seasonal variations in flow. Finally, it should be noted that the Missouri River primarily forms an eastern shore and discharge channel for the state.

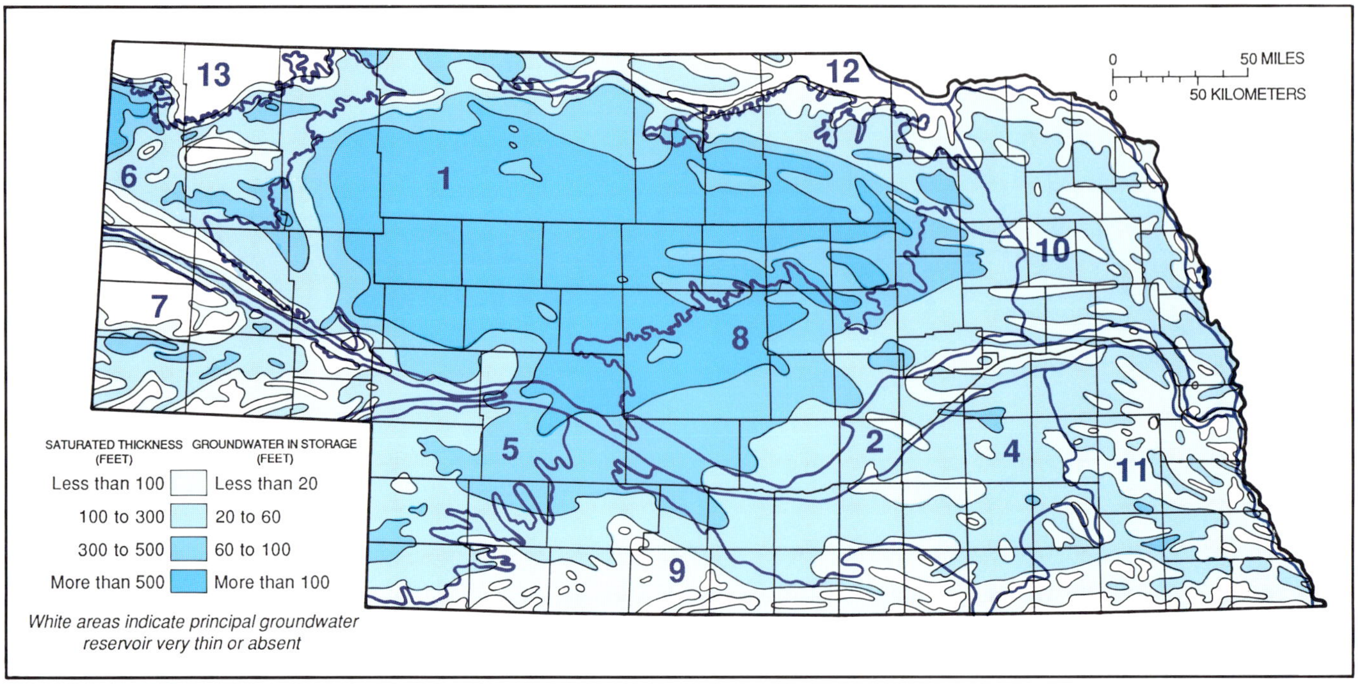

Saturated thickness of the principal groundwater reservoir and groundwater regions

The great Nebraska water resource, once hidden but now highly prized, is underground. A quick calculation of groundwater in storage reveals a quantity that would cover the state to a depth of 34 feet. However, not all of this "gold" is located where it can best be used. This condition is shown in a map that compares groundwater in storage and groundwater regions. But technology has allowed for use of the resource in ways not imagined in the 18th Century, when the first Europeans began to explore the region.

The thirteen groundwater regions described here represent areas where groundwater occurs under similar kinds of conditions (the boundaries represent zones of gradual change). They should help the reader understand the kinds of geology and topography that affect the state's groundwater and access to it. They are:

Region 1—Sand Hills: Large yields of good-quality water can be obtained from aquifers of Tertiary and Quaternary ages, which are largely composed of sand, sand and gravel and/or sandstone. Ogallala Group deposits underlie the entire region but thin out toward the east. Runoff rarely occurs because precipitation readily infiltrates the sandy soils. Most groundwater either is lost to evaporation from lakes and to transpiration from subirrigated meadows or is discharged by seepage into streams known for their uniform flow.

Region 2—Platte River Valley: High yields of good-quality water can be obtained from sand and gravel of Quaternary age in most of this region. The Ogallala Group underlies alluvial (river-deposited) sediment in parts of the North Platte River and South Platte River valleys and the Platte River valley as far east as Grand Island; it is also a source of water supply. Concentrations of dissolved solids in groundwater range from 500 to 1,000 milligrams per liter in the alluvium throughout most of the western two-thirds of the region. The highest concentrations occur in the South Platte River valley. The water quality there is affected primarily by dissolved solids carried from soils by infiltrating precipitation and irrigation water. In some of the Platte River valley, fertilizer applications are affecting groundwater quality.

Region 3—Missouri River Lowlands: Large-yield wells can be developed in sand and gravel of Quaternary age. Groundwater is more mineralized than in the Platte River valley but is still of usable quality.

Region 4—South Central Plains: Abundant groundwater can be found in this region. Sand and gravel of Pliocene and Pleistocene age and the Ogallala Group of Miocene age in the west yield large amounts of good-quality water. Water-quality problems are developing in specific areas. Groundwater levels have risen in the west where water diverted from the Platte River is used for irrigation.

Region 5—Southwestern Tablelands: Sandstone and sand and gravel deposits of the Ogallala Group have been developed extensively for irrigation. Pumping lifts of 200 feet or more are common. Groundwater levels have declined progressively since development began.

Regions 6 and 7—Panhandle Tablelands: Although thick sequences of moderately permeable sediments of Tertiary age underlie the tablelands, wells are usually several hundred feet deep, and those of large capacity are restricted generally to areas underlain by the Ogallala or Arikaree groups. Rocks beside major drainages usually are not saturated. In Region 7, wells also derive water from Brule Formation fractures in the valleys of Pumpkin and Lodgepole creeks and in Sidney Draw.

Region 8—East Central Dissected Plains: Ogallala Group deposits underlie the western two-thirds of the region, becoming thinner toward the east. In the central and eastern parts of the region, Pliocene and Pleistocene deposits of sand and gravel, mantled by loess (wind-blown silt), are thickest within buried ancient valleys. Water quality is generally good.

Region 9—Republican River Valley and Dissected Plains: Sand and gravel underlie the floodplains and terraces of the Republican River and its principal tributaries where they have cut into the uplands. Ogallala Group deposits are thickest in the northern and western parts of the region. Sand and gravel of Pliocene and Pleistocene age fill some buried ancient valleys. Cretaceous deposits of shale and chalk underlie the entire region and either are exposed or are thinly mantled by loess on the south side of the Republican River, where good-quality water is hard to obtain.

Regions 10 and 11—Nebraska Glacial Drift (Till): Glacial till of low permeability mantles most of the region. Pleistocene deposits of sand and gravel in buried ancient valleys and along stream valleys are one source of groundwater, but they are of limited extent. Sandstones of the Dakota Group underlie much of the region, except for the southeastern part of Region 11, where older shales and limestones are the subcropping bedrock. In some locations, groundwater in the Dakota Group is very salty. The Dakota Group sediments are extremely variable, and prediction of yields from this group is difficult.

Region 12—North Central Tableland: The floodplains and terraces of the Niobrara River and its tributaries overlie sand and gravel deposits that generally yield small to moderate amounts of water. The eastwardly thinning Ogallala Group locally yields moderate amounts of water. Where the Pleistocene and Ogallala Group deposits have been eroded away, groundwater is available only by drilling several hundred feet to sandstones of the Dakota Group.

Region 13—Hat Creek-White River Drainage Basin: Because fine-grained sediments of Tertiary and Cretaceous ages occur at the surface and underlie the region, groundwater is hard to obtain here. Small yields are obtained from deposits underlying the river valleys and from sand lenses or fracture zones where the Brule Formation is saturated.

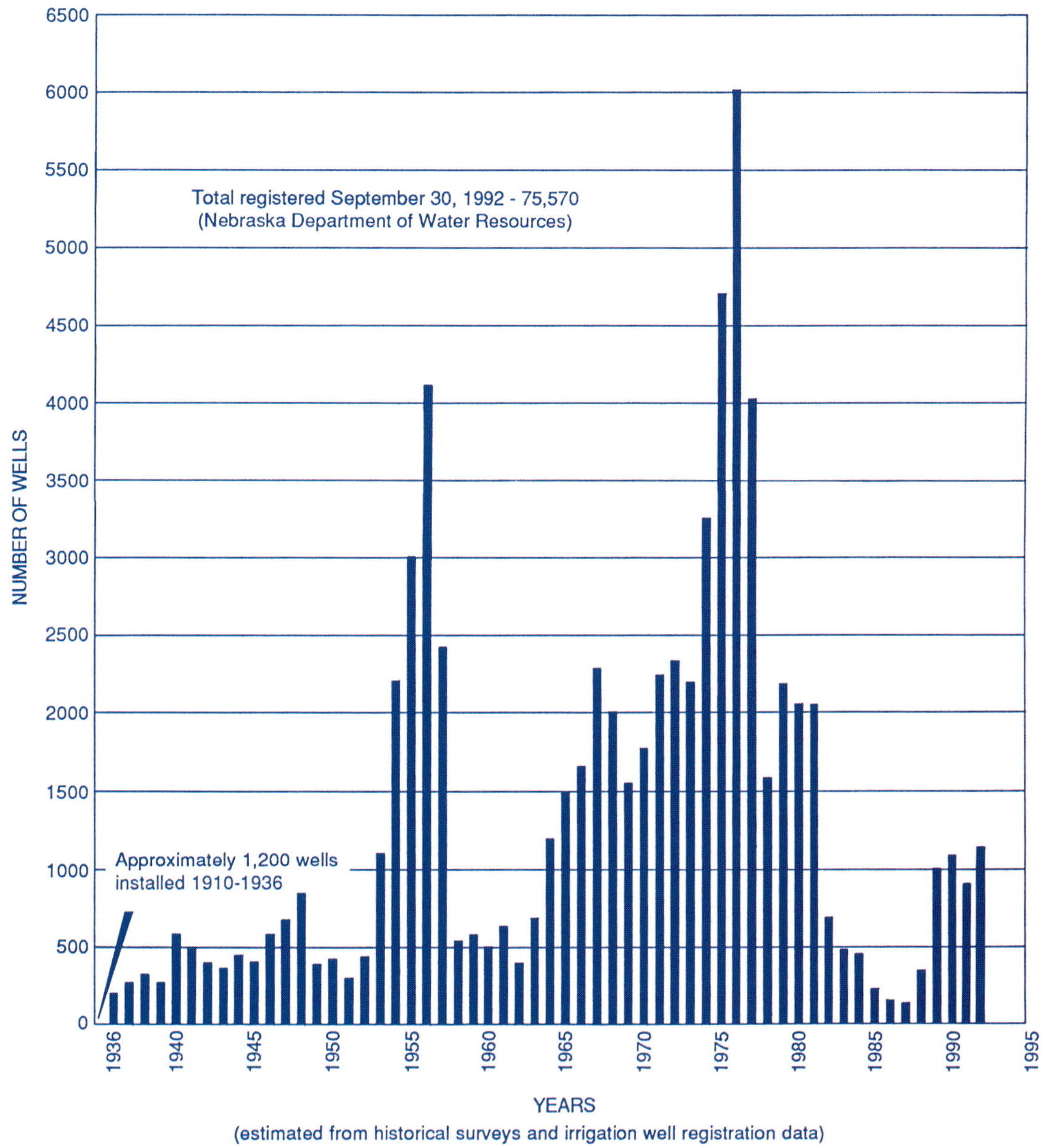

Annual installation of registered irrigation wells

The chart showing annual installation of registered irrigation wells indicates the pattern of well development that followed the technological advances allowing for it and the weather patterns driving it. Particularly prominent are the "spikes" of well installations following the droughts of the 1950s and the 1970s.

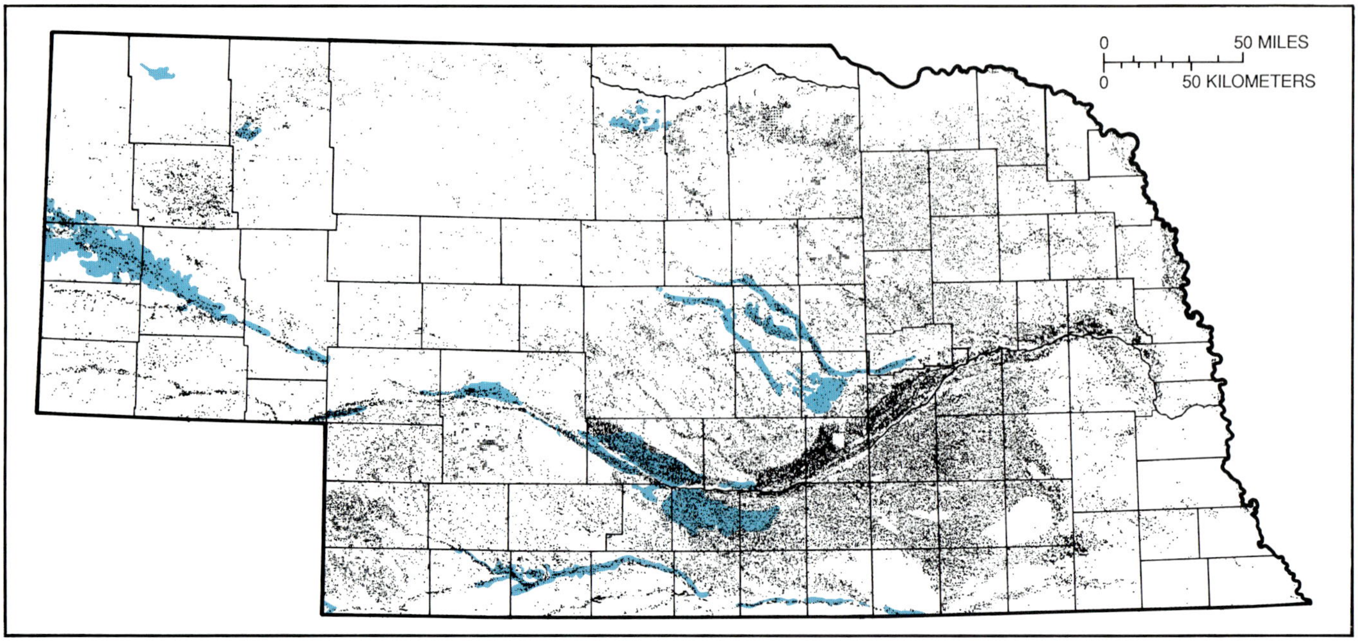

Surface-water development and registered irrigation wells

The map of surface-water development (in blue) and registered irrigation wells (in black) shows the extent of the two main means of irrigating. These have placed the state second nationally in irrigated acres.

Era	Period	Epoch	Ages in millions of years	Group or Formation	Lithology	Water-Bearing Properties
Cenozoic	Quaternary	Holocene	0.01		Sand, silt, gravel and clay	Principal groundwater reservoir, as defined in text; Ogallala is absent in east and northwest. Arikaree is present primarily in west.
		Pleistocene	~2.0			
	Tertiary	Pliocene	5		Sand, gravel and silt	
		Miocene	24	Ogallala	Sand, sandstone, siltstone and some gravel	
		Oligocene		Arikaree	Sandstone and siltstone	
			37	White River	Siltstone, sandstone and clay in lower part	Secondary aquifer in west; water may be highly mineralized.
		Eocene	58	Rocks of this age are not identified in Nebraska.		
		Paleocene	67			
Mesozoic	Cretaceous	Late Cretaceous		Lance	Sandstone and siltstone	Generally not an aquifer; yields water to few wells in west.
				Fox Hills		
				Pierre	Shale, some sandstone in west	Generally not an aquifer; sandstones in west yield highly mineralized water to few industrial wells.
				Niobrara	Shaly chalk and limestone	Secondary aquifer where fractured and at shallow depths, primarily in east
				Carlile	Shale; in some areas, contains sandstones in upper part	Generally not an aquifer; sandstones yield water to few wells in northeast.
			98	Greenhorn-Graneros	Limestone and shale	Generally not an aquifer; yields water to few wells in east.
		Early Cretaceous	144	Dakota	Sandstone and shale	Secondary aquifer, primarily in east; water may be highly mineralized.
	Jurassic		208		Siltstone, some sandstone	Not an aquifer
	Triassic		245		Siltstone	Not an aquifer
Paleozoic	Permian		286		Limestones, dolomites, shales and sandstones	Some sandstone, limestone, and dolomites are secondary aquifers in east. Water may be highly mineralized.
	Pennsylvanian		320			
	Mississippian		360			
	Devonian		408			
	Silurian		438			
	Ordovician		505			
	Cambrian		570			
Precambrian						

Table of water-bearing rock units (hydrostratigraphic table)

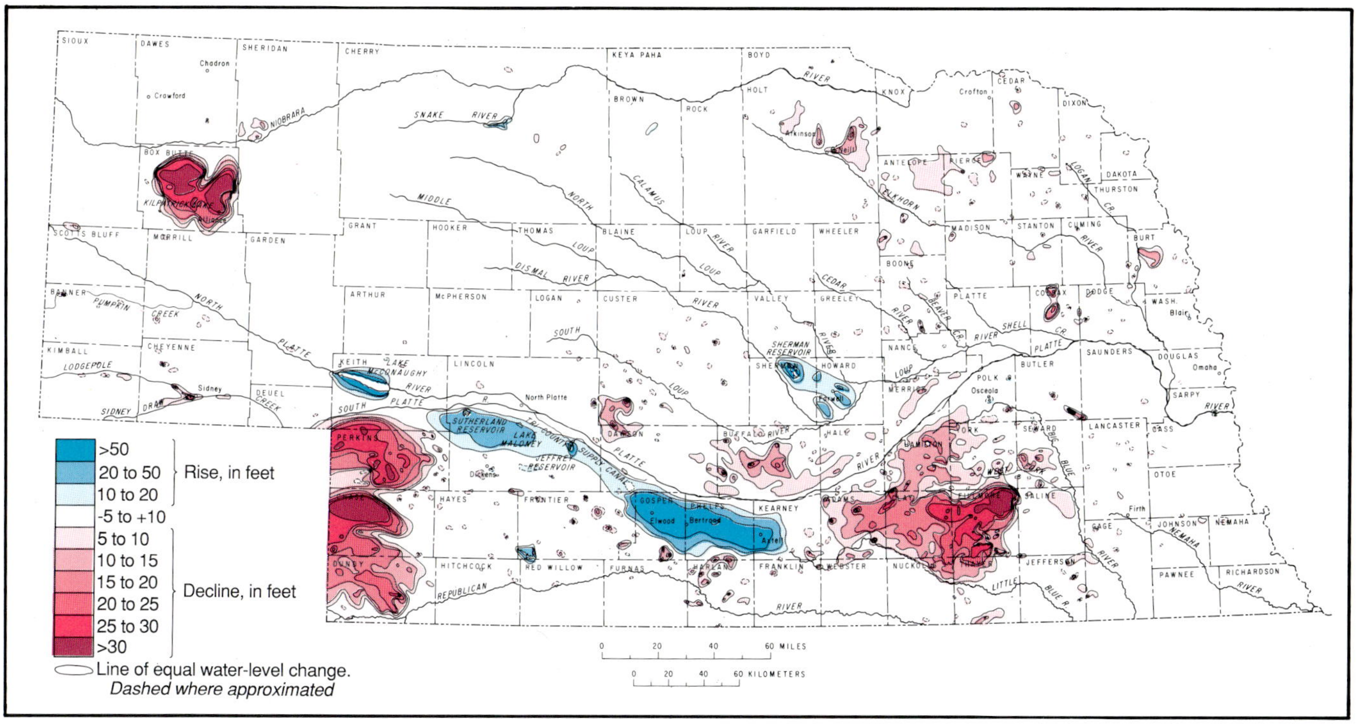

Significant rises and declines in groundwater levels, fall of 1991

Finally, the table showing the units of water-bearing rocks (hydrostratigraphic table) on the previous page describes the properties of the state's rocks as aquifers and the map of groundwater rises and declines above shows what groundwater levels have been altered since the development of the state's groundwater.

The constant battle in the history of the development of the state's water resources was inclusion versus exclusion. This was also the battle in the development of the book. No single volume could possibly identify and credit the many prominent and not-so-prominent people who have contributed to Nebraska's water development, conservation and preservation. This work admittedly falls far short of that impossible task.

We do, however, offer here a list of names, without mention of exact contributions, of those who could have been included or whose contributions could have been given more space. They are: Carl Curtis, former U.S. Congressional Representative and Senator; Hugh A. Butler, former U.S. Senator; Kenneth Wherry, former state legislator and U.S. Senator; Roman L. Hruska, former U.S. Representative and Senator; Edward Zorinsky, former U.S. Senator; J. J. Exon, former Governor and U.S. Senator; Arthur L. Miller, former U.S. Representative; Glenn Cunningham, former U.S. Representative; David Martin, former U.S. Representative; Charles Thone, former Governor and U.S. Representative; Virginia Smith, former U.S. Representative; John Cavanaugh III, former legislator and U.S. Representative; Douglas Bereuter, former legislator and current U.S. Representative; Robert Crosby, former Governor; Frank Morrison, former Governor; Arnold J. Ruhnke; former legislator; Hans O. Jensen, former legislator; Don Thompson, former legislator; Willard H. Waldo, former legislator; Fern Hubbard, former legislator; Richard Marvel, former legislator; Jerome Warner, legislator; George Geddes, former legislator; Rudolph C. Kokes, former legislator; Loran Schmit, former legislator; Richard Lewis, former legislator; Jack D. Mills, former legislator; Howard Lamb, former legislator; Chris Beutler, legislator; Dan S. Jones, Jr., former director of the Department of Water Resources (DWR); Marion E. Ball, former director, DWR; John Nueberger, former director, DWR; Carl Swanson; Cyril Shaugnessey; Henry Lange; Donald L. Wagner; Alfred Drayton; Wes Sandall; Tom Downey and Bill Christensen.

In addition, we tried to get as many state and national groups as we could that were especially prominent in water issues. Groups that might have been profiled but were not include all *local* water policy or educational organizations, as well as statewide or national groups such as the Nebraska Farm Bureau Federation, the Nebraska Farmers Union, the state Grange, the Arbor Day Foundation, the Sierra Club, the American Rivers Society and the League of Women Voters, among others.

Even with these now mentioned, more come to mind. Therefore, the final choices portrayed in the book are offered with some apology for those left out but with hope that most everyone understands that there always will be one more important person, fact or story. These pages offer a sampler, a "reader," as much as an encyclopedia. The goal has been to catch the flavor and to transmit the fervor of Nebraska and its water as seen near the end of the 20th Century.

Acknowledgments

More than thirty people contributed writing to this book. They were dedicated to the task and met nearly impossible deadlines with very few exceptions. The chapter authors are individually identified and profiled in the biographies section near the end of the book. Others, who wrote vignettes, "endpieces," or profiles only, are identified after their bylines or after the individual piece. We thank them heartily.

Their writings were reviewed by many tens of reviewers and perhaps involved hundreds of individual reviews, since few entries and no chapter had less than three independent reviews, many of them two or more at evolving stages of development. Most authors also reviewed the writing of most other authors. Therefore, a roster of the reviewers is impossible to construct. And, it is nearly impossible to adequately thank them.

Vince Dreeszen, Mike Jess, Sue Seacrest, Les Sheffield and Perry Wigley, as members of the project management team, read and reviewed at least one version of each entry, many two or more. They also offered much important criticism and advice along with excellent and consistent management for the project.

Sally Herrin as photo researcher and co-photo editor continually uncovered the necessary visual proof to support this history. She was equally adept at both finding and presenting the impossible. The success of her work is very visible in the pages before you.

As the work moved from draft to final a team of three gave full measure of their many and great talents to graphic support and layout. Thanks to Dee Ebbeka, Jerry Leach and Ann Mack this book is just one more job very well done.

Also deserving of a hearty thanks are Bernice Goemann, Duane Mohlman, Melba Stemm and Karen Stork, who helped compile the index, wrote selected profiles and did much essential but unsung work in typing material into the word processing system. Beverly Martin and Jane Turner, who typed manuscripts, and Marj Seidel, who did much proofreading, also deserve our sincere gratitude.

The project enjoyed the services of two very capable and professional editorial associates. Brad Rundquist helped to manage the basic planning and organization for the book, which allowed it to be completed in a timely manner. He also wrote a multitude of vignettes tirelessly. Devorah Lanner picked up when Brad left and saw the work to conclusion.

Editing such a collage of topics, writers, reviewers and products within the time frame allotted is not a task for the faint of heart or anyone lacking stamina. Our editor, Charles Flowerday, cut his teeth for this type of "magnus opus" on *An Atlas of the Sand Hills*. To his credit he took on *Flat Water* with knowledge that it would be more of the same. He did his job with the very best of the skills that he and his profession can offer. Thank you Charlie!

Major Funding and Support Provided By

Conservation and Survey Division
Cooper Foundation
Nebraska Groundwater Foundation
Nebraska Humanities Council
University of Nebraska Water Center

and

Alfs Well Drilling Company
Central Nebraska Public Power and Irrigation District
Grosch Irrigation Company
Hastings Irrigation Pipe Company
KN Energy, Inc.
Lindsay Manufacturing Company
Lockwood Corporation
Kenneth Morrison
NC+ Hybrids
Nebraska Machinery Company
Nebraska Water Users, Inc.
Nebraska Public Power District
Pioneer Hi-Bred International, Inc.
Reinke Manufacturing Company
Sargent Irrigation, Inc.
T-L Irrigation Company
Valmont Industries, Inc.
Western Land Roller Company

Additional Funding Was Provided By

Ainsworth Irrigation District
Vance Anderson
Burton Well Drilling
Farmers National Company
Farwell Irrigation District
Frenchman-Cambridge Irrigation District
Gering-Ft. Laramie Irrigation District
Henderson Irrigation Company
IMC Fertilizer, Ind.
Industrial-Irrigation Services, Inc.
Madsen's Well Service
Nebraska Coalition for the Environment
Nebraska Farm Bureau Federation
Nebraska State Irrigation Association
Nebraska Water Resources Association
Remmenga Drilling Company
Shuck Drilling Company
Sargent Irrigation District
Pathfinder Irrigation District

Land and Water in 19th-Century Nebraska

"The Desert Shall Rejoice and Blossom As the Rose!"

by Robert Manley[1]

The west is a desert, and we have told ourselves and the world that we have made the desert blossom as the rose. We have told the truth. But we would be wise to remember every moment that roses also blossomed in Mesopotamia and Syria and Tunis and Ur of Chaldees—and they are desert wastes now.

—Bernard DeVoto (1948)

First, the immigrants confronted the land—a rolling, nearly treeless plain west of the Missouri. Explorers who crossed it in the early 1800s were not impressed with what they saw. They called it "The Great American Desert."

It was the Native Americans, however, living in what would become Nebraska who supplied the state with its official name. It was their word for the Platte River, its valley an east-west thoroughfare for water and people that would loom large in the state's history. They called it "Broad, Flat Water."

Unlike the explorers, the immigrants to the West could ill afford a pessimistic view. Along with a few belongings, they brought mostly hope. Such was the attitude, when, in the spring of 1850, some flatboats gave a wagon train safe passage over the swift-flowing Missouri. The wagons then climbed the bluffs, where they circled up. It was Sunday, and after their camp-making chores were done, the folks in the train gathered for worship. Harrison Presson, a Methodist pastor, presided, choosing for his Scripture a passage from Isaiah 35:1: "The wilderness and solitary place shall be glad for them, and the desert shall rejoice and blossom as the rose."

For easterners apprehensive about being at the "edge of civilization," this was a comforting Scripture. But the verse is more important for what it reveals about European attitudes toward wilderness. As Roderick Nash points out, pioneers on the western trails brought with them attitudes toward wilderness that were shaped in prehistory, when human beings began to change their environment as they began to herd and plant. At this time, says Nash, "civilization created wilderness."

That attitude was embellished by the writers of the Old and New Testaments and by the experiences of generations of Europeans who lived along what Nash calls the "frontier of the Middle Ages." English scholars who translated the Bible in the 17th Century believed that the term "desert" meant a remote, treeless, undeveloped land—a wasteland waiting to be conquered and turned to useful purpose. Their translations reflected that view.

So, experience and Scripture spoke eloquently to the Europeans venturing to the New World. True believers, bearers of an expansionist European heritage, were not to be dominated by wilderness. Rather, they were commanded to impose order and value upon "the desert." "Go in and possess the land," thundered the Old Testament prophets, and make the desert "blossom as the rose."

Contrast this European view of wilderness with that of the aboriginal who inhabited the plains. These people knew no wilderness, only an environment with which they had to come to terms if they were to survive. They gave no thought to "conquering" the land. Their world-view and technological limits meant they were more accepting of the land the way it was. And except for periods of crop failure, they created a way of life that suited them and of which they were proud. "Improving" the land was a foreign concept. Their goal was to survive, so they accepted drought and other natural calamities as part of the natural rhythm of the land.

The two cultures agreed upon one point. Life in the wilderness depended upon water. Native Americans erected their villages close to water, as their ancestors had done in prehistoric times. Although the horse, introduced to the plains dwellers by the Spanish, enabled them to range great distances in search of game, even horse-mounted people understood their dependence on water. During dry times they migrated great distances in search of life-giving water and forage. The fact could not be evaded.

This re-created sod house in Sioux County near Toadstool Park evokes the hardships faced by the immigrants and pioneers in the semi-arid West. Even in a land without timber, they could build houses. But, having no way to make it rain when life-giving water was scarce, they soon sought irrigation.

J.W. Goeke, CSD

[1]Identification and biographies of chapter authors are on p. 276. Other authors are identified after their by-lines.

Life in the prairie depended on water. The Native Americans understood. The Europeans were about to find out.

Into the Desert

Although President Thomas Jefferson believed history would vindicate, perhaps even applaud, his decision to buy the Louisiana Territory, the fact is he had no idea what he had bought. In 1804, he sent Meriwether Lewis and William Clark to find out.

These two intrepid explorers and their companions stuck close to the Missouri River as they made their way west. They did not see themselves as plainsmen. But they heard about the prairie country, and they did travel a few miles up the Platte River.

A few years later, Lieutenant Zebulon Pike was ordered to explore the central plains. He had no choice—he became a plainsman, and his journey brought him into the Republican Valley in southern Nebraska before he turned southwest and headed for New Mexico. Pike reported that most of the country he crossed was little better than desert. The region, he predicted, would one day be compared with the Sahara.

Then, in 1820, the War Department dispatched Major Stephen Long to have another look at the plains. It was an important event, as Long's was the first expedition to use steamboats to ascend the river. The steamers didn't work very well, but it was apparent that the age of steam would not be long delayed.

From his base just north of where Omaha now stands, Long and his party struck west, into the Platte Valley. Edwin James, the party's botanist, looked, examined and recorded. In his report James said of the plains country: "Though the soil is in some places fertile, the want of timber, of navigable streams, and of water for the necessities of life render it an unfit residence for any but a nomad population." Long's cartographer named the country, "The Great American Desert." This remote, treeless, semi-arid country would never support an agricultural population, Long and his party thought.

The nation's lawmakers, however, saw a use for the desert. It was a good place to relocate the Native Americans being uprooted by the westward movement of the Europeans. The desert would become, by act of Congress, "Permanent Indian Country."

That policy was short-lived. Long and other explorers demonstrated that the prairie could be easily crossed. The best route west followed the Platte Valley, a superb highway.

Most of the early plains travelers were fur traders. Operating out of St. Louis, these traders traveled up the Missouri River, then "jumped off" to pick up a river flowing into the Missouri from the plains. Some entered via the Kansas River. Others followed the Platte. At any rate, by the 1820s, regular traffic across the plains included fur traders, often joined by tourists, missionaries and artists, all anxious to see the Great West.

Returning from the Rocky Mountains with their loads of furs, the traders hoped to be able to float boats down the Platte River. It seemed like a good idea because the Platte carried an enormous volume of water in the spring, certainly enough to float flat-bottomed bullboats and mackinaws. But appearances were deceiving. There might have been lots of water in the Platte, but it did not flow in fixed channels. It was no wonder, as mentioned earlier, that the Native Americans called the river "Broad, Flat Water," for indeed it spread shallowly over the floor of its valley, meandering, braiding, twisting, defying efforts to float boats.

In 1842 John Fremont met Pierre Papin working his way down the Platte with a flotilla of 11 boats loaded with furs. Papin said that his boats went aground, on the average, 50 times a day. Elisha Perkins, on his way to California in 1849, saw another fleet of barges bobbing and bumping down the Platte. The navigators, all fur traders, were angry and frustrated. They could not keep the boats afloat. It was a pity, observed Perkins, that the Platte "is not deep enough to admit navigation as it runs through a country that one day will be the garden of the world so rich and fertile."

In the same year, 1849, Major Osborne Cross, going west along the Platte with a company of dragoons, similarly observed that the river was not navigable "nor can it be made so." From a commercial point of view, said Cross, the Platte had "very little to recommend it." The river was remarkable only because it had more sandbars, less water and more islands "covered with useless timber" than any other river in the universe, Cross observed.

Washington Irving, a distinguished chronicler of the fur trade, offered this succinct evaluation of the Platte: "It is the most magnificent and useless of rivers."

That view, however, was not shared by the pioneers in covered wagons bound for California, Oregon and other points west. As historian Merrill Mattes points out, the Platte River was a godsend to these travelers. It ran in the right direction, and its valley provided the travelers with fuel, forage and water—the trinity of the trail.

To the overlanders, water was both a plague and a blessing. On the negative side, torrential downpours brought terrible hardship to humans and animals; flood-swollen rivers swept away teams, equipment and people. It must be remembered that the emigrants followed the Platte in the spring and early summer, when prairie streams ran at flood stage. Stream crossings were trials. In June of 1853, Velina Williams, on the road to Oregon, reached Prairie Creek.

An emigrant family fording the Platte River. Lashing logs to the wagons helped, but good-sized logs were few.

Nebraska State Historical Society

Today Prairie Creek, except for flooding, is usually a small, quiet stream. In 1853, it was a formidable barrier. Although Velina's wagon successfully negotiated the crossing, she wrote in her diary: "The banks were very high and steep. Before we started on this journey we should have thought this stream impassable without a bridge, but no accident occurred."

The Platte offered an even greater challenge. Frances Sawyer, a young married lady from Kentucky, was in a wagon train that followed the south side of the Platte. Hearing that there was less disease on the north side of the river, her husband decided to ford the Platte a few miles west of Fort Kearny. The crossing was a terrifying experience for this well-bred lady. "The Platte is a mile wide at this point," she wrote in her diary, "and our wagons pulled very hard in the quicksand. Mr. Sawyer went over in the carriage with me. The water was so deep that our mules had to swim in some places. I was greatly frightened and held on tightly to my husband!"

Drinking water was another problem for these pilgrims. Experienced travelers and guide books told the emigrants to take their water from running streams. Of course, Platte water was muddy and filled with "wrigglers." But a handful of cornmeal, dropped into a bucket of Platte water cleared up most of the mud; and pouring the water through an apron removed the largest of the wiggling critters.

If preparing a "Platte River cocktail" took too much time and effort, the alternative was to draw water from trailside wells. Water in these shallow wells, however, was usually contaminated. E. W. Conyers, going from Illinois to Oregon in 1852, reported that members of his party drank water from a trailside well in central Nebraska. Within days several had died of cholera, the deaths caused, Conyers believed, by drinking water from that well.

While hardships were ever-present and hazards continually challenged the covered-wagon pioneers, there were compensations for taking the Platte trail. First, there was the remarkable scenery. On May 22, 1850, Byron McKinstry, California-bound from Iowa, saw the Platte for for the first time. "I am delighted with the scenery," he wrote in his diary. "The beautiful green bottoms profusely sprinkled with flowers and the picturesque views of the river and its cottonwood isles, among which it [the Platte] flows in several shallow channels. . . ."

Gilbert Cole, on the trail in 1853, wrote

Journal-Star Printing Company

J. Sterling Morton

of the "marvelous beauty of the Platte River islands." Cole said that there were "few of these island beauty spots in the whole wide world. When the Maker of the Universe gathered up his emeralds and then dropped them with careless hand upon a few of earth's waters, He wrought nowhere a more beautiful effect than in the Platte islands of Nebraska."

Overlanders who were farmers invariably took time to examine the soil in the Platte Valley. Most remarked that it was fertile and capable of growing crops. But, as McKinstry noted in his diary, the country did not have enough timber "for farming purposes, though the soil is favorable." That was one serious problem with the prairie—lack of timber. To eastern farmers who used timber to build houses and barns, to make fences and burn in their stoves, the thought of settling in a country with little timber made no sense.

Nonetheless, the overlanders were impressed with the Platte River and its valley. Mr. and Mrs. Langdon, who went west in 1854 had reason to remember their trip along the Platte. On May 30th their wagon train paused long enough to allow Mrs. Langdon to give birth to a "fine son," whom his parents named Thurston Platte Langdon.

Finally, there were the entrepreneurs, men and women who, instead of hurrying along the trail, picked a likely spot and opened road ranches. These were service stations along the trail, where overlanders purchased goods and services. Many ranchers kept herds of cattle, mules and horses for trading; and Joseph E. Johnson, who ran a ranch where Shelton now stands, cultivated a large garden and sold produce. He also cut hay and offered it for sale. Road ranching was one of Nebraska's first important industries—a fledgling form of the tourist industry operating along the wagon trails.

What had happened to the "desert"? As they traveled eastern and central Nebraska, the overlanders seldom referred to the land as a desert. Struggling through the deep sands along the North Platte Valley, however, their tune changed. One fellow from Ohio camped near Chimney Rock wrote in his diary that while the scenery was glorious, no person in his right mind would ever want to live in "this God-forsaken country."

The Town Builders

Nebraska Territory was opened to settlement in the spring of 1854. Town builders made up the first wave of settlers. Eager, expectant, ambitious fellows, the town builders were intent upon finding good town sites, platting towns and selling lots, then growing rich in the process. J. Sterling Morton, one of Nebraska's pioneer town builders, said that he was more interested in platting a town and building a political career than he was in planting and picking corn.

In selecting town sites, the town builders looked for locations near rivers and creeks. Towns on the Missouri River would have steamboat connections with distant markets and business centers by boat. The builders of Brownville, Nebraska City, Plattsmouth, Omaha City and other aspiring river towns boasted of this vital transportation link.

One imaginative town company platted a town on the Nemaha River, several miles upstream from the Missouri. According to tradition, the town builders hired a small steamer, hitched oxen to the vessel and dragged it from the Missouri, through the mud and puddles of the nearly dry river bed, to their wharf, so they could advertise that their town was served by a steamboat.

While fur traders and explorers a generation earlier had concluded that the Platte River was not navigable, hope burned eternal in the breasts of town builders. In 1860 the Nebraska territorial legislature heard John A. Latta promise to run a steamboat on the Platte between the river's mouth and Fort Kearny for the puny price of 20,000 acres of land. Whereupon, the legislators requested from Congress a grant of 20,000 acres of Uncle Sam's vast domain. Latta's proposal deserved serious attention, be-

National Archives, via Nebraska State Historical Society

A photo of the Missouri River snagboat, the U.S. R. E. DeRussy, taken in 1868, the year it began operating

cause the idea that the Platte was unnavigable was "injurious to the interests of our territory and [was] retarding the settlement of the valley," they said. The request insisted that there was "sufficient volume of water flowing in said river, which is one thousand miles in length, to carry ordinary steamboats such as navigate our rivers, and believing that if the channel were once opened and steamboats placed thereon, that navigation would be of great advantage to the general government"—to say nothing of what it would do for Nebraska.

Alas, Nebraska did not get the federal land grant, and Latta's scheme was forgotten. But later an entrepreneur in Fremont built a small steamboat and launched it on the Platte. He wanted to provide service for towns springing up in the Platte Valley. Unfortunately, a trip upstream from Fremont to North Bend, a distance of about 16 river miles, took the better part of a week. So the owner put his craft into service as a ferry. After a dozen or so trips, the craft ran aground and was wrecked.

But the mighty, unruly Platte bedeviled not only navigators. It was a gigantic nuisance to the those who settled in the valley and understood that if their towns were to prosper, the river would have to be bridged.

The history of most Platte Valley towns contains a chapter relating the long, expensive, frustrating efforts to build and maintain bridges over the Platte. Fremont businessmen raised money to build a bridge, then watched ice flows and flood waters smash the structure into splinters. Year after frustrating year, the bridge was rebuilt, only to be washed out. The "outing of the bridge," one Fremont resident said, became part of the community's spring ritual.

In Scotts Bluff County, plans to construct a bridge over the North Platte River became a pawn in the county seat contest. Gering got the courthouse, and the settlers on the north side of the North Platte, where Scottsbluff would one day be located, got the long wooden bridge that allowed them to get to the courthouse without risking life and limb in the swirling waters of the river.

One reason Grand Island enjoyed rapid economic growth is because south of the city the Platte was easily bridged. South of town the Platte was filled with islands, so nine short, substantial bridges were built from island to island, providing a way for settlers from Hamilton and Adams counties to get to Grand Island, "the metropolis of central Nebraska."

The inability to bridge the Platte had an important political impact upon Nebraska. The river literally divided Nebraska into two sections—north of the Platte and south of the Platte—and these two areas quickly became political factions. Sectional political contention had a tremendously negative effect on early Nebraska. In 1857, for example, Congress appropriated $50,000 to "improve" a wagon road from the Missouri River to Fort Kearny. South Platters were confident that the money would be spent to improve the road from Nebraska City to Fort Kearny.

But the federal funds went to improving the trail that originated at Omaha and followed the north side of the Platte. J. Sterling Morton and other furious South Platters met in convention in Nebraska City and announced their determination to secede from Nebraska Territory and join Kansas. Of course, the South Platters were not to achieve their goal, but this political brawl, one among many, illustrates the negative impact the squabbling over the boundary caused by the Platte River had on Nebraska's early history.

Washed-out bridges, sunken steamboats, destructive floods and political fights were facts of life in early Nebraska, but these events should not obscure the important positive role played by rivers in Nebraska's development; for every town builder realized that his town, to grow and prosper, must have at least

Bridge-building was a major enterprise for the first settlements. Many were torn out by spring floods. This postcard of the Kearney bridge is dated 1912.

Nebraska State Historical Society

Nebraska State Historical Society

Flume house and flour mill, West Point, Nebraska

one mill. And in this pioneer generation, most mills were water-powered. So towns were laid out by streams that their promotors insisted provided "inexhaustible water power."

In fact, many early town builders were mill operators by profession. They understood that the site they selected for their mills would most likely develop into a good town site. Mills were extremely important to pioneer towns and to local farmers.

Take the case of Nebraska City. In the 1850s and 1860s, Nebraska City became an important supply point for the Colorado gold camps. In response to miners' needs, Nebraska City shipped enormous amounts of supplies by wagon to the mills. Flour was a big item in that overland trade. Initially the supplies came upriver in steamboats; but Nebraska City businessmen quickly grasped the opportunity to build their own mills—most of them steam-operated—to produce flour for the Colorado trade and keep the profits in their town. In short order, several mills were turning out flour for the Colorado trade. This meant farmers had a handy market for their wheat and corn, and more and more acres were put into crops. The desert was beginning to blossom.

H. W. Parker successfully combined the roles of miller and town builder. He was one of the first settlers in Beatrice, where he operated a mill located on the Blue River. In 1865 he bought land along the Blue River in the southern part of Seward County, laid out the town of Camden—which he expected to become the state capital of Nebraska—and built a water-powered saw and grist mill. Newspaper articles emphasized the town's "superb water privileges." Though it didn't become the state capital, it remained an important rural center for years, largely because of its mill.

If a town didn't have a mill, the question was eventually "agitated." In 1870 the editor of a Fremont newspaper asked why the city didn't have a mill. The Platte River "promised water power of unlimited capacity." Farmers were producing bumper crops, and the local market would easily absorb the mill's products. When would a "good active man with abundant capital" seize the opportunity to build a mill, benefiting not only himself but Fremont as well?

Public meetings in Beatrice in the early 1870s raised the issue of the town's "industrial potential." Although one water-powered mill was running, enough water power was available for many more industries. Why not, one man asked, form a local development company that would bring industries to southeastern Nebraska? Surely a woolen mill, an oil mill, a paper mill and a tannery would prove profitable. Gage County farmers endorsed these ideas. A farmers' convention adopted a resolution stating that "we do gladly offer and pledge our support to a woolen factory or any other kind of manufacture that will advance the interests of the community and farmers generally."

Recognizing the importance of mills, territorial lawmakers adopted laws that regulated the location of mills and protected mill operators from unfair competition. Legislators also adopted the principle of riparian rights, that is, the English common law that held that water users could take water from a stream so long as the water was eventually returned to the stream so that others might also use the water. That principle worked well in England and in the eastern United States, where streams were used principally for power. It worked for a while in Nebraska—until irrigation.

Pioneer mill operators understood that they had a legal right to streamflow. But what concerned them was getting an adequate supply of water. Spring floods were likely to wash away the mill dam. By the time the dam was repaired and the mill back in operation, the streamflow was substantially reduced due to summer drought. A mill known as Schaupp's Mill along Wood River in Hall County illustrates the problems millers faced.

In 1874, Wisconsin mill owner J. G. Schaupp announced that he was going to build a water-powered mill on Wood River. He also promised to bring with him experienced farmers eager to settle in this new country. It is not known if Schaupp brought the farmers, but he did build the mill. Within a year, however, Schaupp was in trouble. Wood River didn't have enough water to power his mill. He went to the Hall County board and asked that they put a bond issue before the voters. Schaupp said that $15,000 was needed to build a canal to bring water from the Platte to Wood River. The increased flow would benefit his mill and also could power factories certain to be built on the river. J. M. Cummings, a Hall County farmer, enthusiastically endorsed the bond issue. "Open the canal," he said, "and a stream of wealth will flow in from the east and make our county the richest in the state. We will have a little New England of our own here in the West, and the busy hum of the manufactory wheels will be heard, making glad the hearts of thousands."

Cumming's vision was not shared by a majority of the voters, who turned down the bond issue. Whereupon, Schaupp closed his Wood River mill and moved to Grand Island, where he built a steam-powered mill.

West Point is another notable example of the connection between water power and town building. A town company made up of Omaha men and led by John Neligh and Uriah Bruner decided to locate a town in the Elkhorn River valley. That river, they believed, possessed tremendous water-power potential. In 1859 a water-powered saw mill was going, and a few years later a grist mill. By now West Point was recognized as one of the up-and-coming towns in northeastern Nebraska, largely on the basis of the water power in the Elkhorn River valley.

Water power did bring factories to West Point. In 1873 an agricultural implement factory produced 100 plows, 1,100 cultivators, 1,100 harrows and 3,000 railroad scrapers. A soda-water factory and a furniture factory also were humming away. To support more economic development, local businessmen formed the West Point Manufacturing Company, much on the order of modern industrial development corporations. Neligh and other West Point businessmen went east to raise money. Upon returning from a successful money-raising tour, Neligh was carried through the streets "by the elated citizens who were overjoyed by the prospect of having West Point a manufacturing center."

The effort paid dividends. West Point got a planing mill, a paper mill, a furniture mill, a huge creamery, a brick works and a number of other smaller businesses. By the late 1870s, West Point was hailed as "The Lowell [Massachusetts] of the West." But West Point's prosperity soon collapsed. Since railroad-rate structures discriminated against Nebraska industries, goods made in West Point could not compete with goods produced in the east.

Nebraska's pioneer economic developers had another serious problem: the difficulty of raising development capital. Part of the problem lay with Nebraska's 1875 constitution, which limited the authority of towns and counties to vote what were known as "mill and factory bonds." Greeley and Stanton counties, to cite two examples, passed bonds to help pay for mills, only to have the issues declared unconstitutional by the Nebraska Supreme Court. So, many mills were financed by local loans and gifts. Some town builders tried to interest eastern investors in building mills and factories. The mill that helped Cedar Rapids (in Boone County) prosper, for example, was financed by a group of Chicago investors. In 1882 the mill produced 150 barrels of flour a day. The $30,000 mill, declared a local reporter, "is beyond question the finest mill in this section of the State."

Despite problems with water supply and financing, 144 water-powered mills were running in Nebraska in 1880. That was the high point. After that the number steadily declined. Steam engines powered the new mills. Moreover, irrigation was diverting water from rivers and streams. A man living in the Republican River valley reported in 1889 that the river was no longer a mill stream. Irrigation and drought had dried up the river during the summer. In 1889 Nebraska had 279 mills, but only a few were water-powered.

The mill operator is a neglected figure in pioneer history. That is unfortunate, for these businessmen were very important in building towns and encouraging industrial and agricultural development. And let's not forget that the local mill pond was a community's recreational center. For swimming, fishing, boating or ice skating, a town's citizens, young and old, went down to the mill pond in every season. These ponds also had a commercial value. They provided ice for local markets and for cooling railroad cars of produce and meat.

The Pierce Commercial Club banquet in honor of the opening of the "modern" flour mill and new creamery

Nebraska State Historical Society

The 1870s: "The Great American Desert Has No Terrors for the Hardy Pioneers."

Many of the town builders who poured into Nebraska Territory were speculators. As W. W. Cox, a Seward county pioneer said, these men came to make a pile of money quickly so they could get back to "civilization" and enjoy their profits.

The nationwide financial panic of 1857 put most of the town builders out of business, and for a few years Nebraskans muddled through in despair and uncertainty. Nebraska needed a new vision, and George Miller, publisher of the *Omaha Herald,* believed he had that vision. "Nebraska is agricultural, or it is nothing," Miller declared. And he urged Nebraska's pioneers to get on with the task of building farms.

As noted earlier, Nebraska's response to Colorado's demand for farm products had a ripple effect on the territory's infant economy. The demand for grain encouraged farmers to expand their operations, and the agricultural frontier by the late 1850s had moved out of the Missouri Valley and onto the rolling prairie to the west. Railroad construction increased also. The rails both followed and often preceded the line of settlement, as trains brought in settlers and opened new markets. A generous federal land policy tempted people to take land claims and, after 1862, homesteads. Prospective settlers were urged to come west and "grow up with the country." In the fall of 1870, the *Plattsmouth Herald* said that settlers were moving into the Republican Valley. Homeseekers followed hard on the heels of surveyors. As soon as the lines had been run and the corner stakes set, eager settlers took claims. "This looks as though the Great American Desert has no terrors for the hardy pioneers," the writer concluded.

Central to the process of developing the prairie wilderness was the homestead law. It is important to understand why this law was written and how it operated.

Nebraska State Historical Society

O. M. Kem, a Populist U.S. Congressman from Nebraska, was the first man from a sod house to be elected to Congress. Many homesteaders built sod houses. Photo dated 1886.

First, it should be understood that the lawmakers wanted to divide up the public domain among honest, sturdy pioneers, anxious to build homes. But they were bothered by the thought of giving away land. That was "paternalism" of the worst sort. Free land might undermine individualism and self-reliance. Finally, however, the proponents of the homestead law concluded that unbroken prairie—wilderness—had no value. Only as the land was "improved" did it take on value; and since claimants were required to make improvements on their claims, they were not receiving a gift from Uncle Sam. By their labor they were "earning" it. Here again the European heritage is at work. So, following enactment of the homestead law, prospective settlers came to Nebraska by the thousands to take a homestead or a land claim and to make the prairie "blossom as the rose."

Having dealt with the threat of paternalism, lawmakers enacted in 1862 a homestead law they believed would encourage small farmers to put down roots and become permanent inhabitants of the land—to expand the number of America's small farmers who were the backbone of the nation and the sinew of a democratic society.

That is not how the law worked out, however. Many homesteaders came west not to settle permanently but to speculate. They took claims, made improvements and then sold their claims to new arrivals and moved on. Please understand that this "claims business" had no legal basis. Those who wrote the Homestead Law never thought that the settlers would act as many of them did. But the homesteaders saw a chance to make money, and it was irresistible. After all, the improvements they made—the houses they built, wells they dug, fences they strung—were what gave value to the land. If they decided to let go of their claims, were they not entitled to compensation for the improvements they made? That was the reasoning behind the claims business.

In other words, homesteaders and other settlers, generally, saw land as a commodity. It existed to be improved and then peddled. Their great grandsons would one day see water the same way.

The settlers in the 1870s looked first for claims in the valleys. The soil was richer, they figured, and water was obtained more easily, either from streams or from shallow wells. They avoided the uplands, if possible, where, it was said, the land was infertile and water hard to obtain.

But since there wasn't enough valley land to go around, many settlers ended up taking upland claims. Those who built homes on the divide between the Platte and the Republican rivers in southwestern Nebraska were hard pressed to find water. Many settlers had to make long trips to get water, returning to their homes with wagons loaded with barrels of sloshing water. Others got water from neighbors with wells. A few dug wells, but in this country it was usually several hundred feet to water, so well digging was both difficult and expensive. Travelers going over a barren prairie divide stopped at farmsteads to get water. One pioneer remembered that the going rate for water was 10 cents a quart for drinking water, 25 cents to water a team of horses, oxen or mules. In northeastern Harlan County, Allen Elliot took a homestead in 1874 right on top of the divide. He put down a deep well and got a good supply of water. When travelers stopped to get water, Elliot refused to take payment. His water was free to one and all. The community that grew up around Elliot's farm was known as Freewater. Today only a cemetery marks the site of this pioneer community with such a wonderful name.

Land agents and promoters, of course, were important in attracting settlers to Nebraska. It was their job to spread "the good news" about the state. Champion Chase, mayor of Omaha, was a typical Nebraska booster. In 1878 he spoke at York's July 4th celebration. He declared that Nebraska was on a par with all other farm states: "With abundant natural irrigation through her spreading rivers, her winding creeks and numerous streams and her never-ending springs, the deep, inexhaustible richness of her broad acres will always be well watered." With sweeping oratorical certainty, Chase insisted that Nebraska was destined to be "The Garden of the West." No wonder settlers in southwestern Nebraska named a county and town for the Omaha mayor. He certainly had a way with words.

Robert Furnas, second governor of Nebraska and an important agricultural leader, was not as eloquent as Chase, but he too was determined to promote the state. Realizing that prospective settlers needed more than glowing words to assure them that it did indeed rain in Nebraska, Furnas systematically collected rainfall statistics for the state. After studying figures for the 1850s, 1860s and early 1870s, he concluded that while Nebraska did not receive as much total rainfall as eastern states, the rain came at the right time, during the spring and early summer growing season.

Remember that Furnas used statistics

'Dutch Joe': A Hero of the Nebraska Frontier

"Dutch Joe" we called him. We were the homesteaders upon the high tables and in the rich black valleys of the sandhills west of Valentine in the eighties. . . . Water was our first necessity and our greatest difficulty. . . . The high table homesteaders hauled their water in barrels, sometimes a distance of seven miles, while they broke out their first fields and laid the foundations for a real farm home.

The first experiments at digging wells on the high table were failures. Some dry holes were sunk two hundred feet and abandoned. It was then that Dutch Joe [Grewe] appeared on the horizon. . . . He was a sturdy fellow of medium height, with a pleasant smile, firmly set, determined lips, and extraordinary muscular development. He undertook to prove that water could be obtained upon the high tables and dug his first wells down through the hard, dry . . . rock over two hundred feet to the abundant underflow of pure, cold water beneath. What a celebration was held when the first Dutch Joe well reached water upon the "German Table." . . .

After that Dutch Joe was in constant demand. Other settlers would break out prairie for him, do his farm work and haul him cedar posts while he dug their wells. In the next seven years he dug over 6,000 feet of wells in the settlement, ranging from 100 to 260 feet in depth. There was no well digging machinery in the region at that time, and the settlers were too poor to import any. Dutch Joe's wells were large, round cylinders, straight as a gun barrel from the gramma grass roots to the gravel underflow. . . .

Nebraska State Historical Society

Joseph ("Dutch Joe") Grewe

One day in 1894 Joe was called upon to go down to the bottom of the first well he had dug in the settlement and clear out some obstruction. From the bottom of the well he gave the signal to hoist a bucket full of loose rock. When the bucket had almost reached the top it slipped from the steel catch which held it to the rope and falling swiftly 200 feet crushed the head of the German hero of the sandhill settlement. . . . Men who risk their lives on fields of battle are justly held as heroes. Those who risk and lose them in the cause of making human homes in what was once a desert are also heroes. Among these I write the name of Joseph Grewe. Let no one who has never dug in the darkness and danger of a deep well dare dispute it [brackets added].

—Addison E. Sheldon; adapted and reprinted from* Nebraska History, *vol. 1, February 1918

gathered from eastern Nebraska. He realized that in climatic terms, western Nebraska was something else. While governor, Furnas was invited to an irrigation convention in Denver. "The western portion of our State particularly is deeply interested in the proposed enterprise," he said in accepting. After this 1873 conference, Furnas toured the irrigated farms around Greeley. He was impressed and saw that irrigation would bring about the redemption of the arid lands of western Nebraska. He didn't keep his opinions to himself, speaking out frequently in favor of irrigation. Later he remarked that he had been "rapped over the head with pen and tongue for stating that irrigation would benefit western Nebraska."

The fact is few Nebraskans gave much thought to developing western Nebraska in the 1870s. They were busy settling the eastern part of the state. But John Wesley Powell, a geologist with the U.S. Geological Survey, was doing a great deal of thinking about the arid and semiarid West. Powell was to study this region of little rainfall, which included western Nebraska, and to make recommendations concerning government policy for developing it.

After field investigations, his findings and recommendations were published in government reports and in popular national magazines. He didn't mince words. Powell declared that conventional farming would never succeed in most of the West. Only a small part of the land could be irrigated, and only where this was possible could crops be raised. Powell recommended a thorough survey of the region so that life-giving streams could be mapped and reservoir sites identified. He insisted that laws written to govern land distribution in this part of the nation must recognize that land without water had little value.

Powell recognized that enormous sums of money would be needed to build the irrigation systems. If this were done, he did not want the government to finance the work; he thought it better for farmers to work cooperatively to bring water to the land. It would take new laws and institutions, concluded Powell, to rescue these lands "from their present worthless state."

Powell was not popular with Nebraska's leaders and promoters (usually one and the same). How dare Powell, that arrogant bureaucrat, suggest that western Nebraska would never be farmed? The Lincoln *State Journal* insisted that Nebraska's agricultural production was comparable with eastern states and was constantly increasing as the farm frontier moved west. The major's "slanderous utterances" were without foundation, the paper said.

While the promoters put up a howl, a few farmers in western Nebraska had already begun to irrigate. Nothing spectacular, but they were turning streamflow onto their land. In the 1850s, several ditches drew water from the Platte, one in Buffalo County. In the 1860s, W. M. Hinman and John Burke, who ran road ranches in Lincoln County near Fort McPherson, irrigated gardens with river water. Produce from the gardens was sold to travelers and soldiers.

In 1870 a group living near North Platte formed an irrigation company. Their goal was water for field crops. This was an important "first," irrigating field crops such as corn. As a local writer said, give the irrigators a couple of years and you would "see our desert blooming as a rose."

The company raised money for a ditch, and water flowed to a few fields. Due to the lay of the land west of North Platte, irrigating was relatively easy. The whole project cost less than $600. But then came several wet years, and the canal was abandoned. Nonetheless, the project demonstrated that irrigation was practical in the western Platte Valley.

Successful irrigation was also practiced at Fort Sidney, where a ditch brought water from Lodgepole Creek to water gardens, lawns and trees. In the early 1870s newspapers carried descriptions of the Fort Sidney irrigation system, and agriculturalists from eastern Nebraska toured the facility and were impressed with what they saw. The post's commandant wrote to Governor Furnas to describe an innovation worked out by the soldiers. Metal "siphon tubes" transferred water from laterals to crop rows and, furthermore, worked well.

West of Fort Sidney, along Lodgepole Creek, other irrigators were operating. Ranchers found that with very little effort and money they could turn water out of a creek onto their hay meadows. A man near Potter named Gunderson displayed considerable ingenuity. He built a small dam across the creek near his place. From the pond he took irrigation water in summer. When he didn't need to irrigate, he used the flow through the dam to generate electricity that lighted his house. In the winter, the streamflow powered his grist mill.

While Nebraska's newspapers carried articles describing these irrigation experiments, the average citizen no doubt regarded the information as interesting but not very important. In the 1870s it was widely accepted that Nebraska's climate was changing—that rainfall was increasing. Tree planting, said the experts, was partly responsible for this important climatic change. In 1869 the *Omaha Herald* set forth the opinion that in a few years western Nebraska would "come under the influence of tree-growing" and "no longer require artificial moisture." J. T. Allan, a well-known horticulturalist, popularized the idea that until tree planting extended into the west, those lands "would be useless except for pasturage." Yes, indeed, echoed the editor of the Lincoln *State Journal,* the arid west was destined to be "conquered by homemade rain produced by homemade trees." To hasten the process, Congress enacted the Timber Culture Act. Introduced by U.S. Senator P. W. Hitchcock of Nebraska, the law gave 160 acres of land to anyone who would plant a part of the quarter section to trees. Hitchcock called the law a major step toward solving the problem of the dry climate of the plains.

The popular idea that "natural laws" brought increased rainfall enabled Nebraska's promoters to deal with a troublesome issue. They wanted settlers to

Nebraska State Historical Society

Samuel Aughey

believe that western Nebraska was, potentially, wonderful farming country. Talk about irrigation retarded settlement, stood in the way of progress.

Fortunately for Nebraska's boosters, help was at hand in the person of Samuel Aughey, professor of natural sciences at the newly established University of Nebraska. Born in Pennsylvania, a Lutheran pastor by profession, Aughey had a passionate interest in natural science, and he was determined to use his scientific knowledge for the benefit of Nebraska.

In January of 1873, Aughey was invited to address the state legislature. He discussed the geology of Nebraska and spoke of ways science could help in the development of the new state.

First, Aughey discussed Nebraska's serious energy problem. Economic and industrial development were impossible without an abundant supply of cheap fuel. Such a supply was at hand, Aughey announced, for he had examined "coal outcroppings" in southeastern Nebraska and had concluded that enormous beds of coal lay beneath the state's prairies.

He then turned to soils. Although he had conducted only a few soil tests, his preliminary findings were exciting. In his opinion, Nebraska's soil was preeminently suited for both crops and fruits. "No other region, except the valleys of the Nile and of the Rhine, can compare with the loess deposits of Nebraska," he declared.

Moreover, due to its organic composition, Nebraska's soil would never require artificial fertilizers. With proper plowing and cultivating, the soil would exhibit a "remarkable power of regeneration." The soil also resembled a sponge. It had the capacity to hold water, releasing it slowly upwards by capillary action as growing crops required moisture. As a result, said Aughey, crop failures would be virtually unknown in the state.

The professor went on to describe Nebraska's superb water resources. The state had "countless numbers" of streams and waters. Running water was likely to be found on nearly every section of land within the borders of Nebraska. Where streams were absent, water could "easily be obtained by digging or boring to a certain depth," normally 20 to 60 feet.

Then Aughey unlimbered his big gun. While Nebraskans believed that in time their state would be an important agricultural region, the future was brighter than the average person realized. For, according to Aughey, Nebraska's rainfall was increasing and the line of increased rainfall was steadily marching westward. It was not a new phenomenon. It had been going on for nearly 20 years because "as civilization extends westward, the fall of rain increases from year to year."

In 1879 Aughey, Nebraska's foremost scientific promoter, was asked by the State Agricultural Society to respond to a plan before Congress to prohibit homesteading west of the 100th Meridian, which ran through Dawson County, Nebraska. Apparently, Powell had persuaded eastern members of Congress that closing off the semi-arid and arid regions to settlement was good policy.

Aided by C. W. Wilber, an itinerant geologist from Illinois, Aughey produced a small pamphlet that attacked the plan to restrict settlement. To hold back the line of settlement, the scientists argued, would be to hold back the operation of those "natural laws" that brought increased rainfall. Wilber later came up with a catchy phrase: "Rainfall follows the plow."

In the face of angry western reaction, buttressed by "scientific findings," Congress backed down. Homesteaders were loosed. The "natural laws" were given free play. Wilber was elated. After all, he declared, "a desert is the result of conditions that can be controlled by the genius and industry of man." He and Aughey agreed that "rainfall is steadily increasing westward, following the pioneer farmer and his plow, which is the primal cause of all these beneficent changes."

Nebraska had a great future. If in doubt, one had but to listen to Professor

Aughey. Read what he said and be convinced. Here is a sample:

> Nearly everyone who comes into [Nebraska] feels a general quickening and elasticity of spirits. The appetite and digestion improve wonderfully. Mind and body are lifted up. All this occurs even with the execrably prepared food eaten in most of the rural districts. For in most of the rural districts, hot biscuit, green with soda, is still the form of bread usually eaten.
>
> Owing to these facts, Nebraska must sooner or later become a health resort. In addition to the health producing properties of the climate, there are in various sections of the State mineral waters of high medicinal value. . . . [Salt water from Lincoln wells] is used in two bathing establishments in the city. This water is also believed to be specific in many cases of dyspepsia, constipation, incipient scrofula, and skin and kidney diseases. . . .

The 1880s: "Bound to Boom!"

For settlers in western Nebraska, the 1880s were pivotal years. In the early years of the decade, rainfall was adequate. Good crops were harvested west of the 100th Meridian. Rainfall did appear to be increasing. In the spring of 1886, The *Omaha Daily Bee* explained it this way:

> Why Rainfall Increases
>
> Observers and scientific men are agreed that the turning up of the soil increased rainfall by retaining moisture in the earth which otherwise would run off rapidly into the tributaries of the rivers. . . . The moisture thus retained is slowly evaporated and taken up into the atmosphere to return as rain in the region where it first fell. This is doubtless one reason why rainfall is increasing. But there are others equally important, for example, the planting of trees has unquestionably brought increased rainfall. It is noted as a singular fact that the building and operation of railroads and telegraph lines, even in advance of settlement, is generally followed by a steadily increasing rainfall. The result is attributed to the effect of the electrical currents running on rails and wires. In the case of the railroads the effect is more probably produced by the disturbance of the atmosphere attendant upon the rushing of the trains. Heavy rainfalls are known to follow artillery battles as the result of the detonation of the guns. The same result takes place through the disturbance of the atmosphere by the trains flying over the iron rails.

But in 1887 drought edged on to the plains. Crops withered and died. H. O. Smith, a Dawson County irrigation pioneer, summed up the plight of the western farmer in the late 1880s: "Sowing in hope, cultivating in doubt and harvesting in despair."

Advocates of irrigation held out hope. Among this small band of visionaries was General John Brisbin. An Indian-fighting officer, Brisbin was well known in Nebraska. After touring Colorado in 1886, he wrote an article for the *New York Tribune* in which he talked about irrigation farming in Colorado. He discussed "a new source of wealth growing up in our country, almost unobserved, but colossal in its nature and extent. I mean the owning of water in the streams and rivers." Formerly, said Brisbin, land was considered the west's most important resource. Now, however, it was evident that land without water had little value. His Colorado tour convinced him that investments in irrigation projects were likely to pay handsome dividends. He also pointed out that digging irrigation ditches did not require expensive, complicated machinery. "The best ditching machine," said Brisbin, "is an Irishman and a wheelbarrow or a plough or scraper."

Then, in 1889, the Nebraska legislature passed the St. Rayner bill, the state's first irrigation law. Under this law people could take water from streams after they had posted notice at the diversion point. Beneficial use was the basis of a person's right to water. And work on the ditch had to start within 60 days of posting the notice.

Congress was also debating irrigation. Discussion centered on what role the federal government ought to take in reclaiming dry lands and how much money Uncle Sam could afford to spend on irrigation projects.

Eastern and southern members of Congress vigorously opposed use of federal money to pay for irrigation. They argued that irrigated farms would add to the farm surplus and further reduce prices. The nation needed no new farms. Since Congress was not in the mood to vote funds for any kinds of "internal improvements," said Wyoming's U.S. Senator Carey, why shouldn't the federal government hand the remaining public lands in the semi-arid and arid regions over to the states. Let the states deal with reclamation and irrigation. But Charles Manderson and A. S. Paddock, Nebraska's U.S. senators, were not enthusiastic about that proposal. They wanted Congress to appropriate money for irrigation. Millions of federal dollars had gone to reclaim eastern swamplands and to build southern levees. Why should not federal funds be used to reclaim these potentially fertile western lands?

As the debate in Congress droned on, a water crisis developed in western Nebraska. In the spring of 1880, Governor Albinus Nance received a letter from Guy C. Barton of North Platte. The South Platte River in Nebraska was nearly dry. Colorado irrigators had taken all the water from the river. This was "a matter of growing importance to western Nebraska," said Barton, and should concern residents in the rest of the state as well. "You are aware," wrote Barton, "that the Platte River is a very wide and shallow stream presenting a very large surface for evaporation. I believe that this evaporation under the influence of our prevailing West winds is the source of no small proportion of the precipitation of rain along the valley east of us and that the total absorption of the stream in Colorado will have a visible effect upon the rainfall in this state."

In the summer of 1887, complaints from the South Platte Valley became more numerous and indignant. For example, John Wilson wrote Governor John Thayer to report that the South Platte was dry again, "causing great inconvenience and loss and suffering" among the settlers. Those who lived on the prairie had not yet had time to dig wells and so depended on the river for water. Water levels in valley wells had dropped alarmingly, and the hay crop was just about a total failure. What were the settlers to do?

Thayer wrote the governor of Colorado to express his concern about the South Platte. Colorado irrigators were taking all the water from the river, declared Thayer, with "disastrous consequences" for Nebraskans in the valley. Thayer implored the governor "in the name of humanity and the sense of fair play" to ask the Colorado irrigators to refrain from appropriating the total flow of the South Platte. Let them know, said Thayer, that they are "unwittingly bringing disaster upon their brethren on this side of the state line. I trust they will be inspired with the sentiment 'live and let live.' "

Colorado's attorney general responded. He said that while Colorado's

Nebraska State Historical Society

Union Pacific land sale advertisement

officials understood the plight of and sympathized with Nebraska settlers, there was nothing the government could do. Colorado water laws permitted farmers to take water from the river "for domestic, agricultural and manufacturing purposes, subject to preference in order and to priority of rights of appropriation." The irrigators were not violating Colorado law, so its elected leaders were powerless.

General Brisbin said that Thayer's letter was the first message of the kind ever sent in the west, "but it will not be the last one." River waters would soon be "in contest," and the federal government would have to step in to regulate the flow of interstate rivers.

This was precisely the argument adopted by Nebraska's political leaders. Federal funds were needed to build reservoirs in the mountains, and federal regulation was needed to see that the water got downstream to Nebraska. In the fall of 1889, the Lincoln *State Journal* reported the alarming news that a huge dam was to be erected near the headwaters of the South Platte in Colorado. The impounded water was to be used to meet Denver's need for water. The Lincoln editor observed that with Colorado irrigators draining off the summer flow of the South Platte, and "Colorado capitalists cornering the head of the stream in the winter," not much was left to dribble into Nebraska. He also warned those who believed that the Platte's "underflow" would continue even if the surface flow were interrupted that they were whistling in the dark. If the South Platte flow were cut off entirely for five months, the "underflow" would vanish. The valley of the South Platte in Nebraska would be as "dry as the Sahara."

Nebraska politicians were well known for their propensity to play upon the theme of rugged individualism. They insisted that Nebraskans were fiercely independent and resented any encroachment on their individualism. Perhaps, but by 1889 the farmers in the South Platte Valley were willing to surrender some of their independence for federal help in keeping water flowing down the channel of the river. One look at a map of the West shows why Nebraska was so vulnerable. The sources of her major river system, the Platte, lay in other states. The prospect of having their water supply interrupted mightily disturbed Nebraskans.

In the eastern and central parts of Nebraska, many rivers flowed within the boundaries of the state. In the 1880s, more and more of these rivers were being tapped to provide the water power for mills and factories. Since coal had not yet been found in Nebraska, rivers provided essential energy.

As noted earlier, Professor Aughey said that Nebraska's coal shortage was temporary. The Lincoln *State Journal* urged Nebraskans to search for coal. "Successful coal mining in Nebraska will revolutionize the commercial and manufacturing status of every important town within Nebraska's limits. We shall build our own agricultural implements, put up our own wagons and carriages, construct our own machinery, make our own paper and reduce by a large per cent. the living expense of every man, woman and child in the community. . . . Let us betake ourselves to coal mining at once."

To encourage the search for coal the legislature passed a coal bounty law. A cash prize went to anyone who found veins of commercial quality coal. Persons submitting claims for payment would have their "strikes" verified by an expert. In 1886, for example, a resident of Cedar County applied for the bonus. He said he had found a vein of coal. Professor L. E. Hicks from the university's geology department was dispatched to inspect the "strike."

Hicks examined the boring and reported that the prospector had tapped into a layer of shale, not coal. However, he noted that the boring had reached a significant flow of water. Perhaps, said Hicks, the discovery of underground water would prove more valuable in the long run than finding a vein of coal.

Hicks was asked to examine other shafts and borings in northeastern Nebraska and elsewhere. In every case he reported that coal had not been found, but in most of his investigations he saw evidence of a considerable flow of underground water. He theorized that Nebraska possessed a supply of underground water that, in his opinion, might be inexhaustible.

While Hicks was examining "coal strikes," other Nebraskans were busily harnessing the state's rivers. Although Nebraska had no coal, announced promoters, the state did possess innumerable mill streams.

Kearney launched a major attempt to promote industrial development by harnessing the Platte River. In 1882 local businessmen set up the Kearney Canal and Water Supply Company. A committee of these men was sent to examine the work being done in Colorado. After their return, a plan was drawn up to dig an 18-mile canal to bring water from the Platte River into Kearney, where it would generate power for factories and mills. The committee also suggested that the water might be used for irrigation.

In 1885 George W. Frank, an Iowa capitalist and promoter, bought a controlling interest in the canal company. Frank predicted that Kearney would soon become the "Minneapolis of the West." A local editor applauded such enthusiasm. "The possibilities for Kearney, on the completion of the canal, are great," he wrote. The city's residents would appreciate "the immensity of the project when they see "a volume of water twenty

Journal-Star Printing Company

W. R. Akers

feet wide and four feet deep pouring down the tailrace." They would then understand that the canal would be an enormous blessing to the town. Kearney would be able to "reach forth a welcome hand to manufacturers." The canal could not help but be "an incentive to the establishment of many large business establishments in this city."

Needing capital to finance his wondrous dreams, Frank persuaded New England investors to put money into this industrial development scheme. And in 1889, plans were announced for the construction of a cotton mill in Kearney's new west side industrial park.

The mill was an imposing building, three stories tall, 408 feet long and 102 feet wide. The machinery was water powered, with a steam engine standing in reserve. The first shipment of cotton goods left the mill in September of 1892, but problems plagued the operation from the outset. The early 1890s were years of drought and economic depression, hardly a time to launch a risky, large-scale venture such as a cotton mill. Water supply turned out to be a problem. The Platte River did not carry enough water to turn the mill's turbines. In fact, the mill operated only when a strong south wind blew water into the Kearney canal! The cotton mill operated for eight years. Then, as one Kearney historian has observed, "The cotton mill, like the paper mill, cracker factory and pressed brick works succumbed to the inevitable." The Kearney "boom," based on "an unlimited water supply," ended

But there were other grand power schemes—many of them. One exuberant promoter proposed a canal that would take water from the Platte and turn factory wheels in Hamilton, York and Polk counties. The canal, the promotor declared, would supply power to meet "all of Nebraska's manufacturing needs for years to come." The canal was never dug.

Even George Miller, a staid, conservative Omaha business leader, caught the water-power fever. He announced that money had been raised to construct a power canal from the Platte River west of Fremont running to Omaha. This "enormous flow" of water would generate more electricity for Omaha than was then in use by any other city in the nation. "As a manufacturing city there would be no limit to Omaha's growth and prosperity," declared Dr. Miller. The canal was never built.

Another interesting scheme was proposed by Joel Hull, the founder of Minden. In the 1880s Hull talked about turning Minden into a prosperous manufacturing town. All that was needed was an "unlimited water power." And that could be provided by bringing Platte River water to Minden via a canal. Hull invited engineers to survey the project's possibilities, and they agreed it was practical. Hull needed no convincing. He was already convinced that the canal was vital to Minden's future prosperity. He also suggested that farmers would be able to irrigate from it. By the time he got around to raising money, however, there was no money to be had because of drought and depression. Then, several years of above-average rainfall convinced farmers that they didn't need irrigation, and Minden businessmen were not interested in financing factories. Nevertheless, Hull continued to talk about the canal until he died in 1914, confident to the end that the canal would one day be built. And he was right. In the 1930s, Hull's canal was finally dug as part of the Tri-County irrigation project.

The Kearney canal, the central Nebraska canal, the Omaha power canal, Hull's canal—these schemes are a fair sample of how Nebraskans regarded water power in the 1880s. There was also, however, considerable talk about irrigation, especially in the state's western counties. But even in this part of the state, where crops were frequently destroyed by drought, irrigators were few and far between. They were not a popular breed. Cattlemen insisted that the land could not be farmed, even with irrigation. Promoters who held vast tracts of land opposed any talk of irrigation, because such talk implied that the country was not productive. But cattlemen and promoters did not stand in the field and watch row crops burn up. Some of the farmers who did began to think about irrigating.

Early in the 1880s residents of North Platte drew up a plan to build a canal that would tap the South Platte River for power and irrigation. The piece of ground to be irrigated lay about 20 miles west of the city and could be irrigated cheaply and with little work. A committee went to Greeley, Colorado, to look over the irrigation works there. While in Colorado, the men from North Platte met Lord Ogilvie, an English entrepreneur who agreed to build the ditch in return for half interest in the canal company and half the irrigated land. The Nebraskans accepted his offer.

The ditch was completed and water began to flow. But only a handful of farmers bought water. The problem was that most farmers knew nothing of farming with "artificial moisture." So the company set up demonstration farms to display the proper way to irrigate. In addition, the company leased 80-acre

Original headgate of the Farmers' Canal about 1889 near Wyoming-Nebraska border

Nebraska State Historical Society

tracts of irrigated land, complete with house and barn, to farmers who didn't have the money to buy land. All the ingredients that one day would go into successful irrigation were present. But it was simply ahead of its time. The experiment folded because of lack of interest.

Farther up the North Platte Valley, farmers were also discovering that they had to irrigate if they were going to produce field crops. Leon Moomaw's family settled in the valley near Chimney Rock in these early days, and he remembered that his father and their neighbors usually planted small fields of crops in the spring. Rainfall was sufficient to start the crops; then came the scorching sun and withering winds of July and August, and the crops burned up. Moomaw never forgot one heart-breaking scene on the family farm: "I was playing about our sod house—mother and father had been out looking at a small field of grain just to the north of the house. I was startled when upon their return, both were crying. Mother said to me with tears falling freely, 'This is the final day of destruction—our crop is burned up. I don't know what we will do.' "

The editor of the Bayard newspaper tried to comfort the discouraged settlers. One day, he said, the North Platte River would be harnessed to run scores of flour mills and factories. But, as Moomaw and the other pioneers realized, it was ridiculous to talk about water-powered mills when there were no crops.

So, who first thought of using North Platte River water to irrigate crops in the valley? According to Moomaw, the decision to irrigate came in 1886 when B. F. Gentry and S. W. Ripley ran water from Winter's Creek into a millet patch. Others followed. Moomaw concluded that the decision came to many at the same time, almost spontaneously.

A. B. Wood, a pioneer newspaper editor, said that the settlers certainly did not bring the idea with them. He admitted that irrigation to him "was a matter of dense ignorance—probably I had a vague idea of what the term meant but so lacking in knowledge were most of us that a vast stream of water flowing down in the midst of this area awakened no serious foresight." Generally speaking, remarked the Bayard editor, the North Platte River was looked upon as "a great hazard for crossing rather than a vast potential for usefulness."

Moomaw and Wood overlooked the fact that there were settlers in the valley who had experience in irrigation. These were the families who had moved to Nebraska from Colorado's irrigated districts. W. R. Akers, destined to become a leader in Nebraska's irrigation movement, was one of them.

Nebraska State Historical Society

South end of the flume across the North Platte River, south side of the Irrigation Land (Company) Canal. Photo undated.

Akers and his family had taken a farm in the North Platte Valley in 1885. Two years later he was busily helping his neighbors organize a canal company. Just as Akers and his friends began to move dirt, they received word that another group of farmers had started a ditch a short distance upstream. Alarmed by the report, Akers sent C. W. Ford galloping off to Sidney, the county seat of this huge territory, to register their water right. Ford arrived in Sidney and discovered that Nebraska had no comprehensive water law. He returned home and sent a copy of Colorado's statute to the Nebraska legislators in Lincoln. In due course, the company filed its water appropriation, the first such filing recorded in Nebraska.

Construction of the Ford-Akers canal, later known as the Farmers Canal, began in March of 1888. Water was delivered to farmers a few months later. It was no small accomplishment. The farmers who dug the canal were destitute. Moomaw said that the men and their families "picked up dry, bleached bones from the prairie and sold them in order to be able to purchase slips and scrapers to begin their excavation work."

That year a number of canals were dug by gangs of hungry, destitute farmers. But they brought water to the land. An 1889 survey reported that Nebraska had 214 irrigators. An irrigated farm averaged 55 acres. The total amount of land irrigated in Cheyenne, Scotts Bluff, Lincoln and Sioux counties was 11,744 acres.

This was not a particularly impressive record, but the irrigators faced many obstacles. Even with water flowing onto their fields, farmers quickly realized that the goal of profitable farming had not yet been achieved. Few knew how to farm under irrigation. Some answers would come from experience, but questions of a more complex nature related to irrigated cultivation and engineering could be answered only by trained specialists. Fortunately, the University of Nebraska (NU) and its experiment station were being staffed by a new generation of scientists and engineers.

It was perhaps prophetic that the first bulletin published by the NU Experiment Station was on the subject of irrigation in Nebraska. It was written by Lewis E. Hicks, the university's geologist, who endorsed irrigation, although his stand placed his university position in jeopardy. He was particularly unacceptable to the "boomers," who regarded him as a "kid-gloved book farmer" who knew nothing about practical farming.

In this bulletin, Hicks urged Nebraskans to look into irrigation. Don't let the controversy with Colorado over the South Platte River keep you from studying Nebraska's total irrigation potential, he warned. Irrigation, particularly in the western counties, would greatly increase the value and productivity of the land.

As for the argument that rainfall was increasing in Nebraska, Hicks observed that the residents of western Nebraska told him that the "line of advancing rainfall" had not reached their area yet. Since he was certain that irrigation would double crop yields, Hicks advised farmers to irrigate. "This result can be secured at once," he said, "without waiting for increased rainfall or anything else." The professor advised those who accepted the idea of increased rainfall to support irrigation, since spreading water on thou-

Nebraska State Historical Society

North end of a flume across the South Platte River west of North Platte, carrying North Platte River water for the South Side Irrigation Land Company Canal

sands of acres of farmland would undoubtedly enhance the conditions that bring rain. Ditches dug for irrigation were a good investment, he said, even if they were used for only a few years. ''It would hasten that change of climate which might in the end make irrigation unnecessary and it would pay to have good crops in the meantime.''

Hicks pointed out that farmers in western Nebraska would benefit immediately from irrigation. He predicted, as well, that in time farmers in the central and eastern parts of the state would also irrigate. At that time eastern land values were low and crop prices were rock bottom. But the time would come, Hicks insisted, ''when those lands will be too valuable to neglect any means of increasing their productiveness and then suggestions regarding irrigation and fertilizers will receive the attention they deserve.''

While Hicks hoped that the experiment station bulletin would encourage Nebraskans to think seriously about surface-water irrigation, he did not attempt to answer one important question in this publication. How could ''supplemental moisture'' be obtained for farms that lay outside the easily irrigated valley lands?

Throughout the 1880s and 1890s, a goodly number of westerners said that the answer to that need resided in artesian wells. (In these wells water rises higher than the top of the aquifer because it is held under pressure by fine-grained sediment; water from these wells sometimes flows at the surface.) They reasoned that wells could be sunk to tap into the underflow, and this water, under high pressure, would flow to the surface without mechanical assistance. It was a great idea.

In 1880 U.S. Senator A. S. Paddock from Nebraska urged his fellow senators to fund an artesian-well study. Only 3 percent of western land could be reclaimed by drawing water from streams, said Paddock. Artesian wells would increase that acreage considerably. Federal funds would pay for locating underground supplies and for developing appropriate drilling methods. The experiments would be carried out on government-owned land. Once artesian wells were proved to be feasible, he expected private capitalists to purchase the land and to pay for the wells.

Congress appropriated funds for the study, and a team of scientists was hired to direct the experimental drilling in eastern Colorado. Samuel Aughey of the University of Nebraska was a member of that team. The editor of the *Omaha Republican* hailed the good news. The investigators would find the best places ''to bore down and tap the subterranean ocean of the plains.'' Artesian wells, once proved to be practical, would make ''the sagebrush lands as rich as the lands belting the Missouri River.''

Unfortunately, the scientists found little to encourage such optimism. Aughey was co-author of the report that con-

Artesian well on the Grathern (?) Farm near Columbus

Nebraska State Historical Society

cluded that the chances of artesian wells providing water for irrigation "are not very encouraging." Aughey and his fellow investigators asked Congress to fund further field work. The important work should be continued, but they also recommended "that the greatest caution be exercised in the location and prosecution of future borings, lest a repetition of failure should permanently discourage all experiments of this kind on the part of the government."

In 1886 the artesian-well scheme reappeared in Congress. Richard J. Hinton was hired to write a survey of irrigation, and the huge volume that Hinton assembled included a lengthy study of California's artesian wells. Hinton suggested that the West had an enormous supply of underground waters—which might be tapped by artesian wells.

Hinton's conclusion placed him in direct opposition to Major John Wesley Powell, who totally rejected the idea that artesian wells would provide irrigation water. Powell's detractors, and there were many of them, both in government and in the scientific community, rushed to support Hinton. They criticized Powell for ignoring artesian wells and for insisting that only costly reservoirs and canal systems would meet the irrigation needs of the semi-arid country. Powell was put on the defensive. He finally agreed to look into artesian wells.

In 1890 Congress appropriated funds for another artesian-well study, and Hinton was named director of the project—a direct slap at Powell. Hinton was given three months to complete his study. Unable to conduct field investigations, Hinton called upon western scientists, including Professor Hicks of Nebraska, to submit information about the prospects for artesian wells in their areas.

Hinton assembled a tremendous amount of information. This study is often cited as the first systematic investigation of the nature and extent of groundwater. In the conclusion, Hinton said that while a considerable supply of underground water occurred in the plains and Great Basin areas, the technology simply didn't exist for raising the water. Artesian wells did not meet that need.

Nebraska editors overlooked Hinton's statements and jumped to their own optimistic conclusions. The Lincoln *State Journal,* for example, declared, "There can be no doubt that the entire area between the Missouri River and the Rocky Mountains is underlain with a sheet of water that is practically inexhaustible . . . the underground waters can be made to supplement the waters of the streams in

Nebraska State Historical Society

Artesian well near Lynch, Nebraska, Boyd County

reclaiming nearly every square mile of the arid region lying east of the base of the Rockies." The *McCook Tribune* examined the report through the same rose-colored glasses. If, said the editor, it should be demonstrated that artesian waters could supply water in sufficient volume to irrigate crops, then "the beneficient results to this bailiwick will be of the most encouraging character—the prospect will be most entrancing. Speed the good work!"

But who would pay for this good work? Having appropriated money to support several studies, Congress did not appear in the mood to throw good money after bad. William Paley, a Frontier County farmer, wrote Governor Thayer in the fall of 1890 and asked him to persuade the legislature to appropriate funds to drill artesian wells. To which Thayer responded, "The State can do nothing in regard to establishing artesian wells, or at least nothing can be done except by an act of the legislature, and I do not believe that body will ever enter into the building of artesian wells."

So farmers who attended an irrigation convention in O'Neill in 1894 turned to Washington. They adopted a resolution that called upon Congress to vote more money in order to discover once and for all, if "the underflow and artesian water is of sufficient volume and can be brought to the surface." Congress did not respond to the plea.

The fact is there were a number of artesian wells scattered over Nebraska. In Seward County, near Beaver Crossing, were a number of "flowing wells." Johnson County also had artesian wells, and there were a considerable number in the Sand Hills. The public fountain in Government Square in Lincoln was fed from a 1,060-foot artesian well. According to Erwin Barbour, water from the fountain was "carried away in bottles and pails because of real or imaginary curative properties."

Yes, there were artesian wells in Nebraska, but scientists offered little hope that such wells could be used to provide irrigation water for parched farmland.

The 1890s: "Water Preaching, Water Teaching and Water Reaching"

In 1890 the U.S. Census Bureau announced that the American frontier had vanished. No longer could a line be drawn between unsettled and developed land, between wilderness and civilization.

Several years later, a young historian named Frederick Jackson Turner published an essay in which he theorized that many of the distinctive traits of Americans, such as individualism, initiative and independence, had been produced by the frontier experience. These traits, however, would not conquer the lands of the West, Turner said. In this region, the successful settler must be "both a capitalist and a protege of the government."

The settlers with claims in the semi-arid and arid regions reached this conclusion too. To bring irrigation to their land, settlers urgently needed federal assistance. Nebraska's Congressional delegation agreed. They argued that federal aid was justified since the settlers had been "lured" into this country by generous federal land policies. Having given the settlers land, the government had the obligation to see that they survived and prospered.

Nebraska Senator William V. Allen also played upon the theme that the nation was running out of room. Irrigated farms would provide homes for the "surplus population of the country which is congesting in our cities and threatening the safety of the nation itself." Since small farmers were the backbone of the nation, as the great Thomas Jefferson had once implied, every effort should be extended to open the west to settlement. Thousands of small irrigated farms would increase the number of independent farmers and so revitalize the American political system. According to Senator Allen, the settlement of the semi-arid and arid West had national implicatons.

Since the federal government did not respond to the urgent pleas of westerners, several members of Congress from the region proposed having the states take over the remaining federal lands and use the income from their sale to pay for irrigation. That idea never had much support. It would take too long to get the money, and westerners were in a hurry. Most Nebraskans, despite their oft-expressed scepticism about federal programs, realized that they needed federal intervention to protect their water rights. Federal laws were needed to supervise the distribution of waters that flowed in the Platte and other interstate rivers. As J. H. Abbott of Big Springs told Governor Thayer in 1891, Nebraska had to have Congress on its side. Colorado and Wyoming had first crack at the North Platte and South Platte rivers, so it was almost impossible, said Abbott, to convince eastern capitalists to invest in Nebraska irrigation projects. These investors knew full well that their investments would be "rendered useless when the time came when these states would use all the waters now flowing into our state."

Democrats and Republicans eventually came to agree that federal intervention was required. The Populists supplied the final adhesive, for this party, made up of farmers and small-town business people, insisted that the federal government take charge of irrigation. In the early 1890s, O. M. Kem, a Nebraska Populist serving in the national House of Representatives, urged that money be appropriated quickly by Congress so that destitute farmers could be put to work building canals and dams. *The Alliance,* a farmers' newspaper, had long supported irrigation as a way to provide employment for idle workers and farmers. "On every hand we find idle, willing men in search of work," wrote the editor of the paper. "Our country has been flooded with all kinds of foreign cheap labor, so that at present rate of wages and scarcity of work it is almost impossible for men to earn a living."

The ideological tide slowly turned. Even conservatives such as J. Sterling Morton must have noticed the shift. In the early 1890s Morton was appointed U.S. Secretary of Agriculture. He sent a message to a convention of Nebraska farmers who had gathered to talk about irrigation. Morton congratulated the farmers for their independence and individualism. These sturdy yeomen would never turn to the federal government for help in building irrigation works. These projects were best done with local funds and local labor. Government "paternalism" would never do, insisted Morton.

The farmers sent Morton a note thanking him for his message. They then set to work on a list of resolutions, each of which asked Congress to finance irrigation projects. Morton, apparently, was out of step.

A terrible drought seared Nebraska in 1890. According to a North Platte reporter, not a bushel of wheat, corn or potatoes was raised in Lincoln County that year. The destruction was total.

Farmers had two choices—they could pull out, leaving their burned up crops behind them, or they could irrigate. The tragedy marked the turning point in the irrigation campaign.

It was during that dreadful year that the first irrigation project was begun in Dawson County. A company made up of non-Nebraska capitalists provided the funds to pay for the canal, but the company met nothing but opposition and indifference. After a year the investors pulled out. As H. O. Smith, a Dawson County implement dealer observed, even Mother Nature conspired against the advocates of irrigation, for in 1891 it rained and rained. Smith said that this "year of mud and moisture" ended talk of irrigation. But only temporarily, for it stopped raining in 1894 and crops burned up again. A handful of determined Dawson County men formed the Farmers and Merchants Irrigation Company. Although afflicted with "an acute attack of financial distress," Smith later wrote, the company secured money from some Ohio investors, and the canal was completed and water released on the land in July of 1895.

Drought spared no part of Nebraska in the early 1890s. The editor of the *Cambridge Kaleidoscope*, writing in the fall of 1890, said that during his 10 years in western Kansas and western Nebraska he had rejoiced with the farmers during good years and "wept with delinquent subscribers because of crop failures." He was convinced that irrigation was indispensable to the development of "this great fertile, but drought-ridden western country." He predicted that before the end of the century, farmers in his district would no longer depend upon "natural rainfall, which so frequently comes only after the crops have been irreparably damaged by drought, and provide facilities for and rely upon irrigation for moistening their broad acres."

On January 3, 1891, the McCook Farmers' and Business Men's Club announced that they were hosting an irrigation convention in their city on January 28. Counties in western Nebraska, as well as counties in Kansas and Colorado, were invited to send delegates. According to a McCook reporter, the convention was a resounding success. The 200 citizens who attended the convention needed no persuasion—they were all "wide awake to the importance of irrigation and alive to the blessings which will follow in the wake of irrigation when it shall be an accomplished fact." One question, and only one question, was before the convention: "How can irrigation be brought about?"

Professor Hicks told the delegates that the question, insofar as the Republican River valley was concerned, could be easily answered. He estimated that the river carried enough water to irrigate 100,000 acres. Tapping into the "underflow" would produce water for at least a million acres. Hicks was overly optimistic, and in the 1895-1896 report of the State Board of Irrigation, Akers said that the question of water supply in the Republican Valley had not been answered. He suspected that there was less water available than Hicks suggested.

But there was no time to lose. The McCook newspaper that carried stories about the irrigation convention also printed the report of the Red Willow County relief committee that was in charge of distributing free food to destitute settlers. During the month of January, the committee reported, dried fruit, tea, coffee, oatmeal, beans, rice, corn meal and flour had been distributed to 211 families, a total of 1,032 persons in the county. The population of Red Willow County, prior to the drought, was about 8,000 persons. The southwestern corner of Nebraska was really suffering. As the McCook editor declared, "It is a condition we confront, not a theory." The *Sidney Telegraph* joined those who said that irrigation was the hope of western Nebraska. The editor wrote:

Irrigation is the problem of Nebraska's development. It should have the support and attention of every man having anything at stake in the future of the state. In the last six months Nebraska has been the victim of a national advertisement of the wrong kind. It is now the interest of everybody having his home or capital here to wipe out the memory of last summer's catastrophe with some grand achievements. The eastern public draws no distinction with this section of the state. It knows that an appropriation of $1,000,000 has been asked of Congress to aid destitute settlers. There is not a merchant, banker, railroad nor real estate speculator who is not interested in preventing a re-currence of the drought. Thus irrigation becomes of vital importance to the state generally. In discussing this question, we are speaking of agriculture pure and simple and this has not been a profitable occupation. The prosperous farmer can stand an occasional affliction, but a continued failure affects the state. The farmers of Western Nebraska are known to be a hardy set of yeomen who can endure untold hardships, yet even this becomes monotonous. If then, the enterprising and energetic men of Nebraska can make the desert a garden and put prosperity where distress now flourishes are there not untold possibilities of growth in the next ten years? . . . The present agitation has resulted in a compact and aggressive organization of the friends of irrigation.

Residents of northeastern Nebraska shared that outlook. A resident of Holt County told Governor Thayer that most of the folks who took claims in that part of Nebraska came "without means and then trusted on Providence, good luck and the county poor fund to get through." The 1894 drought devastated these poor settlers. Arthur Mullen had just opened his law practice in O'Neill. He later said, "No one could have lived through the drought of 1894 without realization that man had to do something to counteract the evil of nature in the dry years." That meant irrigation.

Mullen, who in the 1930s would be instrumental in securing federal funds for Nebraska's power-irrigation system, dug deep in his well-worn pockets and came up with $100 that he invested in a Holt County irrigation project. The Niobrara River was to provide water for the project. But there was a problem. The engineer who laid out the ditches was either incompetent or a fool, for they ran uphill away from the river. Of course, the system was absolutely useless. Although, Mullen recalled, in 1899 and 1900 Holt County was inundated with heavy rains, and the ditches carried off the flood waters from the swamped fields.

Although interest in irrigation was on the rise, particularly in western Nebraska, the movement needed a spokesperson. W. E. Smythe, a reporter for the *Omaha Bee*, turned out to be that person. His editor gave Smythe the task of writing a series of articles about the relief needs of settlers in the western counties. Smythe agreed to accept the assignment, but he asked permission to write another series of articles about irrigation. The editor agreed on one condition. Smythe would sign the articles so that he, not the newspaper, would receive the criticism certain to come.

The editor misjudged the temper of the times and Smythe's ability to present the case for irrigation. Smythe's articles were well received. They sounded a call for an irrigation crusade. Walter Prescott Webb said that Smythe had the "fervor of an evangelist" and played upon "emotion, imagination, pecuniary desire and patriotism" to communicate that irrigation would "conquer" the semi-arid and arid lands. Smythe turned up at many of the early irrigation conferences, and he usually brought with him scientists, such as Professor Hicks, to discuss the technical aspects of irrigation. According to Smythe, Nebraska's western counties, which had suffered so intensely from the drought, could be made "to blossom into assured prosperity by irrigation." Smythe called upon the farmers to organize and to write a practical irrigation program. Above all, those who believed in irrigation would have to lobby the legislature and Congress for needed legislation and appropriations. "Agitation is the mother of legislation," Smythe declared.

The irrigation movement gathered steam. After the McCook convention, another rousing convention was held in Sidney on February 7. The Sidney newspaper said that the meeting was a tremendous success: "It is a clean cut and definite movement aiming at the accomplishment of specific results." Yes, indeed, echoed the McCook paper, "irrigation will soon be an accomplished fact in Western Nebraska—not a theory." From Sidney the irrigationists moved on to Ogallala, where another convention assembled on February 10.

The first statewide irrigation convention was held in Lincoln in February of 1891. Robert Furnas, former governor and long-time advocate of irrigation, was named chairman. Thirty-six Nebraska counties sent delegates who cheered speaker after speaker, each insisting that the time had come to act. Finally, the convention formed a committee to draw up an irrigation bill and submit the proposed law to the legislature.

J. H. Vickry of Gering, an irrigation backer, wrote to Governor Lorenzo Crounse in 1892. Vickry pointed out the problem that he and other irrigators faced. There was no longer any doubt that irrigation was essential. "We cannot make a success of agriculture without the use of irrigation," he told the governor. A steady stream of water poured down the North Platte River, but no one yet knew how to get it on the land. According to Vickry, "The one great task that now confronts us is how are we to raise the money necessary to build irrigation canals of sufficient length and capacity to supply the water necessary for our success." Vickry pointed out the basic problem: "Our laws on irrigation are very imperfect and should be amended. What we want is a law strictly local in its effect and all obligations made under it to be paid by the parties whose land it benefits." Farmers in western Nebraska wanted a law that would create local irrigation districts with power to issue bonds. "We want to keep big enterprises out, which might build needed facilities but would charge dearly for the water." Should the control of water fall into the hands of corporations, Vickry said, "we will have a hard fight for our living where it is obtained by farming."

But Nebraska's lawmakers took no action on the so-called irrigation bill in the 1891 and 1893 sessions. Farmers who owned land in five Panhandle counties—Box Butte, Sheridan, Sioux, Cheyenne and Scotts Bluff—announced that they were ready to secede from Nebraska and join Wyoming because that state already had acceptable irrigation laws on the books.

Irrigators were confused and frustrated. Many farmers abandoned the cause, while western editors criticized both the irrigators and the do-nothing lawmakers. At this critical time, the Business Men's Association of North Platte put out a call for a convention to meet in that city in December of 1893. This meeting turned out to be a pivotal event in the irrigation movement, as the delegates voted to establish the Nebraska Irrigation Association. The new organization was expected to provide continuous, organized, effective leadership. I. A. Fort was elected the first president, and he told the convention that irrigation was no longer a theory. Irrigation was practical, possible and necessary. During the past two years, he said, canals had been built to provide water for 1 million acres of Nebraska farmland. This work was done by farmers who often had to borrow money to buy food for their families and forage for their horses. A North Platte newspaper summarized the work of the convention under the bold headline: "Irrigate or Emigrate."

The Nebraska Irrigation Association held its second meeting in Kearney in December of 1894. The local editor said that "water preaching, water teaching and water reaching" were much in evidence at the session. Meanwhile, Fort was on the banquet circuit, talking irrigation to any group that would listen. For example, he appeared at the state convention of Nebaska bankers in the winter of 1894, telling them that the "question uppermost in the mind of the farmer is, how shall I best insure myself against a repetition of the losses of 1890, 1893 and 1894." The answer: irrigate.

Fortunately, said Fort, Nebraska was well situated when it came to potential water supply. The North Platte and South Platte rivers came from a huge drainage area, and there was little doubt that the supply should remain constant even in dry years. Fort said he was frequently asked if irrigation paid. To questioners, Fort replied, "That if life is worth living *it must pay.*"

He also reminded the bankers that irrigation was not simply a local issue. It had implications for the entire nation. Fort said that they were living in the "thickening up era," that is, the era in which America was becoming densely populated. Millions of small irrigated farms were needed to "relieve our crowded cities and give employment to the idle thousands now roving over the land." Irrigated farms would be an "antidote for anarchy, distrust and disorder." Fort was confident that irrigation "will relieve the evils now threatening society and the Nation."

Sidney hosted the association's third convention in 1895. At this meeting the delegates decided to sponsor an irrigation fair the following year at which irrigation practices and equipment could be put on public display. But 1895 was important for another reason, for in that year Nebraska's lawmakers enacted much-needed irrigation legislation. One law created a state board of irrigation that had the authority to settle water disputes. Another allowed citizens to form irrigation districts with power to levy taxes and issue bonds. State Senator W. R. Akers had a hand in writing and enacting both these landmark bills.

An organized irrigation movement was at last taking shape. Conventions sponsored by the state association publicized irrigation and recruited supporters. Legislation was in place, and articulate leaders kept the issue of irrigation before the public. And by the middle of the decade, another piece of the irrigation puzzle fell into place—science began to make a significant contribution.

O. V. P. Stout was one of the new breed of scientists who came to the University of Nebraska in Lincoln at this time. In 1896 he began a study of Nebraska's streamflow and water supply. Stout said that irrigation could not be profitable unless careful studies were made to lay a proper foundation. In 1896 he wrote: "It is only when we come to the consideration of some one plan or project to water lands that we come to the consideration of how little information is available that is really definite and immediately applicable to the case in hand."

To straighten out the situation, Stout joined with federal and state engineers to gage streamflow systematically and carefully. Streamflow information was absolutely vital to both irrigators and officials regulating irrigation. The work went well, and a few years later Stout reported that Nebraska's irrigators and other citizens possessed "a better knowledge of the capacity of the streams of the state to supply water for irrigation than has been possessed by the inhabitants of any other state at a correspondingly early period of its irrigation development."

Erwin Hinckley Barbour was another University of Nebraska scientist who provided valuable service. In 1892 he began a soil survey of Nebraska. He rejected the idea that rainfall was increasing, and he urged Nebraskans to stop looking for coal. Speaking as a geologist, Barbour insisted that Nebraska's most important resource was water, and it deserved careful examination. He said, "The development and employment of all of the [water] supply flowing on the surface or percolating underground necessitates a study as thorough as that given by other states to coal, iron or precious metals."

Barbour began his water study with an examination of Nebraska's underground water. He observed that the water table in most parts of Nebraska was subject to considerable fluctuation. It rose in wet years, dropped during dry times. He also looked into "sheet water," a misleading popular term that suggested that a vast lake of water lay under Nebraska's prairie. According to the popular notion, this lake would be easily tapped and the supply was virtually inexhaustible. Barbour insisted that this idea was entirely false.

He also studied how to conserve moisture in and on the soil. Progress was being made, Barbour said, in finding ways to catch and hold surface moisture. Farmers had also found ways to conserve soil moisture. He recommended that farmers in the semi-arid zone adopt the so-called Campbell method of dry farming that used "proper cultivation techniques" to produce crops on unirrigated land even during dry years.

As for streamflow, Barbour said that the mistaken notion persisted that rivers "cannot be drained, that as rapidly as water is drawn off as much more rushes in to supply the loss." Although some rivers, such as the Platte, did have "underflows" that might be drawn to the surface "if mechanical difficulties can be overcome," this underflow, too, had its limits. It was not inexhaustible.

Barbour also looked into water pollution, long a problem in Nebraska. As a matter of fact, Samuel Aughey, the university's first professor of natural science, had studied the problem too. In 1882, Aughey said that much of the illness in Nebraska was "caused by well water that has been permitted to become tainted by foul matter." He told a group of Lincoln businessmen that Lincoln had an obvious water-quality problem. Typhoid plagues that regularly swept the capital city were caused by bad water. While "the pure atmospheric conditions of Lincoln" neutralized the effect of impure water, that situation would not last forever. Aughey said that the residents must find an adequate supply of pure water, or else the city's growth would be limited.

Twenty years later Barbour had the same warning. Lincoln obtained its water supply from three well fields, all of which were located within the city's limits. A 60-foot dug well on F Street ought to be abandoned, Barbour advised, because its water had a high salt content. The same applied to the 68 wells in the South Street field. That latter group of wells, Barbour noted, "is virtually condemned." In 1897 Barbour announced that none of Lincoln's wells provided water of acceptable quality. Every well, to some degree, produced water tainted with salt.

1896 Irrigation Fair Intended to Rival State Fair

When the Nebraska State Fair opened in Omaha in late summer of 1896, the people of western Nebraska hardly noticed. Their attention was tuned to the Great Nebraska Irrigation Fair that was about to begin in North Platte.

The irrigation fair, born out of the 1895 meeting of the Nebraska State Irrigation Association, was intended to rival the state fair. North Platte irrigation promoters like Ivan A. Fort, pioneer president of the irrigation association, and William L. Park, who would become president of the Nebraska Irrigation Fair Association, were eager to demonstrate the benefits of irrigation to farmers who still depended on the rain. They invited manufacturers of irrigation equipment from around the country to come and display their wares. Popular events and exhibits were planned as well, primarily as a means to boost attendance.

North Platte residents had but 10 months to prepare for the fair, yet they faced the challenge with enthusiasm. While city workers erected stables, barns, buildings and a grandstand that could seat 2,000, Park and Emil F. Seeberger, secretary of the irrigation-fair association, traveled the Midwest promoting the event, which was to open October 9. Their plans were based, to a large extent, on similar irrigation fairs held previously in Kansas and Colorado.

Not to be forgotten, the women of the city organized a Women's Department of the Irrigation Fair. They held ice-cream socials and dances to raise money to have their own building constructed at the fairgrounds. In it they hoped to display works of fine art.

Interest perked up when area residents learned that Colonel William F. "Buffalo Bill" Cody, a sometime resident of North Platte, would be at the fair with his famous Wild West Show.

Arrangements were made to ship in a steam-powered merry-go-round and the highly regarded fish exhibit from the Nebraska State Fair. Horse and bicycle races were organized and several bands were hired. Anticipation grew.

When opening day finally came, people were more than willing to put down a quarter to pass the gates. They were greeted by windmills of all descriptions turning in the breeze, steadily pumping water from the 10-acre lake that was built specifically for the fair. Also in place around the lake were pumps and ditch-digging devices of all kinds. Men stood by their machines and demonstrated the workings to all who would hear them out. Fair officials wanted, more than anything else, to show farmers how to irrigate to make a better life for themselves, and to boost the well-being of the state. Manufacturers wanted, of course, to make a buck.

Irrigators from several counties in the region, including Dawson, Buffalo, Scotts Bluff and Keith, brought in wagon loads of irrigated fruits and vegetables. One Lodgepole Creek farmer from Cheyenne County displayed an onion weighing 19 ounces that, when cooked, made a fair meal for seven people, according to a report in the Nov. 20, 1896, annual of the Nebraska Irrigation Association. Wagon tours of area irrigation canals and ditches were also offered, free of charge.

The Irrigation Age, a national irrigation magazine of the day, said the fair demonstrated the progress made by Nebraska and the semi-arid region in resolving the issue of agriculture on the Great Plains. The article went on to say, "No longer does Nebraska ask aid for its drought sufferers; no longer is the population decreasing because of unsuccessful farmers leaving the state for more favored regions; they are irrigating with the aid of canals and ditches and with windmills and pumps, and they have again begun to feel that spirit of conquest that filled them when they undertook to make homes for themselves and families on the lonesome prairies."

The largest crowds came, understandably, on Monday, October 12, the day Buffalo Bill's Wild West Show put on two shows. More than 2,000 people went to the first show and 6,000 saw the second. Admission was 50 cents for adults and 25 cents for children under 9 years old.

Local newspapers reported that the most popular attractions were the women's art building, which had barely been completed in time, and the fish exhibit, which boasted a live 125-pound catfish.

When the fair closed on October 15, organizers pronounced it a success. A. G. Wolfenberger, president of the Nebraska State Irrigation Association, said during the 1896 meeting in Lexington, "The brightest dreams of those in whose brains the first conception of this splendid project took ideal form were more than realized in the magnificent display of products of garden, orchard and farm, which surprised and pleased every visitor, no matter from whence he came."

Park reported at that same meeting an increase in irrigation interest among area farmers after the fair. He also urged irrigation-association officials to join the irrigation and fair associations and to hold the 1897 fair in conjunction with the irrigation meeting. Later in the meeting, the irrigation association passed a resolution calling upon the Nebraska State Legislature to recognize the Nebraska Irrigation Fair and to appropriate a reasonable amount to assist it in "its work of educating the people of Nebraska in this method of agriculture."

The legislature failed to appropriate any money, and the Great Nebraska Irrigation Fair, which had occupied the thoughts of most North Platte residents for the better part of a year, died a quiet death.

—Brad Rundquist, weekend and copy editor,* Ft. Dodge (Iowa) Messenger, *and formerly CSD editorial assistant

But, on the positive side, Barbour told about enterprising entrepreneurs who took advantage of Lincoln's supply of salt water. Several companies bottled water from Lincoln wells and sold the liquid as "medicines and curatives for a wide range of human afflictions." Medical doctors ran "baths." Patients who bathed in warm salt water reported dramatic improvements in their physical conditions. This was particularly true of the victims of rheumatism. One commercial venture, said Barbour, deserved special recognition—the resort established on the shores of Salt Lake (now Capitol Beach) west of Lincoln. "Pavilions, bath houses and restaurants have been built, a small steamer and numerous sail and row boats have been added, and the whole bears the rather pretentious name of Burlington Beach," he said.

In the summer of 1897 Barbour embarked upon an interesting field investigation. He hired three students, arranged for teams, wagons and camping equipment and drove from Lincoln to Denver along the Platte Valley. The purpose of the trip was to examine the homemade windmills erected by farmers in the valley. Barbour wanted to gather information about these windmills, then disseminate it "so that people in the arid region could make and use windmills."

A fascinating book was the result of Barbour's tour. In the book he described the various kinds of windmills built by farmers and ranchers—"jumbos," "merry-go-rounds," "battle-axes"—complete with diagrams and photographs. Barbour said that most of these crude contraptions pumped water for stock and gardens. Nonetheless, they provided low-cost service and helped keep people on the land during dry years.

Barbour predicted that windmills would become increasingly important in the near future. As water in streams was

Nebraska State Historical Society

A "merry-go-round" mill near Lincoln, Lancaster County, around 1898

diverted for irrigation, windmills could be used to tap "the underflow of rivers." Windmills of increased efficiency could do the job. Barbour was not alone in this vision. In 1894 Smythe predicted that within 10 years 80 percent of the irrigated land in Nebraska would be "watered by pumps, windmills and other mechanical appliances." In 1898 an Omaha reporter said that Nebraska had 2,000 irrigation windmills whirling away. Many of the mills pumped water that supported farm gardens the produce of which "improved the diets of families. An appetite for fresh fruits and vegetables," said the writer, "will ruin the patent medicine business and drive doctors to change their businesses or leave the country."

But Barbour warned farmers and ranchers not to expect too much from their windmills. Mills in use then had technical limits. "Below a depth of 75 feet," said Barbour, "water cannot be raised profitably by present pumping devices." So, Barbour recommended that studies be conducted that would lead to the production of improved windmills and pumps. He believed that pump irrigation was destined to become an important factor in Nebraska's movement to irrigate.

W. R. Akers agreed. In a speech to the 1896 convention of the Nebraska Irrigation Association, Akers urged irrigators to think about the future, to look ahead. Even with streams tapped for irrigation, even with reservoirs to retain flood waters, even then some land would not be reached by irrigation. That is where technology entered the picture. Akers believed that the "time will come when there will be thousands and thousands of apparatuses for raising the water out of the bowels of the earth and lifting and pouring it out over the surface of the land, and all the irrigator will have to do will be to go to the central station and touch the button and the water will begin to flow and Nebraska will prove herself to be the grandest country on the face of the earth."

It was an incredible vision—the desert made to blossom by machines lifting underground water to thirsty crops. Akers, an irrigation pioneer, had indeed seen the future.

Hasty's Dream

A discussion of water and irrigation in 19th Century Nebraska that focuses solely on legislators, political leaders and scientists misses an important point. Valuable work was done at the local level by persons whose names seldom appeared in headlines. Scores of dedicated people, whom their neighbors most likely called cranks and visionaries, set to work to get water on the land.

One of this band of unappreciated pioneers was Dominicus S. Hasty, who moved to Arapahoe in Furnas County in 1871. The town had recently been platted by a Plattsmouth-based town company. The town site lay near the Republican River. Several creeks flowed nearby and emptied into the Republican. Since a town needed a mill or two, this was a good location; and before long the proprietors of the town had a small water-powered saw mill in operation on Muddy Creek. Since the settlers didn't have enough money to pay for a building, the machinery sat in the open air. But it was an important beginning. The residents of Arapahoe continued to talk about getting a grist mill.

This is when Hasty arrived on the scene. He was a surveyor and engineer by training, and under his supervision, the Muddy Creek dam was strengthened and a dependable supply of water obtained. The saw mill's output increased, and Arapahoe had a mill pond where boys fished in the summer, where couples skated in the winter and where the townspeople secured an abundant supply of ice.

More than ever, the townspeople wanted a grist mill. Hearing that a mill back east was being modernized, several Arapahoe men went east and purchased the discarded machinery. They brought the equipment to Arapahoe and installed

Travis Brothers jumbo, Lincoln, Lancaster County, around 1898; irrigated 5 acres of market garden. Cost—$8

Nebraska State Historical Society

it in a small mill building along the Muddy. They discovered they had made a mistake in putting the grist mill near the saw mill, for the Muddy did not provide sufficient water to power the two mills.

Hasty suggested that a canal be dug to bring water from the Republican River. Such a canal posed no engineering problems and the work could be done by farmers. But Hasty's proposal fell on deaf ears. Then, in 1879 Enos Clarke moved to Arapahoe. Clarke had money, and Hasty talked him into purchasing a mill site and a right-of-way for a mill race. The mill went into operation in 1880. The canal from the Republican ensured a steady flow of water. With the arrival of the railroad and an efficient water-powered mill, Arapahoe became an important commercial center.

Hasty had another idea. Why not use water from the Republican and its tributaries to irrigate farmland? In 1887 he purchased part interest in a Hitchcock County irrigation company. Farmers and ranchers eagerly bought water from the company. Hasty expected Furnas County farmers to do the same. So in 1891 he formed the Arapahoe Irrigation and Improvement Company, filed an appropriation for water from the Republican River and hired men to dig a canal to carry water to farms on the north side of the valley. Hasty's ditch was something of an engineering marvel. He had to build several wooden flumes (channels) to carry water over ravines and creek beds. One flume was more than 100 feet long and rose 25 feet above the creek that it crossed. Local scoffers said the flume would never hold water. They were wrong. The ditch was completed in 1893, and everything worked fine. There was only one problem—farmers did not buy the water. They said that irrigated farming was just too much work. There was another problem. Several times ditch banks collapsed, and water flooded fields and adjacent crops. Owners of the flooded land sued Hasty, who suspected that the farmers had cut the banks to have a court case. Finally, two farmers agreed to take water for the season, but only because Hasty gave them water.

The ditch operated again in 1894. Hasty hired ditch riders to patrol the canal and armed them with shotguns. There were no breaks in the canals that season, and water flowed into Arapahoe, where the townspeople rejoiced. The owner of Hasty's former mill, however, was not happy, for Hasty's canal drained all the water out of the Republican River. His mill stood idle, due to lack of water, while Hasty's canal carried water north of town.

The mill operator brought suit against Hasty. Three years later, the case ended up in the Nebraska Supreme Court, where Hasty's water right was upheld. During the three years Hasty had been in court, however, the canal stood dry. In 1896 Professor Stout reported that Hasty's ditch was dry, flumes were blown down and it was not useable. Stout added that in his estimate the Republican River was not "an ideal irrigation stream."

Hasty was not discouraged, and in 1897 he had his system in working order again. Once again, he offered water to farmers, but there were no takers. He decided to run water into Arapahoe, where residents could use it to water lawns and trees. Townspeople paid for the water, but the farmers remained adamant. Finally Hasty said that if the farmers would maintain the canal, he would provide them water without cost, but still, no takers.

At the end of the 1897 season, Hasty gave up. Though an engineering success, his irrigation ditch was a financial failure. Hasty's experience was not unique. In other parts of Nebraska men tried to launch irrigation projects, only to be met with ridicule and indifference. Hasty and other irrigation promoters were criticized for suggesting that "artificial moisture" was needed to raise crops in Nebraska. But their biggest problem was that farmers quite frankly did not want to irrigate because it was too much work. Repeatedly Hasty heard farmers declare, "If I have to irrigate, I'll go back east to where God provides irrigation." Hasty was a pioneer who showed the way, but there were few as yet ready to follow his pathway to the future.

A New Century

By end of the 19th Century, Nebraskans began to realize that water was a valuable resource. They also understood that irrigation was practical and, in some parts, essential. Farmers began to look to scientists, engineers and irrigation bureaucrats for information about managing and using water. And the Nebraska State Irrigation Association provided a forum where professionals and irrigators met to discuss issues and opportunities.

Thanks to the work of water specialists, more was known about water than ever before. Popular and technical publications carried pages of statistics relating to irrigation practices, streamflow and water use. With data accumulated, the professionals concluded that Nebraska's surface water supply was limited. There was not enough water in Nebraska's streams to bring about the "conquest of arid America" that Smythe had talked about.

Professor O. V. P. Stout, for example, repeatedly warned that the irrigation potential of the Republican River in southwestern Nebraska had been overstated. The supply of water in that river, he said, would not support extensive irrigation. A Dawson County official brought up the same issue regarding the Platte. In 1896 he noted that 14 canals in the county took water from the Platte River. He wondered if the supply of Platte water would be adequate to meet the demand.

This was the basic question faced by irrigators in the 1890s. Even if an irrigator held a valid water right, there was a good chance that he would not get the water he needed when he needed it. Without dams and reservoirs to harness Nebraska's river system, spring flood waters swept through the state, and then by summer, streambeds were nearly dry.

Trying to find a way to adjust to the pattern of seasonal river flow, some irrigators decided to use river water as "supplemental moisture" whenever it was available. That is, let spring flood water flow onto the fields. Allow the moisture to soak into the seedbed where, theoretically, the water would remain until needed by growing crops. In the fall, if water were available, repeat the process. Under this scheme the earth became a reservoir—an inefficient form of storage but the only one available.

The idea of supplemental moisture hung around for about a generation. It is easy to understand why—because dams, reservoirs and irrigation systems were expensive and took time to build. Indeed, at the turn of the century, money loomed as the irrigators' number-one problem. Help was on the way, however, for the long campaign to publicize the need for irrigation in the West paid off. In January of 1901, Senator Francis Newlands of Nevada introduced a national reclamation bill in Congress. President Theodore Roosevelt endorsed the bill. He told Congress in December of 1901:

> It is right for the national Government to make the streams and rivers of the arid region useful by engineering works for water storage as to make useful the rivers and harbors of the humid region by engineering works of another kind. . . . The government should construct and maintain these reservoirs as it does other public works.

Nebraskans had been using this argument for years. No wonder they idolized Teddy.

The Newlands bill, which became law in June of 1902, provided that proceeds from the sale of public land in the West were to be used to pay for irrigation works. The North Platte River valley, reaching from eastern Wyoming into western Nebraska, was selected as one of the first projects under the Newlands law. With the enactment of this federal law, historian Martin Carlson points out that Nebraska had gone through five stages of irrigation development: individual, corporate, district, state and, finally, federal.

As is true of every new piece of legislation, the Newlands act had defects. The biggest problem arose from the question of how farmers, residing in irrigation projects administered by the federal government, were to pay for their water. The issue was not easily resolved, and for years the repayment issue plagued the North Platte project. Some Nebraskans reached the conclusion that future irrigation projects ought to be combined with hydro-electric projects, so that the sale of electric power might be used to help pay for constructing dams, canals, reservoirs and power houses. Along with this idea, there also developed the belief that Nebraska's water resources should be developed under public, not private, control. Both of these ideas had strong backing in Nebraska by the beginning of the new century.

While harboring questions about the best way to pay for irrigation projects, most Nebraskans thought that the water-law issue was settled. Nebraska's laws, based on Wyoming's statutes, appeared to be well suited to Nebraska's needs. According to historian Donald Pisani, however, 19th-Century irrigators never fully understood the problem that would arise in the future under prior appropriation. According to Pisani, Wyoming's legislators adopted prior appropriation to encourage the rapid agricultural settlement of the state. The cattle industry had collapsed due to bad weather and low prices. Wyoming's leaders thought that prior appropriation—the promise of abundant, free water—would bring farmers flocking to Wyoming.

The problem was, says Pisani, that prior appropriation did not take into account the efficient use of streamflow. In the 1880s, for example, rights granted to Colorado farmers in the South Platte Valley, under a modified prior appropriation system, added up to several times the average amount of water that flowed in the river's channel. No wonder the South Platte ran dry in Nebraska!

The problem was recognized by some irrigation officials. In 1902 Frederick Newell, the first director of the U.S. Reclamation Service, said:

> There must come a time when water must be apportioned with justice to all, and a century or more hence we will have it distributed not upon priority rights, but upon technical rights. We cannot have a farmer getting more water than he is entitled to, because his great grandfather or somebody else happened to secure the water right two months ahead of somebody else. Water must ultimately be conserved in the most just manner for the general welfare of all citizens.

In 1900, however, irrigators were not thinking a century ahead. They were thinking about getting water. Most irrigators realized that the supply in streams was limited. They had concluded that artesian wells were not workable, so they began to discuss pump irrigation. They knew that windmills were not suited for large-scale irrigation, so they talked about using gasoline engines and steam engines to power pumps. From the irrigators' perspective, it would not be long until efficient engines and pumps raised water—and until the issues of underground water ownership and management would come up for public debate.

In the early 1900s the irrigation movement was well underway, but there were still problems. For example, while legal and technical aspects of irrigation had been fairly well worked out, there was as yet little understanding of how to farm under irrigation. How much water should be applied to crops? When should water be applied? What crops were best suited to irrigation? Would drainage systems be required? These and other pressing questions had to be answered, and after 1900, irrigators looked to university and government researchers for answers. The University of Nebraska began to offer short courses in irrigation, sugar beet culture and other relevant topics. Supported by both private and public funds, irrigation experiment stations were established where the best practicies in irrigation farming could be demonstrated, tested and refined.

Unfortunately, the answers did not come rapidly enough to satisfy the people on the land. Old-timers in the North Platte Valley have said that by 1912 most irrigators were ready to call it quits. They were bitterly disappointed because the promise of irrigation was so slow to be realized.

Indeed, in 1900 expectations did exceed reality. A down-scaling of expectations was certain to come. Some predictions, in fact, would prove totally groundless, such as the widely held notion that irrigation meant saving the small farm. H. O. Smith, the Dawson County irrigation pioneer who was right about so many things, endorsed this idea. He insisted that his neighbors who opposed irrigation did so because irrigation meant working small farms. To these men, Smith said, "real farming" meant working large fields of wheat and corn. "The greater the acreage, the greater the farmer," was their thinking. Smith said that these farmers would rather work 600 acres and make a "bare living or less," than work 40 acres and "have a bank account." "This idea of farming large areas has made more business for the sheriff than [for] any other branch of industry," he said. In his view, the salvation of Nebraska agriculture lay in small, irrigated farms.

W. E. Smythe, irrigation's best-known promoter, also accepted the premise that irrigated farms would be intensively cultivated small farms. The creation of thousands of such farms would be the salvation of the nation. Irrigated farms would save agriculture; they would stem the flow of families from the land; they would restore the small farmer to political, social and cultural prominence. Western land agents were quick to catch this vision. A Gering newspaper carried this notice in February of 1891: "There are openings for small ranches, hundreds of acres of irrigated land for sale, and dozens of irrigated farms for rent, in the beautiful Platte Valley. You poor, heel-ridden eastern renter, take the advice of many and come west!"

The story of land and water in 19th-Century Nebraska, then, is the story of dreams clashing with reality. It is the story of efforts to develop an entirely new industry—irrigated agriculture. And it is a story with many heroes and heroines. While legislators, scientists and developers played highly visible roles in this drama, there were many people whose contributions are easy to overlook. There were farmers and ranchers who took a considerable risk going into irrigated farming; there were small-town bankers and newspaper editors who supported those who wanted to find a way to make the land pay. And there were inventors and tinkerers who in their dimly lighted shops came up with the machines and paraphernalia needed by 20th-Century irrigators.

Mads Anderson is typical of these tinkerers and risk takers (*see endpiece following ''Farm Family'' chapter*). A Danish immigrant, Mads moved to Howard County, where he ran a store. In 1894 he moved his family to a farm along the Loup River. He built an enormous water wheel that sloshed water out of the river and dumped it on his truck garden. Always alert to opportunities, Mads hired a Danish blacksmith to make a ''land roller.'' This implement was common in Denmark, where it was used to pack the seed bed. The land roller worked well in Nebraska, too, and Mads formed a company to produce it. Later he added other lines of farm equipment, such as grinders and pumps. By the 1920s Mads Anderson's firm, The Western Land Roller Company, was producing irrigation pumps. He clearly saw the future of Nebraska and Great Plains irrigation. Mads Anderson, and thousands of other irrigation pioneers, innovators and risk takers, made Nebraska ''blossom as the rose.''

Sources (and Suggested Reading): *The Great Plains*, Walter Prescott Webb, 1931, Ginn and Company, New York. (An understanding of the plains country begins with this classic study of people and the plains environment); *Wilderness and the American Mind*. Roderick Nash, 1963, Yale University Press, New Haven. (This is a readable and fascinating inquiry into the attitudes of Americans toward wilderness.); *The Great Platte River Road*, Merrill J. Mattes, 1969, Nebraska State Historical Society, Lincoln. (This is an excellent study of the overlanders who followed the Platte River across Nebraska); *Public Power in Nebraska*, Robert E. Firth, 1962, University of Nebraska Press, Lincoln. (This is a survey of how public power came about in Nebraska. The first part is based upon a dissertation written by Gene Hamaker, ''Irrigation Pioneers: A History of the Tri-County Project to 1935,'' in which the story of Joel Hull's canal is told.); *Western Democrat*, Arthur Mullen, 1940, Wilfred Funk, Inc., New York. (Mullen was an important Nebraska political figure who as a young man witnessed the 1890s drought in Holt County.); *Beyond the Hundredth Meridian: John Wesley Powell and the Second Opening of the West*, Wallace Stegner, 1982, University of Nebraska Press Bison Books, Lincoln (a superb, sympathetic portrayal of Powell); *Pioneering in the Shadow of Chimney Rock*, Leon Moomaw, 1966, The Courier Press, Gering, Nebraska (an excellent personal account of early irrigation in the North Platte valley).

For the role of the University of Nebraska in the development of irrigation in Nebraska see: *The University of Nebraska College of Agriculture: The First Century*, Elvin F. Frolik and Ralston J. Graham, 1987, The Board of Regents of the University of Nebraska, Lincoln; *Frontier University*. Vol. I, *The Centennial History of the University of Nebraska*, Robert N. Manley, 1969, University of of Nebraska Press, Lincoln; ''The Development of Irrigation in Nebraska, 1854-1910: A Descriptive Survey,'' Martin E. Carlson, 1963, Unpublished dissertation, University of Nebraska Department of History. (This dissertation contains a full discussion of the origins of the irrigation movement in Nebraska); *Water Powered Flour Mills in Nebraska*, Thomas R. Buecker, 1983, Nebraska State Historical Society, Lincoln (a concise, thorough overview of this important topic); ''Wealth Beneath the Prairie: The Search of Coal in Nebraska,'' Robert N. Manley, *in Nebraska History*, v. 47, June, 1966. (This article discusses the relationship between the unsuccessful search for coal in Nebraska and the discovery of groundwater.); *Wells and Windmills in Nebraska*, Erwin Hinckley Barbour, 1899, U.S. Government Printing Office, Washington, D. C. (If you are interested in Nebraska's water history, you will want to obtain a copy of this fascinating study.) ''Enterprise and Equity: A Critique of Western Water Law in the Nineteenth Century,'' Donald J. Pisani, *The Western Historical Quarterly*, v. 18, January, 1987 (Pisani presents an interesting and controversial examination of the doctrine of prior appropriation.)

In the 1980s, a ''revisionist school'' of western historians rejected many traditional views of western history. With regard to water, they are extremely critical of past policies. In particular, they question irrigation programs. Two examples of this school of writing are: *The Legacy of Conquest: The Unbroken Past of the American West*, Patricia Nelson Limerick, 1987, W. W. Norton and Company, New York; and *Rivers of Empire: Water, Aridity and the Growth of the American West*, Donald Worster, 1985, Pantheon Books, New York.

A well, presumably artesian, in Holt County, probably near turn of the century

Nebraska State Historical Society

Native Americans and Water, Prehistory to 1900

by Charles Flowerday
Editor
Conservation and Survey Division, UNL

The Paleo-Indians: Ancient North Americans Emerge on the Plains

What they thought when they first crossed over is anyone's guess. Most likely, when the "first Americans" set foot on the new continent, they didn't think of it as a momentous event. At first, moving through the Bering strait or its land bridge, they saw a region much like what they had come from.

When they came no one knows. They may have crossed from Siberia to Alaska during a time of lowered sea level as much as 37,000 years ago, according to some experts. And, beyond curiosity, we don't even know for certain what drove or lured them, though it was likely a quest for better hunting or some intertribal strife.

The oldest estimates of migration to North America are based on findings at sites such as one in northern Texas near Lewisville where an ancient hearth—a fire pit—was uncovered accidently by the operator of an excavating machine. Twenty-some hearths were found there about 20 feet below the surface. Radiocarbon dating of plant remains place this occupation at 35,000 B.C., though some dispute this. (Radiocarbon dating measures radioactive decay of any carbon-based—organic—material to determine its age.) More conservative estimates based on surer evidence put the horizon of New World occupation at about 13,000 years ago.

In all likelihood, then, ancient North American peoples roamed, hunted in and occupied the Great Plains before what is generally agreed to be the limit of any substantial evidence of occupation. This certain limit provides an approximate starting point for the *Paleo-Indian* period, the first major archeological period in the Americas—about 12,000 to 8,000 years ago.

One site among dozens providing this evidence is the Lindenmeier Ranch site in northeastern Colorado, dated at about 11,000 years ago, where projectile points (for spears and lances) were found with the fossilized bones of late Pleistocene mammals now long extinct. (The Pleistocene Epoch included the most recent ice advances and spanned from about 1.6 million to 10,000 years ago.) During the latter part of the Pleistocene, generally a time of ample moisture, lush grasses fed herds of mammoths, ancient bison and other large mammals that no doubt lured hunters in an abundance roughly proportional to the amount of game. Numerous artifacts about 10,000-11,000 years old have been found throughout Nebraska.

These people, like all others, doubtless needed frequent and relatively easy access to water. Since they dug no wells, they depended mostly on stream courses for drinking, cooking, washing and ritualistic needs. In addition, though much later, Native Americans came to rely on stream courses and their valleys for moisture for agriculture, as crops were grown in garden-like plots on moist terraces and bottomland made fertile by frequent flooding. Some of these plots even may have been hand-watered during times of drought. That, however, is getting ahead of our story.

The earliest inhabitants of the central Great Plains were surely nomadic hunters and gatherers. The plains provided some of the great routes of travel for aboriginal peoples because they were the paths of least resistance. The region offered abundant game while providing a virtual "ladder" of eastward and southeastward flowing water.

Paths of migration normally followed stream courses or intersected them for encampments. Waterways were crucial to movement in the hunt, often as much for the cover the ravines provided as for meeting the body's needs. They also probably provided cover for war parties and raids, though no real evidence for such warring survives from this period. In addition, animals were probably ambushed as they came to pools or springs to drink.

Evidence also suggests that harvest of aquatic species supplemented hunting very early on, as few sources of food were overlooked. Some fishing, gathering of mussels (freshwater clams) and hunting or snaring of waterfowl are involved here. This would also include hunting of muskrats and probably beaver, otter, turtles and frogs. Since no evidence of nets, harpoon points or bone hooks exists, harvest of catfish, for example bullhead and channel cat—as well as flathead catfish, drum and gar—was likely done in late summer when streams were drying up and they could be clubbed or gathered by hand. Such a method could also yield mussels, reptiles and amphibians.

The Archaic: Toward a "Modern" Ecosystem

Aside from religious significance and ceremonial uses, which will be explored later, the uses above were the main uses of water made by native people in Nebraska up through the *Archaic*—the second major archeological period, extending from about 8,000 to about 2,000 years ago, or up to 0 A.D., according to some experts. In all probability, the main uses of and adaptations to water during the Paleo-Indian period would have persisted through the livable part of the Archaic as well. The Archaic was a time of transition to a "post-ice-age" ecosystem that supported modern plants and animals. But the first part of the period, roughly the first 2,000-3,000 years, was marked by a hot, mostly dry period called the *Altithermal*. Geologic and archeological investigations attest to a prolonged drought that left few remains or decayed plants to be dated.

A relative absence of water, of course, meant a relative absence of people, who probably migrated east or much farther south. Most likely, the western part of the state was so consistently dry that only a few forays for

The bison (buffalo) have been called a "walking supermarket" for Native Americans before reservation life. They could also be called a "hardware and clothing store," as they supplied hoes, scrapers, clubs, sled runners, tool handles, tanned goods such as tipi covers, moccassin tops, shirts, leggings, dresses, robes, caps and mittens. Paleo-Indians hunted larger bison than the modern version, along with other large Pleistocene mammals.

Charles Flowerday, CSD

hunting and gathering were made into these parts during brief periods of improved climate. Most researchers suppose that the bison were greatly reduced if not nearly eliminated and that this produced a greater reliance on foraging. Seasonal, short-term occupation of the western part of the central plains may have occurred where vegetation was available at predictable intervals.

Some scattered evidence of habitation during the Archaic, however, presents an interesting variety of water-related phenomena in and around Nebraska. Sites from the Altithermal do occur along the Front Range of the Rocky Mountains and in the eastern part of the state. One of the most important is the Logan Creek site about 50 miles north of Omaha. An early Archaic site dated about 6,000 years ago, Logan Creek has yielded evidence of much bison hunting but also taking of birds, fish and mussels. Archeologists have even found a bone fish hook, the oldest such artifact in the state.

An unusual angle on water use comes from a series of butte-top middle Archaic habitations in the Panhandle along the North Platte Valley, particularly Signal Butte, that are dated about 3,000-5,000 years ago. Far from surface water due to their elevation, some researchers believe these people caught rainwater in rock-bottomed pits atop the buttes.

Archaic sites have also produced many stone "boiling pits." Early Native Americans would dig a pit, line it with hides, fill it with water and then warm the water with fire-heated stones. When the water began to boil, food was dropped in, a major step forward in food preparation.

Plains Woodland: Settlement and Agriculture Emerge

While hunting and gathering were probably the mainstays of the economies of plains peoples during the Archaic period, by about 0 A.D., the start of the *Plains Woodland* or *Early Ceramic* period, native people began to practice some limited agriculture, more like gardening than anything else. Like the Archaic groups, Woodland Indians in the Missouri River basin still hunted and gathered wild plant foods as their primary foods, accounting for a cyclically nomadic lifestyle. Edible plants particularly suited to floodplain and marshy environments, such as cattails, arrowhead and marshelder, were gathered extensively.

But Woodland peoples also are responsible for the appearance of three important changes. They buried their dead in communal burial mounds, indicating a more settled life. They made simple conical-shaped clay (ceramic) pottery that gives the period its name. And they began to cultivate corn, beans and squash. This first archeological evidence of agriculture in Nebraska comes from a few sites along the Missouri, Elkhorn and Platte rivers dated about 1,000-2,000 years ago that revealed a little corn. Population increases and influences from the southwest and/or east likely stimulated the cultivation of corn and the making of clay pottery.

Agriculture, pottery and a semi-settled lifestyle are all tied together. Among other uses, such as holding water boiled to make stews of meat, grain and roots or tubers, the pots provided temporary, above-ground storage of grain and dried vegetables. Larger quantities were put in longer term storage in below-ground pits. The pottery also represents another, less obvious use of water, as the clay had to be moistened to be worked into a vessel. Pots were also probably used as containers for carrying water, as were skins and bison bladders, to water streamside plots during dry periods.

As settlements occurred, mostly in the eastern part of the state (though some important finds come from the upper Republican River basin), these were invariably by streams, usually on ridges or benches. While such occupation reaches back to 0-500 A.D. in the eastern river valleys, most of the earliest semi-permanent dwellings in the western part of the state date from about 600-900 A.D. Settling near streams also provided timber for poles to support earthlodges, tipis or bark houses, as well as for fuel and other uses. These gallery forests often gave out relatively quickly however, requiring a move to some other spot near water. In addition the large support poles for lodges would rot in about 8 to 10 years, requiring a rebuilt lodge or relocation.

Another interesting aspect of use of water-related materials concerns communal burial spots, called "ossuaries," of a complex extending from south-central Nebraska to central Kansas—the "Keith-phase" Woodland people. Their burial mounds display an abundance of freshwater shell made into disk-like beads on strings, hundreds of beads at some sites, presumably contributed as gifts to accompany the dead into the next world. Freshwater and marine shell pendants also occur in triangular, oblong, crescentic and narrow curved shapes. The marine shell can be traced to the Gulf Coast, indicating extensive trading networks. One ossuary on Prairie Dog Creek near the Kansas-Nebraska border was radiocarbon-dated at about 600 A.D. Evidence from various periods also suggests that clam shells were used as scrapers, particularly for shelling corn, as well as for ornaments.

The Rainbow site near the Floyd Valley in northwestern Iowa is another important Great Plains Woodland site related to agriculture and therefore to water, as stream-valley cultivation is nearly universal. Archeological seed samples offer evidence of occupation from the early to late Woodland period. Regarding the Rainbow site investigations, in *Woodland Cultures on the Western Prairies*, David W. Benn explains, "Annuals such as sunflower, sumpweed, and chenopodium [goosefoot], are highly variable, sometimes rare components of the climax forest, but in soils disturbed by humans these plants explode into prolific resources. . . . Their cultivation, if not always actual domestication . . . is a logical outgrowth of Archaic subsistence patterns [brackets added]." Corn, beans and some squash joined the plants mentioned above after the onset of cultivation, becoming staples in Woodland culture diets.

The Walker Gilmore site near Plattesmouth is also a significant Woodland find. Animal remains indicated exploitation of mammals, aquatic species and migratory waterfowl. Plant remains provide some of the most substantial evidence for agriculture during the Nebraska Woodland period, including cultivation of gourds and squash. Radiocarbon dates place this site at around 1000 A.D.

Plains Village/Middle Ceramic: Agriculture Takes Hold

The Woodland period gave way to the *Plains Village* or *Middle Ceramic* period, which began around 900 and lasted until about 1500 A.D. During this time, more people lived in Nebraska than during any other prehistoric period. The population probably exceeded that of Native Americans during the Historic period. An increased abundance of corn, due to new varieties, and better hunting allowed for larger populations and the development of crafts for domestic and sacred purposes. Two important variants of the Middle Ceramic are the Nebraska phase and the Upper Republican phase. The first involves settlements near the Missouri along the southern two-thirds of the state's eastern border; the second refers to settlements mostly along tributaries of the Republican River as far west as McCook. Another similar culture is the Istkag on the Loup rivers.

Waldo Wedel, in *Central Plains Prehistory*, explains some of the basic patterns of subsistence for these peoples: "There is no archeological evidence . . . that Upper Republican peoples carried out communal hunts and mass kills of bison like those that were common in the northwestern plains. It may be suggested, therefore, that hunting—like the gathering of plant foods, fishing, and gardening—was probably of such a nature that it could be handled, as seasonally appropriate, by individuals, families, or small groups. In much of the Republican basin, notably from Franklin County westward around the Harlan County Reservoir district, and in the environs of the Medicine Creek reservoir (Harry Strunk Lake), the deeply dissected valley margins, the flat-floored, well-grassed canyons, and the deep, narrow tributary valleys would have provided near ideal terrain for hunters to stalk bison by foot."

After about 1000 A.D., the appearance of numerous, large, "bell-shaped" storage pits, 3 to 5 feet deep, marks the advent of real, though small-scale, agriculture. Crops cultivated in small streamside plots included corn, beans, squash, sunflower and marshelder. The homesteads at this time generally consisted of one or two to a half dozen lodges above streams. Another aspect of water-related phenomena that directly affected cultural development was the increase in precipitation that improved the climate for cultivated plants around 850 A.D. Known as the *Neo-Atlantic* episode it brought increased moisture from the Gulf of Mexico (part of the Atlantic Ocean).

This climatic episode lasted until about

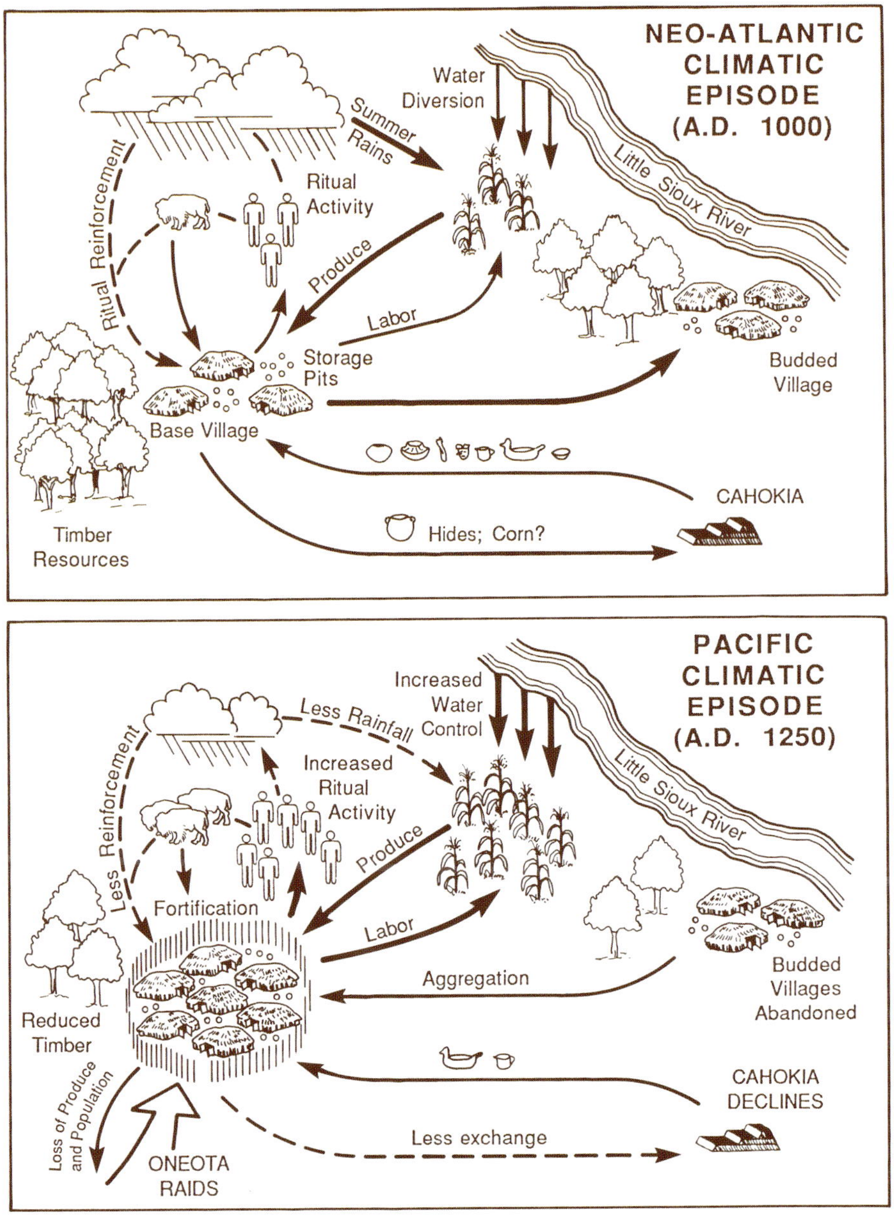

Mill Creek culture of northwestern Iowa probably involved small stream diversions that increased during the Pacific climatic episode (adapted from "Toward a Processual Understanding of the Initial Variant of the Middle Missouri Tradition: The Case of the Mill Creek Culture of Iowa" by Duane C. Anderson in American Antiquity, 52(3), 1987).

1250 A.D., when the Pacific episode began, bringing a higher frequency of drier and cooler western winds. Due to increases in population and food-production during the Neo-Atlantic, particularly in the Missouri Valley drainage, small settlements had "budded" off of larger villages on major valleys and had settled along minor streams and in uplands. With the beginning of the Pacific, many of these consolidated with larger settlements. This evidence comes, among other sources, from the sites of the Mill Creek culture of northwestern Iowa, along the Big and Little Sioux River drainages. Some attempt to divert water from streams, which probably increased during the drier Pacific episode as traditional horticultural methods were under stress, is also in evidence from these sites. As the short-grass prairie moved eastward, and bison followed, bison came to replace deer as the primary source of meat.

From about 1200 to 1400, warfare and drought began to depopulate the plains. This warfare probably occurred between Nebraska village people and those in North and South Dakota. A prolonged drought also played a part in driving out the villagers. The result was that the central plains village culture disappeared.

Late Ceramic: Change, Conflict and the Athabascan Infiltration

Following a two-century gap in the prehistoric sequence in Nebraska, by the beginning of the *Late Ceramic* period (about 1500 to 1800 A.D.), increased moisture on the plains made the region much more hospitable. For native peoples, life began to change much more rapidly than it ever had before. This period includes the first, though initially infrequent, contact with Europeans. Only an occasional Spanish expedition from old or New Mexico visited a few tribes at a time until the early to mid-1700s; then came French traders and trappers from Canada and Louisiana.

But the most significant change during the 1600s, not so much in the eastern submid areas but on the western plains, was the introduction of the Athabascans. Eventually to become Apaches, Navajos and others, they infiltrated the High Plains from western Canada. The ancestors of the Caddoan-speaking villagers (the Arikara of South Dakota, the Pawnee in Nebraska, the Wichita of Oklahoma and later southern Kansas and the Caddo of Texas) had dominated the region, particularly its eastern half. The Athabascan intruders originally were foot-bound nomads who migrated from northwestern Canada. However, through raiding and war, they acquired horses from the Spanish and Puebloan and Caddoan tribes, and thereby, became the first plains Indians to have horses in large numbers.

Most of the evidence for this invasion comes not from the Athabascans themselves but from the villagers' fortifications along their northwestern frontier. This southward expansion stopped at the desert southwest, but like the Caddoan peoples farther east, the Athabascans extended through the High Plains.

This immigration had two important implications for Native American cultures and their relationships to their environment. First, bringing an Old World lifeway evolved through caribou hunting in Canada into the pursuit of bison, they followed the "ladder of rivers" down the plains to expand throughout the west-central United States. They were nomads, the only Indians to be at home on the grasslands in this way from about 1500 to about 1700. Using hundreds of pack dogs per tribe to carry hides and meat, they traded extensively up and down the plains. Doing so, they set in motion a series of cultural collisions, particularly as they began to acquire and trade horses or lost them to raiding, that would reorder the lifeways of most plains peoples.

Secondly, as the Spanish began to take plains Indians as slaves for their mines and Carribean plantations, the Apaches began to raid the Caddoan settlements and carry off the villagers to trade as slaves. Two important developments resulted. Horses made their way north as slaves moved south. And, somewhat ironically, the Apaches became more settled even as horses became more plentiful. Anthropologists believe that as they traded meat for corn early on they developed a taste for the grain; and then, when they began to raid corn-growing villages, they could no longer trade with these same villages and had to learn to grow their own.

In addition, captive Caddoan women probably taught them horticulture and earth-lodge construction. In addition, early contact with Puebloan peoples of the southwest—who

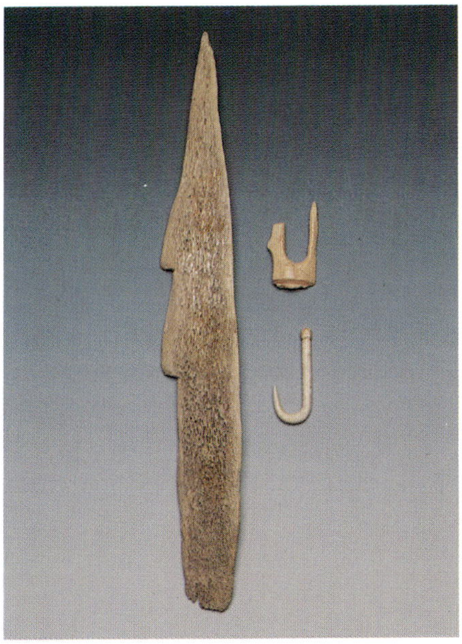

Nebraska State Historical Society

Bone fishhooks (one unfinished) and a harpoon from a late prehistoric eastern Nebraska site

would flee Spanish enslavement and seek refuge with the Apaches, bringing horses, and who temporarily threw off the Spanish, which left much livestock behind—introduced them to horse training, corn growing and, probably, ditch irrigation. In western Kansas, one of the main villages of the Cuartelejo Apaches, and perhaps satellite villages as far north as the Platte River, used some irrigation to water garden crops. The remains of this culture in Kansas and Nebraska are known as the ''Dismal River culture.''

Sites of the Dismal River culture cover an area from the 100th Meridian to the Rocky Mountains and from the Black Hills to northern New Mexico and are dated about 1675 to 1725 A.D. Though less is known about it than the ancestral Pawnee cultures, Dismal River culture is distinctive in that it can be traced to Apache and Pueblo cultures familiar with ditch irrigation. Wedel believes that irrigation methods clearly in evidence at a site in Scott County in west-central Kansas (about 100 miles south of Nebraska) may have extended throughout some villages of the Dismal River culture west of the middle of Kansas and Nebraska.

''The Lovitt [Chase County, Nebraska] and Scott County sites, westernmost of those with evidence of horticulture, are very close to the limits of successful maize [corn] cultivation in the central plains, given a climatic setting like that of today. Here, low and erratic precipitation, high evaporation rates, and frequent but unpredictable droughts require the application of special techniques not so urgently required farther east . . . One of those special techniques was ditch irrigation, clearly indicated at Scott County and there probably a result of plains Apache contacts with fugitive irrigation-conscious Puebloan peoples from Taos and/or Picuris. At this site, spring waters were channeled some hundreds of yards through ditches to supply terrace areas of several acres; this technique was repeated successfully by subsequent white homesteaders who cleaned out and reused the Indian canals,'' he writes in *Central Plains Prehistory* [brackets added].

By the 1700s, however, French traders had supplied the Pawnee and other village tribes with guns, the Utes and Comanches launched offensives from the Rockies and northern Shoshoneans were warring on the Apaches from the north- and southwest. The Comanches especially became known as the best horsemen of the New World, and set about to nearly eradicate the Apaches. This was much more devasting than such fighting might have been earlier because the Apaches had made themselves as vulnerable as their previous Caddoan victims by becoming more settled. This conflict between nomads and villagers is a theme of native peoples' adaptation to environment that will recur, as we will see. As a dominant force, the Apaches had had their day. They fled or merged with other tribes. More and more tribes entered the plains, and by the 1800s, their civilization had survived mostly through a merged tribe called the Kiowa-Apache in Montana, Wyoming and far western South Dakota and its southern band near the Oklahoma Panhandle.

Into the Historic: The Pawnee Flourish and Decline

Before we examine the influx of new people and cultures that characterizes the *Historic* period (beginning about 1750-1800 A.D.) we need to back up a little to look at the history of the Caddoan people in Nebraska—the Pawnee. People known as ancestral Pawnee occupied east-central Nebraska when habitation increased rapidly after the Pacific climatic episode, probably a result of migration from the south. Called the ''Lower Loup'' culture, it resembled the Upper Republican of about five centuries earlier in its basic lifeway, depending on an economy of corn, beans and squash, as well as on bison and other game. The differences included larger villages of a few dozen to a hundred or more dwellings per settlement comprised of bigger, rounder lodges. They practiced agriculture more consistently, producing larger yields from more cultivated acres, probably an acre or so per lodge. More and larger storage pits, dug outside the lodges instead of inside, among other evidence, attest to this. In addition, the hunt took on increasing prominence, especially after the introduction of horses.

In contrast to the Upper Republican and Dismal River cultures, frequent contacts with whites have made the Lower Loup culture much more accessible to us and, ironically, in large part contributed to its downfall. Also unlike the Upper Republican culture, villages were generally located along the main stream valleys. The Skidi or ''Wolf'' band settled on the fork of the Loup river system and the Chaui or ''Grand'' band lived below them on the Platte. Two other bands, the Pitahauerat (or ''Tappage'') and the Kitkehahki (or ''Republican,'' along the river of the same name), contributed to a population that probably numbered as many as 10,000-15,000 by the 18th Century, with some estimates running much higher. As this culture arose and expanded, it became the most powerful nation on the central Great Plains. By about 1800, the Pawnee dominated territory from the Niobrara to the Arkansas rivers and from eastern Nebraska to eastern Colorado.

The Lower Loup phase of Pawnee culture, dating from about 1550 to 1750 A.D., is considered the Pawnee's highest cultural point. Pottery making and other crafts flourished and lodges were well built and uniformly laid out, all of which suggests that religious and ceremonial life peaked as well. By the mid-1800s, however, a mere 100 years later, the culture was in decline. Persistent raiding by the Lakota Sioux, whose expansion onto the plains seriously restricted their traditionally uncontested hunting territory; contact with whites, who brought smallpox, cholera, flu and whooping cough, to which the Pawnee had little resistance; and increasing reliance on trade with whites and other tribes all contributed to disruption of their traditional lifeway. Some decline in horticultural production may have also accompanied this faltering of their society.

Plentiful harvests during good years were stored, of course, and perhaps yielded enough surplus that some was left for the following year. Also, during severe droughts and subsequent crop failures, and during heavy or soaking rains while crops were dried and stored (which could ruin much of the harvest), the communal hunt probably began early to keep all from going hungry. Wild plant foods—vegetables, nuts, berries, fruits, seeds and root or tubers—also helped meet emergency and supplemental food needs.

Especially after the acquisition of the horse, which the Pawnee called ''big dog,'' (the Lakota called it ''sacred dog,'' the Comanche, ''medicine'' or ''god dog,'') the twice-yearly bison hunt became an increasingly important part of the native economy. These usually began after the second hoeing of the corn in June or early July and ended in late August or early September just before harvest. In late October or early November, again the villagers departed and did not return until spring at corn-planting time. While the movement of virtually the whole village twice a year was difficult and sometimes dangerous, it was an essential part of subsistence. For the large permanent villagers, who numbered many hundreds, even thousands, of people, local game and wild plants were soon used up and timber resources likewise. In addition, with large numbers of horses to feed, local grazing also had to be supplemented regularly.

The winter hunts were generally harder, but the summer trips also could be quite rigorous, especially during droughts. Generally the expedition leaders followed the ridges above streams or knew where to intersect them at the right time. Water was also found during droughts by digging a few feet into sandy streambeds if these were dry. However, the main hazard of the summer hunt was the threat of violent acts by other tribes, especially the nomads.

The Semi-Settled Siouans: The Omaha, Ponca and Oto-Missouria

The Omaha

Besides the Lakota Sioux and Pawnee, by historic time, the Omaha were another major tribal force in what became Nebraska. In the late prehistoric period, they lived along the Ohio and Wabash rivers, probably an outgrowth of the Middle Mississippian cultural complex of about 800 to 1400 A.D. According to traditional tribal accounts, they moved up the Mississippi and Missouri valleys, an event captured linguistically in their name, which means "going against the current." The dominant tribe of the Siouan Dhegiha family, the Omaha are closely related to the Ponca, whose language is nearly identical and who were probably originally a clan within the Omaha, as well as to the Kansa, Osage and Quapaw. All except the Quapaw, whose name means "with the current," moved north and west. The Osage stayed near the river that bears their name; the Kansa followed the Missouri eventually to the Kansas river area; and the Omaha and Ponca, along with the Iowa (another Siouan tribe) migrated into the Red Pipestone Quarry region in southwestern Minnesota, probably by the mid-1500s.

They stayed there more than 100 years, living on game and supplementing their diet with corn and wild vegetables. At some point the Ponca split off, heading up the Missouri and then toward the Black Hills. Probably finding competition with the Cheyenne and Lakota too fierce, they returned to rejoin the other two tribes. Soon thereafter, the Iowa crossed the Missouri to live east of it.

Both Omaha and Ponca legends attest to the influence of the Arikara on the combined tribe after a peace that followed hostilities with the Caddoan tribe.The Arikara taught them to make earth lodges, which replaced the bark huts characteristic of a forest people. And apparently they transmitted some methods of cultivation, including certain rites regarding maize. An Omaha legend recounts how the Arikara discovered corn and then invited six tribes to winter with them, the Omaha among them, sharing their crop. It is, however, unlikely the Omaha lost the knowledge of maize altogether. An outside possibility is that they migrated far enough north that its cultivation was difficult or nearly impossible for a time. Nevertheless, "the Arikara were very generous in the distribution of the fruit of this new plant among their guests, and in this manner a knowledge of the plant spread to the Omaha," reads an account of a tribal legend in *The Omaha Tribe* by Alice C. Fletcher and Francis LaFlesche.

This same Omaha legend explains that a young Arikara man was out hunting and became intrigued by a bison bull that frequented a certain bottomland spot by a prairie river. He waited day and night and each morning observed it standing in the same place. It faced each of the four cardinal directions—north, east, south, west—on four consecutive days. After this, the bison vanished as mysteriously as it had come, leaving no footprints. All that was left was a new plant with a fruit that grew in blue, white and yellow kernels and could be eaten without harm.

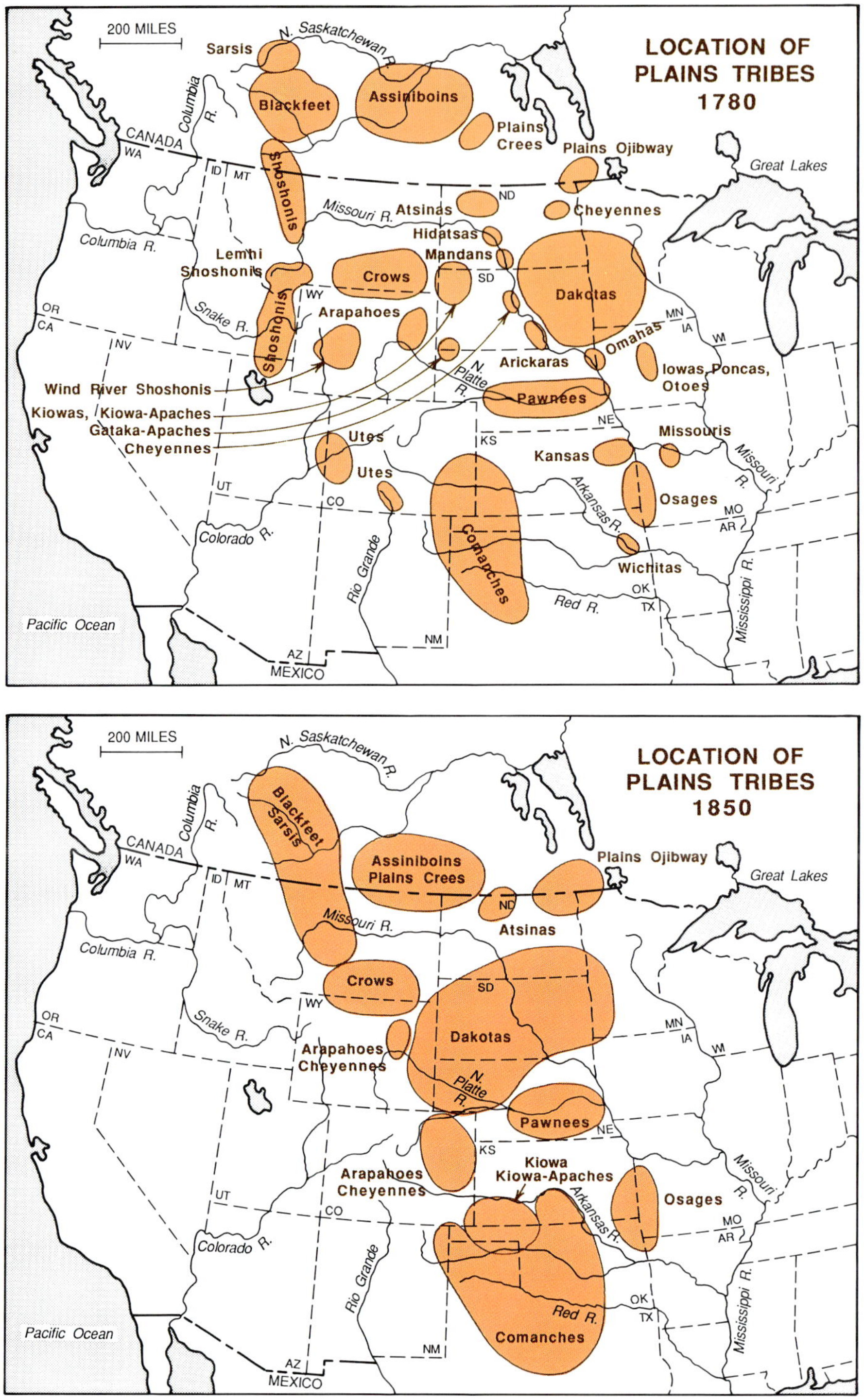

Locations of Native American tribes, about 1780 and 1850, respectively (adapted from Plains Indians, *by Francis Haines, 1976, Thomas Y. Crowell Company, New York and Cleveland, through HarperCollins Publishers, New York ©)*

Such a legend unites the "community of sustenance," as it were: the sun, the prairie, the river, the bottomland, the bison and the maize, along with the human being who watches ardently, inspects intently and partakes gratefully, later worshipping this congregation of plenty.

Another sacred legend concerning the corn plant is also found in *The Omaha Tribe* and comes from a ceremonial song. Interestingly, it bears passing witness to the fact that Omaha *men and women* cultivated corn, not just the women as in most plains tribes: "The maize being one of the greatest of means to give life, in honor of it we sing. We sing even of the growth of its roots, of its clinging to the earth, of its shooting forth from the ground, of its springing from joint to joint, of its sending forth the ear, of its putting a covering on its head, of its ornamenting its head with a feather, of its invitation to men to come and feel of it to open and see its fruit, of its invitation to man to taste of the fruit.

"When maize was discovered the grain was distributed among the people that they might plant and eat of the fruit of their labor, and from that time on it has been the custom to sing the song of the maize and to repeat the distribution of the corn every year at the time of planting.

"The songs [stanzas] are many. They begin with the gathering of the kernels. The people talk of where they shall plant. Then the men select the land and wherever each man selects, he thrusts a pole in the ground to show that now the corn shall be planted."

By the 1700s, the Omaha had moved to a spot on the Big Sioux River south of what is now Sioux Falls. There they built a village near where the river makes a loop and where, as a traditional story goes, "a stream joined the river through two cliffs, like pinnacles, standing at its entrance, through which the wind rushes with such violence as to disturb the water," according to *The Omaha Tribe*.

During this period, the tribe codified its characteristic government and organization, a process accompanied by a wealth of mythic and ceremonial symbology, including a peace ceremony used in one form or another to bring peace to the various tribes of the region. Among the artifacts central to this symbology was the Sacred Pole. Made of cedar, a sweet-smelling, hardy tree characteristic of the plains, it symbolized the power and unity of the tribe.

One sacred legend has it growing near a lake, and the Thunderbirds—cosmic beings who represent the thunderstorm and its rain, lightning and fire, as well as truth—light on the tree and set it ablaze although no one can see this except at night. The son of a tribal elder sees it and reports this encounter with the divine to his father, who views it himself and then tells the leading men of the tribe. In another, the tree gives forth a mysterious light but no heat. A young man of the tribe tells this to his father, who reports it to the tribal elders. In 1888, the Sacred Pole was sent "for safe keeping" to Harvard University's Peabody Museum. It was returned to the tribe in 1989.

About 1735, the Omaha settled on Bow Creek in what is now Cedar County, Nebraska. Apparently an intertribal dispute led to a battle that caused a separation of the tribe and abandonment of the village thereafter called "Bad Village." The tribe eventually reunited and later moved to a site near Dakota City. As of 1765, they had moved south along the Missouri, and about 10 years later, resettled on Omaha Creek near present-day Homer. There they built the settlement they called "Big Village." It was here the infamous Blackbird, a much-feared chief who learned of the powers of arnsenic and put it to use poisoning his enemies, ruled, and under his leadership the Omaha became a powerful military threat. Blackbird died of smallpox about 1800 and is buried on Missouri River bluffs at a spot memorialized by the ethnographic painter Karl Bodmer.

Another well-known chief, Big Elk, became prominent about 1810 because of his eloquence and far-sighted leadership. He had secured treaties with the federal government, had visited Washington and had tried to prepare his people for what he once called "a flood of whites." Wars with the Sauk and Fox drove the tribe to the Elkhorn River in 1819. In 1829 half the tribe moved back to Big Village but were soon driven away the next year by the Sauk and Fox. In 1830 they relinquished rights to land in Iowa in the Treaty of Prairie du Chien. After a smallpox epidemic, they moved back to Big Village in 1834, where they stayed until 1841. There, the Lakota warred against them persistently, and so they resettled at the mouth of Logan Creek on the Elkhorn. After a brief stay again at the Big Village, they moved to Papillion Creek near Bellevue. Conflict with the Sioux and smallpox had reduced their numbers to around 1,300 by this time. What is left of Omaha land is a reservation along the Missouri in Thurston, Burt and Cuming counties, although much of it is leased to whites.

The Ponca

The exact time when the Ponca split from the Omaha is not clear. Some think the Ponca split from the Omaha at the Bad Village site and that the conflict gave the site its name, while others dispute this, but regardless, by about 1800 they occupied a village 8 miles northwest of old Niobrara, Nebraska, in Knox County. They too had learned to build earth lodges from the Arikara. Their activities at this site are described by James H. Howard, in *The Ponca Tribe*, as follows: "The Ponca were quick to apprehend the value of a middleman's position in the trade, and in 1795 they began the practice of stopping and raiding trading craft as they passed up the Missouri. Some of these stolen goods the Ponca then traded to the tribes farther upriver. This piracy was perhaps motivated not only by greed but also by a fear that the upriver tribes, such as the Arikara and Dakota, would acquire guns which would later be turned on the Ponca. The traders, of course, were anxious to deal directly with the upriver groups, since the farther one got from the settlements the less common trade items became and the greater the number of furs that could be secured for them. This Indian piracy, which was practiced by both the Omaha and Ponca, as well as the various Dakota bands on the Missouri, delayed for a considerable period the development of trade on the Upper Missouri. . . ."

The tribe also secured much trade through "legitimate" channels, as their position on the river was strategic for both purposes, and the Ponca village was a traditional stopping place for boats going up or down the Missouri. By the time Lewis and Clark reached them, they were well acquainted with Europeans, their trade goods and their diseases. Most of the famous military men, explorers, traders and ethnographic artists visited them. And from the time of the first treaty with the U.S. government in 1817 and through another in 1826, until their removal to Indian Territory, the Ponca never entered into hostilities with the government, a record of which they were proud. Unfortunately for this relatively peaceful, semi-settled tribe, its corn-raising was regularly disrupted by the Lakota, as with other tribes in the area (the Pawnee, Omaha, Arikara, Mandan and Hidatsa), from about the mid-1800s until the removal in 1877.

The Poncas were supposed to be given a reservation in north-central Nebraska just north of the Niobrara, but a mistake by the federal government also gave this land to the more numerous and powerful Sioux, so the Ponca were moved to Indian Territory (now Oklahoma), where many died of disease. After an attempt by chief Standing Bear to bury his son on their ancestral homelands resulting in his capture, and that of those accompanying him, a much-publicized trial regarding the right of the Indian to full status before the law, and much public outcry, focused attention on the Ponca's plight. The Sioux agreed to return the land. Some Poncas now live in Oklahoma and some in Nebraska.

During the 1950s, as part of the policy of "termination," the northern Ponca reservation was eliminated and the tribe dissolved. Through an aggressive plan of lobbying and certifying lineage, the northern Ponca were again recognized by the U.S. government as an official tribe in 1990.

The Oto and Missouria

The Oto and Missouria, along with the Winnebago and Iowa, belong to a Siouan dialectical family called the Chiwere group. They were probably a single nation living near the Great Lakes in the early 1600s. The Oto and Missouria then separated from the Iowa and Winnebago; the Missouria settled near the Grand River in present Missouri, and the Otos, apparently after an intertribal row, moved up the Missouri to become a distinct tribe.

The Sauk and Fox, persistent enemies, made a devastating attack on the Missouria around 1730, and another that was even worse in 1798. After the second one, the tribe fled and sought refuge with other tribes, among them Osage and Kansa, but most joined the Otos between the Nemaha and the Platte, who had entered Nebraska in the early 1700s. Accordingly, the word, "Nemaha," comes from the Oto meaning "water of cultivation." And some attribute the native term for the Platte, "Nebraska," to the Oto, while others think it is derived from the Omaha. The Oto proceeded to adopt the earth lodge housing of the Pawnee, planted corn, beans, pumpkins, squash and melons and hunted bison twice a year. Village life, however, was more important than the hunt.

When the Louisiana Territory came under Spanish control in 1763 the Oto anticipated good fortune through favorable trading with the Spanish. To be nearer the Spanish, they even moved their village on the Missouri in northeastern Nebraska, where they had fought with the Ponca and Iowa, and returned to the Platte. Eager to assert control over the Missouri River tribes and dominate trading traffic, the Spanish made entreaties to the Oto through

Creole traders from St. Louis. Their goods were welcomed, but the Oto and other tribes opposed their efforts to ascend the Missouri and trade with the Lakotas and other upper Missouri tribes. The northern Sioux wanted to prevent the Spanish from arming their neighbors to the south, and Oto middlemen did not want the Spanish to compete with their trade with tribes further up the Missouri.

Ironically, British merchants, whom the Oto also opposed at first, but bearing many inexpensive goods, including guns, eventually made inroads into trading with the Oto and others. The Oto abandoned plans for a trade monopoly and even welcomed a few British merchants into their village on the Platte. In 1795, the Spanish made a concerted effort to drive out the British traders and reestablish their influence among the Oto, wanting good relations with the tribe because of its strategic position.

James Mackay, a naturalized Spaniard and explorer, met with the Oto and promised that the Spanish would build a fort and trading post near their village. When these plans never materialized, the Oto, reunited with the Missourias by now, made life miserable for Spanish shipping and trading in retaliation, raiding and seizing goods from any Spanish who came near the village. The Spanish responded by forbidding any of their traders to venture up the Missouri, thereby eliminating their own influence. The British, however, were active in the region and willingly supplied the Oto-Missourias with guns. With these, they launched a series of attacks against the Kansas, Spanish allies. The firearms helped the combined tribe keep its enemies at bay and its strategic location secure for a number of years.

Lewis and Clark found the Oto on the south side of the Platte about 5 miles upstream from the Elkhorn River, numbering about 500 people. The Missouria had about 300 members. They hunted in western Iowa, in Nebraska on the land between the Nemaha and the Platte and onto the Great Plains. Explorers and traders observed that the land was "fertile and well watered." Continued warfare with the Sauk produced the intervention of the U.S. government in the Treaty of Prairie du Chien in 1830, in which the tribe gave up their claim to land in western Iowa. They also agreed to donate a small tract along the Missouri between the Great and Little Nemahas as a reservation for mixed bloods from their tribes and various others on the prairie. Differences about the meaning of the treaty meant the Oto-Missouria continued to hunt in western Iowa but came into increasing conflict with the Kickapoo and Potawatomie, who were being resettled from Indiana and Illinois. As game was increasingly depleted in the area, the Oto-Missouria were more and more anxious for white goods. Under these circumstances, much of their land in eastern Nebraska was ceded by 1850 and attempts at farming much smaller tracts were not very successful.

On March 15, 1854, the Oto-Missouria gave up all their land west of the Missouri except about 162,000 acres along the Kansas-Nebraska border intersected by the Big Blue River. Attempts to convert the tribe to farmers generally faltered, and the fertility of the soils and ample rainfall made it attractive farming ground for whites as well. Through questionable land sales negotiated by the federal government, much disputed between more traditional tribal members (called "the Coyote Band") and the more assimilationist members (called "the Quaker Band"), they were removed to Indian Territory in 1881.

The Later Immigrants: The Santee, the Winnebago and the Sauk and Fox

Three tribes who now have reservations in Nebraska were living near the Great Lakes by about 1700, pursuing a lifeway that had more in common with traditional woodland tribes than with plains Indians. While these tribes raised corn, beans, pumpkins and squash and hunted bison, they also built bark or reed-mat houses, gathered much wild rice, an aquatic crop, and hunted woodland game.

Of these, the Santee Sioux and Winnebagos were Siouan-speaking people and the Sauk and Fox were an Algonquian tribe. The Santee, in their woodland lifeway, hunted and trapped game, including the beaver and waterfowl plentiful in their water-rich country. They gathered fruits, roots, maple sugar and rice, as well as raising crops and hunting twice a year on the eastern edge of the bison range. Upon European contact, about 1635, the Santee lived near Lake Superior. When the French fur traders introduced guns to the Ojibway in exchange for beaver pelts, raiding from the Ojibway on the north and the aggressive Sauk and Fox on the south pushed them westward. By the 1840s, they had settled in southern and western Minnesota. The Santee were divided into four bands, the Mdewakantonwan, the Wahpeton, the Wahpekute and Sisseton, who are considered a bridge between woodland and plains culture.

Between the 1830s and 1850s, pressure from settlers, disease and land grabs by the U.S. government pushed the Santee into desperation. In late summer of 1862, the corn crop had been devastated by insects, government benefits reduced and the agency storehouse, filled with food, locked. After a dare over stealing a farmer's eggs and the ensuing argument, Santee warriors killed five settlers. The men responsible told the rest of the tribe and the elders decided make war on local whites instead of merely waiting for federal troops.

Eventually the army rounded up and jailed about 1,800 Santee. It tried and convicted 307, sentencing them to death. Intervention by ministers, citizens and President Lincoln saved all but 33 of them from hanging; those were dispatched in the largest mass execution in U.S. history. From this point on, the tribe's fortunes were determined entirely by the government.

In 1863, legislation banned the Santee, the Winnebagos and the Ojibway from the federated United States. The Mdewakantonwan and the Wahpekute were sent to a reservation on Crow Creek in South Dakota. Three hundred died there in the first few months. Part of the federal policy toward the tribes was to mandate self-sufficiency in farming, presumably to stamp out nomadism, but this was not a lifestyle familar to these people. In addition, except for the bottomland already taken by the Indian agent, the land was barely arable with their techniques, in part because of limited rainfall. Delays in delivering proceeds from the sale of the Minnesota lands, elimination of government annuities (benefits such as blankets and foodstuffs) in retribution for the previous warfare and hostilities with the Winnebagos, also removed to Crow Creek, made their lives there destitute and miserable. Eventually, pleas for better treatment and land resulted in relocation on a reservation in northern Nebraska on the lower Niobrara River. There, as with many reservations, much land is leased to whites.

The Winnebagos also inhabited the Great Lakes region around the early 1700s, living near the present site of Green Bay, Wisconsin. Surrounded by Algonquian neighbors, their culture reflected both Algonquian and Siouan influences. The Winnebagos also hunted bison along the eastern edge of its range, sometimes accompanying their Siouan relatives, the Iowa, Oto and Missouria. In an area with a wealth of lakes and wetlands, they hunted aquatic species such as beavers and ducks, as well as much fish, either speared or taken with weirs (fences in waterways). They used their canoes to stretch nets that caught ducks during molting season, as well as using them as "baskets" for harvesting wild rice.

Thick near the Fox River, wild rice was harvested by most tribes in the area, and they occasionally fought over rice-growing territory. "Poling" with their paddles into the rice, they would bend the stalks over and swat them with the paddles, letting the rice fall onto a cloth spread on the floor of the canoe.

When the Santee uprising was put down, the Minnesota settlers also demanded the removal of the Winnebago, who had moved to central and then southern Minnesota by the mid-1800s. After the federal government sent them to the Crow Creek reservation, the desparate circumstances there caused most of them to flee to the reservation of their Siouan relatives, the Omaha. Others returned to Minnesota. The Crow Creek Winnebagos were subsequently resettled in northeastern Nebraska, in Thurston, Dakota and Dixon counties, north of the Omaha. The Omaha sold them part of their reservation in 1865 and 1874.

Their Indian names meaning "Yellow Earth" and "Red Earth" people, respectively, the Sauk and Fox were confederated tribes that were driven from their home in Michigan and also settled near Green Bay and the Fox River. The Sauk dwelt in permanent villages but also spent months at a time hunting and raiding. They farmed and hunted woodland game and bison, sometimes as far west as Kansas. As a woodland culture, they made extensive use of birchbark and dugout canoes. They believed the earth to be inhabited by magical beings, manitos, who could act for good or ill and represented all of nature.

Along with the Fox, a more hostile tribe, they controlled the trade route from Green Bay along the Fox River, raiding and demanding tribute from traders. Extensive warfare with other tribes and pressure from the French drove them south along the Mississippi, but settlement was increasing there, and the defeat of their British allies in the War of 1812 weakened their position.

In 1832, the militant Sauk chief, Black Hawk, led an attempt to return to Illinois, part of a previous homeland. The Illinois militia, however, defeated them badly. The two tribes moved to Iowa and gave up all right to land east of the Missouri in 1845. Fighting with the Dakotas pushed them farther west to the Osage River Indian Agency in Kansas, an environment that little resembled what they had known, after which, the federal government removed them to Oklahoma in 1886. However, in 1869, the government set up a reservation straddling the Kansas-Nebraska border for a few families that lies near a tract reserved for a few Iowa families, most of whom had also been moved to Indian Territory.

The Horse and the Hoe: The Nomads and the Villagers

For purposes of brevity and discovery, the period after European contact in Nebraska can be broken down into two distinctive lifeways. One is the more horticultural orientation of the semi-sedentary tribes along the western fringe of the woodland ecosystem, which combined the hunt with farming—the Caddoan Pawnee and Siouan Omahas, Poncas and Oto-Missourias. The other involves the nomadic traditions of the more northern and western tribes—the Lakota, Cheyenne and Arapaho.

Both of these groups were affected by and benefitted from European contact. Both were also eventually supplanted by the European settlers as contact with whites proved increasingly devastating to their traditional lifeways. In addition, the conflict engendered between the two lifeways weakened the overall ability of the Indians to fight back the white invasion. This was most true with the horticulturalists, whose more physically stable societies made them ideal hosts for the often fatal diseases the whites brought and easy prey for the nomads' raiding. In addition, with the Pawnee at least, their more politically stable organization made them less physically and politically adaptable as, prior to relocation, they were forced to consolidate bands and villages, with ensuing compromises and conflicts in leadership. Later, the inability of the nomads to present a united front against increasing white incursion on their lands hastened the loss of all but a remnant of those lands.

Preston Holder, in his seminal treatment of this subject in *The Hoe and the Horse on the Plains*, focuses on the Pawnee (and to a lesser extent, the Arikara and other Caddoan groups) as representative horticultural societies and on the Lakota Sioux as a telling example of the nomadic societies. During the Historic period a new way of life emerged for most plains tribes. Some of this stemmed from the introduction of the horse, radically transforming the lifeways of the Cheyenne and Lakota within 50 to 100 years, making mounted nomads out of more sedentary and foot-bound societies.

This lifeway has yielded most of the popular depiction of plains Indians. But the horse also changed the lives of the semi-sedentary societies as well. Other, less obvious changes also came to the plains tribes as contact with Europeans produced a variety of new trading rivalries and alliances, new technologies, new means and measures of wealth and, in general, a leverage on change not possible before. Holder explains: "In native North America stable villages of horticulturalists were a relatively late phenomenon [considering all of human occupation]. Here, as elsewhere in the world, they were the result of some thousands of years of cultural development from simple bands of roving hunters and gatherers. Throughout the Mississippi Valley, as well as in the American Southwest, there is now clear archeological evidence of such a [horticultural] development. In all cases the shift in the pattern of settlement seems to be a reflection of the development of the horticultural practices. Once this development appeared along the Mississippi Valley, a new adaptation to the Plains was possible. Horticulturalists were not long in moving out onto the river bottoms of the eastern peripheries of the Plains" [brackets added].

In this adaptation of much earlier Mississippi Valley and southwestern traditions, a horticultural mythic structure grew up around and enveloped the residual hunting and gathering myths. The Pawnee, in particular, with their star-oriented religion, were quick to connect all life to its cosmic origins and thereby, their culture, including the need for water for a plentiful supply of crops. All things derived from the heavens. No practical or sacred act could take place without the proper ceremony. In sacred mythology, "Mother Corn" was the primary earth-bound, nurturer for the Pawnee, as well as for the Omaha and Ponca.

Such earthly mythology was never far removed from more cosmic powers seen as bestowing life-giving water. The Skidi Pawnee word for a sacred bundle translates as "wrapped up rainstorm." This bundle was also called a "mother," after the two ears of sacred corn wrapped up in each. In fact, in their origin myth, the supreme spirit "Tiwara," the combination of the male forces of the east and the female forces of the west, is reaffirmed in the association of the fundamental male energy with the seminal rains and the female energy with the corn which was planted, tended and harvested by women. This mythological world view was so thoroughgoing that it reinforced nearly all political and social hierarchies within the tribe, was celebrated and/or consulted prior to and after all important hunting and gardening and governed the physical positioning of lodges and their interiors, including the inhabitants.

Beyond the primacy of the Evening Star and the Morning Star, sacred entities and forces were elaborately structured mythically to coincide with the four semi-cardinal directions. (Since the cardinal directions were derived from the cosmos, all other phenomena had to reside in the spaces left, the semi-cardinal areas.) Wind (northwest), Lightning (northeast), Thunder (southeast) and Clouds (southwest) were each associated with a direction, as were sacred wildlife, trees, colors and medicine bundles. Elaborate rituals surrounded the powers of the thunderstorms, and certain plantings, harvests and hunts could not take place without such a ritual and an actual thundering. In addition, during the creation ritual after the first spring thunder, the complete cosmology was rehearsed.

In other ceremonial uses of water, the Pawnee conducted the Great Cleansing Ceremony twice a year before going out on the hunt. It combined the heavenly symbols with the earthly ones from both land and water. The ceremony ended with a race to the river by the whole village. There the priest carried the sacred pipe into the water and symbolically washed himself by dipping a hand, one and then the other, into the water and touching it to his mouth, nose, forehead and down his face. After all the sacred objects and their priestly carriers were symbolically washed, the people waded into the river to clean themselves. While the priests went to a sweat lodge ceremony the rest of the village swam and played in the river.

The Pawnee also revered the animal spirits who were the messengers of Tirawa at lodges set up at sacred sites in their homeland, most of which were related to running water. These were "Dark Island," a Platte River island near Central City; "White Bank," a Loup Fork cutbank near Fullerton; "Water on a Bank," a rounded hill along the Solomon River in Kansas; "Hill Points the Way," (also called "Guide Rock" by whites, lending its name to the settlement near the Republican River in Webster County); and the most important, "Hill Island," (Pa-huk) a point jutting into the Platte near Cedar Bluffs.

The Pawnee's intricate religious system ordered and bound up a social organization that was quite hierarchically structured. These hiearchies supported the horticultural myths and lifestyle, just as the lifestyle gave meaning to the hierarchies. The most characteristic feature was a fledgling, male-dominated class system with hereditary chiefs at the top, who, along with priests and doctors (sorcerers), made up an upper class, and warriors and fraternal orders who were part of the "commoners." The cyclical seasonal, theocratic regimentation of all important activities produced a stable society that dominated the region until about the 19th Century. After this period, the very rigidity of the system—what under other circumstances may have been termed stability—probably contributed to its downfall.

Lakota traditions also embrace the exalted mythical status of water-related phenomena.

The Lakota revered, and sometimes feared the "Thunderbeings," or Thunderbirds, who commune with and/or comprise "Wakan Tanka," the supreme spirit. The Thunderbirds stand for rain, fire and lightning and truth. In one Brule Lakota (Sioux) myth, the Thunderbeings battle the Great Unktehi, the water monster, for the future of the earth. The water monster—an apparent reference to flooding—puffs up the Missouri and makes it overflow. The Thunderbeings take pity on the humans, who have retreated to the highest ground further west, and fire their lightning bolts at the earth all at the same time. This makes the waters boil and dry up. The Unktehi, made up of big and little water monsters, burned up and left only their bones, which turned to the rock of the Dakota badlands. After that, the water power of the Unktehi was ursurped and taken up by the Thunderbirds.

As such a mythology of insurgence might suggest, the bison-hunting nomads were usurpers. They had ventured onto the plains late but with a vengeance. They had supplanted the cultivation of corn with the hunt. In keeping with such an emphasis on opportunism they had a more flexible and resilient social structure, one that depended more on a "meritocracy" based on exploits in war and in the hunt. Their many tribes formed and broke relations quickly, and their relationships to other groups were much more temporary and expedient. Following or not following "leaders," whose authority was never inviolable, was more a matter of individual choice, based on the merits of the case and the leader. Their individualism was more pronounced (while still serving the welfare of the tribe, since food and wealth were shared more equally, not as hierarchically), and their religious system was based much more on the individual vision-quest, often leading to important acts in war or hunting, not on a priestly hierarchy and strict ceremonial observance.

With the introduction of the horse, then the increasingly inconstant world of the encroaching Europeans and, later, the expansion of the Lakota and Cheyenne onto the plains, the dominance of the Pawnee was in danger. Although the Pawnee made good use of the horse, to the extent that they needed to spend more and more time away from their villages just to find grazing grounds for their mounts, the horse opened up nomadism, and the plains, for the Lakota. The flexibility of this lifeway allowed them to avoid prolonged disease-infesting contact with whites, made possible expedient alliances and deft surprise in war and, some food shortages not withstanding, made the semi-arid plains a haven instead of an ecological problem. Permanency of residence along stream courses made the Pawnee vulnerable to raiding by nomads and put them right in the way of white westward expansion, first with the traders and then with the settlers. Nomadism provided the essential, although temporary, adaptation, while allowing the Lakota to still trade meat and hides with horticultural societies, as they generally needed to, for crops. In addition, their social and religious structure rewarded inventiveness and opportunism, most necessary to survival where reduced moisture meant less reliable resources in general and primary reliance on a mobile, grass-eating source of food and raw materials.

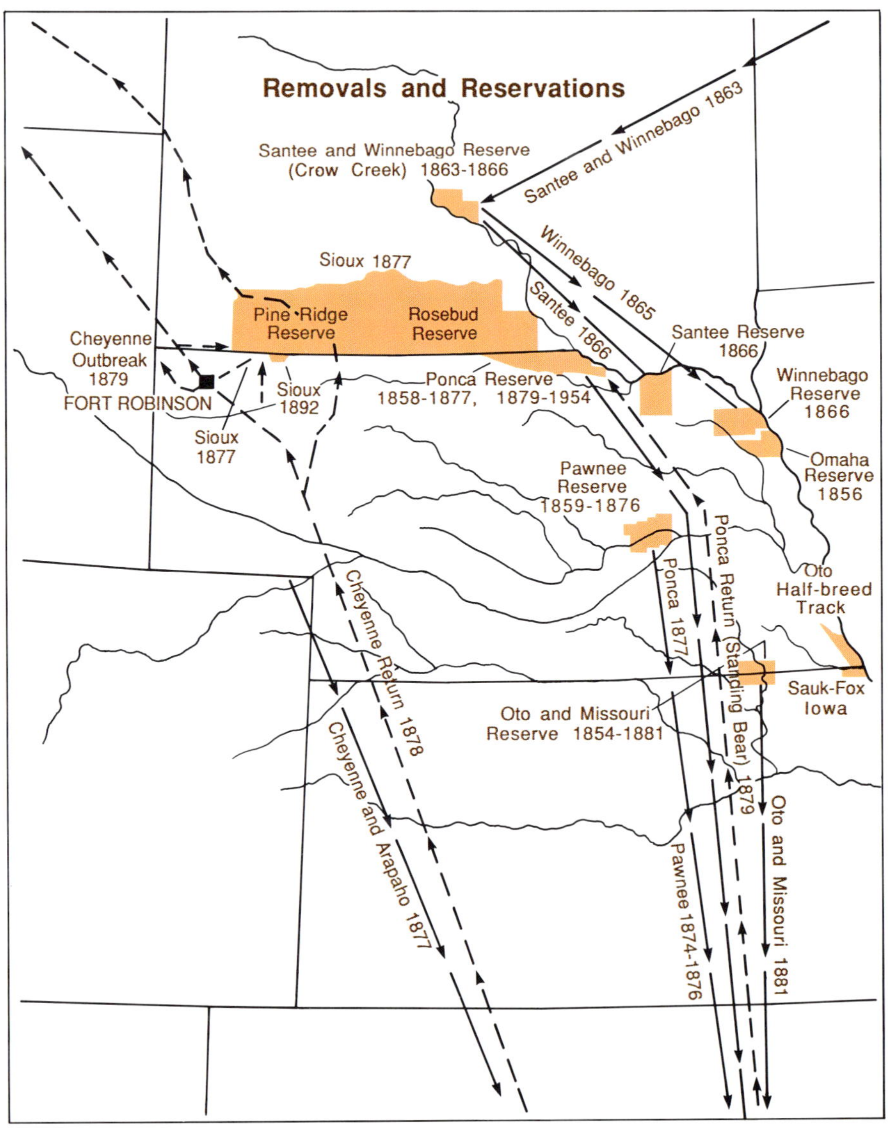

Removals to reservations (adapted from "First Voices," Nebraskaland special edition, Jan.-Feb., 1984, Nebraska Game and Parks Commission)

The Historic: A Fight for Resources—A Collision of Cultures

Through the 1800s, increasing pressure on Native American peoples by Euro-American culture was evident and inevitable. Yet, not until the acquisition of Oregon and California and the discovery of gold in the latter did sustained hostility between Indians and European-Americans really begin. Even though it would be many years before whites would encroach on the hunting grounds of the nomads, the Platte River Road, following the region's largest stream course, made frequent contact, and conflict, unavoidable. The road cut through prime bison-hunting land. Migrating pioneers shot game, especially bison, further scaring away the herds, and depleted the timber and grass. And the passageway itself attracted more whites who wanted to sell to and service the wagon trains headed west. Those most immediately affected were the Pawnee, Sioux and Cheyenne.

The Pawnee, their society increasingly in trouble due to disease, war and food shortages, were more willing to allow whites passage if they could receive government rations and protection from their aggressive Indian neighbors. While some angry braves threatened settlers in eastern Nebraska and stole food and livestock, they mostly became a nuisance to whites by begging and stealing along the road. In 1857, in return for payments of $40,000 for each of 5 years, then $30,000 a year, and promises of protection and training in various trades and farming, they gave up all their land except a reservation that is now Nance County. This did allow them time to begin the transition from a semi-nomadic to an agricultural economy. But after 15 years there, they were eventually surrounded by whites, persistently attacked by the Sioux, misunderstood by crooked or pacifistic agents and were beset with crop failures due to grasshoppers or drought. After one particularly vicious attack by the Sioux, during which more than 100 died, they agreed to move to the Indian Territory.

The nomadic societies, particularly the Lakota Sioux and Cheyenne, were to play out an extended "endgame" in the often bloody match with the expanding United States that took most of 30 years to resolve. Much of this contest of cultural wills involved a struggle over land that mostly could not be farmed without irrigation but contained minerals, often gold, silver or copper, that the whites sought eagerly. Perhaps because these tribes had fought so hard to make their way onto the plains—being driven from the Great Lakes region and the north-central plains—they fought desperately to hold onto the hunting territory that sustained them. In this "war," two river roads featured prominently, those following the Platte and Powder rivers.

Primarily because of the threat of attack by either tribe, both of which were extremely proud, warrior societies, the army established Fort Kearny on the central Platte, near the present city of that name, and Fort Laramie on the North Platte about 100 miles into Wyoming. Nebraska was opened for settlement in 1854, but settlers stayed near the Missouri at first. By the mid-1860s, the government would no longer allow pioneers to leave Fort Kearny until a wagon train had formed large enough to defend itself. Councils and peace treaties were made with selected groups of nomads in 1851, 1866 and 1868, but none of these were very effective because of the relatively egalitarian nature of nomadic culture, meaning that some agreed to observe the conditions and some did not, and because after each one, white encroachment would abate briefly, then continue. The Indians had become accustomed to trading for guns, ammunitions and other metal products, so were sometimes open to negotiation, but generally wanted government payments and benefits without giving up any of the range or manner of their traditional life.

After the Civil War, as the army had much more ready manpower and the nation was eager to expand, the building of the Union Pacific Railroad, again along the major east-west stream course in the central plains, the Platte, was an important undertaking for the reunified nation. This especially aggravated the nomads. They raided railroad workers and even once, derailed a train. Keeping the workers safe provided the justification for an ongoing series of raids and reprisals against the Cheyenne and Lakota Sioux. Once built, the railroad signaled the death knell for the traditional lifeways of these people. It opened new areas to settlers, reduced the favorable bison-hunting territory and provided an easy route of supply for army bases. With each treaty, more Indians accepted "agency" life, where they were given rations and were allowed for a while to hunt during the summer. But a sizable number refused to accept reservation life and retreated to a hunting range in eastern Montana and Wyoming and the western Dakotas. When the government attempted to open and maintain the Powder River road through this territory to allow for exploitation of the region's precious minerals, coal and rangeland, the Indians became increasingly resolved to keep their land or die trying.

In 1876, Custer's famous "Last Stand" at the Little Bighorn in Montana, which involved a large number of Nebraska agency Sioux, became the classic example of the tenacity of the nomadic cultures in defending what remained of their land. And the massacre of more than 150 Lakota who were disarming at Wounded Knee, South Dakota, in 1890 became the enduring symbol of white aggression. Crazy Horse surrendured in 1877 and was killed while being arrested. Sitting Bull fled with a group of non-agency "roamers" to Canada for a few years but the Canadian government refused to accept them, and with the bison herds shrinking, they returned to accept agency life in 1881. The Lakota were placed on two reservations just across the state line in southwestern and south-central South Dakota, the Pine Ridge and Rosebud reserves. Much of this land is leased to ranchers who live off the reservation. The people now living there are the poorest per capita in the nation.

Estimated Population of Plains Tribes, 1780

Tribe	Tribal family	Secured horses	Population
Wichitas	Caddoan	1680	3,000
Osages	Siouan	1700	6,200
Missourias	Siouan	1700	1,000
Kansas	Siouan	1724	3,300
Pawnees	Caddoan	1720	10,000
Iowas	Siouan	1730	1,200
Otos	Siouan	1730	900
Poncas	Siouan	1730	800
Omahas	Siouan	1730	2,800
Cheyennes	Algonquian	1750	3,500
Dakotas	Siouan	1750	25,000
Plains Ojibways	Algonquian	1770	3,000
Plains Crees	Algonquian	1770	4,000
Assiniboines	Siouan	1730	10,000
Blackfeet	Algonquian	1730	5,000
Sarsis	Athabascan	1760	700
Arikaras	Caddoan	1740	3,800
Mandans	Siouan	1745	3,600
Hidatsas	Siouan	1750	2,500
Atsinas	Alqonquian	1750	3,000
Crows	Siouan	1730	4,000
Arapahos	Algonquian	1730	3,000
Kiowas	Algonquian	1730	2,000
Kiowa-Apaches	Algonquian-Athabascan	1730	500
Utes	Shoshoni	1660	4,500
Comanches	Shoshoni	1700	12,000

—Adapted from *Plains Indians*, by Francis Haines, 1976, Thomas Y. Crowell Company, New York, and *American Indian Almanac*, by John Upton Terrell, 1971, World Publishing Company, New York and Cleveland

Sources: *American Indian Almanac*, John Upton Terrell, 1971, The World Publishing Company, New York and Cleveland; *American Indian Myths and Legends*, Richard Erdoes and Alfonso Ortiz, 1984, Random House, New York and Toronto; *Black Elk Speaks*, John G. Neihart, 1932, William Morrow and Company, New York; *Central Plains Prehistory*, Waldo R. Wedel, 1986, University of Nebraska Press, Lincoln and London; *Ceremonies of the Pawnee*, James R. Murie and Douglas R. Parks, 1981 Smithsonian Institution Press, Washington, D.C.; *The Cheyenne Indians—Their History and Way of Life*, George Bird Grinnell, (1962 Cooper Square Publishing) 1972, Bison Books reprint, University of Nebraska Press, Lincoln and London; *Corn Among the Indians of the Upper Missouri*, George F. Will and George R. Hyde, 1917, Bison Books reprint, University of Nebraska Press, Lincoln and London; *The First Voices*, Nebraskaland Magazine special issue, vol. 62, no. 1, January-February 1984; *The Hoe and the Horse on the Plains—A Study of Cultural Development Among North American Indians*, Preston Holder, 1970, University of Nebraska Press, Lincoln and London; *The Lost Universe*, Gene Weltfish, 1965, Basic Books, New York; *The Omaha Tribe*, Vols. I, II, Alice C. Fletcher and Francis LaFlesche, 1972, University of Nebraska Press, Lincoln and London; *The Otoe-Missouria People*, R. David Edmunds, 1976, Indian Tribal Series Publishers, Phoenix; *The Otoes and Missourias*, Berlin Basil Chapman, 1965, Times Journal Publishing Company, Stillwater, Oklahoma; *Ponca Indians*, Joseph Jablow, 1974, Garland Publishing Inc., New York and London; *The Ponca Tribe*, James H. Howard, 1965, Smithsonian Institution Bureau of American Ethnology Bulletin 195, U.S. Government Printing Office, Washington, D.C.; *Prehistoric Agriculture on the Great Plains*, Mary J. Adair, 1988, University of Kansas Publications in Anthropology No. 16, University of Kansas, Lawrence, Kansas; *This Land was Theirs—A Study of North American Indians*, Wendell H. Oswalt, 1978, Wiley and Sons, New York; "Toward a Processual Understanding of the Initial Variant of the Middle Missouri Traditions: The Case of the Mill Creek Culture of Iowa," Duane C. Anderson, in American Antiquity, v. 52, no. 3, 1987; *Red Cloud's Folk—A History of the Oglala Sioux Indians*, George E. Hyde, 1937 University of Oklahoma Press, Stillwater, Oklahoma; *Red Cloud and the Sioux Problem*, James C. Olson, 1965, University of Nebraska Press, Lincoln and London; *Uses of Plants by the Indians of the Missouri River Region*, Melvin R. Gilmore, 1914, Bison Books reprint, University of Nebraska Press, Lincoln and London; "Woodland Cultures on the Western Prairies: The Rainbow Site Investigations," David W. Benn, ed., with David A. H. Baerreis, in *Woodland Cultures on the Western Prairies*, 1990, Iowa Office of the State Archeologist, Report No. 18;

Interviews with: Robert Bozell, research archeologist, Nebraska State Historical Society; David Wishart, professor of historical geography, University of Nebraska-Lincoln.

Profiles:

William R. Akers: 'Akers Law' Set Up Water Rights and Board of Irrigation

In 1882, William R. Akers moved to Fort Collins, Colorado, to practice law. Shortly after his arrival, Akers, a Civil War hero from Ohio, was elected to the House of Representatives of the Colorado legislature.

In Colorado he witnessed the progress of area irrigators and recognized the chance to use irrigation for his own good and for the good of farmers struggling in other dry regions. He therefore spent two years investigating the technology and then left for lands unwatered.

Akers set out for Nebraska in 1884 but instead landed on a farm about 10 miles inside of the Wyoming border. He lived on that farm along the Platte until 1887. Then, he filed for a homestead in Morrill County. His homestead is now part of the city of Morrill.

Soon after settling, he and a partner became interested in starting an irrigation ditch company. Akers and Charles W. Ford, also a resident of modern-day Morrill, secured a point of diversion along the North Platte River a little more than a mile inside the Nebraska state line.

In a few months, the men took for partners several former Colorado residents and incorporated the Farmers' Canal Company. Akers, serving as company president, registered the diversion in September 1887 and thus became the first to file an irrigation claim in Nebraska.

The group of men sold the Farmers' Canal Company before the planned 10-mile ditch was complete. The group they sold it to suspended construction in 1898 and a Nebraska court ordered the sale of the canal in 1901. It was purchased in 1904 by the Tri-State Land Company, which extended it to 96 miles by 1905. The canal was later renamed the Tri-State.

In 1893, Akers was elected in Morrill County to the state Senate, largely on his campaign promise to draft water-regulation legislation. The water-appropriation bill he pushed through, patterned to a large extent after the Wyoming Irrigation Code and proposed legislation drafted at a state-wide irrigation meeting held in Lincoln in 1891, established the "first in time, first in right" doctrine and cleared the way for the formation of irrigation districts. In later years, the district laws gave irrigators the legal and financial means to develop large-scale projects. The "Akers Law" also set up the State Board of Irrigation. The governor, the attorney general and the commissioner of public lands made up the board's membership.

At the end of his second term in the Senate in 1896, Akers was selected assistant secretary of the State Board of Irrigation. A few months later, he replaced Robert B. Howell in the position of board secretary and, at the same time, was made state engineer. He held those positions for only a year, as he was appointed receiver of the Alliance land office by President William McKinley in 1897.

In 1905, Akers retired and moved from Alliance to Scottsbluff. There he bought a farm four miles from town and faded from public light. Akers died Jan. 16, 1915 while visiting a brother in Chicago.

Akers once said, "We have a practical water supply possible to water almost all of Nebraska. . .Nebraska is a grand river flowing on its way waiting to be utilized by the intelligence of the people."

—Brad Rundquist, weekend and copy editor, **Ft. Dodge (Iowa) Messenger,** *and formerly CSD editorial assistant*

Ivan A. Fort: Irrigation Missionary

After the Civil War, Ivan A. Fort looked for opportunity in the expanding west. The 21-year-old Massachusetts native came to Nebraska in 1866 and went to work as an engineer for the Union Pacific Railroad, a job he held for less than a year. For several years he made his living as a hunter, selling game and furs to railroad employees. In 1871, he bought a farm in Lincoln County near North Platte.

Later, Fort found an interest in real estate and surveying and was elected Lincoln County surveyor in 1889. As surveyor he saw different ways in which farmers were bringing water to their land. Fort quickly recognized the benefits of irrigation to crop yields and land values, and dreamed about what the state could do if all farmers used the technology. He set out to teach them and became, in a sense, an irrigation missionary.

He began to organize irrigation associations in western Nebraska in the dry year of 1893, when he saw people leaving the state for regions of more dependable rainfall. Fort was instrumental in the organization of the Nebraska State Irrigation Association. He was elected association president at the group's first meeting in December 1893.

In 1894, on his own time and with his own money, he traveled more than 30,000 miles throughout the semi-arid region promoting irrigation and helping to organize 75 local irrigation associations. He left an additional 25 associations in his wake during the first two months of 1895.

Fort served two terms as president of the Nebraska State Irrigation Association and worked non-stop for irrigation through 1896. That year, as state lecturer for the association and national lecturer for the National Irrigation Association, he delivered 25 speeches across the Midwest.

He remained active in the state association after he stepped down as president. In 1896, he helped organize the Great Nebraska Irrigation Fair, which was held in North Platte.

Speaking at the irrigation association meeting in November 1896, Fort said, "I am pleased to see the interest in irrigation is growing everywhere, and steadily increasing. . . . I am pleased to say that at points where water cannot be obtained by means of canals, the people are devising ways and means for securing water by means of the underflow, [using] windmills, pumping plants and machinery of that character, and it will only be a question of time until on every farm west of the 100th Meridian, yes, and I will say east of the 100th Meridian, there will be a windmill or some form of machinery at work pumping water which will insure crops to the irrigator regardless of rainfall."

—Brad Rundquist, weekend and copy editor, **Ft. Dodge (Iowa) Messenger,** *and formerly CSD editorial assistant*

William E. Smythe: His Irrigation Articles Spawned a Movement

William E. Smythe was born the son of a shoe factory owner in Worchester, Massachusetts, in 1861. He set out for the West in 1888, arriving that year in Kearney. There, at age 27, he became editor and publisher of the *Kearney Enterprise*. In 1890, he accepted an editorial position at the *Omaha Bee*.

In January 1891, the paper's founder, Edward Rosewater, asked Smythe to write appeals urging the people of Omaha to contribute food, money and seed to drought-stricken farmers of western Nebraska. Smythe, however, wanted to help farmers help themselves. He proposed irrigation.

At the time, though, irrigation was an unpopular notion in Nebraska. To admit the need for additional water was to admit the land was not suited for agriculture. Rosewater was aware of popular sentiment against irrigation and didn't think it best to promote the idea in his newspaper. But when Smythe said he would take full responsibility for his opinions, Rosewater told him to proceed.

Two years before, Smythe had seen irrigation at work in California and Colorado. In a series of six articles, he told of the prosperity of Colorado irrigators and described "the greatest wheat crop that was ever known" harvested in the San Luis Valley of California. Such success could be Nebraska's, he said, if only irrigation could be adopted.

Smythe's articles told where irrigation was needed, where the water could come from, how much money it would take to build irrigation works and how they could be financed.

He quit his job after the last irrigation article was published and appointed himself leader of the movement to reform the social order of the state through irrigation; he spent the remainder of his life promoting settlement, possible through the virtues of irrigation, in Nebraska.

Smythe traveled the state organizing local irrigation meetings in places like McCook, Sidney and Ogallala. He helped bring together delegates at a state meeting held in Lincoln on February 11-12, 1891. At that first statewide convention, Smythe helped draft proposed legislation to encourage irrigation as had been promoted in his articles. Delegates sent the proposed law to the legislature, but it was not passed until 1895.

Later in 1891, Smythe founded the *Irrigation Age*, which became the nation's most important irrigation journal. Smythe also helped organize the first National Irrigation Congress, held in Salt Lake City in 1892. He was elected president of the national organization at the second annual meeting, held in Los Angeles.

In 1900, and again in 1905, he published an irrigation history titled *The Conquering of Arid America*. The book has been called a classic in the history of conservation.

—Brad Rundquist, weekend and copy editor,* Ft. Dodge (Iowa) Messenger, *and formerly CSD editorial assistant

Ella Sumner, Nellie Richardson, Effie Arnett: Early Irrigation Promoters Weren't All Men

Dr. Ella P. Sumner was one of the lady delegates from Nebraska to the Phoenix session of the National Irrigation Congress. She was born in New York, but lived and received most of her education in the state of Vermont. She came west to Nebraska with her husband in 1882. She took a three-year course in the Omaha Medical College, receiving her degree in 1894. She was the second women in the state to pass examination successfully before the State Board of Pharmacy. Dr. Sumner was a close student of current events, and a woman of refinement and culture, cheerfully cooperating in all movements for the advancement of the masses.

* * *

Nellie M. Richardson, of Lincoln, Nebraska, was a lawyer and regular practitioner in the capitol city. She received a seminary education in the east and a musical course of instruction in Chicago. She studied law, and was admitted to the bar in Scotts Bluff County in 1888, and to the bar of the Supreme Court two years later. After some valuable experience in practice she entered the College of Law, University of Nebraska, graduating in 1894, with the degree LL.B. She was a member of the Women's Relief Corps of the Grand Army of the Republic, serving as senior vice-president of the department of Nebraska (a powerful position in the major organization for Civil War veterans), and was prominent in the women's club and city improvement work during her residence in Lincoln. She had valuable experience in irrigation and devoted considerable attention to that branch of the law. She was a delegate to the National Irrigation Congress in 1894, and represented the Commercial Club of Lincoln in the State Convention of 1896. She was there chosen assistant secretary. Mrs. Richardson rendered valuable assistance in the production of the Irrigation Annual for 1897, performing a large share of the supervisory and editorial work.

* * *

Mrs. Effie Arnett represented the Lincoln Commercial Club at the Fifth National Irrigation Congress, in Phoenix, Ariz., and was also a regular delegate to the same organization at the Fourth National Congress held at Albuquerque, N.M. She accompanied her husband, Lee Arnett, in attendance to every session of the National Irrigation Congress save one since that body was organized. Her study and observation have been such as to give her an insight into the science and art of irrigation possessed by very few persons, even among practical irrigators. In both of the congresses first above referred to she served upon important committees, and has become one of the best known lady delegates of this national organization. Her study of the great problems of the arid regions is prompted by her patriotic and benevolent impulses as an intelligent American woman.

—Reprinted from the first and second Irrigation Annuals, Nebraska State Irrigation Association

Nebraska State Historical Society

Old Dam at Milford, Seward County, 1892

Nebraska State Historical Society

Mill and water power, Ponca, Dixon County, 1904

Climate and Hydrology

by Ann Salomon Bleed[1]

The state of Nebraska sits on the eastern edge of the Great Plains. This location has shaped the history of the state and its water use. In his book *The Great Plains*, western historian Walter Prescott Webb emphasizes that the physical forces in the Great Plains are

> a permanent factor in the interpretation of history, one that must be understood. If the Great Plains forced man to make radical changes, sweeping innovations in his ways of living, the cause lies almost wholly in the physical aspects of the land. A study of these physical aspects . . . not only illuminates the later historical development, but in large measure serves to explain it.

Webb defined the Great Plains as a treeless area with a relatively flat surface "where rainfall is insufficient for the ordinary intensive agriculture common to lands in a humid climate."

Later, he described the Great Plains as "a semi-desert with a desert heart." He said it had "a soul too dark" to be converted, arguing that it would be "folly to exhaust the treasury trying to tame over the West in the image of Illinois."

If, according to Webb, the 98th Meridian separates the timbered East from the grassy West, then Nebraska is truly where the West begins. The state straddles the 98th Meridian; one third lies to the east, two thirds to the west. In fact, the tree line, where native forests used to grow beyond any stream valleys, lies further east, along the Missouri River valley *(see map of Natural Vegetation of Nebraska in "Environmental Quality and Changes in Ecosystems," p. 163)* Between the Missouri River and the 98th Meridian, irrigated agriculture begins in earnest.

Bernard DeVoto, another prominent historian, put is this way: "The West begins where the annual average rainfall drops below 20 inches." In years of normal rainfall, that definition places the beginning of the West somewhere in western Nebraska. In dry years, however, the line moves into eastern portions of the state *(see map of normal annual and growing season precipitation in "Introduction," p. 2)*.

Like other Great Plains states, Nebraska is characterized by cold winters and hot summers. Average monthly temperatures in January and February range from 20 to 30 degrees F. Northeastern Nebraska, the coldest part of the state, has experienced temperatures below 0 degrees F for periods of 30 consecutive days. The average annual number of days with maximum temperatures less than 32 degrees F ranges from 53 days in the northeast to 32 days in the southwest *(see map of average annual days with low temperatures, p. 46)*. Nebraska summers are hot. South-central Nebraska averages more than 50 days a year with temperatures above 90 degrees F. Yet the northern and northwestern parts of the state generally experience less than 20 such days *(see map of average annual days with high temperatures, p. 46)*. The average annual freeze-free season ranges from 180 days in eastern Nebraska to less than 120 days in the west *(see map of average annual freeze-free season, p. 46)*.

Wind is another prominent feature of the Great Plains. With large flat areas and no mountains to slow them down, winds attain high speeds, especially in the early spring when warm and cool air masses mix frequently. In *Mollie—the Journal of Mollie Dorsey Sanford in Nebraska and Colorado Territories, 1857-1866*, a pioneer woman wrote of her first experiences with the Nebraska wind in the southeastern part of the state.

> We have been treated today to a genuine Nebraska wind storm. We just don't do anything during these storms, but gouge the dust out of our eyes and ears, and wish ourselves back in civilization again.

Sanford also wrote:

> Sabbath, April 20th, the second one spent here. The wind blew so hard we did not venture to go to church. Our neighbor, Mrs. Boulware, came in a while. She says if we wait for calm weather we will never go, or not until June, so it is some consolation to know there is a limit to the wind.

The prevailing wind direction is from the north and northwest in the winter and the south and southwest in the summer. The combination of high temperatures and forceful winds creates high rates of evaporation and plant transpiration. (Transpiration is the uptake and emission of water by plants). Surface evaporation plus plant transpiration, or evapotranspiration, is usually higher in the summer than the rate at which water penetrates the soil and reaches the groundwater. As important as these aspects of climate are, the most distinguishing characteristic of Great Plains

A thunderstorm rolls in over Swan Lake in Garden County, summer of 1992.

Jim Swinehart, CSD

[1]With special thanks to Vincent H. Dreeszen, who contributed significant parts of this chapter

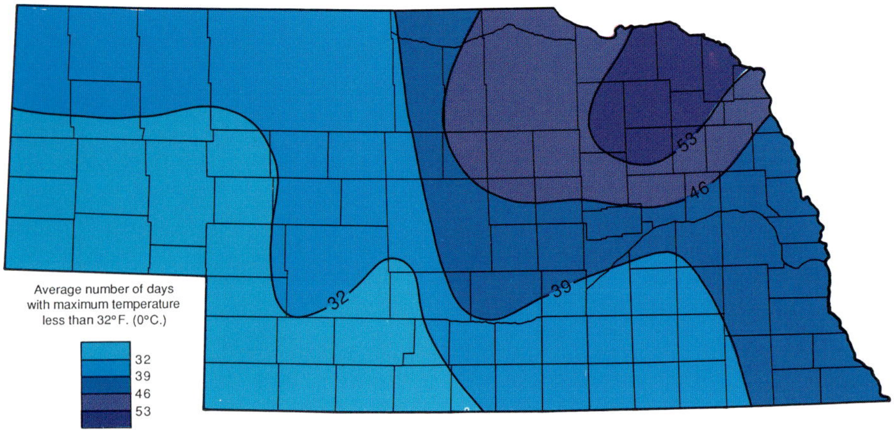

Average annual days with low temperatures

climate is the frequent deficiency of the most essential climatic and biologic element: water.

On the earth's surface, water can be considered a renewable resource. Unlike many other natural resources, for most practical purposes, it is neither created nor destroyed but moves and changes through the hydrologic cycle *(see diagram of the hydrologic cycle, p. 47)*. If so, then why do people talk about the "consumptive" use of water regarding domestic, industrial or agricultural use? And why is irrigation called the major consumptive use of water in Nebraska? What they actually mean is that the water either has changed into a form not easily used or has moved to where it is not readily accessible. Pollution, for example, is one way water can be made unusable. However, when water is diverted for irrigation, it is not destroyed. Much of it is transpired—released as a gas into the atmosphere through crop plants. Some evaporates, and some remains in the plant, not to be released until harvest. Excess water either runs off the land or soaks into the ground, where it may or may not be easily accessible to the next user.

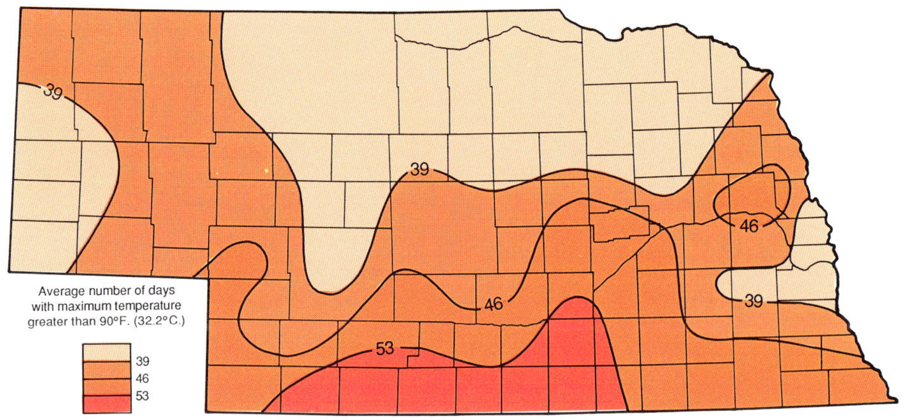

Average annual days with high temperatures

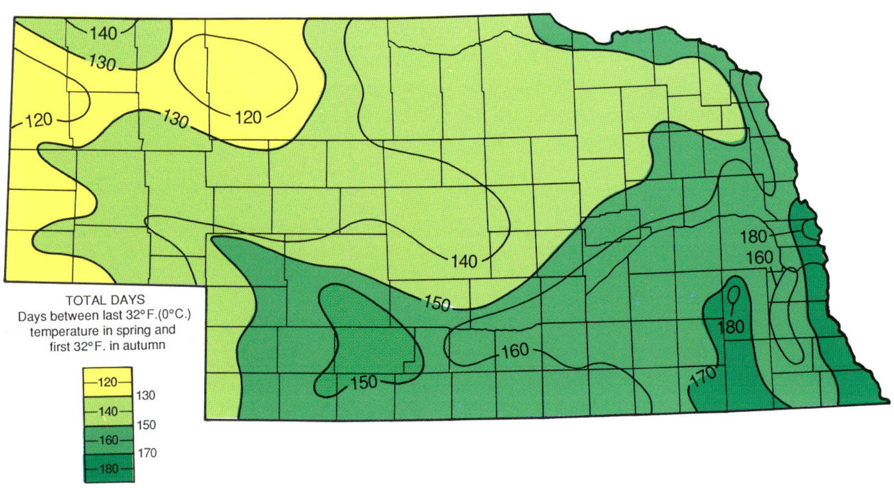

Average annual freeze-free days

The fact that water is neither created nor destroyed is a tremendous advantage to human society. However, its changeable nature and constant movement make it a tricky substance for humans to handle. At a given time or place, our ability to predict the occurrence or amount of water may not be much better than our ability to predict the weather. Water does not stay put like coal or other minerals. Rather, it moves across international and interstate boundaries, creating legal and political difficulties in the process.

When water is in a gaseous form in the air, it may move as fast as hundreds of miles per day, carried by the prevailing winds. As a liquid, water may move tens of miles a day, as in a river, or only a few hundred feet a day, as in a lake or pond. Water that percolates between the sands and gravels or other pore spaces underground moves much more slowly than it does on the earth's surface, often just a few feet per year.

Where Does the Water Come From?

One obvious source of water is precipitation. The average annual amount of precipitation in the state ranges from 35 inches in the southeastern corner to 14 inches in the west. This range, from sub-humid to semi-arid, is comparable to the change in average annual rainfall from the east coast to Chicago *(see map of normal annual precipitation in "Introduction," p. 2)*. Much of this precipitation comes as rain during the growing season. However, the averages belie the tremendous variation in annual amounts of precipitation *(see average annual departure from normal precipitation in "Introduction," p. 3)*. Such variation makes being a farmer or rancher in Nebraska a risky business.

Rivers that flow across the state's boundaries, primarily from the west, also bring an average of about 2 million acre-feet of water into the state *(see map of streamflow in "Introduction," p. 3)*. (An

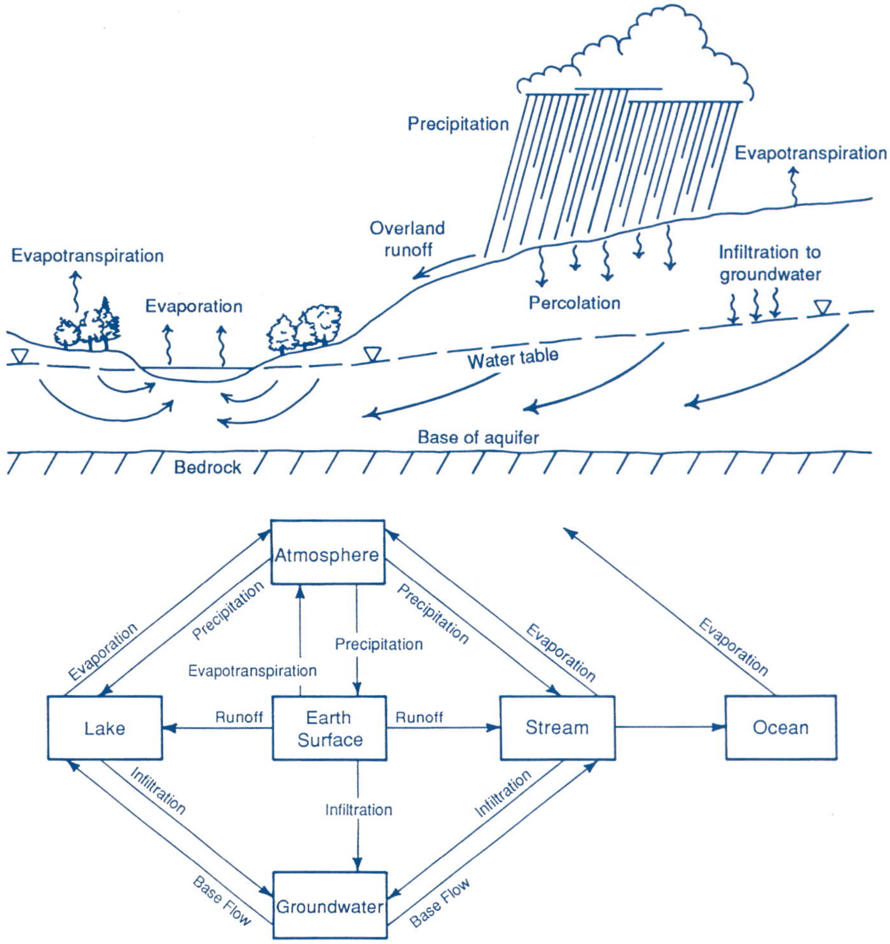

The hydrologic cycle

acre-foot of water is the amount of water that would cover one acre to a depth of one foot.) The Platte River system, consisting of the North Platte River, the South Platte River and the Platte River, carries a major part of these inflows. Much of this flow comes from snowmelt in the Rocky Mountains. Tributaries in the foothills of the Rockies also contribute to the Platte system, as well as contributing to the Republican and Niobrara rivers.

Finally, the state receives water from groundwater that flows across the state line. Accompanying this chapter is a map showing the water-level or water-table contours of the state's groundwater *(see p. 48)*. In general, water flows perpendicular to the contour lines from higher water-table elevations to lower ones. A look at the map shows that water enters the state through groundwater flow from the north and west.

Where Does the Water Go?

Much of the precipitation that falls on Nebraska runs off overland, eventually reaching a stream, and flows to the east and south out of the state. A larger amount infiltrates into the ground and eventually reaches the groundwater. Nebraska is very fortunate to have soils and geological formations that readily allow the infiltration of groundwater. Because water moves slowly within these formations, water that reaches the groundwater (called *recharge*) may stay there a long time, creating groundwater reservoirs that, in some cases, are hundreds of feet in thickness *(see map of saturated thickness in "Introduction," p. 4)*. The thickest of these reservoirs is in the deep sand and gravel deposits under the sandy soils of Nebraska's Sand Hills. Two billion acre-feet of usable water are estimated to be stored in the state's groundwater reservoirs. This supply is 700 times that stored in all of its surface-water reservoirs.

Although much of the water that recharges Nebraska's groundwater reservoirs stays there for long periods of time, some water moves from the groundwater back to the earth's surface. Nebraska has a number of natural springs and flowing artesian wells (wells that flow without pumping because of water under pressure), where groundwater flows to the surface. More commonly, the groundwater slowly seeps from the reservoir to a stream or lake. In the water table map *(see p. 48)*, any location where the groundwater contours form "V's" that point upstream, water will be moving from the groundwater to the stream.

A large amount of the flow in many of Nebraska's streams comes from groundwater. The map of streamflow in this book's introduction *(see p. 3)* shows how flows in the Niobrara River increase, picking up groundwater as the river runs along the northern border of the Sand Hills. The Loup River system also drains the groundwater reservoir under the Sand Hills. These flows feed the lower Platte River below the city of Columbus and provide much of the drinking water for cities along the lower Platte, including Fremont, Lincoln and Omaha. The Republican River, the Big and Little Blue rivers and the Missouri River, though primarily fed by water from overland runoff, also carry some groundwater. Like the Platte and Niobrara, they carry some of this water beyond the state's boundaries. On the average, 9 million acre-feet of surface water flows out of the state.

Some water crosses the state line as groundwater flow. By far the largest amount of water leaves the state through transpiration and evaporation. Water evaporates from the state's many rivers, lakes and reservoirs, as well as from the

Ann Bleed, DWR

Nebraska's western grassland (beyond the timbered east), where trees are mainly in stream valleys, is seen in this photo looking west at the confluence of the North and South forks of the Dismal River.

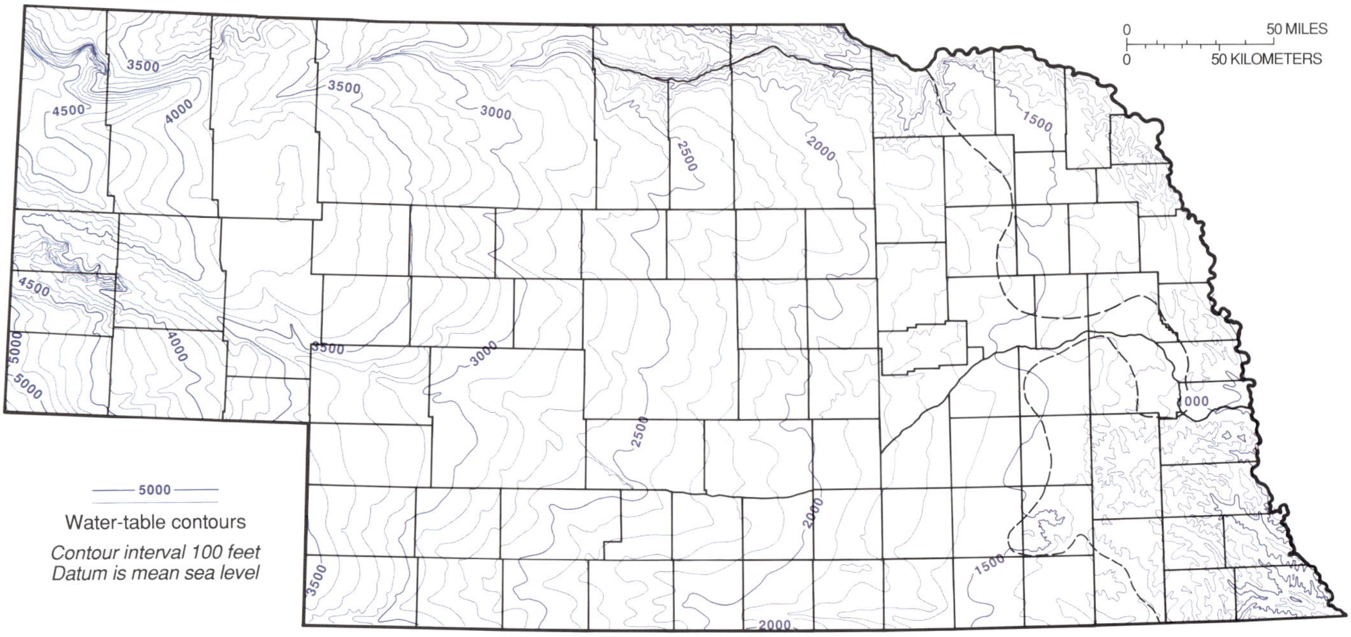

Configuration of the regional water table. (East of the dashed line, some data were taken from perched water tables or water confined under artesian pressure.)

ground after it rains or snows. In addition, water is lost from the subsoil either directly or through plant transpiration. Many of Nebraska's native plant species have adapted to the water supply by developing roots that reach deep and tap into groundwater supplies. Evapotranspiration accounts for the movement of large amounts of water into the atmosphere, especially in sub-irrigated meadows, wetlands and along rivers.

Having the right amount of water at the right time and place is critical to the success of all living populations. Those that are successful have adapted to the seasonal and yearly variations in the availability of water. Human societies, though more able to manipulate the timing and location of water sources than other creatures, are not completely free from such difficulties.

In Nebraska, irrigation is the largest "consumptive" use of water. East of the 98th Meridian, where rainfall is more plentiful, crops can be grown with little or no supplemental water (irrigation) in most years. Toward the west, the chances of producing a crop without supplemental water decrease markedly. Nebraska's early farmers learned this lesson the hard way. As wet cycles encouraged optimism, those with doubts were kept somewhat quiet out of a fear that the reputation of the state as a place for a good life would be ruined if the need for supplemental water for crops were known. However, the dry years of the 1890s convinced those in the west that irrigation was necessary. At first irrigation was considered an insurance policy against drought, but later it became an integral part of the farming operation.

The first irrigators looked to Nebraska's rivers for a source of water. Some of the most fertile lands in the state occur along the Platte, Republican and Big and Little Blue rivers. Unfortunately, these rivers do not maintain a steady flow. Fed primarily by overland runoff, they may have high flows in the spring or after a rainstorm, but the flows generally drop off sharply in the summer. They also vary greatly from one year to the next. Several consecutive drought or low-flow years are not uncommon. Early irrigators who built canals to divert these flows often found that, though plentiful in the spring, they were inadequate in the summer at the time crop water needs were the greatest.

Above its confluence with the Loup River, the Platte River system, a major

A spring feeding the North Fork of Birdwood Creek, McPherson County

Ann Bleed, DWR

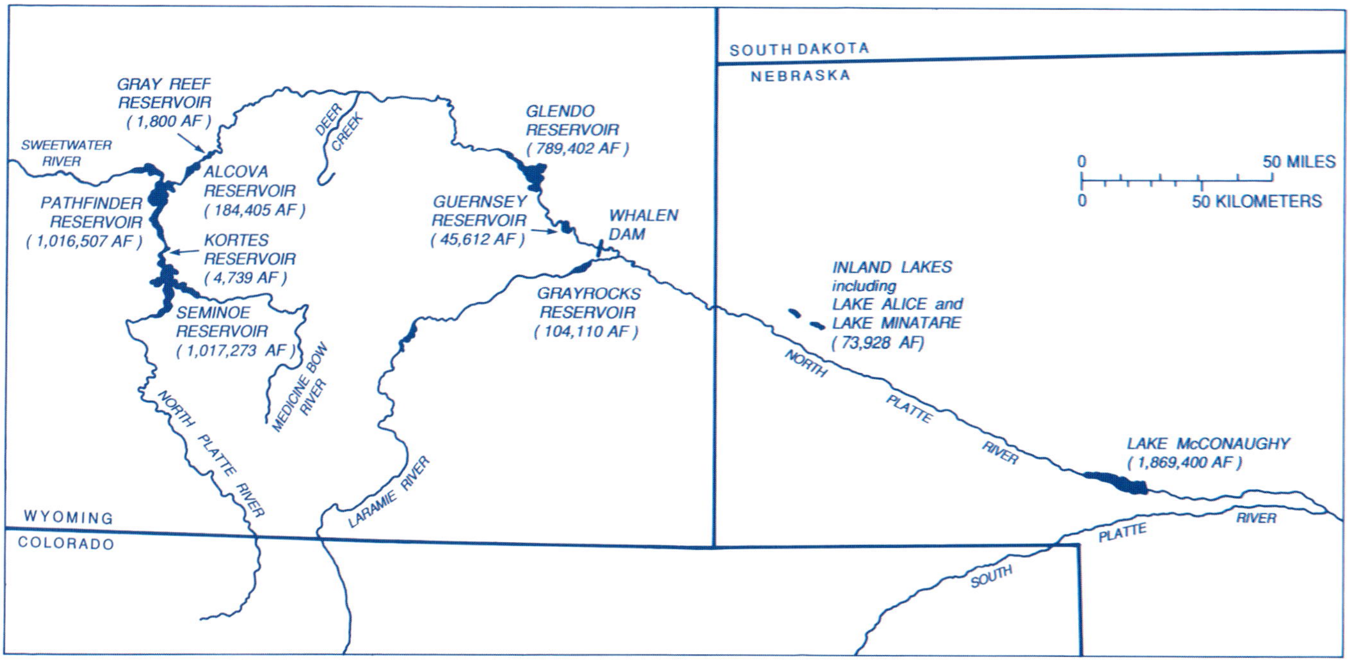

Reservoirs along the North Platte and Platte rivers in Wyoming and Nebraska

source of water for crops since the beginning of irrigation in Nebraska, is the ultimate of runoff streams. The North and South Platte rivers originate from snowmelt in the Rocky Mountains of Colorado and Wyoming. With few natural tributaries to replenish the supply from the state line to Columbus (near its confluence with the Loup), Nebraskans irrigating out of the Platte system depend greatly on snowfall in the mountains. The snowmelt flows in the Platte usually occur in May and June. By July, when irrigation requirements are at their peak, the flows decrease.

The Platte's extreme variations in flow caused many problems for early irrigators. For this reason, 27 canals, built on Nebraska's part of the Platte system when water was plentiful, were abandoned by 1910. To solve these timing problems, reservoirs were built to capture flows during the non-irrigation season and store them until the crops needed water. The larger of these reservoirs are on the North Platte River *(see the map of the reservoirs on the North Platte and Platte rivers, p. 49)*. The first of these large reservoirs was Pathfinder Dam, completed in 1909 on the North Platte River in Wyoming *(see "Politics and Policy," p. 207)*. Six additional reservoirs in Wyoming and Nebraska's Lake McConaughy were built on the North Platte River for a total storage capacity of nearly 5 million acre-feet.

This storage of water, together with irrigation diversions, changed the flow system of the Platte River. Similar storage and diversion projects also changed the nature of the Republican River. In both cases the river width and expanse of open, unvegetated sandbars narrowed, and the islands and banks became overgrown with willows and cottonwoods.

Just how much these developments changed the hydrologic system of the river is unknown. Unfortunately, reliable records of streamflow and climate do not exist before 1900. Gaging stations increased in the 1940s, but even today, streamflow records, though much improved, are often inadequate. The wide and shifting riverbed makes it hard to get accurate flow measurements. Nevertheless, few dispute that water storage has decreased peak spring flows on the North Platte and Platte rivers. The South Platte River, with no large dams, still exhibits relatively high uncontrolled peak flows. Most also agree that the diversion of water has contributed to dry streams in the summer, but it is unknown just how dry the Platte was before the irrigation diversions. On the other hand, the release of storage water from reservoirs and return flows from irrigation projects have added water to the river in the summer and fall. The end result is a new, highly managed river system.

Irrigation return flows include water from surface-water and groundwater sources. Water that is diverted but not applied to the fields is usually spilled back to the river. Return flow also includes groundwater inflows to the river derived from water that seeps through the walls of the canal and water that percolates through the soil and eventually reaches the water table. Unlike surface-water "spills" (returns), which return to the river in a matter of hours or days, groundwater return flows may not reach the river for several weeks or months. Additional groundwater storage, built up over time in areas where surface-water projects move or store water, may con-

(Continued on p. 54)

R.H. Willis, water commissioner for the Platte River from 1895 to 1919 and chief of the Bureau of Irrigation, Water Power and Drainage after that, stands beside what is probably the Platte at high flow. Photo undated.

Department of Water Resources

Understanding Groundwater Starts with the Rocks: Geologic Framework, Groundwater Occurrence and Irrigation Development in Nebraska

Nebraska is blessed with abundant groundwater. This groundwater is found beneath much of the state, principally in the sediments and sedimentary rocks that blanket the ancient and largely crystalline rocks geologists call Precambrian *(see block diagram of Nebraska's geology, p. 50).* These younger deposits contain pieces of evidence such as fossils and sedimentary features that reveal the tremendous changes in the environment that have gone on over the last 570 million years, during which these deposits accumulated. *Change* is the name of the game in geologic history, just as it is in human history. Fundamental changes in geologic history across the Great Plains states, particularly during the last 66.5 million years (the Cenozoic Era), have had a dramatic role in and effect on the creation or destruction of the state's water-bearing sediments and rocks.

From the Cambrian Period through the Cretaceous Period of geologic time (570-66.5 million years ago), Nebraska frequently was covered by a shallow inland sea or was a coastal plain where sediments were deposited by rivers and wind. The sediments deposited in the sea range from land-derived sands and muds laid down mostly near shore to lime muds that accumulated either farther offshore or near-shore but away from river mouths, where sands and muds entered the sea. Sediments deposited on the coastal plain when relative sea levels fell were river channel, floodplain and delta deposits, as well as wind-deposited dune sands and silt blankets. Stream and wind erosion periodically outstripped deposition, resulting in the removal of some of the previously accumulated sediments. Land plant cover on the coastal plain aided in developing ancient soils in the upper parts of the sediments there. Through time some of the sediments were hardened by compaction or by the introduction of natural cements that precipitated out as waters moved through these sediments. Other sediments remained relatively unaffected. Countless geology students have remembered this span of earth history generally by the adage, "the seas came in, the seas went out. . . ."

Some may react to the changes just sketched out and ask, "So what?" But the reader whose imagination is still intact can appreciate their enormity. If you have driven across Nebraska, you have a feel for how big the state is. Now imagine covering all of it with a sea perhaps hundreds of feet

(continued on next page)

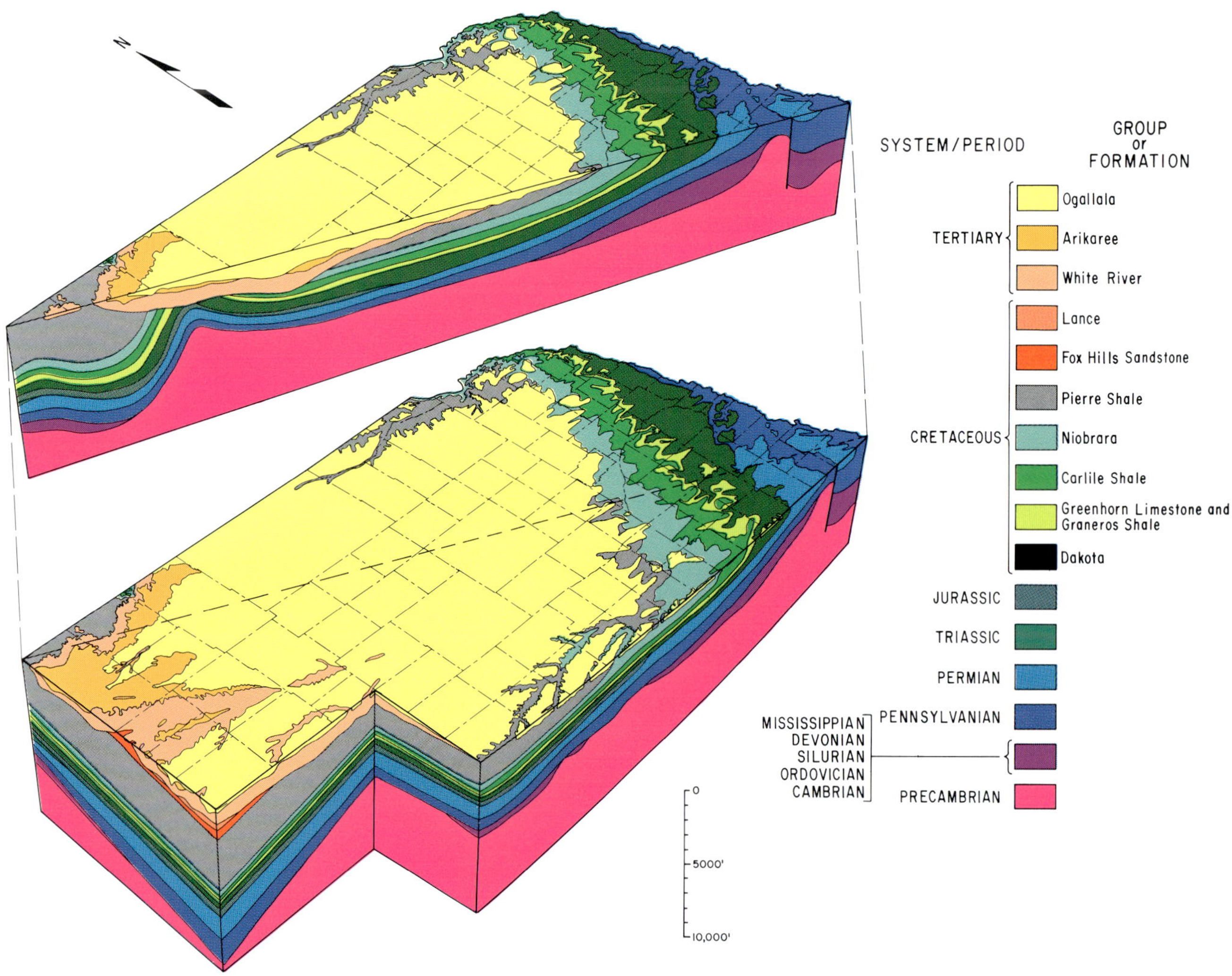

Block diagram of Nebraska geology showing distribution of strata (layers) of different geologic ages and major folds and faults (from The Groundwater Atlas of Nebraska, *Conservation and Survey Division, IANR, UNL)*

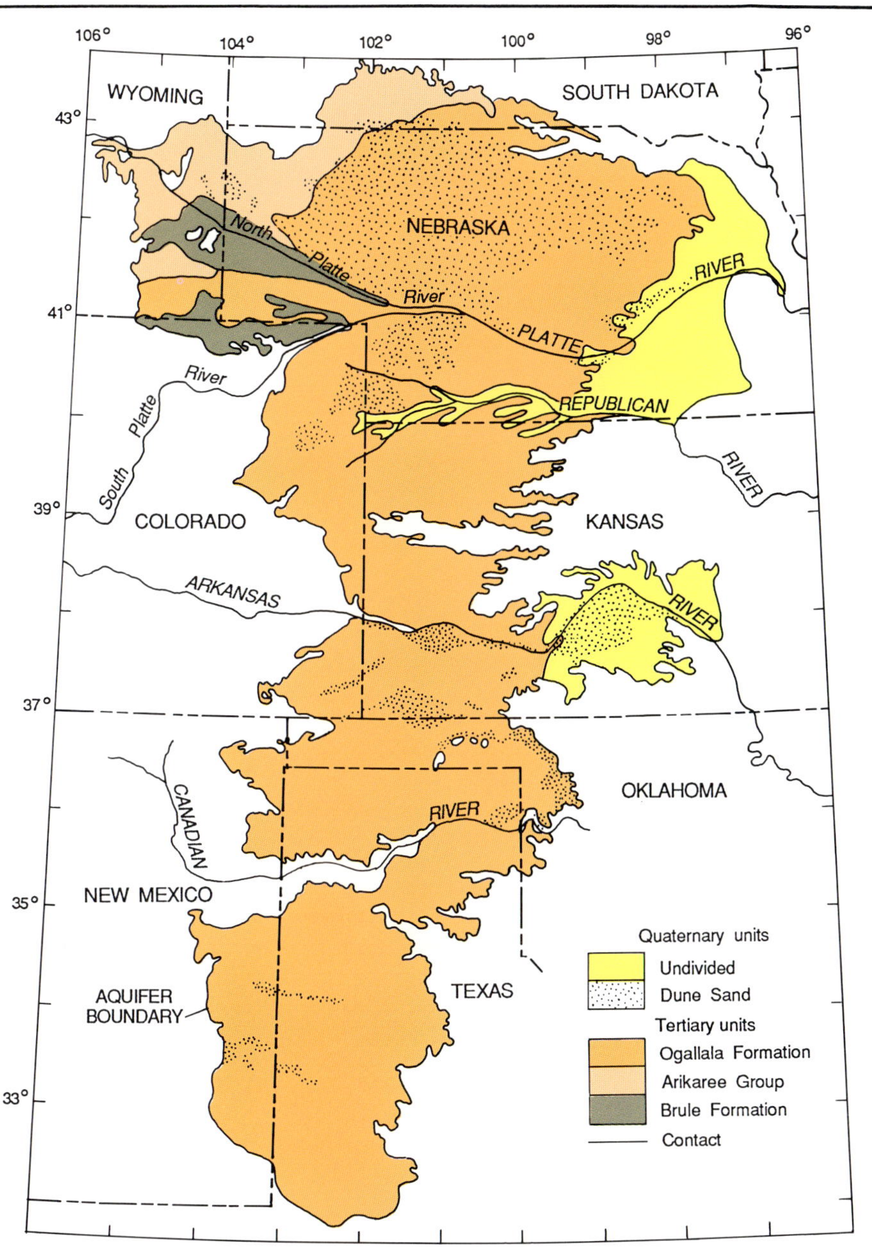

Distribution of principal geologic units in the High Plains Aquifer (from U.S. Geological Survey Professional Paper 1400-A, fig. 1)

deep. That sea extends from the Appalachians to the Rocky Mountains, or even from coast to coast, and from the Arctic Ocean to the Gulf of Mexico. Now drain off part or all of that sea and then bring it back again to cover parts or all of the former flooded area. Repeat that mental image time after time, and you begin to get a better appreciation of the fantastic things that were taking place across Nebraska. That's change writ enormous! And if that's not enough for you, then from time to time have a big meteorite strike the earth and imagine the damage it could do.

More than 30 years ago, Bob Dietz, a highly respected geologist, coined the term *astroblemes* for the impact structures made from such a hit. There is a likely crater site about 67 million years old in west-central Iowa near the town of Manson. Rocks buried under the younger sediments beneath the farm fields there are intensely deformed. A piece of extraterrestrial rock perhaps several billion cubic feet in volume fell out of the sky and hit that spot, creating a huge deformed area of about 300 square miles and blasting up enormous volumes of rock debris into the atmosphere and sending out a shock wave of tremendous force in all directions. Everything standing would have been leveled for hundreds of miles around the site and perhaps for years sunlight would not have been able to penetrate the debris cloud that appears to have surrounded the earth, according to geologists at the Iowa Geological Survey and the U.S. Geological Survey. The climatic effects of this and possibly other nearly simultaneous impacts in other areas of the world may have led to the extinctions of dinosaurs and many other life forms at the end of the Mesozoic Era (about 66.5 million years ago).

The seas really ''went out'' about 66.5 million years ago across most of what is now North America and have not returned to Nebraska or the rest of the Great Plains since. With the exception of a few scattered small sediment remnants of intermediate age, the geologic record in Nebraska is missing from 66.5 to about 38 million years ago. Starting at about 38 million years ago, volcanic eruptions from the Rocky Mountains west and south into Mexico produced vast clouds of volcanic ash that rained down on Nebraska periodically and produced most of the deposits known as the White River and Arikaree groups of rocks (*see block diagram of geology*). These eruptions became less common beginning about 17 million years ago. Then rivers transported proportionately more sediments eroded from Colorado's and Wyoming's Rocky Mountains. These sediments began to be deposited across Nebraska, forming the initial parts of the High Plains (Ogallala) Aquifer.

The High Plains Aquifer underlies most of Nebraska and parts of Colorado, Kansas, New Mexico, Oklahoma, Texas, South Dakota and Wyoming (*see map of High Plains Aquifer, p. 51*). The U.S. Geological Survey has included all of the sediments and sedimentary rocks of the White River, Arikaree, Ogallala and most younger sediments beneath the area shown in the aquifer map. The thickest parts of the aquifer and the greatest quantities of water are in Nebraska. This water is generally most abundant in the Ogallala Group and younger Pliocene and Quaternary sediments beneath the Great Plains.

In Nebraska and adjacent areas, the Ogallala Group is composed mostly of mixtures of gravels, sands and silts in varying proportions deposited in complex valley fills. Time after time, rivers eroded valleys and filled them with sediments carried from the mountains to the west, then shifted their courses and began new cycles of cutting and filling. Now and again volcanoes to the west of Nebraska erupted and sent volcanic ash into the atmosphere that settled on the accumulating Ogallala deposits. These deposits were sometimes preserved. Preserved in them sometimes were fossils of ancient life. For example, at the Ashfall Fossil Beds State Historical Park in Antelope County, many ancient rhinoceroses and other animals were buried by ash. Sediments also accumulated in lakes as the sediment pile called the Ogallala accumulated and soils rich in calcium-carbonate formed on land surfaces. To the south, in west Texas, much of the Ogallala is wind-deposited sediments, indicating drier conditions in that area than in Nebraska.

Since the end of Ogallala deposition, riv-

(continued on next page)

ers continued to cut valleys across parts of Nebraska and to fill them with sediments. General pictures of the changes in river positions through time were worked up in 1990 by three colleagues at the Conservation and Survey Division, V.L. Souders, J.B. Swinehart and V.H. Dreeszen *(see map of postulated evolution of Platte River and related drainages, p. 52)*. The oldest of these river systems crossed Nebraska at about the time that the oldest glacial ice sheets in North America reached eastern Nebraska *(see Rolling Hills on topographic regions map, p. 53)*. Ice sheets also affected that area in the Illinoian glacial advance *(see postulated drainages map)*. Imagine the effect that these river-blocking ice masses at least several hundreds of feet thick had on rivers flowing up to their edges, perhaps ponding those rivers, certainly changing their courses and causing them to deposit sediment upstream. The glacial ice and rivers issuing from the melting ice sheets caused enormous changes, particularly in eastern Nebraska. The former west-to-east drainage pattern was changed to a more north-to-south pattern that is still present. Ogallala- and Pliocene-aged aquifers were mostly destroyed in the glaciated part of the state by glacial erosion, and many cubic miles of sediment were washed by glacial melt waters to the Gulf of Mexico. In place of these aquifers, thick deposits of impermeable or poorly permeable glacial till or boulder clays, with a few interbedded river silts, sands and gravels, were left behind.

Ice sheets only made it as far south and west as northeastern Kansas. Rivers to the south were not affected in the same ways as those in Nebraska. River deposits are not so thick and do not cover such a large area there.

Overlying the river sediments are wind-transported silts (loess) beneath the land surface on the Plains, Dissected Plains and parts of the Rolling Hills areas of the state (*see topographic regions map*). Sand covers the Ogallala and younger river deposits beneath the dunes of the Sand Hills areas. J.B. Swinehart has done extensive studies of the dunes and believes that the dunes we see today formed less than 10,000 years ago in several arid periods separated by wetter times.

Much of the loess and sand blanketing most of the state is less than 50,000 years old. The present river valleys are all very young geologically. River deposits buried directly beneath their floodplains are all probably less than 10,000 years old. During the last 10,000 years the river systems have eroded their valleys more than once and these each time have been filled in places with river-, lake- and wind-deposited sediments. These gravelly fills in ancient, buried valleys are good water-bearing sediments where they can be found. River system erosion across all of Nebraska during this time was responsible for dewatering some of the older parts of the High Plains Aquifer that formerly were saturated but were breached by that erosion.

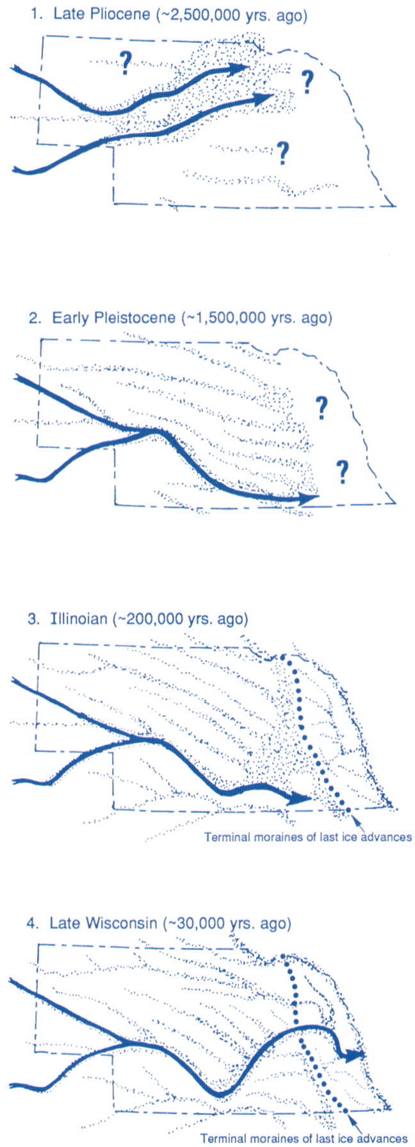

Postulated evolution of Platte River and related drainages

To further complicate the geologic picture, the sediments and rocks of Nebraska have been folded and faulted in places (*see block diagram of geology*). Since the geologic bedrock map is such a small-scale map, many of these structures have been omitted. Most of these occur in the Panhandle and in eastern Nebraska. Some of this deformation has been so recent that even sediments as young as the Pliocene have been deformed; even younger deposits may have been too. These folds and faults are responsible for the position and development of at least some Nebraska river valleys and valley fills, and perhaps most of them.

The picture just painted is essentially without life and could lead you to see the Great Plains area as lifeless or as unaffected by events in other parts of the world. But neither of these things was true. The seas and lands teemed with life that migrated, died out locally or became extinct, or originated and expanded to occupy other areas after the environment changed there. These environmental changes were produced by changes in elevation of lands due to mountain uplift or land subsidence, by physical events resulting from volcanic eruptions or meteorite impacts, by changes in the amount of energy received from the sun and by changes in life forms and their impacts on the places where they lived. Life and the non-living earth have affected one another throughout the time that life has been present on the earth, and extraterrestrial forces have affected both as well. Even the formation of our aquifers and river systems here on the Plains were tied inextricably to the interplay of these forces.

Irrigation and Aquifers

For the most part, the Cambrian to Cretaceous sediments and sedimentary rocks beneath Nebraska are not important aquifers for one or more of the following reasons: they are often deeply buried and are overlain by better post-Cretaceous aquifers; their water is poorer in quality than that in overlying younger aquifers; or they do not yield any water to wells. The Cambrian to Pennsylvanian sandstones, limestones and dolomites have been used as water sources in parts of eastern Nebraska. The Dakota Sandstone of Cretaceous age is an important aquifer in parts of eastern Nebraska and is used in public, private, irrigation and industrial wells.

For obvious reasons, irrigation development in Nebraska is mostly in the areas underlain by the High Plains Aquifer (*see map of registered irrigation wells; compare with postulated drainages map and topographic regions map*) with the exception of the Sand Hills. If the soils in the latter area had been better suited to farming, that area would probably be more heavily developed than it is (that it is not suited is an accident of geologic history).

Many people believe that the regional aquifer in Nebraska (or anywhere) is some kind of vast underground lake. It is not; such an idea could not be further from the truth. In reality, most of our aquifers are uncemented sands and coarser grained river deposits. In these the groundwater is present in the spaces between adjacent grains called pores. The porosity formed at the time of sediment deposition and is thus called primary and intergranular *(see diagram of types of porosity, p. 53)*. After the rocks have formed, secondary porosity occurs by fluids dissolving either crystals (intercrystalline porosity), fossil shells (moldic), or fine-grained lime muds (vug porosity) from limestone or dolomites or dissolving cements from sandstones, all of which create cavities. After the sediments have been deposited, larger scale openings

(continued on next page)

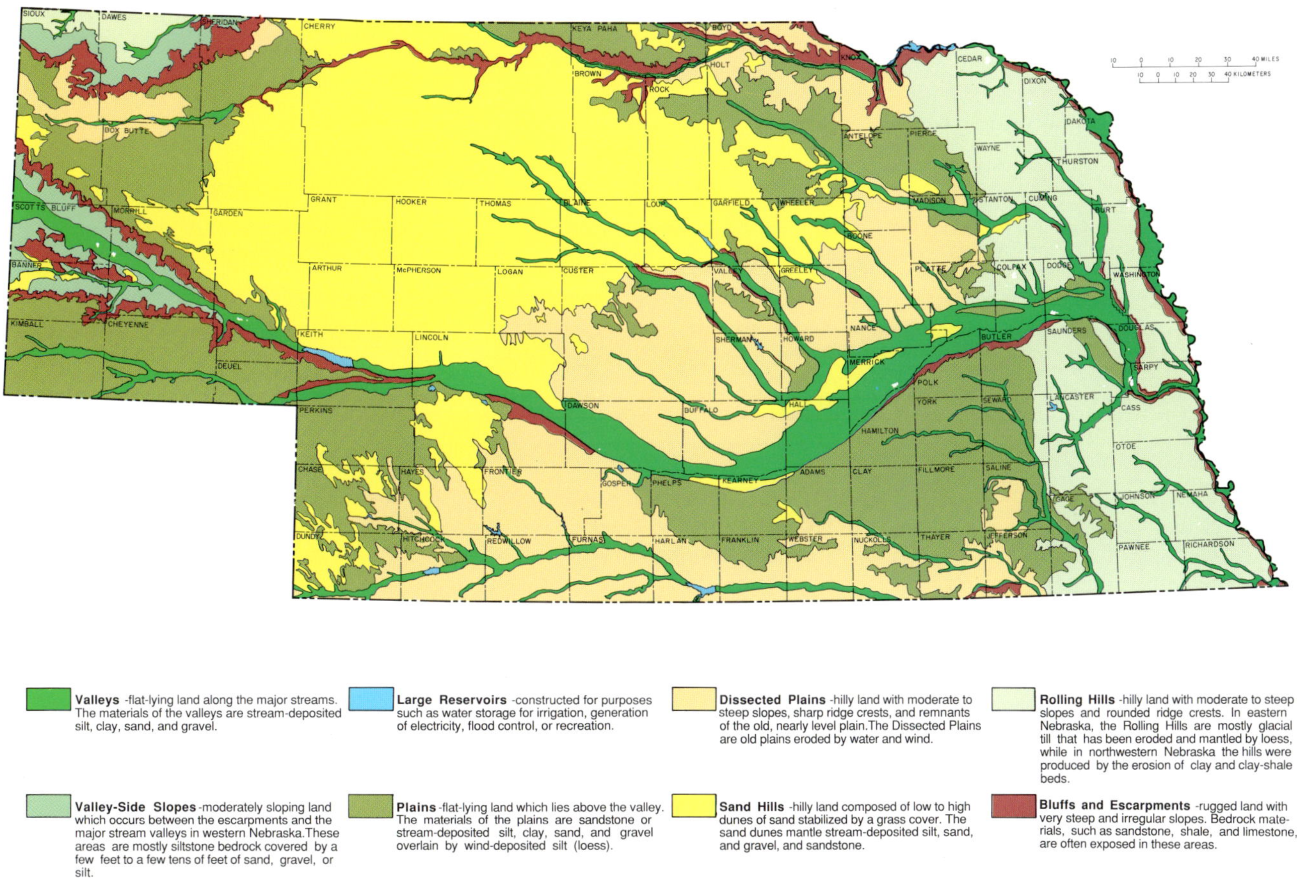

Topographic regions of Nebraska

are produced by fracturing or by solution. Recently, springs exiting from the tops of cylindrical conduits several feet in diameter and many tens of feet deep have been described along the Dismal River by geologists Ann Guhman and Darryll Pederson. These larger scale openings pass through unconsolidated sands and other sediments. Piping or development of narrow conduits through soils by erosion of the earth materials by percolating waters can also produce passages for water movement. If interconnected pores have diameters too small to allow water passage, the aquifer may contain water but is said to be not permeable.

Intergranular porosity occurs in some Cambrian to Permian sandstones, in some Cretaceous sandstones, in White River Group silts, sands and sands and gravels, in Arikaree Group sands and coarser deposits and through most of the Ogallala and younger sediments and sedimentary rocks. Intercrystalline, moldic and vug porosities occur in limestone and dolomites, while porosity due to dissolving cement may occur in sandstones of any age. Fracture porosity occurs mostly in White River Group and older rocks, large-solution cavities in parts of the Niobrara Chalk of Cretaceous Age, and conduits in the post-Ogallala sediments beneath the Sand Hills. Piping and fracturing occur in loess and in some parts of the weathered White River Group. Of all of these, the intergranular porosity mentioned above is the most important in Nebraska's principal aquifer, the High Plains Aquifer.

All of the aquifers, formed from rocks and sediments containing one or more of the porosity types described above, origi-

(continued on next page)

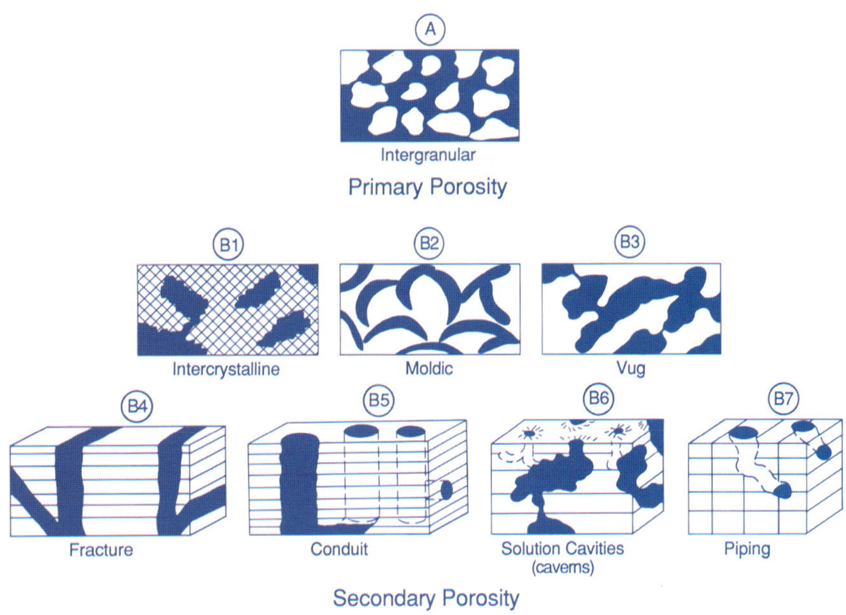

Types of porosity (pore space in black): a) primary porosity: small-scale intergranular; b) secondary porosity: small scale (< 1 in. in diameter): B1—intercrystalline; B2—moldic—fossil shells dissolved, rock matrix undissolved; B3— vug—parts of rock matrix dissolved; large scale (usually > 1 in. long or wide): B4—fracture; B5—conduit; B6—solution cavities (caverns); B7—piping in weathered sediments.

nated as a result of the interplay of forces and events carried out over lengthy periods of geologic time. A few seemingly minor changes in these events could have resulted in great changes in our aquifers and in the distribution and flows of our rivers.

—by R.F. Diffendal, Jr., CSD research geologist

Sources: "Appraisal of Data for Ground-water Quality in Nebraska," R. A. Engberg, 1984, U.S. Geological Survey Water-Supply Paper No. 2245; "Boiling Sand Springs, Dismal River, Nebraska—Agents for Formation of Vertical Cylindrical Structures and Geomorphic Change," Ann I. Guhman, and Darryll T. Pederson, 1992, in Geology, v. 20, p. 8-10; "Cenozoic Paleogeography of Western Nebraska, J. B, Swinehart, V. L. Souders, H. M. DeGraw, R. F. Diffendal, Jr., 1985, in R. M. Flores and S. S. Kaplan, eds., *Cenozoic Paleogeography of the West-central U.S.*, Society of Economic Paleontologists and Mineralogists, Rocky Mountain Section, Rocky Mountain Paleogeography Symposium 3, p. 209-229; "The Groundwater Atlas of Nebraska," Robert Kuzelka and Darryll Pederson, project leaders, Charles Flowerday, ed., 1986, Conservation and Survey Division, Institute of Agriculture and Natural Resources, University of Nebraska-Lincoln, Resource atlas no. 4; "The Hydrogeologic Framework of the Paleozoic Aquifers of the Papio Natural Resources District, Nebraska," J. W. Pipes, unpublished master's thesis, University of Nebraska-Lincoln; "Older Groundwater Reservoirs in Nebraska," 1992, Marvin P. Carlson, Steven S. Sibray, and Franklin L. Mills, contract report for a cooperative project between the Conservation and Survey Division and the Nebraska Department of Environmental Control; "Overview of Nebraska's and the Regional Aquifer-Part 1," R. F. Diffendal, Jr., 1985, Proceedings of the 1985 University of Nebraska Water Resources Seminar Series; "Plate Tectonics, Space, Geologic Time, and the Great Plains—a Primer for Non-geologists," R. F. Diffendal, Jr., 1991, Great Plains Quarterly, v. 11, p. 83-102; "Summary of the High Plains Regional Aquifer—System Analysis in Parts of Colorado, Kansas, Nebraska, New Mexico, Oklahoma, South Dakota, Texas, and Wyoming," 1988, J. B. Weeks, E. D. Gutentag, F. J. Heimes, R. R. Luckey, U.S. Geological Survey Professional Paper 1400-A; "Water-bearing Formations of Nebraska," G. E. Condra and E. C. Reed, 1936, Nebraska Geological Survey Paper No. 10; "Wind-blown Deposits," J. B. Swinehart, 1989, in *An Atlas of the Sand Hills*, Ann Bleed and Charles Flowerday, eds., Conservation and Survey Division, Institute of Agriculture and Natural Resources, University of Nebraska-Lincoln, p. 43-56.

(Continued from p. 49)

tinually release water to streams. Limiting the amount of this return flow is one of the major goals of canal operators, but one that is easier said than done *(see vignette on Wes Sell, p. 70).*

The impact of irrigation return flows can be significant. A number of tributaries to the North Platte River between the state line and Bridgeport were once ephemeral streams, flowing only during spring runoff. Return flows from irrigation canals, diverting water from Wyoming, have turned these into perennial streams, which today not only support irrigation but also a trout fishery.

This impact of return flows was noted early on in the development of the river. In 1927, R.H. Willis, water commissioner for the Platte River from 1895 to 1919 and then chief of the Bureau of Irrigation, Water Power and Drainage (later to become the state Department of Water Resources), commented on the impact of reservoir releases and return flows. At the time he wrote, Pathfinder Reservoir, completed in 1909, and Guernsey Reservoir, just being completed in Wyoming, were the only major reservoirs on the river.

> The quantity of water returning to the North Platte river, after being diverted for irrigation, is amazing to those who have not observed the habits of this stream. . . . In 1909 there were probably fewer than 50,000 acres under irrigation in the valley west of Bridgeport. . . . By 1919 the total had reached 233,000 acres. At present [1927], the irrigated area in the valley comprises approximately 350,000 acres.
>
> Since 1910 there have been two main factors affecting the habit of the river, namely, the storage of flood water in the Pathfinder Reservoir, and the increased diversions from the stream. The reservoir serves as a direct retard of the flow of the stream from the non-irrigation to the irrigation season. The effects which flood storage and return flow have had on the lower section of the North Platte river are shown graphically in Fig. 1 *[See accompanying graph on p. 54]* It will be noted that since the construction of the reservoir the mean flow for May and June has been approximately 50 percent less than the mean prior to that time. For July, the mean flow has been 30 percent less; *for August, 95 percent more; for September, about 225 percent more; and for October, 215 percent more than*

(Continued on p. 56)

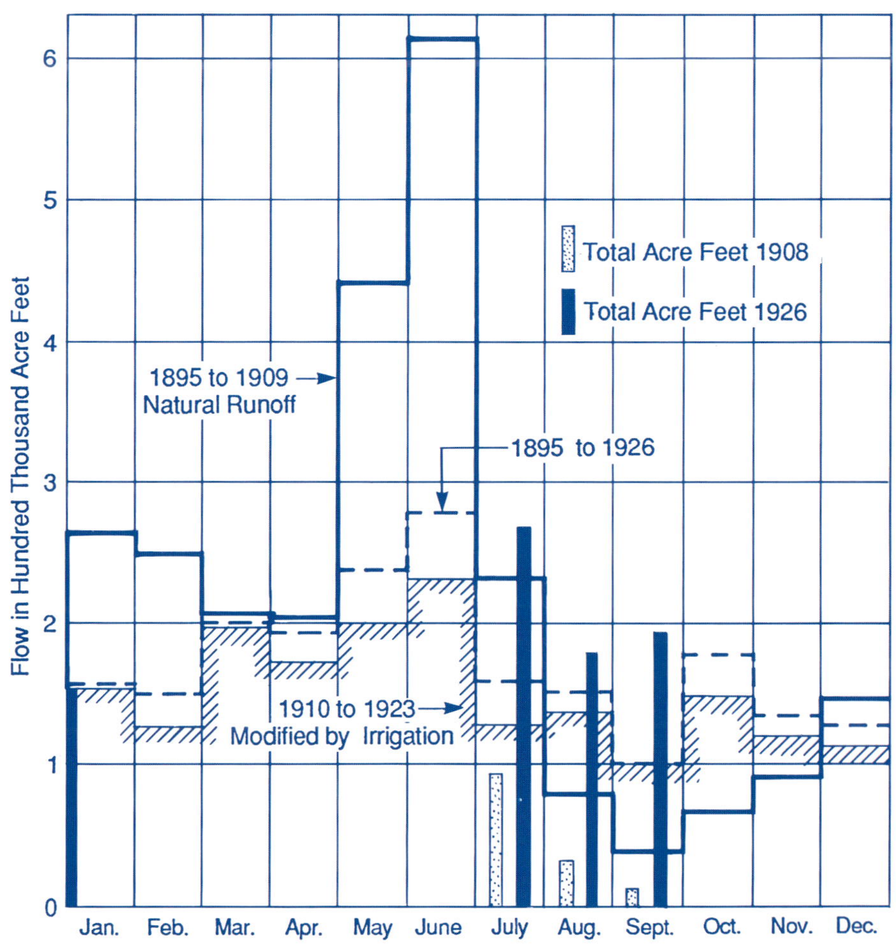

Mean monthly discharge of North Platte River at North Platte from beginning of record-keeping to 1927. Adapted from "Return Water and Drainage Recovery from Irrigation," a symposium by R.H. Willis, Esq., and D.W. Murphy, in American Society of Civil Engineering Paper No. 1730, vol. 94, 1930, p. 327-343.

Global Climate Change and the Great Plains

The Historical Climate

Geologic history shows us that the current climate system is quite different from the historic past. Over the past million years, North America has been dominated by periods of extensive continental glaciation. Sea levels were much lower than today and thick ice sheets extended down from northeastern Canada into the plains. The climate of the Great Plains consisted of ice-cap conditions to the north and east and tundra conditions to the west and south. Heavy mountain snowfalls pushed alpine glaciers down the front range of the Rocky Mountains. However, they did not extend very far out onto the plains. During this last ice age, a time known as the "Pleistocene," several periods of glacial advance were followed by short periods of glacial retreat known as "interglacials." We are currently in one of these interglacials that began about 12,000 years ago.

Although it might seem that the earth is not obviously glaciated and that the term interglacial is inappropriate, it must be remembered that vast amounts of ice still remain today on Greenland and Antarctica. During this current interglacial, the climate system has varied significantly from current conditions. From the onset of this interglacial to about 3000 B.C., the earth warmed considerably, reaching temperatures much warmer than the current climate. This period of time, extending from about 3000 B.C. to about 100 B.C., was called the "Climatic Optimum" and brought tropical heat and desert dryness to the Great Plains. The sand dunes of the Sand Hills region of Nebraska were active and blowing during this time period. Many researchers have pointed out that this glimpse of the past might also offer a view of the future should the forecasts of global "greenhouse warming" occur.

The "Climatic Optimum" ended, and the climate system gradually cooled for several hundred years only to warm again during what has been termed the "Little Climatic Optimum" of 1000-1200 AD. The climate of the plains then became much wetter and colder during the next six centuries. This time period known as "The Little Ice Age" persisted until the mid-1800s, when the earth's climate system began to revert to a full ice age. Glacial advances were common in mountainous regions, and in some European mountain regions, glaciers began to overrun some settlements.

It is interesting to note that the harsh climatic conditions of that time were well documented in the literature and art of the time. The small settlements that had existed in Greenland completely perished at this time (not solely, but perhaps to a large extent the result of harsher climatic conditions). Several noteworthy explorers set out at this time to find a "northwest passage" through the Arctic, and many met with disaster due to the harsh climate. European settlers first arrived in North America and began to establish colonies. The history books relate that the winters were harsh and many of the colonists perished. Considering "The Little Ice Age," one understands why climate was such a limiting factor on these early settlements. Toward the end of this period, the first scientific weather records began in the United States, and many of the existing record low temperatures for much of the United States date back to the start of this record.

The Climate During the Past 100 Years

The "Little Ice Age" slowly loosened its grip and eventually came to an end by the 19th Century. The climate system warmed until the 1930s and 1940s culminating in the dramatic event known as the "Dust Bowl." This was a period when the highest temperatures on record were established. All time record highs in the range of 115 degrees F to 120 degrees F were established up and down the Great Plains. Unlike today's climate, where temperatures of 100+ degrees F normally last for only a few days at a time, heat waves of the 1930s produced weeks of consecutive 100+ degrees F temperatures. These heat waves occurred with some of the lowest precipitation records for the Great Plains and caused a significant emigration of plains farmers, as so well presented in John Steinbeck's novel, *The Grapes of Wrath*.

Mercifully, the climate system began to cool down and precipitation began to increase extending up to the mid-1970s, a period of time in which scientists began to warn of an impending new "Little Ice Age." The three winters of 1976, 1977 and 1978 were three of the coldest winters in the past 100 years. The impact on our nations's economy and our collective need to conserve energy was summed up by then President Carter as "the moral equivalent of war." Snow cover in the 1970s commonly covered the Great Plains for several months at a time. The rivers were ice-covered each winter and spring ice jams were a common phenomenon. Great Plains blizzards were a common climatic winter phenomenon. In searching for a human-induced cause, Professor Reid Bryson of the University of Wisconsin determined that humans in industry and in farming were putting vast amounts of dust into the air and coined the phrase "human volcano." He maintained that this blocking of the sun's energy was causing the rapid cooling in the climate system. His comparison of human activities to a volcano was grounded in the concept that increased volcanic activity in the past has been related by scientists to some of the cooling trends in climate. But then, one more time, the climate began to change.

Current Climate Trends

In the early 1980s, the U.S. Environmental Protection Agency, which had been monitoring the role of pollution and the climate system, began to warn of an impending "global greenhouse warming" that might be caused by increases in carbon dioxide and to some degree other "greenhouse" gasses such as methane and chlorofluorocarbons. Greenhouse gasses do not hinder incoming solar radiation but tend to trap outgoing radiation of heat from the earth's surface. These scientists recognized that the amount of dust being put into the air by the "human volcano" phenomenon was decreasing but greenhouse gasses were rapidly increasing. At first, the warnings of warming were virtually ignored, until heat waves and droughts began to occur across the southeastern United States, the Midwest and western United States. By the mid-1980s the media were reporting that more and more scientists were finding that the "signal" of climate change in the form of global greenhouse warming had begun. This opinion was and still is not universally accepted by all climate researchers. There are at least two dissenting points of view, the first arguing that the current warming trend is a random cycle and does not represent human-induced change and the second that the warming trend is the result of an urban bias in the collection of temperature data (as cities grow, so should the urban "heat island," the reasoning goes).

As the decade of the 1980s drew to a close, climatologists were telling us that seven of the warmest years this century for the nation had occurred in the 1980s. Nineteen-ninety followed as the warmest year of record for the United States, and 1991 was just behind as the second warmest year on record. A drought in California was in its seventh year, and the winter of 1991-1992 was being called the "year without a winter" as mild weather persisted across much of the nation. Along with these mild conditions came record amounts of cloudiness and fog throughout much of the Great Plains. It is interesting to reflect on the fact that a little over a decade ago, the media were flooded with scientific predictions of a new "Little Ice Age." Although it is possible that the current predictions of continued global warming may also be preempted in the near future with a reverse trend to cooling, the current highest probability is for continued warming, so it would make sense to plan accordingly.

(continued on next page)

What About the Future?

With all of the dire statements about our changing climate and warnings of global warming, it would be only natural to worry about a potential return to "Dust Bowl" conditions across the Great Plains. We are somewhat more prepared to withstand the onslaught of a "Dust Bowl" type drought due to better irrigation technology (compared to the 1930s); however, the external pressure for access to finite water resources continues to grow. For the answer to these questions, we can take a look at what are called global circulation models (GCMs) and their attempt to "model" the future climate.

Although some believe there are certain cycles to climate, such as the so-called "22-year drought cycle" in the Great Plains, most climate cycles are limited in their geographic extent, and they all have a degree of non-reliability. The climate system is far too complex to be neatly modeled using a cycle or two as the basis for a climate forecast. Instead, climatologists have turned to high powered computers and have designed computer models, called "simulations," to understand the dynamics of the climate system and how it might change. The advantage of this "simulation" approach is that it allows the climatologist to put all of the known information about our earth-atmosphere system into the computer model along with the myriad of mathematics that describe the various physical laws, and then have the model "simulate" what the climate might look like in the future using various scenarios. These scenarios can include current trends; past events not occurring now, such as vast volcanic activity; or projected conditions such as increased "greenhouse gas" concentrations. Although it should be noted that these models do not have the refinement to offer information at the scale of county size, they are excellent methods of examining regional (for example, Great Plains) climate conditions.

It is generally agreed that the primary greenhouse gas, carbon dioxide, continues to increase rapidly in our atmosphere. Using the current exponentially rising trend in global carbon dioxide, climatologists anticipate that global carbon dioxide should double by about 2050 A.D. However, this assumes that there will be minimal pollution abatement during the coming decades. As a scenario, this doubling has been put into the various climate GCMs. As might be expected, there is some disagreement among the various models as to what our future climate might look like. However, there is strong agreement about several aspects of our modelled future global climate. The following is a brief summary of these major anticipated changes in climate:

The Great Plains should see a significant decrease in precipitation during the growing season. And the Great Plains should have greater loss of moisture in winter due to higher temperatures and greater evaporation rates. With decreased precipitation and cloud cover during the growing season, the probability of heat stress due to extremely high maximum temperatures should significantly increase. The Environmental Defense Fund physicist, Michael Oppenheimer, using the NASA Global Circulation Model, estimates that there will be a radical increase in the number of days over 95 degrees F during the next 50 years. His estimates for the number of days above 95 F include, for example, Louisville, Kentucky, soaring above 95 F for 45 days, up from 5 days now; Houston to 117 days, up from 30 now; New York City, 48 days, up from 15 now, and Omaha, 80 days, up from 22 today.

Outside the Great Plains, the greatest amount of warming is expected to occur throughout the Polar latitudes, where it is anticipated that there will also be a huge increase in cloud cover and precipitation (especially winter snow falls).

Important to note is the fact that the various computer models show that the entire earth is not expected to warm and in fact regions such as eastern Canada, forecasted to cool, have already begun to show significant cooling. A large controversy associated with this is the belief among many climatologists that the warming of the earth will produce enough cloudiness that the climate system will regulate itself and begin a gradual cool-down, preventing the forecasted dire consequences from occurring. If, however, the forecast of continued rapid increases in global warming takes place, it appears that we in the Great Plains may have the most to lose with a forecast of dramatically lower precipitation amounts along with higher temperatures and evaporation rates.

Although we can anticipate being one of the hardest hit areas in this future scenario of global warming, there are some agricultural areas that should benefit from these changes. As forecast in these climate models, Canada and central Asia, for example, stand to benefit from a longer growing season and increased precipitation. This, obviously, hints that there could be drastically changed roles in the world grain market during this change in climate. It is also reasonable to expect that there will be severe negative effects on wildlife in the Great Plains with the acreage of wetlands drastically decreasing.

The question of what to do to ensure availability of water resources in the future is predicated upon the belief that in fact global warming will continue. If the current trend were to reverse and the global climate were to begin cooling with increased precipitation amounts over the Great Plains, our worries of water shortages will be unfounded. However, if the warming continues, decreasing precipitation in the Great Plains will occur concurrently with increased pressure for access to the dwindling water resources. It would seem appropriate at the current time to begin the process of prioritizing the uses of water resources in our area along with developing contingency plans for the development of additional wetlands and perhaps dam development.

—Kenneth F. Dewey, vice chair, Department of Geography, University of Nebraska-Lincoln

(Continued from p. 54)

> *the mean flow prior to the beginning of storage*. . . . This increased flow during the months cited is the result of the application of water to the irrigated area between Whalen [in Wyoming] and North Platte [italics and brackets added].

Today these streams, together with wetlands created by seepage water, provide habitat for waterfowl and other wetland species and support a cold-water trout fishery. Their flows are an important source of water for irrigators between the state line and Bridgeport, and their fall flows supply a significant part of the water for Lake McConaughy. Though these return flows enter the North Platte River in Nebraska, they have their origins in Wyoming. If diversions in Wyoming for canals in Nebraska were cut off, the return flow would be severely diminished. The importance of the return flow to irrigation on the North Platte is not only recognized by water scientists but also forms the basis for dividing up water between Wyoming and Nebraska in the U.S. Supreme Court Decree for the North Platte River. This 1945 decree is perhaps the earliest official recognition of the connection between surface water and groundwater in Nebraska and their potential for conjunctive use, that is, the integrated management of groundwater and surface water.

Unfortunately, although reservoir storage and irrigation return flows helped to provide water to meet crop water demands in mid-summer, the demand on rivers in the Platte system and on the Republican River was, and still is, greater than the supply. To avoid conflict, in 1895 the state passed a water-rights law. This law allocates water to the holders of the oldest permits first and then, if water is still available, to the younger permits. The Nebraska Department of Water Resources (DWR) and its predecessors have administered this system.

In 1929 R.H. Willis set forth some of

Gaging station on the Castle Rock Canal, looking upstream. R.H. Willis stands beside the shelter. Photo undated.

Department of Water Resources

the problems involved in water administration:

> In order that an irrigation farmer may enjoy the benefits of his water supply, while not depriving others of an equal privilege, the irrigation world ordains that the rules of orderly distribution be observed.
>
> It is on our statutes and in the constitution that the appropriator *first in time is first in right*. The water users can not enjoy the benefits of those rights unless the available water supply of the basin is lawfully administered. To do otherwise would retard development of our irrigation institutions. . . . And beside, there would be no peace and security for those who have already established their home.
>
> As a matter of fact, practically all of the normal flow of the stream in which Nebraska is concerned is over appropriated to a more or less degree during July, August and September, or at least a part of that period in some seasons and on some streams. Hence we are confronted with the problem of orderly distribution.
>
> The administration of the water supply brings the State officials in contact with various problems . . . [such as] storage water, natural flow, diversion dams, good, poor and indifferent headgates, pumping plants, acreage reports, measuring devices, observer's reports, evaporation, Water Commissioners, Interstate rights, temporary injunctions, cooperation, joint ditches, several priorities in the same canal, riparian rights etc. . . .
>
> [On the North Platte River] extending from the Pathfinder Reservoir in Wyoming to Kearney, Nebraska, a distance of more than five hundred miles, there is considerable variance in the daily climatic conditions. Rain may be falling in one part of the basin and [it may be] dry with high temperatures at another. Next day, the conditions may be reversed. . . .
>
> It is [a] difficult matter to close or reduce the flow of junior canals just to the amount necessary to supply senior rights, when the temperature is high, perhaps hot winds blowing, and a shower here and there, without the danger of some water passing the headgate of the last canal downstream diverting water. The loss by evaporation varies each day; it may be high one day or less the next, which will affect the computations of the amount of water needed to supply rights in the lower sections of the streams.
>
> Administering the water supply of the streams within the State does not end at the boundaries of the State. We have several interstate streams of greater importance than our intra-state streams: the Republican, South Plate, Horse Creek, White River and the North Platte Rivers. The latter has more complications than any of the other streams [brackets and italics added].

To administer water rights, DWR also needs to know what river flows can be allocated on any given day. From this need arose the department's stream-gaging program. Michael Jess, current director of the department, presented this program in writing to the Nebraska State Legislature in 1977 in the "Proposal For an Expanded Stream Gaging Program."

> The stream-gaging program in Nebraska has evolved through the years as federal, state and local needs for surface-water data have increased. Although some records have been collected since 1891, a systematic collection was not begun in Nebraska until 1894. Formal cooperative gaging activities between the U.S. Geological Survey and the State were started in 1906 but lasted only until 1914. State cooperation resumed in 1931 and has continued since.
>
> Early stream-flow records were obtained strictly for irrigation needs. By 1912 several small plants were using water power for generating electricity. When additional development was planned, lack of stream-flow records for this purpose was recognized as a deterrent to future development. In 1912 this need prompted the State Engineer to suggest the need for additional stream-gaging sites.
>
> Major floods of 1935 in the Republican River Basin, and those of 1940, 1944, 1947, 1950 and 1951 in river basins throughout eastern Nebraska caused local residents to seek flood-control projects. This resulted in accelerated stream gaging to obtain data on which to plan flood-control works.
>
> Many found the records invaluable for such activities as irrigation development, flood-plain zoning, water-resources planning and recreation development. After stream-flow data became available, the Department of Roads began using flood stage, discharge and frequency statistics in designing roads and bridges. The Department of Environmental Control [now "Environmental Quality"] (and previously the Department of Health) utilizes the chemical quality records in evaluating the natural and altered environmental conditions of many streams. Finally, use of the records is made for purposes of research and interpretive evaluation of the resource [brackets added].

Although surface water was the first to be used for irrigation, the limited quantity and unreliability of that supply, as well as problems moving water from the river to distant fields, restricted its use. Groundwater, on the other hand, is an abundant and reliable source in many areas of the state. Once data was developed showing where groundwater occurred and how it could be tapped, it became the predominant source of irrigation water in the state.

Groundwater pumping began soon after surface-water irrigation in the valleys of rivers such as the Platte. Some farmers began to use windmills to pump water to irrigate small plots of land. However, major irrigation development had to wait until better drilling equipment and pumps were developed and until more was known about the state's groundwater.

One of the earliest groundwater in-

(Continued on p. 60)

1935 Republican River Flood Killed 110, Caused $100 Million in Damages in Today's Dollars

Nebraska Farmer

Part of the aftermath of the 1935 Republican River Flood

May 1935 brought heavy rain across the entire Republican River basin. Between May 12 and 22, precipitation ranged from 2 to 5 inches, with another 1 to 3 inches falling between May 26 and 29. The soil had reached the saturation point. Some localized flooding had been reported along many southern Nebraska and northern Kansas streams.

But the worst was yet to come. A high pressure system carrying an extensive amount of moisture from the Great Lakes was moving in from the northeast. At the same time, a low pressure system, with its counterclockwise rotation, collected warm, moist Gulf air as it approached from the southwest.

The two storm fronts collided over the headwaters of the Republican River near eastern Colorado and southwestern Nebraska on the night of May 30. Although there were no U.S. Weather Bureau gauges in the area in 1935, numerous local residents reported rainfalls of 15 to 20 inches. One report of 24 inches of rainfall came from a Newton, Colorado, rancher.

The very wet month of May, coupled with the cloudburst of May 30 and 31, caused the worst flood in Nebraska's recorded history. To make matters even worse, the storm headed east, generally following the Republican River's drainage area. The storm lasted until June 2, dropping an additional 2 to 7 inches of rain over the basin. This added rainfall and eastward movement along the basin meant that the flood crest, which was heading down the Republican River, would meet with already high flood stages on the Republican and its tributaries.

In the headwaters area, where the great storm of May 30-31 occurred, the flood came at night with little or no warning. Forty lives were lost there. It came with a roar that was heard for miles. Perhaps the best first-hand account of the onrushing torrent is provided by J.H. Mullison, then-manager of the Southern Nebraska Power Company near Bostwick, Nebraska. As described by the U.S. Geological Survey publication on the flood, his experience watching from high ground was as follows: "While the flood was still several miles distant, he heard its roar, and when it appeared the advance was a mass of tumbling water and debris several feet in height, quickly filling the main channel, which at the point of observation was about 7 feet deep. Spreading out behind the advancing waves the flood quickly overflowed the bottom lands, giving the effect of an oncoming spearhead. Within a very short time the great quantity of tumbling debris blotted out the sight of the water itself. The flood made short work of bends and elbows in the old channel, taking the shortest course, which resulted in the formation of many new cut-off channels."

The 350-mile Republican River valley—eastern Colorado to Junction City, Kansas—was overwhelmed with a river that was now up to a mile and half wide in places. At Cambridge, Nebraska, where Medicine Creek joins the Republican River, the Republican Valley overflowed to a 2-mile width. This was the largest streamflow ever recorded within the state's borders: 280,000 cubic feet per second. For comparison, the average flow of the Missouri River from 1929-1950, before construction of any significant dams, was 33,350 cubic feet per second.

As a result of its location on the confluence of Medicine Creek and the Republican River, Cambridge was devastated. Seventy-five percent of the homes suffered flood-related damage, and only 20 remained on their foundations. Seventy-six people were rescued by boat from submerged homes.

Farther upstream, near Trenton, ample warning was given. But the warning fell on many deaf ears. Several did not realize the unprecedented depth of the water and refused to believe their homes were in danger. In addition, it was still night. As a result, 60 people lost their lives along this part of the river. The largest death toll along this stretch occurred in and around Oxford.

The Oxford paper reported: "Dwellers on the lowlands were warned of their danger, and in some cases trucks were waiting to take them to higher ground, but they had lived there so long and seen so-called high water so many times without trouble that all warnings and proffered help were rejected, with perilous results. When the houses began to fill up and were finally toppled over by the force of the swift current, shrieks for help could be heard, but shrouded by darkness, nothing could be seen. As daylight broke, those on shore had been pushed back by the ever-rising waters until the scene of the disaster was invisible except by the aid of field glasses, and all sounds were hushed by the roar of the raging river. Taking in the 5-mile stretch west of Oxford, two families suffered the heaviest fatalities. According to the survivors, as the water rose in the houses, the inmates [sic] 'clumb up into the attic, kicked in a hole in the roof, and climbed outside.' Presently the houses began to float but only went a short distance before the bottom struck trees or other objects and began to roll. Their human loads were thus dumped off into the swirling waters. Sometimes they would be successful in grabbing some objects, only to be brushed off like flies by floating debris. Thus it was a continuous fight until overcome by sheer exhaustion, rendering them an easy prey for the angry water."

The major flood waters did not arrive farther downstream until daybreak. As a result, the loss of life there was not as great. Those lives lost were due mainly to lost time trying to save personal property and livestock.

There were many reports of families climbing trees for safety all along the flood route. Families who sought such safety found they had another battle to fight. Many of these trees were already harboring snakes, rats and various other wildlife.

McCook, the largest Nebraska city on the Republican River, is situated on high ground and experienced only minor flooding. However, the city's water system was stopped when the local power plant was flooded. Forty of the power plant's employees spent a harrowing 24 hours on the

(continued on next page)

roof of their plant waiting for the flood waters to recede. Only then could rescue boats safely reach them.

Adding to the melee was a tornado about 6 miles west of McCook on the afternoon of May 31. The tornadic winds caused a thick duststorm over the McCook area that lasted about 30 minutes.

The damage results in the three-state basin (Nebraska, Colorado and Kansas) were staggering: 110 killed; 20,593 livestock killed; more than 275,000 acres of farmland damaged (most of it growing crops); about 1,000 miles of highways and railroad tracks damaged or destroyed; and close to 600 highway and railroad bridges damaged or destroyed.

Nebraska's share of the storm damage was severe: 94 killed; 8,100 livestock killed; 46,500 poultry killed; 341 miles of highway damaged or destroyed; 307 highway bridges damaged or destroyed; and 57,000 acres of farmland damaged (42,000 acres in crops).

Burlington Northern lost more than 40 miles of its main track. Temporary service was finally restored on June 23. Reports stated that the force of the water was so great that rails were actually twisted. Other reports detailed how some sections of track were uplifted and resembled a picket fence. Burlington alone suffered $1.5 million in flood-related damage.

The estimated total property value lost in Nebraska was more than $7.5 million. Today, this would amount to more than $100 million.

The huge numbers of livestock killed in the flood was a serious health hazard. This was especially true near Bartley, Nebraska, where a tree-lined bank in the bend of the river captured numerous carcasses. Adding to the health menace was the problem that such "log jams" also caught great numbers of live rattlesnakes.

After the flood, the destruction of the carcasses was a top priority. To assist local workers, the federal government deployed 1,600 conservation corps and relief workers. When their task was complete, attention turned to the massive clean-up of the landscape and restoration of public works. For this, the federal government extended $434,000 in aid to Nebraska.

Since 1935, numerous reservoirs have been built. Not only have they helped prevent another such flood, but they provide sources of irrigation and recreation. Some of the reservoirs built in Nebraska include: Swanson Lake and Harlan County Lake on the Republican, Enders Reservoir on Frenchman Creek, Harry Strunk Lake on Medicine Creek, Hugh Butler Lake on Red Willow Creek, plus countless farm ponds on smaller tributaries.

Flood Damage

	Nebraska	Kansas	Colorado	Total
Persons killed	94	10	6	110
Livestock killed	8,100	12,193	300	20,593
Highways damaged*	341	484	5	830
Highway bridges	307	202	6	515
Crops damaged-acres	42,000	221,507	NA	263,507
Farmland damaged-acres	57,000	202,615	15,000	274,615
Railroads*	NA	171	NA	
Railroad bridges	NA	83	NA	

* in miles

Table from U.S. Geological Survey Water Supply Paper 796-B

—Duane Mohlman, CSD editorial assistant and data systems coordinator

Sources: "Compilation of Surface Water Records, Missouri River Basin Below Sioux City, Iowa, Through 1950," U.S. Geological Survey Water-Supply Paper 1310; "Flood on the Republican and Kansas Rivers, May and June 1935," U.S. Geological Survey Water-Supply Paper 769-B, R. Follansbee and J.B. Spiegel; Personal notes of R. Kutz, U. S. Bureau of Reclamation; Personal notes of R. Hotchkiss, Department of Civil Engineering, University of Nebraska-Lincoln; "The Power Plant and the Big Flood," Nebraska Light and Power Company, McCook, Nebraska, 1935.

The Conservation and Survey Division (state geological survey) drills a test hole 4 miles northwest of Columbus with a Dempster rotary-hydraulic rig near the Loup Canal, 1935.

Conservation and Survey

Harry Burleigh (left) pumps water into one of two nearby mud pits. Miller (first name unknown) stands beside the rotary-hydraulic Dempster rig that drills a test hole. In the mud pit, clay is mixed with the water to form a mud solution. The mud solution is then taken from the pit by the rig's suction hose and pumped down the drill stem. The mud hardens and acts as a casing to help prevent cave-ins and brings samples to the surface. This return solution flows into the second mud pit, where the cuttings settle out. Samples are laid out for record-keeping on the wooden table in the foreground.

(Continued from p. 57)

vestigations was done by Nelson H. Darton in 1896 in southeastern Nebraska. A few other studies by Darton, G. E. Condra, O. E. Meinzer and R. C. Cady followed. In 1930 a cooperative investigation of groundwater resources between the U.S. Geological Survey (USGS) and the state of Nebraska began on the lower Platte Valley and adjacent uplands; and in 1934 Nebraska became the first state to embark on a statewide well-observation program. Declining water levels during the drought of the 1930s near Grand Island were investigated and published by L. K. Wenzel in 1940.

The real growth in groundwater investigation and use would not occur until after World War II. Before that Nebraska farmers would endure the terrible drought years of the 1930s. Vince Dreeszen, former director of the University of Nebraska-Lincoln Conservation and Survey Division (CSD), remembers those years well. He remembers, for example, his father's inability to produce marketable corn on the family farm near Palmyra in eastern Nebraska. A little farther to the east, corn got up to 4 feet high and jobs shucking corn could be found. However, to the west of Lincoln, the corn wouldn't get higher than 1 to 2 feet. One of the few bright spots in Nebraska during these years was the Sand Hills. Here groundwater was close to the surface, and subirrigated meadows continued to produce hay. Ranchers from surrounding areas sent their cattle to these productive pasture lands. The region, however, was not completely spared. Some Sand Hills lakes went dry.

The uncertainty of trying to farm in the 1930s and the fascination with rocks exposed by wind erosion sparked Dreeszen's interest in geology. After serving in World War II, he came back to Nebraska to study geology with the intent of exploring for oil and gas. Meanwhile, as part of the wartime effort to secure a stable food supply, Eugene Reed, CSD geologist and, later, its director, initiated a statewide systematic test-drilling program to assess groundwater supplies. Early investigations concentrated on the valleys of the Platte, Nemaha and Republican rivers. The test drilling in the Republican Valley was done cooperatively with DWR.

In the early 1940s, Reed also started working with domestic well drillers to develop irrigation wells. Equipment to drill irrigation wells in Nebraska's unconsolidated sands and gravels quickly and effectively just was not available. Drillers in the Shickley and Alliance areas began building their own drill rigs. Most of the early irrigation wells were drilled in the river valleys because of the limitations of the technology. Reed was convinced, based on the few scattered test holes already drilled in the uplands, that the potential for large-capacity wells existed there. Former state Senator Maurice Kremer of Aurora, also a believer and risk-taker, put in one of the first upland irrigation wells in Hamilton County *(see profile on Kremer, p. 234)*, but for the most part the use of wells on the uplands did not catch on until much later.

The scope of water-resource investigation in Nebraska was much expanded in 1945 by the USGS as part of the U.S. Department of the Interior's program for the development of the Missouri River basin. At the same time, Reed, then associate director of CSD, redirected the test-drilling program toward a systematic grid pattern.

Dreeszen joined forces with Reed in 1947, and together they developed a database on groundwater availability throughout Nebraska, working to promote the use of groundwater for irrigation, as well as for cities and industry. Still, groundwater irrigation was slow to develop. Pump equipment did not function well; a ready supply of fertilizers did not yet exist, and most banks were reluctant to loan money for irrigation. The few banks and other lenders that did wanted assurance that water would be available for the 20-year term of the loan.

All this hesitation disappeared during the drought of the 1950s. Farmers literally lined up at Reed's and Dreeszen's doors at CSD wanting to know where they could get water. By the end of 1954, about 10,000 total wells had been drilled. Reed and Dreeszen were involved in the first major irrigation development in nearly every county in the state.

During the same period, nearly every city and town upgraded its water supply with new wells, storage tanks and distribution systems. Most domestic wells also were replaced as rural water demands increased.

In the wetter period of the 1960s, well development slowed. However, during the drought of the mid- to late 1970s, spurred in part by high crop prices and the technological improvement represented by center-pivot sprinkler systems, the demand for irrigation expanded rapidly, and another spurt of drilling occurred. Mostly, the 1980s were wet years, and well drilling slowed again, only to pick up in the dry years of the late 1980s and early 1990s. As of 1991, nearly 75,000 irrigation wells had been registered in Nebraska. Dreeszen, justifiably, displays great personal pride when he points out that Nebraska was one of the first states to form a well drillers association to promote the construction of efficient, sanitary wells.

During this time also, CSD and the UNL Cooperative Extension Service actively promoted the formation of county irrigation associations to provide a forum for education about wells, pumps and water-use efficiency. At that time, annual withdrawal of 48 inches of water was not uncommon. But by degrees this effort was successful, as extension specialists used a variety of educational methods at their disposal, including humor. For example, to encourage more efficient use of water, one irrigation engineer coined the term "splash irrigation" to advocate, jokingly, driving back and forth on county roads to splash back on the fields excess water flowing across the roads.

Although CSD's efforts promoted groundwater irrigation, Reed and Dreeszen also were concerned that unrestricted pumping could deplete water supplies. Because of this concern, CSD and the USGS greatly expanded the statewide network of water-level observation wells. Dreeszen, along with Reed, promoted irrigation well registration and legislation on well spacing that would have required 2,000 feet between irrigation wells.

According to Dreeszen, Reed felt that the long-term water supply in the uplands could not be sustained if more than one well per section were installed. Other hydrologists and many legislators shared this concern, and CSD was directed to propose boundaries for groundwater regulatory districts. None were established but the efforts led to legislation in 1959 establishing groundwater conservation districts; eventually groundwater control areas were established in areas with serious rates of decline *(see "Politics and Policy," p. 220)*. As the law was voluntary, several districts were authorized to adopt, administer and enforce rules and regulations, but none did. The tax base was inadequate to provide administration, since cities and towns opted to stay out of the districts, and irrigators were reluctant to regulate themselves. Comments often heard in meetings across the state were "water levels in my well are the same as when it was drilled," "I only use as much water as I need," and, when asked about the cost of water, "My water doesn't cost me anything."

Obviously, the number of irrigation wells in many areas has greatly surpassed Reed's suggestion of one well per section. Groundwater declines have occurred at a rate in the uplands that is considered by some to be a concern. Declines are most apparent in Box Butte County, most of the Upper Republican Natural Resources District and in parts of the Big and Little Blue River basins. The rates of decline, according to Dreeszen, have been somewhat predictable but significantly less than what might have been expected. In these areas, the amount of water remaining is still large. The groundwater control areas created in the Upper Republican NRD and more recently in the Upper Big Blue NRD have been the only parts of the state where groundwater withdrawals are currently restricted.

Most attempts, including computer modeling, to predict rates of water-level declines have been overly pessimistic, according to Dreeszen. Examples of where declines were expected but remain minimal include Polk, York and western Butler counties. The reason is not fully known. In the areas of significant declines in the two Blue River basins and elsewhere, Dreeszen suggests that hydrologists have underestimated natural recharge and return flows from irrigation. He explains the minimal declines in southeastern Hall, western Hamilton and northern Adams counties as loss in flow from the Platte River.

One area where recharge is much greater than anticipated is along the Platte River. In his 1929 article on water administration, Willis described how irrigation projects were already changing the river.

> It is not known, at this time, how many junior appropriators should be closed to supply the Kearney Canal with 162 second-feet of water if permitted to flow down the bed of the Platte River in hot, dry weather. If evaporation loss only were to be considered, probably 162 second-feet of 400 second-feet passing North Platte would reach the headgates of the Kearney Canal. However, there is a stream bed loss east of the confluence of the North and South Platte Rivers. When the flow east of the North Platte is low, the bed loss is very noticeable. The water table during these periods drops below the river bed to the extent that no water flows on the surface. In the year 1910, the water table was about ten feet below the river bed at Kearney in the month of August. However, this condition seems to be improving. The water table has not dropped as low as in the former years. In a few more years the water table, which I believe is the surface of a large underground reservoir, will be stabilized at a higher elevation in July and August by the increasing return flow coming from irrigated areas in the upstream portion of the basin.
>
> In recent periods of water shortage it is less difficult to administer the distribution of the available water supply as far as Cozad. In other words, it seems that the water table has reached a stage of equilibrium at Cozad; beyond that point the water is below the river bed. Perhaps it is a futile hope to plan on the water table rising to the point where water will flow on the river bed at the Kearney headgate during low water periods.

One of the consequences of surface-water development not often recognized was that water levels would rise along canals and under project lands. Benefits from increased return flows to valleys and streams included greater well yields, shorter (and more economical) pumping lifts and an increase in number and size of small streams and wetlands. Other examples of where groundwater levels rose as a result of surface-water use include Dawson County; other irrigation canals and ditches in the valley to the west, as well as the Elm Creek and Kearney ditches to the east; in the Middle Loup, the Sargent and Farwell projects. Other projects where lesser rises occurred are along the Republican River and in the lands of the Mirage Flats and Ainsworth projects.

However, the most striking example of water-level rise associated with surface-water projects is the groundwater mound created by the combined projects of the Nebraska Public Power District and the Central Nebraska Public Power and Irrigation District (CNPPID). This rise runs south of the South Platte and Platte rivers through central Keith County, southwestern Dawson and most of Gosper, Phelps and eastern Kearney counties. Loss of water from Sutherland Reservoir and through the canals, lakes and project lands of CNPPID has resulted in water-level rises of more than 50 feet in some areas. The mound stretches from Sutherland to Minden, a distance of more than 150 miles. Various estimates of the amount of groundwater that is stored in the mound suggest that it is at least 7 million acre-feet. One of the results of this mound was that some irrigators switched from using surface water to using only groundwater; others use both groundwater and surface water, depending on surface-water availability. Though unplanned at first, this project evolved into an economically beneficial conjunctive-use system.

The contribution of return flows to the South Platte and Platte rivers and their valleys from the large mound probably cannot be calculated accurately. High water-table conditions have been largely corrected by extensive ungaged drains; large amounts of water are withdrawn from the area by wells, and, as yet, stream gaging is not adequate. Perhaps the most difficult quantity to document or even estimate is the amount lost to evapotranspiration. The CNPPID project also unintentionally affected the movement of water from the mound south into the Republican River valley tributaries and east into the Upper Big Blue and Little Blue River systems.

Another impact of surface-water development seldom considered in project planning is water rising to the land surface. Most, if not all, surface-water projects that include irrigation have created conditions requiring the construction of extensive drainage systems, often at a high cost. In some areas, as in Dawson County, ditching was done in the early 1900s. Later installation of wells to supplement ditch water resulted in a new equilibrium and conjunctive-use system.

Where groundwater is not replenished from delivery of surface-water, intensive groundwater withdrawal has resulted in groundwater declines. Groundwater withdrawal along some valleys also has affected streamflows. Examples in the western part of the state include Lodgepole and Pumpkin creeks in the Panhandle and in the southwest, Frenchman Creek, a tributary of the Republican River. Impacts on streamflow of Beaver and Sappa creeks, and also on tributaries to the Republican River, can be noted.

Other streams whose beds are near or slightly below the regional water table have also been affected by groundwater pumping. These streams include Wood River and the upper reaches of tributaries of the Big Blue River. Elsewhere in the state, including the Sand Hills, the impact of groundwater pumping on steamflow is not easily noted.

Another problem that can result from settlement and development is pollution. For the most part Nebraska's water supplies are low in dissolved solids and can be used with little or no treatment for human consumption, However, the presence of feedlots and the use of industrial and agricultural chemicals have contaminated water supplies in several areas of the state.

Nitrates from the use of fertilizer are the major cause of water pollution in the state, particularly well-documented in the central Platte Valley area. To date, because there has been little agricultural or commercial activity in the Sand Hills, groundwater in this region is relatively free of significant contamination. However, the ease with which water and contaminants can penetrate the region's sandy soils and groundwater makes this region very vulnerable.

More than plants and animals, human societies have been able to manipulate water supplies to meet their needs. The question is to what extent we have developed a sustainable system and made peace with the physical forces of the Great Plains. In cases where we have fallen short, can we find ways of managing our water resources to maintain the life we have? Or will we have to make radical changes in our ways of living in order to avoid depleting the treasury of both natural and financial resources? The challenge is to find a reasonable and fair system of management that will provide a sustainable, quality habitat for all trying to make it in this part of the Great Plains.

Sources: Agricultural Meteorology, Department of, Institute of Agriculture and Natural Resources, University of Nebraska-Lincoln; An *Atlas of the Sand Hills*, 1990, 2nd ed., Ann Bleed and Charles Flowerday, eds., Conservation and Survey Division, University of Nebraska-Lincoln; *Climatic Atlas of Nebraska*, Merlin P. Lawson and others, 1977, University of Nebraska Press, Lincoln and London; *The Great Plains*, Walter Prescott Webb, 1959, 2nd ed., University of Nebraska Press, Lincoln and London; *The Groundwater Atlas of Nebraska*, Robert Kuzelka and Darryll Pederson, project leaders, 1986, Conservation and Survey Division, Institute of Agriculture and Natural Resources, University of Nebraska-Lincoln; *Groundwater-level Changes in Nebraska*, 1991, Gregory V. Steele and Perry B. Wigley, 1992, Conservation and Survey Division, in cooperation with the U.S. Geological Survey, Institute of Agriculture and Natural Resources, University of Nebraska-Lincoln; *Mollie—The Journal of Mollie Dorsey Sanford in Nebraska and Colorado Territories, 1857-1866,* 1976, University of Nebraska Press, Lincoln and London; ''Natural Vegetation of Nebraska,'' Robert B. Kaul, 1993, Conservation and Survey Division, Institute of Agriculture and Natural Resources, University of Nebraska-Lincoln; Nebraska Department of Water Resources, 1992, State of Nebraska, including predecessors: State Board of Irrigation, State Board of Irrigation, Highways and Drainage (Department of Public Works), Bureau of Irrigation, Water Power and Drainage (Nebraska Department of Roads and Irrigation).

Interviews with: Wes Sell, canal manager, Farmers' Irrigation District; Vincent H. Dreeszen, hydrogeologist emeritus and formerly director, Conservation and Survey Division, IANR, UNL.

In the early 1940s, Nebraska City had water quality problems because the city obtained its water directly from the Missouri River. Nebraska City hired Henningson Consulting Engineers of Omaha to find a suitable water supply for the city. Henningson contacted CSD for help in drilling test holes. Ellis Gordon (third from left), CSD, was sent to oversee drilling. A barge was hired to move the drilling rigs to a small island on the Missouri River.

Conservation and Survey

To Make A Record of This Land: The Geologist as Public Servant

by Vernon L. Souders
Research Geologist, CSD

The director of the Conservation and Survey Division is *ex officio* the Nebraska state geologist, which means, among other things, that this person manages the research and record-keeping agency designated by the state to be a repository of unbiased knowledge regarding its groundwater and surface-water resources, as well as its geology, soils and some aspects of its vegetation.

This division and this designation for its director have their roots in the Nebraska Geological Survey, one of four present Conservation and Survey programs. The story of the Conservation and Survey Division (CSD) closely parallels Nebraska's water history. And since no history of Nebraska's water issues would be complete without the history of its state geologists, that story deserves a telling.

The present geological survey evolved from two sources, the earliest dating back to 1871, four years after statehood and the year the University of Nebraska opened its doors to 20 college and 100 preparatory students. That year Samuel H. Aughey, a Lutheran minister and natural scientist operating independently in Nebraska since 1864, was appointed by the regents to the chair of natural history at the new university. The only scientist on a faculty of five, Aughey somehow acquired the title of "Honorary State Geologist" at the same time.

The title carried no stipend but presumably provided the practical-minded people of Nebraska a focal point for public and private questions about geological resources while at the same time saving the state a considerable amount of money. Possibly the most noted member of the university faculty during the 1870s, Aughey probably toiled harder for the people of the state than for his students, becoming a well-known Nebraska booster. He promoted the richness of the state's soils, the idea of increasing precipitation by cultivating more acres, and the prospects of coal for fuel in this state of few trees.

Conservation and Survey

E. H. Barbour

Aughey resigned in August 1883, apparently because the regents were dissatisfied with the amount of time he spent in the Wyoming Territory investigating coal deposits. Rumor and the press suggested Aughey's resignation was due to his issuing a paper with forged endorsements, but an investigation exonerated him. The regents asked Aughey to withdraw his resignation, but he refused and migrated to the Wyoming Territory.

Lewis E. Hicks joined the faculty the following year as professor of geology and apparently was given the honorary geological title that Aughey had held. A student of the noted French-American naturalist, Louis Agassiz, Hicks was more than a teacher and academic. Like Aughey, he was a practical scientist concerned about the development of the natural resources of the state, and his special interest was groundwater resources.

Critical of the federal policy of spending to investigate only surface-water resources, Hicks became an early advocate of the need to survey groundwater to use it for irrigating crops. Robert N. Manley, in his *Centennial History of the University of Nebraska, Vol. 1—Frontier University (1869-1919)*, states that from Hicks's investigation in the 1880s "the people of Nebraska learned of the great water resources in the state" and that "as both a scientist and a promoter of irrigation Professor Hicks deserves the gratitude of Nebraskans."

However, another geologist, Erwin H. Barbour, also had arrived at the university in 1891 and, on his way to statewide renown, became head professor of geology and acting state geologist the year he arrived.

Barbour came to Nebraska with impressive credentials—a Ph.D. from Yale, where O. C. Marsh and J. D. Dana were two of his teachers, field experience with the U.S. Geological Survey and two years teaching at Grinnell College. Although his principal interest was vertebrate paleontology, Barbour recognized the commercial importance of geology, and the rudiments of a state geological survey slowly assumed shape.

Governor Thayer had appointed Barbour acting state geologist the year he arrived, and in 1893 legislation was enacted that made the head professor of geology at the university, Barbour, the geologist for the state.

No money was designated for the geological survey by either the regents or the legislature. But Barbour did conduct some field work at his own expense in 1891 and 1892. In 1893, Regent Charles H. Morrill, who also was interested in paleontology, began financing expeditions to collect fossils and other geologic samples and continued to do so until 1902. Information was at last being obtained on a systematic basis with private money.

The first direct state aid to the survey came in 1901 when $1,200 was appropriated by the legislature for the 1901-1902 biennium and probably provided funds for the first publications of the geological survey in 1903. Slightly larger biennial sums were voted until 1911 when legislation was passed that enabled the state survey to cooperate with the U.S. Geological Survey.

After 20 years of labor, Barbour appeared to have established a stable geological survey with a financial base. However, the same legislature passed two bills that must have shaken Barbour's confidence in the future of the survey as it was then organized. One was an act that empowered the regents "to appoint a member of the teaching staff to be state geologist," a departure from the 1893 act that automatically made the head of the department state geologist. The other was an appropriation of $6,000 for another ad hoc university unit with the unwieldy name of Nebraska Conservation and Soil Survey.

More changes came in the ensuing decade. Most of them centered on the Conservation and Soil Survey and its dynamic director, George Evert Condra, the second source of the modern Nebraska Geological Survey.

Born in 1869 on a farm near Seymour, Iowa, Condra came to Lincoln in 1892 to teach science and coach athletics in the city's high schools while continuing his education at the University of Nebraska. He completed

Conservation and Survey

Dr. George E. Condra

Dr. G. E. Condra (right) and professors Pool and Barker break for lunch during Sand Hills field work, August 1909.

G. E. Condra had a passion for documenting Nebraska's people, places and use of resources. This photograph of an early corn-husking machine is part of his collection.

his doctorate in 1902; his Ph.D. dissertation, done in consultation with E. O. Ulrich, then at Kentucky, and R. S. Bassler of the National Museum, constituted part one of the second volume of reports published by Barbour's geological survey in 1903.

The big, bull-voiced athlete joined the university faculty in 1902 and began espousing resource conservation, a popular national movement of the early 20th Century. On this issue, Condra became to Nebraska what Theodore Roosevelt was to the United States—a highly visible public figure leading the conservation cause.

Invited to a governors' conference on conservation called by Roosevelt in 1907, Condra returned from Washington and set to work. In 1908 the biennial legislature was in recess, but Governor Sheldon, on his own initiative, created a Nebraska Conservation Commission. The following year the legislature appropriated $1,000 for the commission and required that it be spent by Condra as director of the Conservation and Soil Survey under the control of the regents. In 1911, the indirect appropriation for the fledgling survey was 60 percent of the appropriation for the state geological survey. Finally in 1913, the legislature created the Conservation and Public Welfare Commission, consisting of the governor, the chancellor of the university, the director of the Conservation and Soil Survey and two additional state officials.

One of the principal duties of the commission was to "serve as an advisory board for the various state surveys." This meant that Condra was one of five advising himself and Barbour. During the same session, legislation was enacted to formalize the existence of the Conservation and Soil Survey within the university and to define its work. One of the duties was to survey the natural resources of the state with the following resources being outlined by statute: soil, water, water power, potash, forests and road materials. The appropriation for Condra's survey doubled to $12,000 while Barbour's survey received $10,000, the same amount that was voted in 1911.

A clear but probably covert rivalry had developed between the two men and their surveys. Barbour, long a prominent figure at the university, had other duties to attend to in the geology department and the state natural history museum. Condra could cultivate his single crop.

Barbour continued to be state geologist, but by now there was no longer a state geological survey. The legislature in 1919 ended the rivalry for all practical purposes by enlarging the duties of the Conservation and Soil Survey to include some geological activities while not calling these activities geological, by appropriating $25,000 to this survey, by ignoring the state geological survey and by discontinuing the Conservation and Public Welfare Commission.

The final act of this stage of evolution of the modern Nebraska Geological Survey was written in 1921. Legislation was passed that did away with the Conservation and Soil Survey but created a new, broad-based organization within the university: the Conservation and Survey Division. Among a wide range of other things, the law specified that the division would include a geological survey, that the director of the division would be appointed by the regents, and that the 1913 law that provided for the appointment of a state geologist be repealed.

Erwin H. Barbour and George E. Condra loom as the two greatest figures in the history of Nebraska geology and they cast long shadows. Both were of statewide renown by 1921, and both would linger at the university for years, pacing known territory but never venturing too far into new fields. Barbour continued as head of the geology department until 1934 and director of the state museum until 1941. From 1919 to 1929 he also served as chairman of the geography department. Acquiring the additional title of dean in 1929, Condra continued as director of his survey until 1954.

The early Condra years involved an era of ever-escalating field research, for the most part under primitive conditions—sleeping under the stars, eating sandwiches by the side of the road, and always with insufficient money and insufficient staff. Condra cultivated the art of "courting" the legislature for funds to carry on the mandated research of the survey.

Few scientists trained today have had the opportunities to absorb knowledge in so many natural-resource fields as did Condra. He was well informed in the fields of botany, zoology, chemistry, forestry, mineralogy, paleontology and hydrology. He was one of the first to recognize the tremendous opportunity for the use of groundwater and water from streams to supplement agricultural production. He helped organize the Nebraska Well Drillers Association in 1929 to encourage professionalism, the sharing of ideas and, most importantly, sanitary well construction.

Motivation for good quality rural and urban water supplies resulted from the death of his son Evert at 7 years of age. Evert was a victim of typhoid fever, induced from drinking contaminated water from a farm well while accompanying his father on a short field trip. In the years that followed, many of the home, stock, irrigation and municipal wells drilled in the state were constructed only after consulting the extensive hydrologic and geologic data compiled under Condra's direction.

Groundwater data collection and research were accelerated in 1903 through a cooper-

Conservation and Survey

Eugene C. Reed

ative program of water-resource investigation with the U.S. Geological Survey initiated by Condra. He was also instrumental in establishing one of the first statewide programs of test drilling and water-level measuring in the nation.

In 1954, Condra reluctantly stepped down as director. His pioneering days of innovative research were over, and it was time for a man of new vision and new goals to assume command. This lot fell to Eugene C. Reed, the first native Nebraskan to be recognized, even if *ex officio*, as state geologist. Resembling Barbour more than Condra, the angular, quiet-spoken Reed was born in Holdrege in 1901. He acquired his bachelor's degree at the university in 1923, then spent the next 8 years as a petroleum geologist in Mexico and Venezuela. Returning to Nebraska, Reed completed his graduate studies in 1933 and went to work for the Conservation and Survey Di-

G. E. Condra gets ready to shoot motion-picture footage of Nebraska's land and water from the air. Since about 1972, CSD has continued that tradition by storing and processing aerial and satellite imagery.

Nebraska State Historical Society

vision as a geologist. Becoming associate director in 1944, he toiled at the survey during Condra's declining years before being named director in 1954, without Condra's title of dean.

Reed, along with Condra, stressed the importance of cooperative work with the U.S. Geological Survey in order to get something done in Nebraska. This cooperation was particularly critical for groundwater surveys and the monitoring of groundwater resources. After World War II, Reed, with the help of a young Vincent H. Dreeszen—who eventually would succeed him as director—devised a systematic, long-term drilling program for investigating groundwater and the significant post-Cretaceous deposits (generally water-bearing and above bedrock) of the state in cooperation with the federal survey.

The 1921 act that established the Conservation and Survey Division created, on paper, a grandiose organization for surveying natural resources. The depression, war and flagging energy of an aging Condra in fact prevented realization of this kind of organization. When Reed became director in 1954, the division had a very small professional staff working primarily with groundwater resources and stratigraphy and a loosely associated group of individuals from other parts of the university and federal agencies. Reed increased the geological staff, especially those mapping the deeper subsurface features, and during his tenure from 1954 to 1967 the division probably acquired an identity more nearly akin to a geological survey than to a broadly arrayed natural-resources survey. Applied geology, using subsurface information and techniques, set the general tone of the division.

Conservation and Survey

Vincent H. Dreeszen

Vince Dreeszen became acting director upon the retirement of Reed in 1967 and was appointed director by the regents two years later. Born in Palmyra in 1921, he received his bachelor's degree at Peru State Teachers College in 1942 before entering the navy. Returning to Nebraska after the war, he worked part-time for the division and completed his graduate studies at the university in 1949, the same year he joined the division staff as a geologist-hydrologist.

Named assistant director in 1959, Dreeszen continued shepherding the test-drilling program and studying his principal interests, the Cenozoic rocks of Nebraska and their critical importance to the water supplies of the state. On becoming director, he gradually expanded the scope of the division to more nearly coincide with the provisions of the 1921 act, especially with regard to the survey of water and land resources. The geological survey, while still emphasizing applied geology and relying strongly on subsurface techniques, was enlarged and broadened.

Dreeszen served the division under Condra and Reed and knew Barbour and his work. Over a 95-year span each of these leaders established for the survey a statewide and national reputation for scientific credibility and achievement. The record of public service to the state established during their ten-

Restoration of the rhinoceros-like Diceratherium, *the most numerous of the fossils found at the Agate Fossil Beds in Sioux County, the result of the early digging of the Nebraska State Museum/Geological Survey and part of its early collection; 1906.*

Nebraska State Historical Society

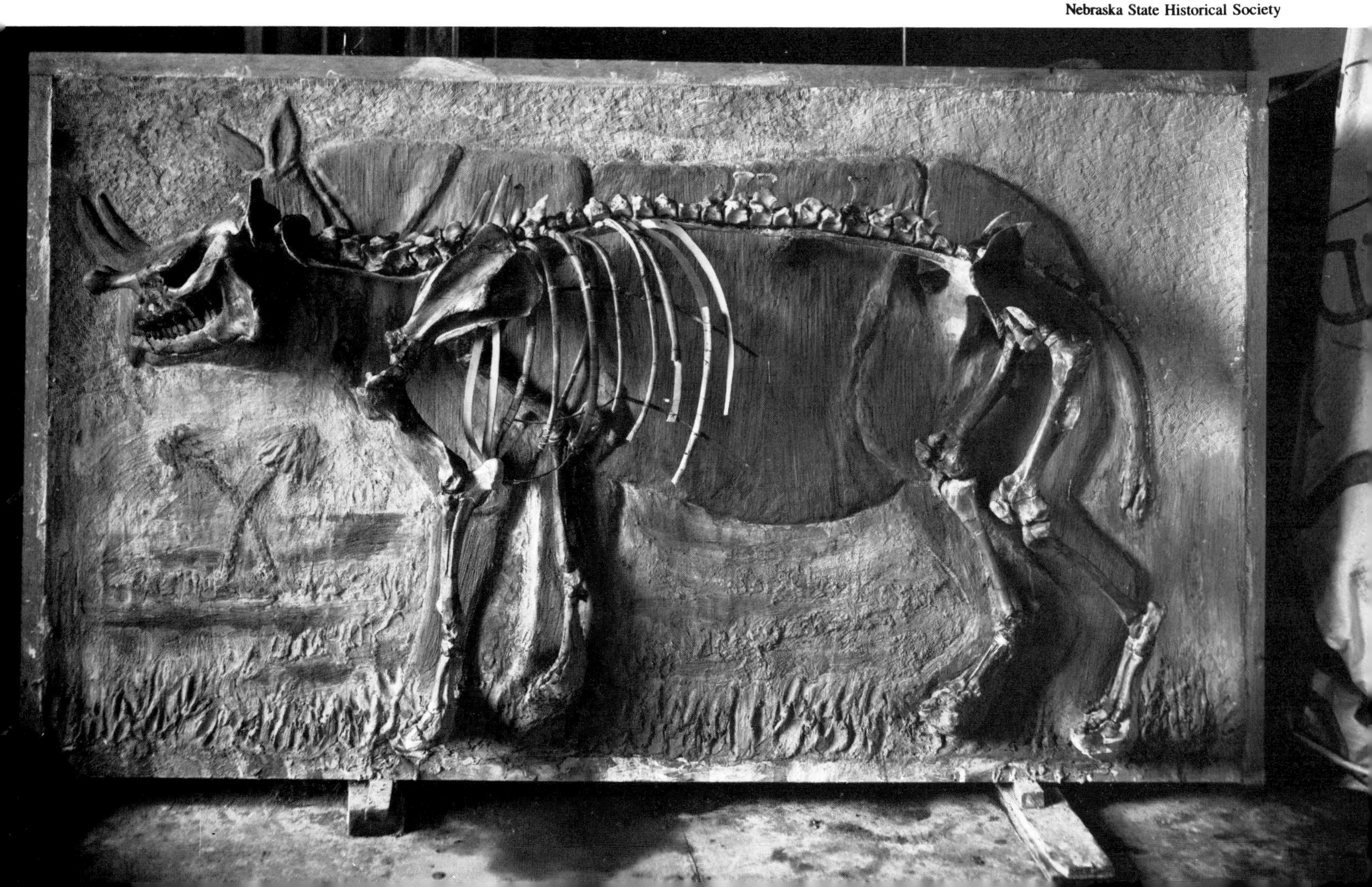

Conservation and Survey

Dr. Perry B. Wigley

ure is one of strong influence on public decisions in matters related to natural-resource use and management.

Dreeszen stepped down as director March 1, 1987. Replacing him is Dr. Perry B. Wigley, experienced in basic and applied geology in the public and private sectors. Wigley, who has been a professor at Eastern Kentucky University in Richmond, Kentucky, assistant state geologist with the Georgia Geological Survey and a petroleum geology consultant based in Houston, Texas, has ushered in another era for the division that will likely see it into the 21st Century.

[During Wigley's tenure, the Conservation and Survey Division has continued to expand into new areas while maintaining service to Nebraskans through basic geological, hydrogeological and soils survey. Examples of this expansion include remote sensing and geographic information systems, a strengthened hydrogeologic research program and a dramatic increase in outside funding.

Geographic information systems (GIS) represent a major change in how people use maps and other spatial data. The technology affects agriculture, natural resources management, environmental protection, wildlife management, land management and many other areas. The division has developed statewide GIS databases on land use, terrain elevation, transportation, water, soils, landslides, ecoregions, wetlands and other subjects. Through the Center for Advanced Land Management Information Technologies (CALMIT), the division has become recognized nationally as a major center of excellence in GIS.

In hydrogeologic research, the division is attempting to elevate the evaluation of the state's water resources from a generally qualitative and descriptive approach to a more quantitative assessment of the geologic, hydrologic and chemical properties of groundwater and surface-water systems and their interrelationships. The many water-related problems currently facing Nebraska will require a quantitative understanding of the physical, chemical and biological interactions that occur as water moves through the various parts of the earth's hydrologic cycle.

Finally, these new initiatives have been aided by a dramatic increase in funding from outside sources. A year after Wigley arrived in fiscal year 1988, outside funding for research stood at about $400,000. Through a team effort of researchers and administrators, in just 5 years that total has tripled to $1.2 million. These funds enable division scientists to further explore basic and applied research questions and help solve the various natural resources problems facing the citizens of Nebraska.—Ed.]

***Adapted and reprinted from* Resource Notes, Vol. II, No. I.**

Profiles:

Nelson Horatio Darton: His Surveys Laid the Foundation

No serious researcher studying the development of water resources in Nebraska, or for that matter in the central Great Plains, can overlook the works of Nelson Horatio Darton, the first geologist to study and publish about those resources in any detail.

Darton originally published the names and characteristics of many hydrogeologic units beneath much of Nebraska, now used by groups as diverse as scientists, well drillers and the general public. These works are listed below. As an indication of the value of these early efforts, I have used these works for my research and can testify to their accuracy and reliability.

N. H. Darton was born in 1865 in Brooklyn, New York, according to accounts of his life by P. B. King (1948) in his *Memorial to Nelson Horatio Darton* for the Geological Society of America and that of W.H. Monroe (1948) for the American Association of Petroleum Geologists, as well as those of other workers. His father was a seaman and later owned a shipyard. Darton graduated from grade school when he was 13 and went to work for his uncle as an apprentice at the uncle's pharmaceutical factory. For the next 7 years he worked in the business and took over more and more of the duties. During this time he also published the results of chemistry studies, learned about minerals and geology by studying his uncle's mineral collection, and, perhaps most interesting to those wondering how he ended up studying water, analyzed water from wells in the New York area for evidence of contamination. He also found time to start a card catalog of articles on the geology of New York.

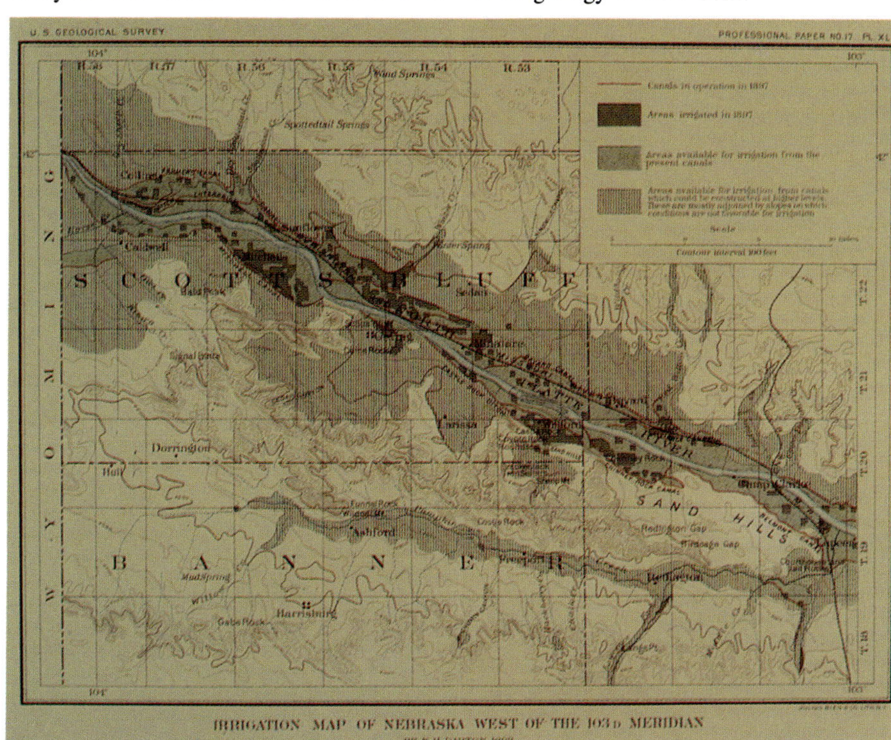

This latter work caught the eye of G. K. Gilbert, a geologist with the U.S. Geological Survey, who wanted a catalog prepared on the geology of the Appalachian Mountains. Darton was hired for the job and went on to do research on the geology of the Appalachians and the Atlantic Coastal Plain. While at work on the Coastal Plain study, he collected data on artesian wells in that area from Long Island, New York, to Georgia and used some of that information to help Long Island find a water supply. After a three-year assignment with the New York Geological Survey, Darton returned to the U.S. Geological Survey and, in 1895, began his work on the underground waters of the western United States.

Darton, on foot or horseback or in a wagon, sometimes alone, sometimes with one or two other researchers, prepared preliminary topographic and geologic maps of major parts of the West in about 11 years. He recorded and mapped irrigation development and studied the geologic framework of southeastern and south-central Nebraska from Lincoln to Cozad. He published similar but more expanded works for the Nebraska Panhandle and coined the terms *Ogallala, Arikaree, Gering, Brule* and *Chadron* formations, and the *White River Group*. He drew maps of the irrigation developments along the North Platte and Niobrara rivers in the late 1800s *(see the map of irrigation west of the 103rd Meridian, p. 67)*. He also prepared a general geologic map of Nebraska.

During that 11 years, he and a few coworkers also published geologic maps of South Dakota, New Mexico, Arizona, and parts of Texas, Colorado, Kansas, Wyoming, Montana, Minnesota and Iowa. And all of the research for these maps and the accompanying reports was done using the transportation mentioned above while wearing a coat and tie! In Nebraska and these other states, all subsequent knowledge has been built on the framework developed by Darton, a true giant in his field.

Works by N.H. Darton on Nebraska Resources and Geology

1. Underground waters of a portion of southeastern Nebraska: U.S. Geological Survey Water-Supply Paper 12, 56 p., 1898.
2. On the Tertiary of South Dakota and Nebraska: *Science*, new series, v.7, p. 359, 1898.
3. Preliminary report on the geology and water resources of Nebraska west of the one-hundred and third meridian: U.S. Geological Survey, Nineteenth Annual Report, part 4, p. 719-785, 1899.
4. Relations of Tertiary formations in western Nebraska region: *American Geologist*, v.23, p. 94, 1899. Also in *Science*, new series, v.9, p. 103.
5. Description of the Delrichs quadrangle, South Dakota and Nebraska: U.S. Geological Atlas 85, 6 p., maps, 1902.
6. Preliminary list of deep borings in the United States, Part II, Nebraska and Wyoming: U.S. Geological Survey Water-Supply Paper 61, 67 p., 1902.
7. Description of the Camp Clarke quadrangle, Nebraska: U.S. Geological Survey Geologic Atlas 87, 4 p., maps, 1903.
8. Description of the Scotts Bluff quadrangle, Nebraska: U.S. Geological Survey Geologic Atlas 88, 5 p., maps, 1903.
9. Preliminary report on the geology and water resources of Nebraska west of the one-hundred and third meridian: U.S. Geological Survey Professional Paper 17, 69 p., 1903.
10. Description of the Edgemont quadrangle, South Dakota and Nebraska (with W.S.T. Smith): U.S. Geological Survey Geological Atlas 108, 10 p., maps, 1904.
11. Preliminary report on the geology and underground water resources of the Central Great Plains: U.S. Geological Survey Professional Paper 32, 433 p., 1905.
12. The Great Plains of the central United States: *Scottish Geographical Magazine*, v. 32, p. 9-18, 1906.
13. Cement materials in the Republican Valley, Nebraska: U.S. Geological Survey Bulletin 430, p. 381-387, 1910.
14. The structure of parts of the central Great Plains: U.S. Geological Survey Bulletin 691, p. 1-26; 1918.

—by R.F. Diffendal, Jr., CSD research geologist

Sources: "Memorial to Nelson Horatio Darton," P. B. King, 1949, in Geological Society of America Proceedings for 1948, p. 144-169; "Memorial to Nelson Horatio Darton (1865-1948)," W. H. Monroe, 1949, in American Association of Petroleum Geolgists Bulletin, v. 33.

R. H. Willis: First Irrigation Chief Faced Threats, Shotguns

R. H. Willis and unidentified man at the headgate of the Dundy County Ditch, undated

Department of Water Resources

In the early days of irrigation in Nebraska, a farm family's very survival rested on a dependable flow of water. But there wasn't always enough water to supply all who wanted it, and sometimes junior water rights had to be shut off. When that happened, the man controlling the supply faced scorn and, all too often, shotguns.

Robert Henry Willis controlled that supply in one capacity or another for more than 58 years. Willis, who went on to head the Nebraska Bureau of Irrigation, Water Power and Drainage from 1919 to 1951, began his career in 1895 as a water commissioner for Banner, Cheyenne and Scotts Bluff counties. His job was to enforce the new "first-in-time, first-in-right" appropriation law, which made him quite unpopular at times.

During the dry summer of 1901, for instance, Willis was ordered by the secretary of the State Board of Irrigation to divide water in the Lawrence Fork Creek, located in Banner and Morrill counties, according to the priority system. Some of the newer canals had to be shut down to supply those with prior rights.

Willis closed the junior canals. The next day, it was reported to Willis that the gates had been reopened. He immediately went back to close them and found at one an effigy of himself hanging by the neck from a tree. Written on a card next to the dummy was, "Dead men will hang from these trees before the ditches are closed."

Willis closed the gate and filed a complaint with the county attorney. The gate owner was fined.

On the same creek several days later, Willis was told that a water user was supplying his ditch against orders. Willis went to investigate and found that a lumber dam had been constructed in front of the diversion. The dam was high enough that water spilled into the diversion ditch.

Willis began to dismantle the dam, then saw two men approach. They informed him they would throw him into the creek if he removed the dam. He did, and the men made good on their promise.

Yet another incident occurred in 1896 when Willis was surveying for the construction of an irrigation ditch. One landowner who didn't want the ditch to cross his land confronted Willis with a shotgun and told him to leave his property. Willis told him to go ahead and try to stop him. The man finally backed down.

Also, as head of the Bureau of Irrigation, Water Power and Drainage, Willis ordered the closing of the Mitchell Irrigation Ditch in Scotts Bluff County in 1935. Water users disobeyed the order and, eventually, the governor had to send in the National Guard and declare martial law *(see vignette on 1935 Water War, p 208).*

R. H. Willis's life began uniquely. On March 22, 1869, he became the first male white child born in Cheyenne, Wyoming. Shortly before his birth, Willis' father, John G. Willis, had opened a general merchandise store. The log-built store burned twice in two years, so John moved his family to Omaha in the fall of 1870. John was familiar with the city, as he had many years before operated an ox-mule freighting company that ran from Cheyenne to Omaha to Denver. In Omaha, John opened another general store. It prospered and remained in business until 1884. Willis Avenue in Omaha is named after him.

Robert Willis attended Omaha High School but refused to take the required Latin and Greek courses. As a result, he had to complete his last year of high school at Creighton University.

His father had made enough money in merchandising to send him to college on the east coast. In 1886, Willis entered Rensselaer Polytechnic Institute in Troy, New York. He spent his summer working at the U.S. Weather Bureau office in Omaha in exchange for help with geometry from the bureau's director.

His education at Rensselaer ended with the drought and financial chaos of 1890. Willis returned home to work in the city engineer's office. He was soon hired by the Union-Pacific Railroad as a surveyor and sent to Idaho.

When the surveying job ended, Willis returned to Omaha and took a job in the county engineer's office. A Republican, he was forced out when a Democratic administration took over in 1894.

Department of Water Resources

R. H. Willis

In September of that year, Willis was hired as an engineer to help build the Belmont Irrigation Canal in the North Platte Valley. Willis' brother-in-law, an Omaha attorney, was financing the project.

He arrived by train in Sidney later that month and set out for Bridgeport with a group of Swedish immigrants. For his first wife and himself he built a sod house near the Camp Clarke Bridge, which is near modern-day Bridgeport. Bridgeport and Scottsbluff didn't exist until 1900.

While living there, Willis came to know William R. Akers and L.B. Carry, state senators who helped usher through the legislature the 1895 water-appropriation law. The measure established the State Board of Irrigation, with the state engineer as its director. At the time, the state engineer was Robert B. Howell. As the first secretary of the board, Howell appointed Willis water commissioner for Cheyenne, Banner and Scotts Bluff counties in 1895.

In 1901, Willis moved his family to Bridgeport. In addition to his duties as water commissioner, Willis served as Cheyenne County surveyor. When Morrill County was established, he served two years as that county's surveyor. At the same time, Willis served on the local school board and ran a private engineering firm, a 160-acre irrigated farm, a grocery store and, later, a cement-block factory.

Willis bought the *Bridgeport Blade* newspaper from Senator Carry in 1903. Later, he paid for construction of a two-story building to house the paper on the first floor and his private engineering office on the second. He sold the *Blade* in 1907 and became part owner of the *Bridgeport News-Blade*.

Willis was appointed superintendent of Division No. 1 in 1910. That division included all the irrigable lands of the state drained by the Platte rivers and their tributaries lying west of the mouth of the Loup River. It also included lands lying south of the Platte and South Platte rivers that were or could be irrigated with groundwater. Willis was, therefore, responsible for enforcing water-appropriation laws for more than half of the state, including the North Platte, South Platte and Platte rivers and their tributaries, as well as those of the Blue and Republican rivers.

State engineer George E. Johnson appointed Willis assistant engineer in charge of all irrigation in Nebraska in 1918. He accepted the job only after being assured he could remain in Bridgeport and not move to Lincoln. Working out of Bridgeport his entire career, he traveled by train to Lincoln about once a month.

The State Board of Irrigation, Highways and Drainage was moved in 1919 to the Department of Public Works, and became the Bureau of Irrigation, Water Power and Drainage. Willis became chief of the bureau.

Willis had been involved in the Nebraska State Irrigation Association for several years, and was made secretary of the group in 1919. His efforts led to the publication of association proceedings, which remain one of the most important sources of information on the history of irrigation in Nebraska. He also devoted a great deal of time to researching the history of irrigation in the state, writing many papers and presenting his findings to several groups. He served as association secretary until 1949, and as secretary emeritus until 1954. *(see vignette on the Nebraska State Irrigation Association, p. 157).*

The Bureau of Irrigation, Water Power and Drainage became part of the new Department of Roads and Irrigation in 1933. Willis remained chief of the bureau. It was the bureau's duty to control the appropriation of water in Nebraska's streams and rivers. To do so, accurate records of streamflow were recorded and published each day.

For several years in the 1930s, he studied and recorded return flows below the Guernsey Reservoir. They revealed that further irrigation development in Nebraska was possible. The planners of the Sutherland and Tri-County projects used Willis' study to demonstrate the feasibility of and get construction funds for those projects.

Willis was awarded the first Nebraska Builders Award, the highest non-acedemic honor conferred by the University of Nebraska, in 1946.

Willis retired at 81 from the Department of Roads and Irrigation in 1951 as the oldest state employee. He remained with the board as a consultant until his death on Jan. 4, 1954.

—Brad Rundquist, weekend and copy editor,* Ft. Dodge (Iowa) Messenger, *and formerly CSD editorial assistant

Wesley Sell: Canal Manager is a Ditch Rider's Ditch Rider

Wesley Sell, current manager of the Farmer's Irrigation District, has been with the district since 1949. He manages a 75-mile-long canal that delivers water to more than 63,000 acres. The canal diverts water from the North Platte River just east of the Nebraska-Wyoming state line. Water also runs into the canal from several tributaries to the North Platte River and from return flows from the Pathfinder Irrigation District to the north.

Ann Bleed, DWR

Wesley Sell and his willow-weaving machine

Sell, who's been applauded as one of the best canal managers around, has a multitude of stories to tell about the trials and tribulations of ensuring a water supply for his irrigators. Maintaining diversion dams and headgates, controlling bank seepage, erosion and weeds, maintaining roads and keeping good records are just a few of the tasks that must be done. Sell's biggest headaches, however, are related to the unpredictable nature of the water supply itself. Sometimes there is not enough water to go around and irrigators have to go on a rotating schedule of six days on and three days off, or forgo early irrigation of alfalfa. Other times there is too much water and the canal must be closely watched and managed to avoid canal breaks and flooding.

During the peak of the irrigation season, ditch riders keep a constant watch on the system. When water is short, headgates, which control the delivery of water to the fields, have to be checked daily and sometimes closed. During the 1960 drought, one of the worst droughts Sell has experienced, a few headgates had to be locked. Sell still remembers one irrigator, who had been caught cheating, begging literally on bended knee and with tears in his eyes for Sell to deliver some water to his field. Sell would not budge. That year he established his reputation for being firm but fair and there were record yields for sugar beets. Most irrigators, Sell claims, are understanding and cooperative when water is short. In fact, Sell gets fewer complaints during a dry year than when there is a full supply of water.

Inadequate water supplies produce problems, but the biggest headaches for Sell come when there is too much water and the canal spills water over the top or breaks. At the height of the irrigation season, when the canal is filled to capacity, a 4-inch rain storm can produce enough runoff to overtop or break through the canal. Whenever a ditch rider observes the water rising quickly, there is a race between rider and water to get to a spill gate to avoid an uncontrolled spill.

In August of 1950 a 7.5-inch rain caused the canal to break, flooding the town of Bayard. Lawsuits followed. Increased heights can contain the canal and the use of car radios has increased the odds of the rider winning the race to the spillways and made the control of flooding easier. Still, sudden rain storms can produce flooding. In 1987, 30 people worked all night to prevent a flood, but still the canal broke.

When the irrigation season ends in September, the work continues. The rest of the irrigation year is spent maintaining equipment, roads and the canal itself. Clay is placed in the canal to protect banks and prevent seepage. In the spring weeds must be removed. Sell describes cutting the weeds with a scythe or even pulling them by hand in the early days of his career. He talks of the miles of ditch that had to be walked to reach headgates because of the lack of roads. If you're lucky, he will show off the district's willow-weaving machine.

During the winter the district used to employ four men to go to the river bank to cut willows and then, using the machine, wire them into 30 foot long mats. These would be used to prevent erosion along the canal banks. The task was repeated every winter because some of these mats would be destroyed every spring when weeds were burned.

Although modern machinery, good roads and car radios have eliminated much of the work and some of the problems of managing a canal, Sell still worries about the future. Flooding and droughts will always be problems, but in addition there are now new environmental regulations that add costs and complexity to canal management. To Sell the biggest challenge will be providing water for fish and wildlife. He thinks farmers may have to irrigate every year as they do in years when the water supply is short and has to be carefully allocated. Sell is quick to point out, however, that the district had excellent crops in the water-short years of 1990 and 1991.

—Ann Bleed, state hydrologist, Nebraska Department of Water Resources

Industry

by Cindy Evert

In a state where rainfall is scant, the road to economic self-determination and achievement has not been easy. In fact, it is difficult to imagine how a state that so thoroughly failed to impress early travelers has come so far in the past century.

Although much of the Great Plains was termed the "Great American Desert" early on, water was instrumental in Nebraska's early settlement and the growth of agriculture and economic development, and it remains one of the most valuable resources in the state today.

Much of the basic sustenance for the Native American tribes was met by the wild game, wood and water obtained in or nearby the river valleys in which they established their villages. Upon the Europeans' discovery of the region, the state's streams and their valleys became commercial "highways" and lanes of exploration.

The river valleys supplied the early trappers with animals for pelts and, so, were an immediate source of income-generating resources. They were soon followed by settlements along the Missouri.

By the mid- to late 1800s, running beside the river, the Platte River Road provided a natural route to the West for the Oregon and Mormon trails. The Pony Express run, stagecoach routes, telegraph lines, railroads and later—mostly after 1900—highways also paralleled the Platte and other rivers. By the time the steamboats started chugging up the Missouri, river towns began to thrive.

The abundance of rivers, lakes and underground water in Nebraska has provided the basis for survival, settlement and development. Developing an agricultural economy in a land some thought not fit for farming was not simple. However, by about 1900, the stage was set for extensive water development as the expanding population realized its need for irrigation and water power. Agriculture's growth to the state's number one industry and its continued domination have reflected the tenacity, hard work and ingenuity of the settlers.

An ad for well supplies and pumps from the late 1890s or early 1900s

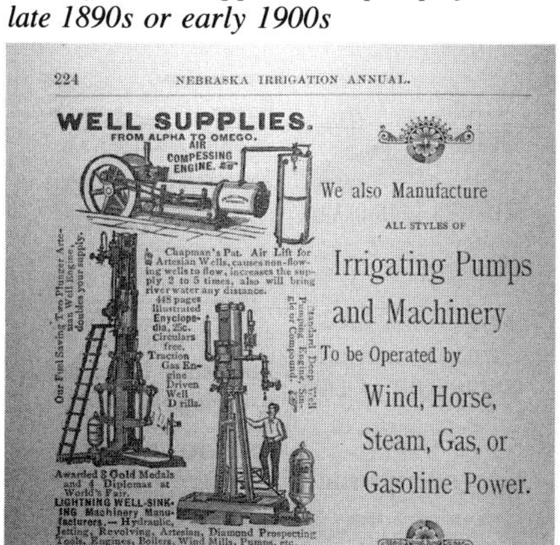

Nebraska State Irrigation Association

An ad for Dempster water-well products from the late 1890s or early 1900s

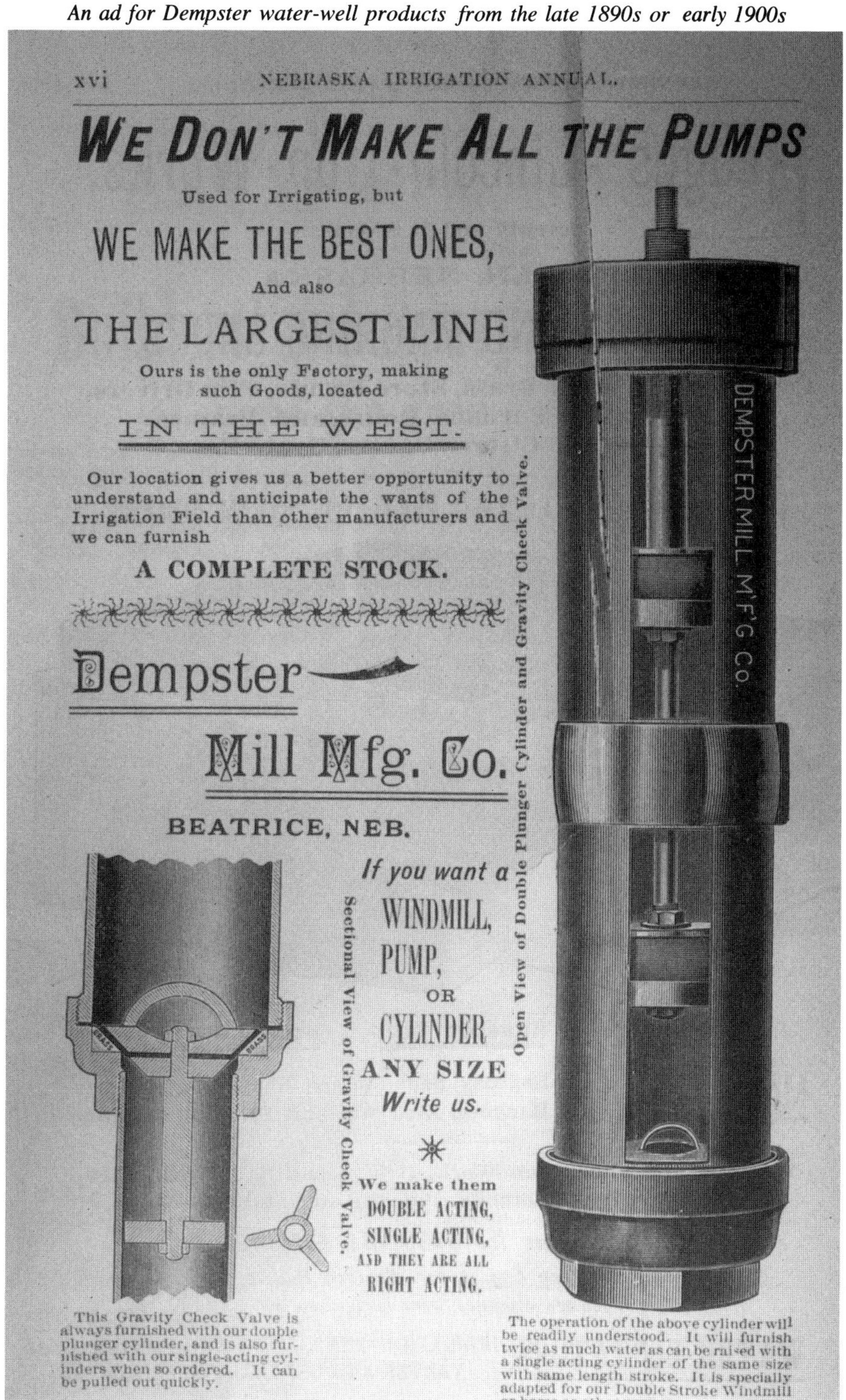

Nebraska State Irrigation Association

Harnessing Water Power

Besides farmers, Nebraska drew many investors bringing "eastern money," looking to make more money by intro-

Nebraska State Historical Society

A saw mill "believed to be operated by the Millers on Long Pine Creek from 1883 to 1892," according to State Historical Society information

ducing new technology to the frontier. They often saw their fortunes reflected in water development.

As settlers discovered that Nebraska's fertile soils were suitable to growing cereal grains, the need for mills to process the local crops (mostly wheat) grew as well. Water provided power for the early mills. By 1867, the time of statehood, flour milling was the leading industry in Nebraska. In 1889, an estimated 279 mills operated in the state. As communities often sprang up around the mills, they became central to the settlement and development of the state.

With seemingly unlimited water power available in Nebraska, promoters stressed the potential not only to grind grain but to power other manufacturing operations. One of the early industrial parks was organized around the "best water power in the state," then producing up to 500 horsepower. In this park, the West Point Manufacturing Company built an impressive complex including a sawmill, planing mill, paper mill, furniture factory, condensed milk operation, and brick works around the flour mill.

A Revolution: Irrigation

The economic impact of irrigation on agriculture and the state's economy has been tremendous. Not only has irrigation increased the number of acres of available farmland in the state, but it has elevated Nebraska's agriculture and agriculturally related business to prominence in the world market.

In the late 1800s, interest in irrigation came and went from one year to the next. During the years of adequate rainfall, farmers weren't interested in it and during times of drought, they had little money to invest. The earliest projects along the North Platte River in Wyoming and Nebraska were the work of individual farmers. Some ventures were soon forgotten after a year of adequate rainfall. Despite the collapsed dreams and failure of early projects, the vision of water development in Nebraska wouldn't die.

In 1891, William Smythe, a writer for the *Omaha Bee* newspaper, determined irrigation was the means to guarantee industrial independence and thereby save the West. "Nebraska's best days are before her," said Smythe, who helped organize what became the Nebraska State Irrigation Association in 1893.

The 1902 Reclamation Act authorized several water projects; construction began in 1903 on one of the first, the Pathfinder Dam built on the North Platte River southwest of Casper, Wyoming. By 1980 Nebraska farmers irrigated more than 100,000 acres of the Panhandle using North Platte River water, and the Pathfinder Reservoir continues to be a key storage facility today.

In 1914, the Tri-County Association was formed to meet the dire need for irrigation by channeling Platte River water through a series of canals and reservoirs. However, it would not come to fruition until 1941.

Floods in 1927 spurred interest in Congress for flood control, while drought re-emphasized the need for irrigation, and depression created a need for jobs. The idea of irrigation became more accepted when coupled with the idea of power generation. Hydroelectric plants had a much wider appeal than irrigation alone, and the power revenues were needed to finance and operate projects.

Early electricity was either municipally owned or controlled by mill owners. However, in 1933, the Central Nebraska Public Power and Irrigation District, a public agency organized as a corporation, was permitted by the legislature to receive federal funds to form public power and irrigation districts. The Platte Valley Public Power and Irrigation District received loans and grants from the Public Works Administration (PWA) for the construction of reservoirs and miles of canals in the North Platte area, creating about 2,000 jobs. The Central Nebraska and Loup River Public Power District received similar PWA loans.

Soon after the formation of the Central Nebraska Public Power and Irrigation District, members received funds to build Kingsley Dam on the North Platte as part of the Tri-County Project. This massive structure, constructed to hold up to 2 million acre-feet of water in Lake McConaughy, has 35,000 acres of surface water and stretches 22 miles up the valley along 105 miles of shore line. Until the construction of the main-stem dams on the Missouri in the 1940s, Kingsley Dam, built in the 1930s, was the second-largest earthen dam in the world.

Water from Lake McConaughy cools the coal-fired Gerald Gentleman plant in

Neligh Mills employees hauling power poles, Neligh, Antelope County. Neligh Mills first produced power in 1900.

Nebraska State Historical Society

Nebraska State Historical Society

C. B. Purdy of Minatare, Scotts Bluff County, stands beside his irrigation works.

Sutherland, operates North Platte's hydroelectric plant, and serves six irrigation districts and companies. Below Kingsley, water is diverted from and returns to the river, being used and reused down the valley for 150 miles. The diverted Platte waters 175,000 acres with gravity irrigation and meets a significant share of the state's total electrical needs with various power plants along the river.

Enabling this use of irrigation water was the develpment of new technology from 1930 to 1950. Western Land Roller was one of the first companies to expedite the irrigation process. In the 1930s, this Hastings company brought state-of-the-art drilling equipment to the area and leased and later sold the equipment to customers. Around that time, the company also developed an irrigation well turbine similar to types it produces today. By the early 1990s it employed about 245 people.

In 1947, Frank Zybach of Colorado began working on an automated sprinkler system that would forever change the agricultural landscape of the state and later be exported to various parts of the world. Zybach and A.E. Trowbridge of Columbus entered a partnership in 1953 and began building center-pivot irrigation systems.

Only 19 systems were produced during the first two years because of the difficulty of selling a new, untried method of irrigation. In 1954, Zybach and Trowbridge sold Valmont Industries of Valley the manufacturing and sales rights, receiving a 5 percent royalty.

Shortly after the patent expired in 1969, the center-pivot industry took off; by 1988, 27,617 pivot systems were operating in Nebraska, more than in any other state. In 1990 of the more than 8 million acres irrigated in Nebraska, nearly 4.2 million were irrigated by sprinkler, 3 million by center pivot. Valmont, now the oldest and largest producer of center-pivot irrigation systems, employed about 4,300 people in the early 1990s. As an innovator in the center-pivot area, the company was one of the first manufacturers to export its systems to other countries, including China, and the first to sell manufacturing rights to the former Soviet Union.

Valmont is one of seven firms in the U.S. making and selling center-pivots. Five of those seven companies are in Nebraska and together, with Valmont, as of 1992, they served 90 percent of the market in the United States and 90 percent of the export market. Lindsay Manufacturing Company in Lindsay employs 438 people; Lockwood Corporation of Gering, 400; Reinke Manufacturing Company Inc. of Deshler, 176; and T-L Irrigation Company of Hastings, 122.

More than 40 companies in Nebraska are directly related to the manufacturing of irrigation equipment. In addition to pumps and gated pipe, items made range from rain shut-offs, flood gates and gaskets to heat exchangers, gear boxes and electric wire. Companies making irrigation-related equipment in the state employ from four to more than 1,100 people. In the town of Henderson with a population near 1,000, 40 jobs in three companies are directly related to the irrigation industry, a significant number of jobs in a rural area.

Indirect Economic Effects of Irrigation by Industry for 1985 ($1,000s)

Industry	Impact
Manufacturing	$ 224,724.25
Transportation, Communications, and Utilities	96,179.36
Trade	154,358.70
Finance, Insurance and Real Estate	144,662.14
Services	160,910.43
Other	53,069.00
Personal Income	476,441.62
Total	$1,310,345.50

Indirectly, irrigation stimulates markets in agricultural equipment and machinery, animal feed and health products, chemicals and pesticides, prefabricated metal and wood buildings, silos, lumber and hardware, diesel fuel, machine shops, transportation, meat packing, and other food processing companies. It is estimated one-half of Nebraska's work force depends on agriculture and the

Cutting dryland alfalfa by horse on the P. L. Tomlinson Farm in Burt County. Agriculture was fundamental to Nebraska's economy early on and still is today. This meant farmers depended on adequate supplies of water, either from precipitation, or from "artificial moisture," irrigation.

Nebraska State Historical Society

Les Sheffield

Lake McConaughy behind Kingsley Dam

many related industries for employment. In 1985, the indirect impact (which accounts for the "multiplier effect" of money spent and respent locally) of irrigated crops amounted to more than $1.3 billion, in addition to a direct impact of $980 million. Approximately $476 million was added to Nebraska's personal income in 1985 as a result of irrigation. Irrigation has provided more economic opportunities and a higher standard of living for both farm and non-farm people in Nebraska. Increased agricultural prosperity and trade from irrigated land not only contributes to the state economy and world markets but helps rural communities build a tax base with which to maintain schools, roads and medical facilities, as well as providing additional income that can be contributed to churches and charities.

The Public Power State

Mill owners were responsible for bringing the first electric power to many Nebraska towns by adding turbines and pulleys at the mill site to generate and transmit power. At first, electric current generated from water power or steam provided only street lighting, but as the incandescent lamp was perfected and costs were reduced businesses and later homes began using electricity as well. Early tries at electrical power generation were hit and miss, sometimes with years of lapses between service. But as people saw how lights and electricity added convenience to their lives, they wanted more.

Under the 1933 law, the formation of the Loup River Power District, the Central Nebraska District, and Platte Valley Power District set the stage for the creation of the Nebraska Public Power District, the Omaha Public Power District and 31 rural power districts.

In 1939, Consumers Public Power District was formed to buy the properties of various private power companies operating in the state to provide a market for the publicly owned hydro-electric and other power produced by the districts. The Nebraska Public Power System, created in 1940 by joint agreement of the districts, acted as a marketing agent for wholesale power and operated high voltage transmission facilities.

The merger of portions of the Platte and Loup districts, Nebraska Public Power System and Consumers in 1970 resulted in the formation of the Nebraska Public Power District. As the major power supplier of 88 percent of the state, NPPD, along with the Omaha Public Power District, generates and transmits much of the power used by other districts. In 1991, NPPD paid more than $13.3 million for lease agreements to 201 cities and villages it serves at retail, and provided $5 million to government subdivisions in 70 counties in lieu of tax and gross revenue payments. In addition, the district has more than 2,200 employees statewide with an annual payroll of almost $90 million.

Although hydroelectric power generated in Nebraska was once significant, it is no longer, contributing less than 5 percent of the total production. The result of the years of attempting to bring dreams of irrigation and electrical power generation to fruition is unique to Nebraska—the only state where the power industry is wholly owned by the public.

Industries Related to Water

Groundwater Pumped by Windmills

Because groundwater is more prevalent than surface water, early settlers needed to devise ways to bring water up from the ground. In a region where the wind seems to blow constantly, it became an obvious power source for pumping groundwater.

Windmills of all sizes and shapes were built on the Great Plains to provide water for limited irrigation in the late 1800s, but it was not until the development of centrifugal pumps in the early 1900s that groundwater became a significant source of irrigation water.

With the later development of vertical centrifugal pumps and turbine-style pumps, steam or electrical power enabled water to be drawn from greater and greater depths.

With the big demand for windmills in this region, three companies came to the forefront of windmill and pump production. Dempster Manufacturing Company of Beatrice became one of the largest makers of windmills in the United States. For a while, the Fairbury Windmill Company and Geneva Iron and Windmill Factory also were producers. The 1890s were the peak years for windmills as prime movers of irrigation pumps. By 1898, 2,000 systems were in use in Nebraska. As of 1992, in addition to still making about 200 windmills a year, Dempster sold parts; also at this time, Muller Industries of Randolph offered parts and rebuilding, and Titan Industries of Paxton also provided parts and remanufacturing.

Food Processing Begins with Mills

In 1855, the editor of the *Nebraska Palladium* said, "We trust the establishment of a flouring mill will be the means of encouraging the farmers to raise a large amount of corn and wheat so as to prevent . . . the importation of breadstuffs. This course will keep a large amount of money in this country which has been sent to St. Louis."

Processing local crops to flour, meal and livestock feed initially saved settlers the expense of importing expensive flour and other food items. Later, as crop pro-

(Continued on p. 76)

Once Life Givers of the Plains, Windmills Still Water a Thirsty Land

Nebraska State Historical Society, E.H. Barbour photo

Matthew Wilson's battle ax mill near Overton, Dawson County, 1898. It pumped water for stock and cost $1.50

Dotting the rural countryside, sometimes entwined with morning glories, they remind passers-by that in the late 19th Century they were an answer to a farmer's prayer for water, water for his family, his livestock and his crops.

Windmills were introduced in the United States about 500 years after they appeared in Holland. There they drained the land by pumping water into the sea from behind the dikes.

Their legacy spread through Europe. Three hundred years ago, Don Quixote's epic, "The Man of LaMancha," described the mythical pursuit of windmills, which Quixote imagined as battle-worthy foes on the austere plateau of La Mancha in central Spain.

But while tilting at windmills has been symbolic of a futile fantasy, in reality the windmill, very early on in Nebraska's history, was a symbol of life.

Some of the first recorded information on windmills comes from a meeting of the Royal Society in London, where on June 14, 1759, John Smeaton read a paper "On the Construction and Effects of Windmill Sails." This later led to research in Batavia, Illinois, on a device called a "wind wheel." This research, published in 1883, provides some of the first recorded data about wind velocity, sail area, spacing and friction that helped perfect the windmill.

Major John Wesley Powell, director of the special committee of the U.S. Senate on the Irrigation and Reclamation of Arid Lands, reported in 1890 that Nebraska's small farms could be irrigated by means of windmills at a cost per acre that would not exceed the usual charge for water rights under ditches, or about $10 to $15 per acre.

The early drought years of the 1890s, especially 1892 through 1895, proved that windmills would increase a crop grower's profits by pumping the abundant water in from "practically inexhaustible stores that underlie a great part of the state." Furthermore, because surface water and Nebraska's topography limited gravity irrigation, and Nebraska had the wind velocity needed, at least 15 miles per hour for more than a third of the time, the windmill offered a small amount of security to water-deficient farmers.

Further documentation comes from Erwin Hickley Barbour, University of Nebraska professor of geology and state geologist from 1891 to 1921. He published Water-Supply Paper No. 29 with the U.S. Geological Survey in 1899, entitled "Wells and Windmills in Nebraska." Barbour, who came to Nebraska with a Ph.D. from Yale University, was the third state geologist and head of the Nebraska Geological Survey, a precursor to the Conservation and Survey Division at the University of Nebraska-Lincoln. Later, he became the director of the NU State Museum and its natural history collections.

He noted that due to the drought from 1892 through 1895, "the whole water table was lowered" and many springs, ponds, streams and wells failed. Barbour described methods of "hoisting water" to provide for human and animal needs.

"The windmill has long been the popular means of raising water for domestic use and for watering stock, and now, in addition, it comes to enjoy the recognition it deserves as one means of irrigating land where other means are wanting," he said.

"The whirling mills are necessary accompaniments of Nebraska scenery. Their work is admirable, though impossible feats are often expected of them. As the matter of windmill irrigation is agitated, the receptive mind of the progressive farmer seizes the idea."

(continued on next page)

Henry Mathiesen's stationary turbine mill near Grand Island, Hall County, 1898. Cost—$0.32.

Nebraska State Historical Society, E.H. Barbour photo

Barbour foresaw windmill irrigation as one of the important factors in the development of Nebraska's agricultural resources. "Already farmers are irrigating with profit from 1-acre to 50-acre tracts by means of one or more windmills," he noted.

A windmill of some kind was "a necessity on most of the 115,928 farms in Nebraska," according to the 1896 "Nebraska Irrigation Annual" appendix.

Windmills emerged above hand-dug wells throughout the semiarid and arid West. In fact, the nation's westward expansion was greatly aided by windmills and barbed wire.

Bringing the Nebraska windmill up to date, Marty Muller, of Muller Industries at Randolph, said windmills are still used in the western part of the state. Muller Industries has manufactured and stocked windmill parts since the early 1900s.

"I would estimate that about 10,000 windmills are used in Nebraska today," he said. "Windmills are used mostly to provide range water for cattle in both eastern and western Nebraska."

In the spring and summer, Muller Industries moves about 300 to 400 rebuilt windmills. He said it's a seasonal demand and it all depends on the cattle market.

Dempster Industries, Inc. of Beatrice has manufactured windmills since 1878. It sells about 200 windmills each year, mostly to owners of small acreages. Some go to such places as Colombia, Venezuela, Equador, and the Caribbean Islands, as well as to Chad and Sudan. And a few go to California for cattle watering.

At the peak of the windmill era in the late 1920s and early 1930s, Dempster Mill Manufacturing Co., made about 10,000 to 15,000 windmills annually and ranked about tenth in the production of windmills among the 100 firms in the nation that produced about 250,000 a year. *(see vignette on Dempster, p. 77).*

But as the countryside became hooked up to electricity through the Rural Electrification Administration, the need for wind power began to decline. Now, Dempster also produces water system pumps and fertilizer and chemical applicators for the agricultural industry.

Wind, domesticated animals, livestock, people, and a thirsty land were the ingredients that made the windmill a necessity. Consequently, many old-timers are sentimental about their windmills. And at present, windmill sales and repair by "windmillers" are generally for symbolic, ornamental structures that may have electrical wires directly over them. A recent advertisement by a national chain of drug stores advertised "American Style Windmills"—8 feet tall— for $39.99, or a 4.5 foot model for $29.99.

The symbolism of the windmill can be observed at Cordoba, population 130, about 8 miles south of Interstate 80 at the Utica exit in Seward County. A windmill, the town symbol, stands majestically in the village square, a reminder of its role in local prosperity.

However, one of the most predominant monuments to the former life-giving windmill in Nebraska is Windmill State Recreation Area, at the Gibbon I-80 Interchange between Kearney and Grand Island. Today travelers who stop to use the camp facilities adjacent to this 14-acre reservoir can appreciate windmills from the past: a railroad windmill that is more than 60 feet tall; a Colorado windmill from about 1890 that was moved to Kansas and later found a home at York; a Waupans windmill erected in 1902 at Arcadia; and a 1910 Dempster Model 9 windmill, all in working condition.

Historically known as Windmill Crossing, Pawnee Indians crossed the Platte River here during their annual buffalo hunts. Operated by the Game and Parks Commission, it has been a stop-off point for generations of travelers.

—Pat Larsen, communications specialist, Water Center, University of Nebraska-Lincoln

(Continued from p. 74)

duction, mills and mill technology increased, the market for Nebraska's flour changed from local-only consumption to exports to markets in western states. The gold seekers in Colorado created a need for milled products that kept some mills running night and day.

The processing of food products, adding value before food items leave the state, became an increasingly important industry. Successors of the early millers contribute to the state's economy today. An estimated 11 companies produce flour and grain-mill products, four make breakfast cereals, 14 produce dog and cat food, 107 make animal feeds, 13 make bread or bakery products and five produce macaroni or pasta.

Each of these companies contributes to the billion-dollar food processing industry in the state, employing about 31 percent of all manufacturing employees in Nebraska.

With grain markets constantly in flux, farmers continue to look for additional, sometimes unique crops to raise, and for ways to process or create new food products. For instance, Nebraska farmers harvested 82,324,000 pounds of popcorn in 1989 on 53,000 irrigated acres. This "alternative crop" makes Nebraska the number one popcorn-producing state in the nation and number one region in world production.

An estimated one-half of the products from growers or processors in the state reaches national or international markets. The increasing world distribution of Nebraska's food products adds to the growth and stability of the economy in the state.

The Ice Industry

For many years around 1890, cutting ice from lakes and rivers was a well-established, important winter industry. The ice blocks were used in homes, businesses, refrigerated railroad cars, and the meatpacking industry. The largest operation was run by Crete Mills, where initially ice was cut from the Big Blue. Demand became so great two lakes were created for cutting and an ice house was built for storage. About 75-100 men were employed during "ice season," harvesting when the ice reached a thickness of 10 inches or more. During a particularly cold winter, as many as 100 railroad cars of ice were shipped daily from Crete.

The five or six ice companies in Nebraska went out of business shortly after World War II with the advent of mechanical ice production plants, electric refrigerators and tighter health standards.

Modern-day Nebraska is home to a few commercial ice manufacturers and distributors (among them are Valley Ice Co. of Lincoln and Omaha Ice Company of Omaha), as well as to wholesalers and dealers of ice-making equipment and machines. In the same vein, refrigeration and ice-machine service companies employ hundreds of people statewide.

Other Water-related Industries

Eastern businesses funded many of Nebraska's business enterprises related to water. Some ventures paid off; some did not; but all contributed to the settlement, economy and history of the state.

The first public utility to be developed in Nebraska was public water. Public water supplies were often contaminated, typhoid was common, and at times fires burned entire business districts due to the lack of water to quench the flames. Private companies built water works in Nebraska cities, hoping to make a profit; few did. Usually, the system was sold or given to the municipality.

In the western Sand Hills, several brackish lakes supported a potash industry that briefly became one of the

(Continued on p. 78)

Dempster Industries: One of the Region's Oldest Water-well Firms Sends Products Around the World

One of the Midlands' oldest and most respected names in the water-well business is still going strong in Beatrice.

After a period of revolving door ownerships and economic problems, Dempster has restored its honored position as a major force for readily available pure water, not just in the central United States but across the country and around the world.

The 114-year-old company is manufacturing a broad offering of pumps, from the world's sturdiest hand pump to giant turbine models that meet the demands of industry, municipalities, irrigation and other substantial water needs. And it still is turning out the water machine that made the company famous and made the prairies inhabitable—the windmill.

In the Nebraska water-well industry, Dempster has become particularly prominent in the past several years through an aggressive marketing tie-up with Roberts Pump and Supply Company of Grand Island. It is not coincidental that Roberts's 12-year stay in Nebraska has matched the comeback period of Dempster. That period also began with the arrival in Nebraska of David Suey as chief operating officer and executive vice president of the investor-owned Dempster Industries, Inc.

Suey and the Roberts family—brothers Jack and Ray, who now run Roberts Pump and Supply, and their father, LaVern—all came to Nebraska from key jobs with the Aeromotor Company. Jack headed the company's Fort Wayne, Indiana, office and Ray the Rockford, Illinois, branch. LaVern worked for Aeromotor in Nebraska. They started their supply company in Cairo, Nebraska, in 1980 and moved it to Grand Island in 1981.

Suey, who grew up "in the corn fields" near Decatur, Illinois, was president of Valley Pump Company, which acquired Aeromotor, before he joined Dempster in late 1979. He spent about 10 years in the company's corporate offices in St. Louis after operating its branch in Conway, Arkansas

Now Roberts Pump and Supply is the sole distributor of Dempster water products in Nebraska; and Dempster is Roberts's major line. Before the Roberts connection, Dempster sold directly to dealers throughout Nebraska.

Dempster itself is the story of water development in Nebraska and the Great Plains and of a large manufacturing firm adapting itself to the conditions and needs of changing times. Its current slogan, "From Wind Power to Air Power," says it all.

"Wind power" refers to windmills, the chief stock in trade of the firm since it was started in 1878. Dempster is the only company in the country that has made windmills continuously since their advent in the 1800s. The only other current manufacturer of windmills in the United States is Aeromotor. Another brand, Fiasa, is imported from Argentina. As late as the 1920s, there were several hundred windmill manufacturers in the United States.

"Air power" identifies the latest in Dempster's manufacture of state-of-the-art farm equipment, a self-propelled fertilizer applicator that disperses the material with high-velocity air through an electronic monitoring system.

[In addition, perhaps one of the most significant impacts on the water-well industry made by Dempster was the invention, or at least the development, of a hydraulic-rotary drilling rig about 1902. That development, improved over time, revolutionized the industry by providing a method of drilling wells quickly and efficiently in Nebraska's soft sedimentary rocks.

The drilling rig was self-contained with gears to rotate the drill pipe and bit and pumps to remove drill cuttings from the hole. The original rig used "horse" or "mule" power. Later, power was provided by an internal combustion engine. The trailer rig was mounted on a wooden frame. How many were manufactured (probably hundreds) and for how long is not known because company records were destroyed.—Ed.]

These products highlight the history of the 114-year-old business, once one of the largest manufacturing firms in the state, possibly the largest outside of Omaha and Lincoln. The years span several economic crises that threatened the firm but failed to close it, as well as a recent period of ownership changes, which brought substantial cutbacks in employment and production.

Now, under a new investment group that acquired the company in 1985, Dempster is "back on track," Suey says.

In its glory years, the company regularly had well over 500 workers. Now it is employing about 100 in the manufacture of a diversified line of pump and water-system equipment, as well as specialized farm implements. Sales are split about evenly between the two divisions. Dempster has a sales and service branch in Springfield, Missouri, and a pump manufacturing and marketing plant in Lubbock, Texas.

Still in the same facilities it has occupied for much of its existence, Dempster recently added computerized lathes and is updating several other areas of its manufacturing plant.

For more than 80 years after it was started by 25-year-old Charles B. Dempster in 1878, the company was owned and operated by members of the Dempster family.

In 1960, it was sold to Omaha billionaire investor Warren Buffet, who, as Suey puts it, "quickly liquidated whatever assets he could." Within two years, Buffet sold the company to a group of California investors. In 1963, an investment group including Charles Dempster and Doris Dempster Crocker, descendants of the founder, acquired what was left of the company.

It was sold seven years ago to the new group of investors, headed by Don Clark of Fredericktown, Ohio, as chief executive officer and major stockholder of Dempster Industries, Inc.

While it closed its foundry in 1982, the company has never been out of the manufacturing business in its 114 years. It now buys castings from other foundries and turns out a broad line of pumps of all kinds—submersibles, jets, utility, centrifugal and windmill—plus well cylinders and water-system accessories. Its well-water products are in use around the world.

This Nebraska company has become a key contributor to a global effort to bring a sanitary water system to every village in the world, a goal that grew out of the 1980s "World Water Decade." Undertaken by the United Nations and other international organizations, the effort addressed the problem of shortened life spans due to contaminated water.

"Many villages of 500 people or more are still drinking from the same water hole in which they bathe, water their animals and wash their clothes in," Suey says.

Much of the drive to clean up water depended on the old-fashioned hand pump, which could provide groundwater where no power sources existed. Hundreds of thousands of hand pumps have been made throughout the world, of which Dempster has manufactured a significant number. It is not unusual for their pumps to be in operation 24 hours a day.

Pumps originally designed for the job failed because they could not stand up to rigorous use, rubbing of camels and other hazards that would break off the cast iron handles. In some instances, the new pumps were sabotaged by witch doctors and other tribal leaders who feared that the advent of pure water could dilute their powers over the people.

In answer to the problem, Dempster created a particularly durable hand pump, called the Mark II, designed to strict industry standards. It is very hard to tear up and is simple to repair, Suey says. The Nebraska product now is a major weapon in the global battle against human consumption of contaminated water.

Its biggest users of water equipment are municipalities, industries, irrigators and home owners, again serving a wide variety of needs. Suey cites paper mills, which use

(continued on next page)

huge amounts of water, as among the company's many customers.

Suey's fascination with pumps and the water business is a natural outgrowth of the little retail pump and windmill store founded by Charles B. Dempster with $37 of his own and $300 borrowed. Settlers pouring into Nebraska in the late 1870s found the river banks and brook sides pre-empted by earlier arrivals. They needed a source of water from their own land and the Dempster Windmill Company provided and installed the machinery to do it.

Six years after founding the company, Dempster and a brother, A. R. Dempster, began to manufacture windmills in addition to selling them. Soon they built a two-story factory and foundry powered by the Blue River.

—Jack Hart; adapted and reprinted courtesy of Water Rights and the Nebraska Well Drillers Association.

(Continued from p. 76)

leading industries in the state, employing thousands of men for about six years.

In 1911, the Potash Production Company (and later Potash Reduction Company) filed a mineral claim in southern Sheridan County. A few years later, as many as seven plants had been constructed in the area and two in other parts of the state. However, Nebraska potash couldn't compete with high grade German potash. After World War I ended and the German product came back on the market, all Nebraska potash companies closed down.

Not funded by "eastern" money, but still hoping to make a profit in the new territory, were water witchers. Witching, whether hoax, entertainment or art, was and is still a water-related business. Many people made their living trying to divine the location of water, particularly in the dry years prior to irrigation and better well drilling equipment. For their part, witchers may have worked to keep the dream of water development alive.

Soil Conservation Service

Children playing at an unidentified reservoir

Water-related Recreation Increases

Americans and Nebraskans place a high value on the outdoors, especially areas of high quality water, accessible recreational lands and clean air. Recreation opportunities have become central to their lives and to the livelihood of communities and regions. Yearly, 500,000 people, or roughly one third of Nebraska's population, will take part in water-related activities.

The popularity of water sports has increased dramatically in the last 50 years, so much so that leisure-time activities have become big business in Nebraska. In 1987, Nebraskans spent $1.2 billion on outdoor recreation in their state. Growth is most evident in park and recreation areas managed by the Nebraska Game and Parks Commission. Nebraskans and visitors from out of state made more than 9.2 million estimated visits to state park and recreation areas as they enjoyed water sports, sightseeing, camping, hunting and fishing in 1990. Attraction to water-related activities has helped tourism earn the third largest amount of revenue from outside the state, creating about 36,000 full-time jobs.

Many of the irrigation, flood-control and power-generating projects developed in the last century have provided excellent opportunities for outdoor recreation. In Nebraska, a total of seven reservoirs licensed by the Federal Energy Regulatory Commission (FERC)

Burlington Beach near Lincoln, a former resort now known as Capitol Beach

Nebraska State Historical Society

Soil Conservation Service

A trout fisherman on Nine Mile Creek, Scotts Bluff County

now provide 43,000 surface acres of water, as well as canals for fishing and other water activities. Lake McConaughy is nationally recognized as one of the premier recreation areas in the United States. In addition, return flows from irrigated land have transformed some former dry washes into perennial streams that fish now use for spawning.

Also included in water-related recreation, the U.S. Army Corps of Engineers and Bureau of Reclamation were instructed in the 1930s to recognize all phases of water development, including recreation and fish and wildlife, as part of their flood control and irrigation projects. Today, in addition to Lake McConaughy, Nebraskans enjoy Harry Strunk, Swanson, Harlan County, Enders, Hugh Butler, Sherman, Merritt and Calamus reservoirs—all created as a part of these water-management plans. Additionally, Nebraska's 23 natural resources districts, created to streamline and coordinate the interests of many soil and water groups, are assuming greater participation in fish, wildlife habitat and recreation development. Outdoor recreation areas have developed at many of the lakes and ponds built and operated by the NRDs.

The multiple-purpose water projects of the early 1900s have created limitations for recreation in some cases, however. During irrigation season, when recreation is at its peak, reservoir drawdown can drastically reduce the water available for boating and water activities. For example, the Calamus Reservoir experienced a 14-foot drawdown during its first operating season.

In general, all state parks or state recreation areas (SRAs) are defined as those with primary value for day use, but include overnight facilities or potential for such. In 69 state recreation areas, 69.2 percent or 80,778 acres, are water, and .9 percent, or 1,019 acres, are wetlands. Ninety-three percent of all SRAs contain water suitable for water-related activities and access to a river or lake. The natural resources districts and the USDA Soil Conservation Service also maintain and/or operate reservoirs.

Use of recreation areas is hard to gauge since in Nebraska, many state parks have multiple accesses, and attendance at more than half the state recreation areas is unmonitored. Knowledgeable estimates and 52 reporting areas contribute to most of the visitor counts, and boat registration and hunting and fishing permits provide additional participation figures.

Before lake development in this landlocked state, boating was a special endeavor saved for vacation trips, not a quick excursion to a neighboring lake or river on a weekend. In Nebraska in 1955, 1,600 boats were sold; by 1990, total boat registrations were 57,969, a dramatic increase (though registrations include boats bought earlier and not all boats registered were sold in the state). In 1969, the primary purpose of 8.1 percent of people at water-based recreation areas was boating. A 1986 estimate was that more than 32.4 percent of the state's population took part in some form of boating—canoeing, sailing, kayaking, rowing, float or power boating, and jet or water skiing. A state Game and Parks Commission study shows the number of people participating in boating activities may not be rising, but the number of boating trips per year is increasing.

Recreational boaters spend $140 million a year in Nebraska on supplies, gas, maintenance and other items. This is estimated at about $540 a year in individ-

Lou Haberstroh, Harry Thiessen and George Steige fishing at Neligh Mill Pond, Antelope County, 1919

Nebraska State Historical Society

ual boat maintenance and at least $33 per trip. With boaters taking their crafts out an average of 11 times a year, boating is not only a leisure pastime but, for some, a devotion.

Fishing is pursued by one out of three people over age 16 in Nebraska. An angler spends about $450 a year on equipment, permits, fuel, bait, meals and lodging. An average small game or bird hunter spends $175 per year on that sport. At this rate, boating and fishing add $642 million to the economy each year.

Many Nebraska streams, such as the Missouri, Niobrara and Calamus rivers, are gaining importance as exceptional canoeing routes. In recent years, the Niobrara has become the symbol of ca-

Nebraska Game and Parks Commission

Melvin Steen, former director of the Nebraska Game and Parks Commission, and his dog

Estimated Park Visitor Counts in Water-related Areas at State Recreation Areas (SRAs), Historical Parks (SHPs) and Parks (SPs), 1991

Bluestem SRA	48,500
Box Butte SRA	85,000
Branched Oak SRA	972,400
Bridgeport SRA	85,817
Calamus SRA	182,600
Champion Mill SHP	13,750
Conestoga SRA	115,800
Dead Timber SRA	62,000
Enders SRA	50,011
E. T. Mahoney SP	620,688
Fort Kearny SRA	115,350
Fremont SRA	790,225
Indian Cave SP	137,339
Johnson Lake SRA	360,600
Lake Maloney SRA	143,300
Lake McConaughy SRA	509,405
Lake Minatare SRA	186,965
Lewis & Clark SRA	285,900
Louisville SRA	219,529
Medicine Creek SRA	51,013
Merritt Reservoir SRA	131,178
Mormon Island SRA	301,500
Niobrara SP	125,775
Olive Creek SRA	48,400
Oliver Reservoir SRA	28,930
Pawnee SRA	744,000
Platte River SP	354,585
Ponca SP	203,217
Red Willow SRA	19,462
Schramm Park SRA	100,311
Sherman Reservoir SRA	71,624
Stagecoach SRA	23,600
Summit SRA	42,100
Sutherland SRA	41,600
Swanson Reservoir SRA	72,380
Two Rivers SRA	275,653
Victoria Springs SRA	39,520
Wagon Train SRA	80,600
Wildcat Hills SRA	57,401
Willow Creek SRA	115,625
Windmill SRA	43,950
35 Non Reporting Areas	340,000
	8,297,605

Lake McCounaghy attracts visitors from far and wide camping out in all kinds of portable shelter. Cabins near the lake also provide lodging for tourists. Water-related recreation is an increasingly important industry in the state.

Nebraska Game and Parks Commission

noeing at its best, selling the enjoyment of the sport to Nebraskans and others. The growing impact of the pastime is evident when one realizes that, in 1972, about 300 trips were taken by canoe down the Niobrara; by 1982, nearly 15,000 total trips had been made down the river. With more interest in the outdoors and conservation, the number of canoeing trips is likely to continue to grow.

Until a debate over a proposed irrigation project in the 1970s drew state and national attention to the Platte River and its annual avian and human visitors, few Nebraskans realized that about 80 percent of the continent's sandhill cranes stopped by the Platte. Each year since, the spectacle has brought state, national and international media, visitors, researchers and conservationists to the area.

Today, as nearly 500,000 cranes are squeezed into a shorter length of river, so too are the crane-watchers, who come to observe the largest concentration of cranes in the world. Upwards of 12,000 people per year have visited the Grand Island and Kearney areas to watch cranes, and the numbers are expected to increase.

A 1987 Game and Parks Commission survey showed $200 million of the $1.2 billion spent on recreation in Nebraska in one year went to viewing wildlife. Although it is difficult to assess accurately, the dollar value to the state of the impact of the cranes, including multiplier effect, was estimated by the Whooping Crane Trust to be $1.26 million in 1988. And the impact is predicted to increase dramatically in the next few years as more visitors come to see the cranes.

With the numbers of visitors increasing tremendously each year, problems will naturally be experienced in the state and the region, not the least of which involves disturbing the cranes. Awareness of the cranes increased the need to protect their habitat, but the attention drawn to the situation has created new problems. Whether the solution is found in an interpretive visitors' center or better methods of managing humans in wildlife settings, people of all interests will have to work together to ensure this economic boon does not destroy the very habitat that has created it.

Nebraska Game and Parks Commission

Oxen ''Neb'' and ''Raska,'' former mascots of the state Game and Parks Commission with handler Bud Geistlinger. The oxen were named by the director of the state Department of Water Resources, Mike Jess, in a contest while he was in high school.

Sources: *Agricultural Atlas of Nebraska,* James H. Williams, Doug Murfield, Merlin P. Lawson, 1977, University of Nebraska Press, Lincoln and London; Biennial Report to Governor and Legislature on the Nebraska Resource Development Fund, 1990, Nebraska Natural Resources Commission; *Economic Atlas of Nebraska*, Richard E. Lonsdale, Merlin P. Lawson, 1977, University of Nebraska Press, Lincoln; ''Gerald Gentleman, Public Power in Nebraska,'' 1966, Nebraska Public Power District; ''Report on the Accomplishments of Irrigation in the North Platte Project Area, Wyoming-Nebraska,'' U.S. Bureau of Reclamation, 1956, U.S. Department of the Interior; ''Nebraska Public Power District—A Vision Fulfilled'' (speech), Donald Schaufelberger, October 12, 1983, national annual meeting of the Newcomen Society, Lincoln, Nebraska; Nebraska Resources Development Fund Guidelines, May 1990, Nebraska Natural Resources Commission; Nebraska Wetlands Priority Plan (draft), adapted and expanded for inclusion in the 1991-1995 Nebraska Comprehensive Outdoor Recreation Plan, Nebraska Game and Parks Commission; ''Outdoor recreation—Platte River Basin, Nebraska,'' 1975, Missouri River Basin Commission; ''PWAs Problem Child,'' Neil McClark, in Saturday Evening Post, September 25, 1937; ''Recreational Aspects of Three Nebraska Lakes,'' Edgar Z. Palmer, December 1960, Bureau of Business Research, University of Nebraska-Lincoln; State Comprehensive Outdoor Recreation Plan, Nebraska Game and Parks Commission, 1980-1990; State Parks Division Annual Report, Nebraska Game and Parks Commission, 1991; ''User Survey of 15 Recreation Areas in Nebraska,'' 1967, Nebraska Game and Parks Commission; *Water in Nebraska*, James Aucoin, 1984, University of Nebraska Press, Lincoln.

Water Availability and Use

by Vincent H. Dreeszen
Hydrogeologist Emeritus
and Formerly Director, CSD

"Water, Water Everywhere . . ."

Common, ordinary water, so abundant on earth but so rare elsewhere in our solar system, is truly an amazing substance. Water has been called "the miracle drug" because of its many unique qualities that are vital to life processes. It slakes our thirst, makes up most of our bodies, carries and distributes food within them and those of plants and other animals, and because it evaporates from the liquid to vapor phases, it cools us. We slip and slide and skate on water in its solid phase.

Another of its unique qualities is as the "universal solvent." More than any other single substance, it has the ability to dissolve other substances. Because of this quality, water can hardly ever be said to be "pure." Water's capacity to dissolve and absorb is very useful to us because we use it to clean our bodies, clothes and accessories.

But by being the "universal solvent," it creates some of the real problems associated with water use—possibly as great if not greater than the problems arising from shortages as we use and abuse water. Water has the capacity to dissolve all minerals, including glass and quartz (silicon dioxide). It dissolves and disintegrates rocks and becomes mineralized even in its natural occurrence. This quality of water renders most of the world's water supply unfit for many uses. Although oceans support an abundance of life useful to us, about 97 percent of water on earth is in our salt-laden seas. And in Nebraska, although we have a tremendous supply of good quality groundwater, as much or more water than that of good quality is highly mineralized in deeply buried rocks.

Water, too, is very vulnerable to contamination because of our uses and abuses of it. Water dissolves and carries chemicals that may be harmful, as well as bacteria and viruses that cause infectious diseases. Perhaps the most serious problem of water pollution occurred during the early settlement in Nebraska and elsewhere in the late 1800s and early 1900s. Epidemics resulted in the deaths of many thousands of people from typhoid, cholera and diphtheria who drank contaminated water from poorly constructed wells and surface supplies. And, more recently, in the late 1940s and early 1950s, outbreaks of polio were linked in some instances to water sources.

All in all, water has many qualities that are useful, but it can be a problem when polluted or a nuisance when we build or farm in the pathways of water. There are substantial costs to us to treat water that we drink or dump into the wastewater stream.

There are substantial costs to society in both damage from floods and in building dams and levees to "fight" floods. (Floods, as viewed through the eyes of an earth scientist, cannot be controlled.) Many, now rare, infectious diseases were water borne, and over several decades, regulations and public acceptance of such regulations provided solutions. Research into solutions and public acceptance of such solutions to the problems of pollution also may come over time.

Water Costs

Water and water use can be viewed in several ways. Like air, water is vital to all life on earth. Viewed as an industry and measured in tonnage as a commodity, providing water is by far the biggest industry in this country. Estimates of magnitude have ranged as high as seven times as large as all other industries combined. Measured in dollars, water systems that store, transport and treat water represent an investment at least equal to all manufacturing.

Water is big business in Nebraska, yet the cost to the consumer compared to many other products we use is quite low. Because of its abundance and generally ready availability, many users view water as belonging to the "commons" and, therefore, as free. Actually, to most water suppliers, the water itself is free; the cost is in buying or leasing land, storage and withdrawal facilities—whether from behind dams or from wells—and in the transportation systems—whether a canal "pipeline" or the smaller pipes that deliver water to the point of use. Other costs to the user include the purchase of a distribution method. In the home, that might be plumbing, faucets and shower heads; on the farm, that might be center pivots or gated pipe. Another cost to communal systems is that of obtaining engineering services to plan the facilities and legal services to obtain water rights.

Though agricultural use has dominated the spotlight in Nebraska's water past, municipal use may play an increasingly prominent role. This photo shows Ogallala's "space ship" water tower.

Nebraska Farmer

The cost, then, to the consumer is that of providing services. Those services also include the cost of the power to pump or deliver the water and, in the case of public supply, commercial use or home use, the cost of treating the water to meet health and manufacturing standards or personal tastes, for example, softening to remove hardness or iron.

There are three other costs of water associated with use, and they are the cost of testing water, of providing facilities and treatment to properly dispose of wastewater and of providing storage and capacity to meet peaking needs. Electricity and water have similar characteristics with respect to the latter cost in that both industries must "overdesign" to meet the demands of Nebraska users in the summer months. Our maximum use of electricity is in the summer for cooling homes and businesses and to pump water for crops. Peaking needs for water also occur in the summer and are for irrigating crops, watering lawns, plants and golf courses and as cooling water for coal-fired steam and nuclear power generating plants.

There are costs to the user for water. It isn't free, but compared to other bulk commodities we use, it is cheap.

Water Use

Nebraska is a big state, stretching mostly east to west in the central plains. That helps account for the diversity of our landforms, climate, soils, land use and even our population density. This density also accounts in large part for the differences in the pattern, amounts and types of water use in the state. The size of the state also is a major factor in determining the supply of water available on an annual basis. That amount is mostly dependent on precipitation, the ultimate source of our supply.

If one uses the total area of the state, about 48,949,000 acres, and generalizes precipitation information presented in the chapter on climate and hydrology, estimates of the annual supply can be made. In a "wet" year, the amount of water falling as precipitation is as much as 100 million acre-feet, in a "dry" year less than 80 million acre-feet and in an "average" year probably about 90 million acre-feet. In addition, some water flows into the state via streams, canals and as groundwater. Excluding the Missouri River, that quantity of water, mostly streamflow and storage releases from the North and South

Platte rivers, also is variable. It ranges from less than a million acre-feet to more than 3.5 million acre-feet. Leaving the state in a dry year is about 4 million acre-feet and in a wet year, as much as 20 million acre-feet. The average is about 9 million acre-feet *(see streamflow map in "Introduction," p. 3)*. A logical question is: what happens to the 92 million or so acre-feet of water of average annual supply?

When one examines land use in Nebraska, the answer becomes fairly clear. Of the 48,949,000 acres of land in the state, about 7 percent is dedicated to urban, highway, state and federal institutions, parks, wildlife and other minor uses. About 2 percent of the land is forested. Of the remaining 91 percent, 43 percent is grazing land, leaving about 23,380,000 acres or 48 percent of the total land area as cropland and ranches. Cropland includes row crops, alfalfa and native hay. Since 1985 the area planted to crops and used for hay has ranged from 16.7 to 18.5 million acres, the variability being due largely to active participation in government acreage-reduction programs.

As indicated above, about 9 million acre-feet leave the state as streamflow or an average of about 10 percent of the total annual water supply. The rest leaves the state vertically as evapotranspiration (ET). Transpiration by native vegetation and crops and evaporation from water surfaces, soils and wetlands (evapotranspiration, when combined) accounts for about 80 million acre-feet or about 90 percent of the total annual water supply.

The use of water by vegetation is a consumptive use, as is the evaporation of water from any surface. The water is considered "consumed" in that it enters the atmosphere as water vapor and is no longer available to us. One can only speculate as to the impact we have had on the ET part of the hydrological cycle and on the annual water supply. Although humans have altered the supply by withdrawal of water from streams, reservoirs and from groundwater supplies, we probably have not had much impact on the total supply. The quantity of water available to us on an annual basis is essentially the same as before settlement by European immigrants.

The use of water can be measured in two ways. When we speak of water withdrawal, it is water either diverted by canals or a pump from a stream or water pumped by wells. Irrigation is a "consumptive use" of water, but the quantity withdrawn is not equal to the consumptive use. That is because water applied to cropland is supplemental to precipitation, and some of the water returns to streams through runoff, tail-water canals and drains or the groundwater reservoir. The common term to describe this phenomenon is return flow. Some of the flow may be used again for irrigation or other purposes, such as hydropower, cooling or for maintaining instream flow and wildlife habitat.

The largest use of water, possibly best appreciated by those in the central and western parts of Nebraska, by far is for irrigation. However, the amount of water used for irrigation, or for any other purpose, is at best only an estimate, whether in the 1850s or the 1990s. The same is true for acres irrigated. No precise figures are available for present-day use or the use in the past. In the case of irrigation several problems emerge. One is that of definition. Some crops are irrigated several times a season, others perhaps only once or twice. In some relatively wet years, particularly in the central or eastern parts of the state, even if the water supply is available, the crop may not be irrigated. Another problem of estimating either acres irrigated or water withdrawn is that many irrigated acres are served by water from both surface-water and groundwater sources.

Most stream diversions and reservoir releases are measured or gaged, as are some stream pumpers. The Nebraska Department of Water Resources grants the rights for storage and diversion for a designated quantity and number of acres. Both the amount diverted or pumped varies annually, as can the number of acres served. A major task of the department is in monitoring diversions and acres irrigated, including ongoing adjudication of water rights. Since water-use reports from well irrigators are not required, that use can only be estimated. However, meters are required on all large-capacity wells in the groundwater control areas of the Upper Republican and Little Blue natural resources districts. The Little Blue NRD has initiated action to require meters on wells in its groundwater control area for the 1993 season. The only other groundwater control area is in the Upper Big Blue NRD, and that NRD board is also considering meters.

Sources of information on irrigation water use include publications of the U.S. Geological Survey and the Conservation and Survey Division, University of Nebraska-Lincoln. Titles are in print for 1975, 1979-80, 1985. The Nebraska Agricultural Statistics Service reports the number of acres irrigated each year and has done so for many years.

The drought of 1859 and the early 1860s spawned the first attempt to irrigate crops. Early direct diversions from streams include: in 1860 on the North Platte River at North Platte and along the South Platte; at Maxwell in 1866 and on Lodgepole Creek in 1868; on Wood River in 1873; on the Republican River in 1877; on Cedar Creek in 1882; on the Platte River in 1882; on the Niobrara River in 1883; and on Frenchman Creek in 1886. R. H. Willis quotes the U.S. Census Report as saying, "In 1889 there were 9,026 acres being irrigated in Nebraska." By 1890 there were still only about 12,000 acres irrigated in the state.

The drought years of 1889-1895 really focused attention all across the state on the need for and importance of irrigation. Interestingly, although the drought was lengthy and severe, 1891, the year the first State Irrigation Convention was held in Lincoln, was a wet year.

There was one important understanding that emerged from the early attempts to divert water from streams—the drought of the 1890s demonstrated the limited capability of streams to supply sufficient water to supplement precipitation. Where the need to irrigate was greatest, the supply was the least. Out of these early experiences, legislators and water leaders saw the need for laws to equitably distribute the limited supply and the need to store water when it was available for later release.

Irrigation as a practice increased rapidly

Biological Systems Engineering, UNL

Test pumping, date unknown, perhaps 1940s

during the next ten years and by 1900 as many as 150,000 acres were irrigated from streams and possibly as many as 1,000 acres were irrigated from wells. The power was supplied by manufactured and ingenious homemade windmills that supplied water mostly to small tracts of less than 5 acres. This early attempt to irrigate from the groundwater reservoir, although limited and primitive, was a portent of the development of the large groundwater resources of the state that would occur.

Estimated Acres Irrigated

Year	Surface Water	Groundwater	Total
1860	9,000	–	9,000
1890	12,000	–	12,000
1900	150,000	1,000	150,000 (circa)
1920	450,000	1,000	450,000[1](circa)
1940	500,000	40,000	540,000
1950	600,000	470,000	1,070,000
1960	850,000	1,650,000	2,500,000
1970	1,000,000	3,000,000	4,000,000
1980	1,000,000	6,200,000	7,200,000
1990	1,000,000	7,000,000	8,000,000

[1]The totals represent acres that have wells or surface-water rights that could be irrigated if conditions warrant. Some duplication after 1940 occurs because many acres served by surface water also receive well water.

Surface-water use continued to dominate as the source of water for the next four to five decades. Surface-water development was given a tremendous boost by the passage in Congress of the Reclamation Act in 1902 and the creation a few years later of the U.S. Bureau of Reclamation. The first major irrigation project in Nebraska was the North Platte Project, with water from the North Platte River supplied by storage in the Pathfinder Reservoir in Wyoming. By 1920, about 450,000 acres were being irrigated, still almost entirely from surface-water sources. Water storage was now a major component of surface-water development and use. Oliver Reservoir was constructed on Lodgepole Creek in Kimball County in 1911, and in Scotts Bluff County, Lake Alice, in 1913, and Lake Minatare, in 1915, were built. Bennett Reservoir, also in Kimball County, was constructed in the early 1920s, and Whitney Reservoir on the White River in Dawes County was constructed in 1923. Bennett Reservoir was abandoned and Oliver was converted to use primarily for recreation by the South Platte Natural Resources District in the early 1980s.

Water provided from surface-water sources continued to be the principal supply for irrigation use until the mid-1950s. Sutherland and Maloney reservoirs became operational in 1935 to supply water from the North Platte River for irrigation and water power in the North Platte River valley. Lake McConaughy stores water behind Kingsley Dam, completed in 1941 by the Central Nebraska Public Power and Irrigation District. Johnson and Jeffrey reservoirs were also constructed in 1941 as a part of the development. Other large reservoirs in the state that supply water for irrigation were built after World War II as components of the Pick-Sloan Plan.

Most urban Nebraskans generally tend to think of water use in terms of their domestic needs, whether for drinking, bathing, flushing the toilet, fire protection, or for watering their plants and lawns. Such uses generally are classified along with industrial and commercial use within city and village limits as domestic and municipal use. And, these uses, however important to each of us, are minimal—the proverbial drop in the bucket—compared to other uses made of water. Some of the uses of water important to most of us include swimming, skating, fishing and boating or just plain viewing as we watch water flow or trickle by in a stream, or observe an expanse of water whipped by breezes and reflecting the colors of the sun and sky. These latter uses are known as aesthetic uses and are a kind of bonus that results, at least in part in this region of few natural lakes, from projects which have been built in the last 50 years or so.

Probably less well understood and appreciated is the role that water plays in providing us with transportation, routes of travel and power for our homes and industries. The early fur traders, explorers and settlers relied heavily upon the rivers and their valleys in their travels. The importance of the Missouri River in bringing the early settlers to their staging point for overland travel west is well documented by historians. The Missouri River today carries barge traffic up and down the river. Many railroads and a number of our roads and highways follow the river trails. And, prior to settlement, one can speculate that the natives and hoofed and pawed creatures first used the river paths.

Dr. Robert Manley, historian with the Nebraska Department of Economic Development, has documented the problems that early settlers faced in developing power for grists and other industries. Coal production beyond a few buckets or wagon loads could not be found. No wonder we turned to water for power. The term generally used for waterpower units, usually mills, during the first half of the 20th Century was either "hydro powers" or "water powers." Dr. George E. Condra in 1953 reports, "About 257 water powers have been built in Nebraska but many of them were used to develop mechanical power for the operation of grist mills. Most of them were too small for electric generation and were abandoned, largely because they obstructed drainage and caused floods, and some of them were removed by floods. However, 44 hydro powers are now in operation."

Change in Water Use

The pioneers who traversed the state in the middle 1800s would be greatly amazed if they could see the transformation of the prairies that to them had seemed so short of water. By the turn of the century, however, dreamers and visionaries began to see that the future of Nebraska depended on the use of water for power and to supplement the undependable rainfall for agriculture. The dreamers could envision the cities and factories powered by water and supplied by a huge, irrigation-based agricultural enterprise. That water would come from reservoirs or storage underground. Clearly, the objective of the dreamers, the state officials, the railroaders and the journalists was economic development. Generally, little thought was given to water-based recreation, to the need for water for wildlife habitat or even to water for domestic and stock purposes. The latter use, however important, was quite small and was measured by bucketfuls, powered by hand and windmill pumps, and in the towns, distributed by a faucet or two in the house and a watering trough for horses downtown. The rain barrel and cistern were sources of "soft" water for homes in town and on the farm as late as the 1940s.

Although much of Nebraska has an abundant water supply, some parts of the state, largely along the margins, have not been so blessed. Individual wells, ponds and streams have not provided a dependable supply for home and livestock during droughts. Some farm homes were abandoned and livestock herds decimated and sold because of an inadequate water supply during the droughts of the 1930s and 1950s. A solution slow in coming but relatively quick in acceptance was the establishment of rural water districts. The Nebraska State Legislature authorized their formation in 1967 and in legislative action creating natural resources districts (NRDs) stopped the creation of any new rural water districts as a separate and distinct political entity after July 1, 1972. Rural water systems after that date could be created as improvement projects by the NRDs. The state has not participated in the funding for rural water systems. But the federal government has since 1969 through the USDA Farmers Home Administration by providing loans and grants for rural systems in Nebraska. The cost of planning, construction and management to provide water to scattered-site rural residents has been large. The process of creating a rural water system has not been smooth because of funding difficulties and the objections of some detractors. Some potential users were concerned about cost and did not sign up initially; leaders in some communities objected because they feared the stigma of a water shortage in the area; and some well drillers objected because of anticipated loss of business.

The first rural water systems formed in southeastern Nebraska in 1969 and the early 1970s. Since then, many rural water systems have been constructed *(see map of rural water systems, p. 85)*. Others are being planned by the NRDs. The impetus for the creation of rural systems was a shortage of water. Now, consideration of water quality is at least an equal factor in the decision process. Water quality may be the major factor in the future, changing the pattern of water use in the state from individual and public water supply systems to more broadly based and distributed rural community water systems.

The story of water use is one of change in use and of scale. Water use for power progressed from the stream and small dam water wheels for milling in the late 1800s and early 1900s to the low-head hydropower dams of the early to mid-1900s, then to become the huge hydropower dams of today on the Missouri and North Platte rivers. These uses, except for evaporation of the impounded water, are non-consumptive uses. Water for cooling fuel-powered electric power plants has also become a major use of water and, whether for cooling a coal-fired plant, a nuclear reactor or even a small-town diesel generator, is a consumptive use because large quantities of water evaporate. Another major need for cooling water was the boilers of steam engines, particularly for the railroads. The water tanks located every few tens of miles along the railroads were one of the first sights seen by travelers but now can be seen only in museums. Information developed by the railroaders in their search for boiler water provided much of the first available knowledge about groundwater quantity and quality in the state. The railroad well often was also the village or town source of water.

The scale and use of water for irrigation, like that for electrical power generation, probably far surpasses even that envisioned by the early advocates. Possibly, though, their dreams of economic viability and stability have been met by use of water from reservoirs on the North Platte River in Wyoming, on several rivers within the state, at some off-stream sites and from the groundwater reservoir.

The bonus of water-based recreation was not envisioned, nor was the considerable economic impact of the use of water for boating, water skiing and fishing generally considered. The 1991 annual report of the Nebraska Game and Parks Commission states that some 189,378 acres of standing water are available for lake and reservoir fishing in the state and that nearly 60,000 boats are registered in Nebraska. This total acreage of water surfaces includes farm ponds, grade stabilization and flood-retarding structures, natural lakes and gravel pits. Many of the parks in the state are centered on water-based facilities. In effect small cities have sprung up around some lakes both as summer and permanent homes.

The use of water for domestic purposes also has expanded greatly. The per capita use has doubled, tripled and quadrupled over the last hundred years. Automatic dishwashers, washing machines and lawn and golf-course watering, particularly since the late 1940s, help account for present-day use of 200 gallons per day per capita in eastern Nebraska and about 400 gallons per day in the western part of the state. The electrification of the rural areas by the Rural Electrification Administration played an important role in evening out water use as farm homes were modernized. As recently as the late 1940s, the rural user was asking his or her well driller for a well supplying two to five gallons per minute. The demand since then for a farm well, partly for fire protection, is often 20 to 50 gallons per minute.

The impact of the use of water goes far beyond the monetary returns from power generation, increased crop production and

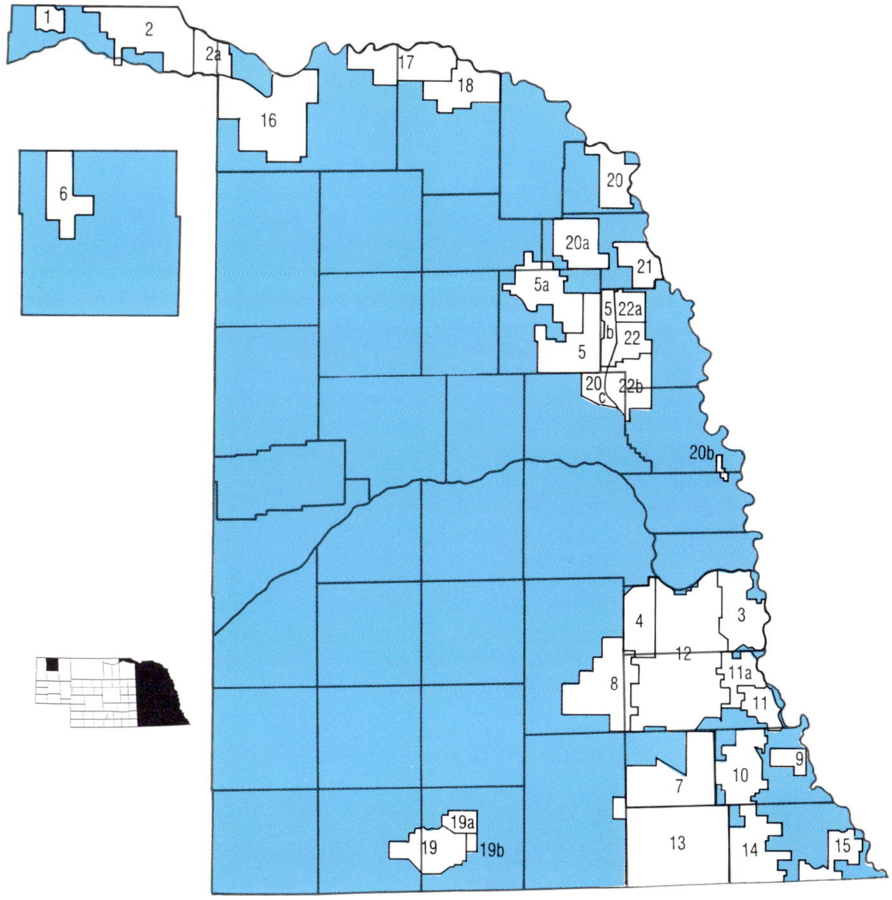

NO.	NAME	YEAR	SOURCE
1.	RWD #1, BOYD CO.	1970	WELLS
2.	RWD #2, BOYD CO.	1978	WELLS
2a.	RWD #2, BOYD CO.	1984	
3.	RWD.#1, CASS CO.	1975	PLATTSMOUTH
4.	RWD #2, CASS CO.	1981	WELLS
5.	RWD #1, CUMMING CO.	1976	WELLS & WESTPOINT
5a.	RWD #1, CUMMING CO.	1980	
5b.	RWD #1, CUMMING CO.	1989	
6.	RWD #1, DAWES CO.	1976	WELLS
7.	RWD #1, JOHNSON CO.	1973	WELLS
8.	RWD #1, LANDCASTER CO.	1974	WELLS
9.	RWD #1, NEMAHA CO.	1970	WELLS
10.	RWD #2, NEMAHA CO.	1974	WELLS
11.	RWD #1, OTOE CO.	1971	NEBRASKA CITY
11a.	RWD #1, OTOE CO.*	1974	
12.	RWD #3, OTOE CO.	1977	WELLS
13.	RWD #1, PAWNEE CO.	1975	RWD #1, JOHNSON CO.
14.	RWD #1, RICHARDSON CO.	1974	WELLS
15.	RWD #2, RICHARDSON CO.	1975	FALLS CITY
16.	LOWER NIOBRARA NRD	1984	WELLS
17.	LEWIS & CLARK NRD	1981	LEWIS & CLARK LAKE
18.	LEWIS & CLARK NRD	1991	
19.	LITTLE BLUE NRD	1979	FAIRBURY
19a.	LITTLE BLUE NRD	1980	FAIRBURY
19b.	LITTLE BLUE NRD	1980	
20.	PAPIO-MISSOURI RIVER NRD DAKOTA CO.	1979	DAKOTA CITY
20a.	PAPIO-MISSOURI RIVER NRD THURSTON CO.	1983	PENDER
20b.	PAPIO-MISSOURI RIVER NRD WASHINGTON CO.	1979	OMAHA
20c.	PAPIO-MISSOURI RIVER NRD	1993	
21.	OMAHA INDIAN TRIBE	1981	WELLS
22.	LOWER ELKHORN NRD	1991	WELLS
22a.	LOWER ELKHORN NRD	1992	WELLS
22b.	LOWER ELKHORN NRD	1993	

*Previously RWD #2 Otoe County

Rural water systems in Nebraska (from Farmers Home Administration)

commercial or recreational use. Many businesses share in the economic gains and to some degree each Nebraskan also benefits from the amenities provided by the increased tax base and by the relative low cost of food and electricity.

Other generally unforeseen impacts resulting from water use have occurred. One impact has been in the alteration of the hydrologic system; that is, changes in stream regimens and flow characteristics and in groundwater storage. Water levels have declined in some parts of the state and risen in other parts. Wetlands and drainage problems have resulted from some surface-water projects, and in some areas wetlands have been altered or destroyed by land and water use. Use of water in some river valleys has changed surface water-groundwater relationships. In some areas as along the South Platte, North Platte and Platte rivers, although the peak flows were reduced by storage and use, the low flows have been enhanced by water release and return flows from drains and groundwater mounds. Stream depletion caused by diversion is reasonably well documented, but the depletion resulting from land conservation practices such as watershed dams, terracing and contour plowing—all of which hold water on the land—is recognized but not well-researched or documented.

Another major impact resulting from surface-water use has been the alteration of terrestrial and aquatic habitat. Significant changes in plant and animal communities have occurred in the river valleys. Trees have invaded river stretches as the result of change in streamflow characteristics in the Platte and Republican river systems. Riverine habitat has been reduced for some species and enhanced for others. Data collection and research is needed to document and evaluate the very complex changes that have occurred and to develop management practices to accommodate the often conflicting needs of plant and animal life, as well as the needs of the various water-user groups and interests. Perhaps more than any other single factor, the future of water use in the state will rely upon resolution of real and perceived conflicts among water users of today.

The one thing about the water use of the future that can be stated with certainty is that use patterns will continue to change. The cost to water users will increase, and the cost of treating waste water and of treating or maintaining the quality of the water supply will likely exceed that of providing the quantity needed by users. Almost surely, too, another change in water use will be that of greater demands upon the water supply by municipalities, manufacturing (including industry and power generation) and recreation and wildlife interests at some expense to irrigated agriculture. Irrigation as a practice in Nebraska will continue into the future but may be constrained by how much water and when and where it will be applied.

Could it be that the changes in water use by the end of the next century will be as dramatic as they have been in the past 100 years? And would the citizens of Nebraska today be amazed if they could return in 100 years to see those changes? The future of Nebraska, its economic vitality and the well being of citizens will be as intimately interwined with the use of water as it has in the past. Water storage from surface and groundwater reservoirs has been the key to water use in the past and will be in the future. Some intentional recharge of and storage in groundwater reservoirs in certain locations may help management of the water supply. The challenge will be that of integrating the use from groundwater and surface-water supplies to meet future needs.

Profile:

Melvin O. Steen: Dynamic Leader of Game and Parks Changed Nebraska

Dedicated, dynamic, capable, hard-headed visionary leader—the late Melvin O. Steen was all of these. Following an extensive career in conservation, in 1956 he became director of the then Game, Forestation and Parks Commission. Steen was a driving force in the restoration of habitat in the state, espousing his principle that wildlife can only survive where there is adequate habitat, and no amount of stocking will help if habitat is lacking. He returned the wild turkey and antelope to their historic ranges, and during his tenure, Nebraskans saw deer hunting expanded statewide. Unlike many states that have large public lands, Nebraska is mostly privately owned, and Steen quickly realized the need for more public hunting and fishing areas across the state. In trying to rectify that lack, he said, "I scraped the bottom of the money barrel until I got slivers under my fingernails to get funds for the development of public-use facilities."

His proudest accomplishment (among many) was the creation of the "Chain of Lakes" along Interstate 80, which paralleled the westward trek of the pioneers. He envisioned a superhighway dotted with small lakes to beautify the freeway, create outdoor recreation for Nebraskans and travelers alike and to slow the headlong dash of visitors from border to border. He presented a plan to the Nebraska Department of Roads (DOR) to pump gravel and sand from strategic locations along the construction route. Because of the high underground water table along the Platte River, these pits would fill with water, creating lakes. When the highway work was done, hundreds of tiny "oases" would be left behind that could be stocked with fish and converted to recreation and wildlife areas. The DOR saw the far-reaching merits of his proposal and set about implementing changes in construction plans and even amended existing contracts to accommodate his proposals.

Steen, too, saw the crying need for water recreation in populous eastern Nebraska and pressed the U. S. Army Corps of Engineers to change their plans for flood-control impoundments in the Salt Valley watershed. Through his efforts, the Corps was prompted to raise the water levels of these reservoirs to make them suitable for outdoor recreation, as well as flood control. Combined, these recreation areas now account for the highest visitation totals in the state parks system.

Born in Pierce, North Dakota, in 1896, Steen graduated from Concordia College at Moorhead, Minnesota, and embarked on a career in conservation that would span half a century. He was well-prepared when he came to Nebraska, having spent 14 years with the North Dakota Game and Fish Department, 10 years with the U.S. Fish and Wildlife Service, and 12 years with the Missouri Conservation Commission. He fostered much sweeping legislation on both the state and national scene that benefitted conservation, including federal legislation that created the Soil and Water Conservation Fund and the creation of the Bureau of Biological Survey (forerunner of the U.S. Fish and Wildlife Service).

His honors, titles, offices, citations and awards are too numerous to mention. In fact, at his retirement dinner (he retired in 1970), it was mentioned that in his office, one could see no walls—they were completely covered with plaques, certificates and awards. In an interview at that time, he was asked to list his proudest accomplishments. Steen penned this partial list: Chain of Lakes, acquisition and restoration of Buffalo Bill's Scouts's Rest Ranch in North Platte, restorations at Fort Kearney and Ash Hollow, the Salt Valley lakes, Fort Robinson, Plattsmouth (Schilling) Waterfowl Management Area, Rock Creek Station State Historical Park near Fairbury and the Oregon Trail there, southwestern reservoirs and Lake McConaughy developments, acquisition and creation of Indian Cave State Park in southeastern Nebraska, creation of Two Rivers State Recreation Area near the mouth of the Elkhorn on the Platte and its trout lake, acquisition of Fort Atkinson, and many other state wildlife and fishing areas and projects.

Mel Steen was sometimes controversial but never dull. The impacts of his vision and the foundations he laid are benefitting Nebraskans today and will continue to do so for generations to come. In a speech, former Governor Frank Morrison once said. "Mel Steen is one of the best things that has happened to Nebraska."

—Elizabeth (Liz) Huff, public information officer II, Nebraska Game and Parks Commission

Technology

by Leslie F. Sheffield

At the turn of the century, irrigation technology for on-farm deliveries of water was not greatly different than it had been for centuries. By 1900, nearly all the land irrigated in Nebraska was watered by diversions from streams or rivers. The water flowed by gravity from diversion structures to canals, to laterals and then to on-farm irrigation ditches.

Unless water could flow from the diversion structure to the canal by gravity, and from the canal to the highest point in the field, some type of pump was required to lift the water. Deep-well turbine pumps weren't available before the early 1900s. And in the few cases where centrifugal pumps were used, they were powered by a small steam or internal-combustion engine. Since centrifugal pumps couldn't lift water more than 20 feet efficiently, they were used only where water was near the surface, primarily in river valleys. (On-farm electricity for pumping didn't become available until Congress created the Rural Electrification Administration (REA) in the 1930s, which provided electricity to rural areas.)

To deliver water from the high end of the field to the rest of it, the farmer or hired hand cut the ditch open with a spade at varying intervals. Water erosion varied from one cut to another and more water flowed into some furrows than others, so uniform water distribution was a serious problem. To achieve greater uniformity, the irrigator had to be present day and night. Irrigation not only was hard and tiring work but required long hours in the fields. Given these hardships, the drive to find more effective and efficient technologies to move water, the subject of most of this chapter, isn't hard to understand.

Some of the back-breaking work needed to manage gravity irrigation before lath boxes and siphon tubes is evident in this photo of a sugar beet irrigator in the North Platte Valley in 1930. Before these innovations, ditches were opened with a spade.

Nebraska State Historical Society

An irrigator in the North Platte Valley opens a ditch to allow water into a furrow between rows of what are probably sugar beets. This photo is undated.

Nebraska State Historical Society

Large Irrigation Projects in Nebraska

U.S. Bureau of Reclamation Projects

The Reclamation Act of 1902 provided federal funds to construct large dams, reservoirs and irrigation projects in 17 western states, including Nebraska. One of the first projects authorized was the North Platte Project. It provided for construction of dams and reservoirs on the North Platte River in Wyoming to supply water to eastern Wyoming and western Nebraska. The North Platte Project was authorized by Congress on March 14, 1903, and construction began later that year.

The building of the Pathfinder Dam, an engineering achievement great enough to be listed in 1971 in the *National Register of Historic Places,* was a saga of great difficulties overcome. "Work crews were constantly menaced by flooding, storms and other natural disasters. Horse freighters hauled cement 45 miles from Casper. Transportation and communication were constant problems. The dam, modified many times from the original design, was completed on June 14, 1909. Completion of Pathfinder and subsequent dams and reservoirs on the North

Platte River made possible a stabilized irrigation water supply for farmers of eastern Wyoming and western Nebraska,'' a Bureau of Reclamation booklet explains.

The Bureau of Reclamation has built and/or services the following irrigation projects in Nebraska:

—North Platte Project, with 263,017 irrigated acres in Nebraska and 68,801 in Wyoming, from dams and reservoirs in Wyoming and western Nebraska;

—Mirage Flats Project near Hay Springs, which began diverting water to 11,662 acres in 1946. Box Butte Reservoir north of Alliance is the principal water supply;

—Frenchman-Cambridge Project, with 45,000 acres irrigated since 1952. Trenton Reservoir and Harry Strunk Reservoir supply the water;

—Nebraska Bostwick Project on the Republican River below Harlan County Dam, built by the U.S. Army Corps of Engineers; it began diverting water to 22,787 acres in Nebraska in 1952;

—Frenchman Valley Irrigation Project, which depends on Enders Reservoir for its water, began deliveries to 9,600 acres in 1959.

—Hitchcock and Red Willow Project, also depending on Enders Reservoir for its supply, began deliveries to 11,490 acres in 1961;

—Farwell Project, dependent on Sherman Reservoir, first began water deliveries to 50,051 acres in 1962;

—Ainsworth Project, dependent on Merritt Reservoir, began deliveries in 1965 to 34,539 acres; and

—Twin Loups Project, which began deliveries to 34,012 acres in 1990. It will supply 53,000 acres by 1995 and depends on the Calamus Reservoir and the Davis Creek Reservoir.

Among the Bureau of Reclamation projects authorized by Congress but not built due to lack of funding and/or local support are the Mid-State Reclamation Project on the central Platte River near Grand Island and the Norden Dam on the Niobrara River, which was to irrigate land in north-central Nebraska. Both of these projects were fought by local and national environmental groups. They were challenged in the courts and eventually lost political support in the state and in Congress.

U.S. Army Corps of Engineers Projects

The U.S. Army Corps of Engineers supervises dams and reservoirs on the Missouri River in Montana, North Dakota and South Dakota and also regulates barge traffic on the Missouri River. On the Missouri along the eastern border of Nebraska, the Corps has built and maintains levees and coordinates dredging operations to maintain a suitable channel depth for barge traffic. These modifications are known as channelization. Hydro-electric power generation at the main-stem dams on the Missouri River is a very important source of both power and revenues, with part of the revenues used to fund irrigation projects under the Pick-Sloan Plan.

U.S. Bureau of Reclamation

An aerial view of Trenton Dam near Trenton, Hitchcock County

The Corps, operating under the Pick-Sloan Plan *(see ''Politics and Policy,'' p. 212)*, built the Harlan County Dam on the Republican River in the late 1940s. It provides flood control and irrigation water to the Nebraska Bostwick Project and Kansas Bostwick Project. As this book goes to press, serious disagreements exist between irrigation interests and recreational and wildlife interests about how much water should be kept in storage. Irrigators want more water released during the irrigation season while recreation and wildlife interests want more to be stored for boating, fishing and wildlife.

Public Works Administration Projects

The only major irrigation projects in Nebraska funded by the Public Works Administration are the Central Nebraska Public Power and Irrigation District (CNPPID) and what began as the Platte Valley Public Power and Irrigation District but is now part of the Nebraska Public Power District (NPPD). In *Irrigation Pioneers: A History of the Tri-County Project to 1935,* Gene E. Hamaker tells of the trials and plights of the people who worked on the ''Tri-County Project,'' as CNPPID was called. After many fits and starts, the project was approved by Secretary of the Interior Harold Ickes and President Franklin D. Roosevelt on Sept. 26, 1935 *(see ''Politics and Policy,'' p. 206).*

From its 75-mile-long main supply canal, CNPPID provides water to about 115,000 acres in Gosper, Phelps and Kearney counties, with limited deliveries to Lincoln and Dawson counties. The canal flows from the diversion dam at the confluence of the North and South Platte rivers just east of North Platte to Johnson Lake near Elwood. This project has provided tremendous benefits, not only to landowners receiving water since 1943, but also to the businesses and communities in the area.

The former Platte Valley Public Power and Irrigation District, now part of NPPD, was built in conjunction with CNPPID. Both projects rely on water stored in Lake McConaughy behind Kingsley Dam north of Ogallala. After water is released from the lake, it flows downstream in the North Platte to the Keystone Diversion Dam, from which part of the flow is diverted into the canal and through a large siphon tube. The tube carries the water under the east edge of Paxton, state Highway 30, the Union Pacific Railroad and the South Platte River. There, it connects with the Korty Canal, which carries water diverted from the South Platte River by Korty Dam. After the two sources of water merge, the water flows to Sutherland Reservoir south of Sutherland and from there to Lake Maloney south of North Platte. In 1992, NPPD provided water to about 80,000 acres operated by farmer-owned irrigation districts between Brady and Overton.

Changes in On-Farm Irrigation Technology

Pump and Power-Unit Technology

Irrigation Pumps

In 1902, when Mahlon E. Layne, a well driller from South Dakota, was working in the oil fields in Louisiana and Texas *(see endpiece on well-drilling technology, p.108)*, he and O. P. Woodburn, a young engineer, drilled four large-diameter irrigation wells near Pierce, Texas, for a rice planter named Milner. They dug pits 8 to 10 feet in diameter and usually 30 to 40 feet deep to water. They then drilled a hole about 2 feet in diameter at the bottom of the pit to a depth of 100 feet or more, until they reached the deeper strata of water-bearing sediment. Layne installed horizontal centrifugal pumps in three wells and a vertical centrifugal pump in the other.

The three horizontal pumps were made by Fairbanks-Morse and powered by that firm's oil-burning engines. Rope, used as belting in grooved pulleys, ran from the engine over idlers and down into each pit. Several 50-pound weights dangled at the top of the pits, providing rope tension. Woodburn recalled vividly the extreme hazards of servicing the pumps while they were running. During the summer of 1902, while climbing in and out of the pits, Layne conceived of the "pit-less" pump. His new idea involved substituting a rotating steel shaft to provide the power from the engine to a pump in the well. Layne's differed greatly from earlier pumps because its shaft was prealigned and enclosed in a pipe that was flanged (attached by a separate part at a right angle) onto the pump. The whole unit was enclosed in a steel casing, which could be installed in a drilled well rather than a pit.

Department of Water Resources

R.H. Willis stands on the sluice gate of the Paxton Canal, October 1929. Writing is on photograph.

Nebraska State Historical Society

Model of the "morning glory" spillway at Lake McConaughy

Layne's new pumps were lubricated by oilers at the surface, and the entire unit could be raised or lowered from the surface, a vast improvement. After trying turbine pumps from a Chicago manufacturer, he went into partnership with Woodburn and salesman P.D. Bowler to develop a better pump. Layne continued to improve his pump and in 1906 filed for a patent. In 1907, the partners formed the Layne and Bowler Company in Houston.

By 1907, the firm was marketing pumps that could lift great volumes of water. By 1909, Layne and Bowler boasted that more than 500 of its pumps were irrigating more than 75,000 acres of rice. Cost of the first pumps was about \$500, not including the cost of the well and power plant. Many other firms began building and selling the new and improved turbine pumps, including one in Nebraska. By the 1970s, Western Land Roller Company, now a division of Ingersoll-Rand at Hastings, was one of the largest manufacturers of turbine pumps. Their turbine pumps were marketed through a network of irrigation well drillers located in several states, some of whom they helped set up in business.

Power Units

The lack of adequate and dependable units to power turbine pumps was a problem early on. Steam engines were used for the early centrifugal pumps, many adapted from the oil industry. Donald Green, in *Land of the Underground Rain,* notes that in 1900 near Mesa, Arizona, an 80-horsepower steam engine was installed to power the centrifugal pump of S. J. Murphy. The Murphy plant needed four people to maintain it—two engineers to service the engine and two firemen to cut wood for fuel. The daily cost for crew, oil and maintenance parts was \$9.25. Cost for wood, obtained from one of Murphy's neighbors, was an added expense. All totaled, costs for the Murphy plant were \$2.27 for each acre-foot of water.

By 1910, electricity was used to drive some pumps. Green explains that the Northern Colorado Power Company solicited irrigation farmers in its area to use electricity. The company recognized that irrigators pumped primarily in the summer months, when its urban customers used little electricity. (Of course, after rural areas received electricity and air conditioners became common, summer power cost more.)

Improved internal combustion engines were developed and became far more common as automobiles, trucks and farm tractors were mass produced. Heavy-duty gasoline, natural gas and diesel engines also became available for irrigation power units. Before then and before the REA delivered electricity to rural areas, many of the pump engines were adapted from those used in oil fields.

After the REA was established, it was only natural that many farmers turned to electric motors. They required minimal maintenance and lasted longer than internal-combustion engines. Therefore, they can be depreciated over a much longer period of time. Recently, as electric motors account for the bulk of the summer load of REA districts, special incentives have been developed to encourage "interruptible service." This is especially true for farmers with pumps providing water for center-pivot irrigation systems. Farmers can sign up for "interruptible service" in exchange for a lower rate per kilowatt-hour. This technique helps the REA districts reduce their peak power needs during the summer and minimize penalties on users for highest peak demand power.

Many farmers with turbine pumps have preferred diesel over gas or liquid-propane (LP) engines because of their longer

Biological Systems Engineering, UNL

Plastic siphon tubes move water from ditch to furrows. A one-inch siphon is used for each row. A one and one-half or two-inch siphon may be divided among two or three rows. Photo probably taken during 1940s.

life and lower maintenance costs. Today, many different irrigation power units are found on farms. An estimate published by Dean Eisenhauer, a University of Nebraska-Lincoln (UNL) agricultural engineer, in the January-February 1992 issue of *Irrigation Journal* indicated the following percentages for use of different types of power units to power irrigation pumps on an estimated 75,000 irrigation wells in Nebraska: electric motors—35 percent; diesel engines—27 percent; natural gas engines—18 percent; LP gas engines—18 percent; gasoline engines—2 percent (used only for shallow lifts with small tractors or old car engines).

On-Farm Delivery Methods

Lath boxes

Somewhere, at some time, an ingenious farmer, tired of the problems of cutting the ditch bank and its uneven water distribution, came up with "lath boxes" made from a cheap, common construction material. Lath was used in the building industry as backing for walls that were then plastered with a cement-sand mixture. In the summer of 1943, my first experience with irrigation involved cutting the thin lath strips into 2-foot pieces and making them into a long, narrow box on a farm east of Axtell owned by the late George E. Johnson, chief engineer and first general manager of the Central Nebraska Public Power and Irrigation District. Paul Harris, the tenant farmer on this land, taught me how to build these lath boxes.

The lath boxes were then "mudded" into the ditch bank so each row of corn received only the water each box could provide. The level of the water in the ditch was established by use of canvas dams (suspended on a pipe with a small bend in the center to allow excess water to flow to the next dam). Lath boxes meant that the volume of water flowing down each row was far more uniform than by the bank-cut method. The principal disadvantage was that they sometimes would wash out, allowing most of the water to flow down one or two rows. After siphon tubes became available, most farmers abandoned lath boxes for the plastic tubes until gated pipe became popular because of its labor savings.

Siphon Tubes

The widespread use of siphon tubes for irrigation was due to their development in 1934 and subsequent marketing by Milo German, a Cozad farmer and entrepreneur *(see vignette on German in "Farm Family," p. 242)*. An irrigator, German had worked hours and hours to achieve a uniform distribution of water, which often meant being in the fields day and night.

German's first siphon tubes were made from discarded boiler flue pipes, but the steel pipes were heavy and cumbersome. In 1942, with the advent of plastic tubing, he began building siphon tubes out of plastic.

In 1945, German incorporated his siphon-tube business under the name, Nebraska Plastics, Inc., using "COZAD" as the trade-name. (This firm is now operated by two of his sons.) For the next 16 years, he improved the original design and worked diligently to market this new irrigation method. At the same time, he managed his farms and a John Deere equipment business in partnership with Bill Rice. In addition, he started a car dealership, an airplane dealership, a potato growing and processing company, an alfalfa buying and processing company, a fertilizer business and an aerial spraying business.

Such endeavors resulted in German traveling to other parts of the country. As he piloted his small airplane (a Beechcraft Bonanza) to major cities, he met and became friends with people like William Lear, then with Motorola and later builder of Lear jets; Walter Beech, founder of Beechcraft; Bill Odom, the well-known aviator, and a host of other colorful personalities.

Gated Pipe

The Ames Irrigation Company of California was the first firm known to have produced gated pipe. Paul Hohnstein, president of the Hastings Irrigation Pipe Company of Hastings, Nebraska, founded the second firm. The Hastings firm is now the largest single manufacturer of gated pipe in the United States, with plants in Hastings and Madera, California.

In the late 1940s, Hohnstein, then a young machinist working out of his garage, had an idea for an irrigation gate in a metal pipe that would improve accurate delivery of water to gravity-irrigated row crops. After perfecting his idea, he filed patent applications, and Hohnstein's Hastings Speciality Manufacturing Company entered the irrigation business. In the early 1950s, the firm incorporated as the "Hastings Irrigation Pipe Company" and moved to a larger building. While his business and many of its buildings have expanded, the firm's main office and Hastings plants are still at the same location.

At the time of his invention, Hohnstein's manufacturing experiences were with the Hastings Air Conditioning Company and the U.S. Naval Ammunition Depot, where he had worked primarily with a relatively new metal, aluminum. He saw many advantages in replacing the galvanized steel pipe then in use with the much lighter aluminum alloys. Aluminum could be produced in longer sections and was much easier to handle. Hohnstein's firm produced gated pipe in three sizes, 8, 9 and 10 inches in diameter. Six to eight thousand feet of pipe per week was then considered a good week's work for the company. By the early 1990s, it could produce several times that amount at its two plants. The standard length of gated pipe now is 30 feet.

Les Sheffield, UNL

Gated pipe used on irrigated pasture at the University of Nebraska West Central Research and Extension Center at North Platte

The mid-1950s brought a major drought in the Great Plains and a large expansion of irrigation. Most of the growth came from wells drilled in the High Plains (Ogallala) Aquifer. At that time, most irrigators were buying gated pipe for gravity irrigation systems. Expansion of gated pipe occurred not only with new wells but also from converting the much more labor-intensive siphon tube and lath box methods, which also eliminated the need for an open ditch at the head of the field. In the late 1970s, 1980s and early 1990s, many farmers began converting leveled land that had been gravity-irrigated to the highly automated center-pivot systems.

Hand-moved Sprinklers

With the development of pressurized city and town water systems, sprinklers were developed to water lawns and gardens. Sprinkler irrigation on a larger scale began with city parks and golf courses and then moved to nurseries and high-value crops such as fruits and vegetables.

Sprinkler irrigation received a significant boost in 1933 when Orten Englehart of Glendora, California, combined a piece of copper tubing and a spring wire from a broom handle to develop the first spring-activated rotating sprinkler head. His friend, Clem LaFetra, put this invention into production, founding the Rain Bird Sprinkler Company of Glendora, one of the principal manufacturers of irrigation sprinklers. The rotating impulse-type sprinkler heads first used on hand-moved sections of pipe made it possible to do light watering of uneven terrain. Application of the technology first occurred in California, where the first systems were made.

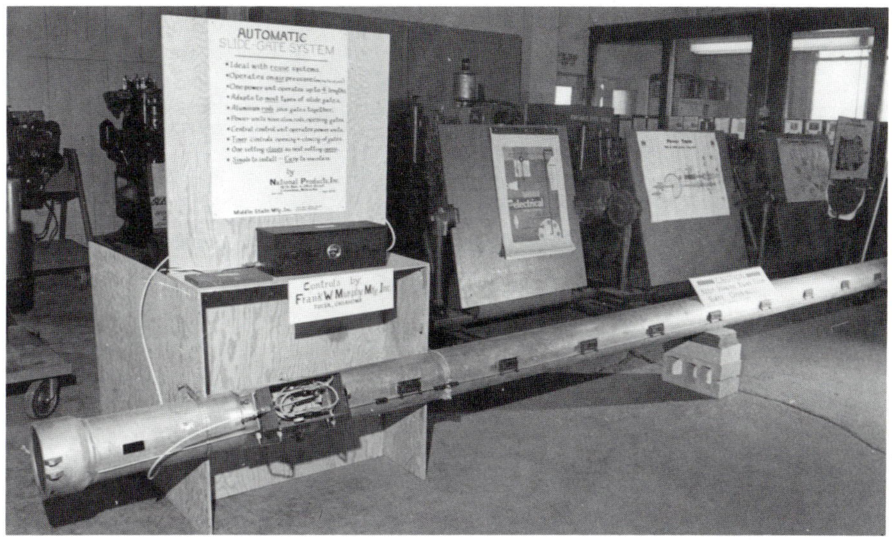

Biological Systems Engineering, UNL

An automatic slide-gate system for gated pipe operated by air pressure. Sign notes that a timer controls the gates, opening one as another closes. Aluminum rods move the gates.

After World War II, the Rain Bird Company decided to expand the market for their hand-moved sprinkler-irrigation pipe to include general crop production in the Great Plains. At the time, the only type of sprinkler irrigation available was with hand-moved sections of aluminum-alloy pipe. These were produced in 30-foot lengths with riser tubes of varying heights, depending on the crop being irrigated. They required much labor to assemble and move from set to set but offered a new method of irrigation.

To move pipe over tall corn, farmers used three-tine pitchforks with the middle tine broken out. One person would get on each end of the pipe, lift a section with the pitchfork and carry it over the tall corn to the next set. It was very hard work, especially during the hottest, most humid days of summer. In the mid-1950s, speaking at an irrigation field day, Alva

Biological Systems Engineering, UNL

Hand-moved sprinklers used on alfalfa in Nevada

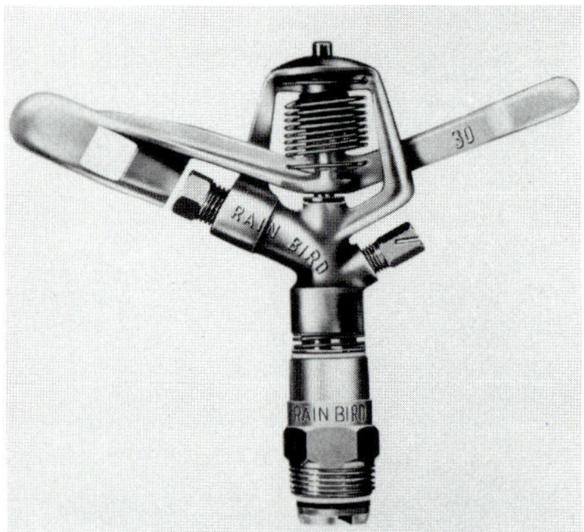

Biological Systems Engineering, UNL

Rain Bird sprinkler head

Hecht, an extension agent in York County for many years, said that in his experience even the hardiest of farmers would give up trying to move the lines after no more than 3 years.

Tow-line or Skid-tow Sprinkler Lines

The first tow-line sprinkler system was invented about 1950 in Portland, Oregon, by W. H. Stout, also the first promoter of side-roll sprinkler systems in 1949, though he didn't invent them. Tow-line systems modify hand-moved sprinkler lines with heavy-duty couplers and stabilizer bars (to keep sprinklers on risers from being damaged when they were moved to another set). They can then be moved with a tractor. In the early 1960s, one of the first manufacturers of tow-line systems in Nebraska was T-L Irrigation of Hastings, whose founder and president is LeRoy Thom.

The modification of hand-moved lines to tow-lines caught on fast. Many small firms began making the tow-line systems or converting hand-moved lines to tow-line systems. The key to their success was the stabilizers that held the risers and sprinklers upright, along with heavy-duty couplers that would withstand towing a quarter-mile line from one set to another.

One firm that began converting hand-moved lines to tow lines began as a small welding shop and garage operated at Lindsay, Nebraska, by Paul Zimmerer. Zimmerer got his start in the irrigation business in 1955 by selling and installing irrigation motors and hand-moved sprinkler lines. After beginning the conversions in 1958, these labor-saving tow-lines caught on immediately with farmers, and in 1961, Zimmerer built a facility for the production of tractor-tow sprinkler lines.

In 1963, he and two sons, Art and Bernie, formed a partnership they later called the Lindsay Manufacturing Company. That same year, they doubled their physical plant, as the demand for tow-line sprinkler systems continued to grow. *(See the section on center pivot companies in this chapter for more about Lindsay Manufacturing Company.)*

Tow-lines, sometimes called skid-tow lines, are still used in parts of Nebraska and around the country since it is a low-cost method of irrigation. But it is much more labor-intensive than center-pivot or lateral- (linear-) move sprinkler systems. These tow-line systems are popular with farmers who want to keep their equipment costs low.

Les. Sheffield, UNL

Alfalfa irrigated with tow-line system north of Mead, Nebraska

Rotating Boom-type Sprinklers

Boom-type systems were first marketed in the late 1950s. This system consisted of a long rotating boom mounted on a wide-axle four-wheel trailer, with sprinklers on the boom to distribute the water in a circular pattern. Most of these used a cable support system to hold the boom as it rotated. When the system wasn't operating, the booms could be lowered by releasing tension on the cable. Boom sprinklers were top-heavy, so one of the hazards of operating these systems was their tendency to tip over in strong winds, especially on hilly terrain.

At first, the system operated only when standing still, and after irrigating in one spot, a tractor would move the systems to another set by adding or removing sections from the pipeline supplying water, which usually came from a well. The trailers had pipe racks to carry extra pipe sections; these provided extra weight to help stabilize the trailer.

Boom sprinklers were available in various lengths from 80 to 250 feet. For irrigating row crops such as corn, travel lanes of eight to 12 rows of a short-growing crop, like soybeans or grain sorghum, were planted so the system could be moved from set to set. The lanes were spaced from 180 to 300 feet apart. Tow-line travel lanes were usually about 60 to 80 feet apart.

One of the principal manufacturers of rotating boom systems was the Vermeer Company of Pella, Iowa. A major Nebraska manufacturer of the systems was Duncan Irrigation in Columbus, which quit operating several years ago. Vermeer was one of the first to offer a traveling boom system that towed a flexible supply hose and used a cable and winch to move the system forward as the cable wound on a large reel mounted on rubber wheels. These systems cut the labor considerably compared with stationary booms.

While a sizable number of firms made boom-type systems mounted on trailers, by the 1980s, the demand for such systems was almost non-existent. Traveling volume-gun sprinklers took over the role that once was filled by rotating boom-type sprinklers.

Volume-gun Sprinklers

The first traveling volume gun sprinkler with a guidance system that pumped water from an irrigation ditch was developed by A. W. Krueger (formerly a county extension agent) and his son, Russell, who then lived near St. Paul, Nebraska. The guidance system allowed the traveling gun, marketed as "Big Squirt," to follow a curving ditch on fields with uneven terrain. Russ Krueger later moved to Longmont, Colorado, where he founded the Irrigation Motor and Pump Company that continued to manufacture the "Big Squirt" traveling gun sprinklers for several years.

During the 1970s, traveling volume-gun sprinklers gained great popularity with farmers, especially for irregularly shaped or long rectangular fields. Volume guns vary widely in size and capacity, ranging from 200 gallons per minute (gpm) to 750 gpm. The amount of pressure varies with the capacity of the gun and the width of the lane needing irrigation.

Most volume-gun systems are mounted on small trailers that are pulled through the field by a cable-winch system, usually operated by the system's water pressure. On most units, a flexible high-pressure hose 4 to 6 inches in diameter supplies water, usually from a well. The standard hose length is 660 feet, so the system can travel a quarter of a mile between sets—660 feet on each side of the water source. Where a volume gun is to be used over several seasons, irrigators often install a buried pipeline with risers at fixed intervals to supply water.

Since the distance of throw from the volume gun is critical in determining the width of the wetted area, these systems require high pressure for maximum lane widths. The cost of the high-pressure hose, which must withstand pressures of 100 to 150 pounds per square inch (psi), is an important factor in the price of a system. Each pound of pressure equals lifting a given volume of water 2.31 feet, so a traveling volume-gun sprinkler operating at 100 psi will have the equivalent of 230 head-feet of lift for the water plus the lift of the water from the well or river.

Some volume-gun sprinklers have a large reel for handling the hose as the system moves. While more expensive than the "hose-drag" systems, the reel system helps reduce energy costs since the hose isn't pulled along the surface. The reel also made the system easier to move from field to field since the 660-foot length of hose was heavy and required a tractor to pull it from set to set.

When the cost of energy rose sharply in 1973 and increased through the early 1980s, the level of interest in volume-gun sprinklers decreased because of their energy costs. Farmers looked for systems using less energy.

Center-pivot Self-propelled Sprinkler Systems

In 1947, Frank Zybach, a tenant wheat farmer near Strasburg, Colorado, east of Denver, went to a demonstration of hand-moved sprinkler irrigation pipe, then the main type of irrigation system. Zybach was intrigued by the potential of the new irrigation method since it simulated rainfall. However, he told the neighbor who invited him to the demonstration, "There has to be a better way to irrigate" *(see Zybach endpiece after "Economics and Finance," p. 132).*

Les Sheffield, UNL

An example of a volume (or traveling) gun sprinkler. These are mounted on a small trailer and pulled through the field by a cable-winch system. Such systems require high water pressure, 100-150 pounds per square inch.

Such a comment was understandable after he watched the laborious task of connecting the pipe sections, running water through them, then walking through the mud to move them and then reconnecting them to the water source to irrigate again. Obviously, irrigation with hand-moved sprinkler lines involved a lot of labor, but it was a totally different method of applying water, compared to gravity irrigation.

Having an inventive nature (with several patents to his credit), Zybach began work on an automated self-propelled sprinkler system in the summer of 1947. He did his early work in his farm shop, where he had installed a forge and other metal-working tools. (Zybach, who died in August of 1980 at 86, was really more a blacksmith, metal worker and "tinkerer" than a farmer at heart).

By the next summer, he had built a small, two-tower experimental center-pivot system. He started out using metal skids at each tower under the pipeline but later used old rod-weeder steel wheels at each tower, which required less energy. To propel the system, he used a small amount of the water in the main pipeline, which flowed into the hydraulic water cylinder. The water pressure in the cylinder forced the piston to the top of the stroke, at which point the

Les Sheffield, UNL

Frank Zybach's first center-pivot irrigation system, near Strasburg, Colorado. Note the low pipeline. Zybach later raised it to irrigate corn and other field crops.

water was expelled through a valve that opened at the bottom of the cylinder. He used a system of levers connected from the water cylinder to a trojan bar, which would catch the lugs on the steel wheels and moved the system forward, one lug at a time.

He also worked out a method to keep the towers and the pipeline (suspended about 2-3 feet off the ground) in alignment as it moved in a circle. Unable to buy any, he built two-way water valves to keep the system aligned. These valves were connected by control wires zigzagging from tower to tower to open and close the valves at each tower as needed for proper alignment. Since the outer tower had to move the greatest distance, it was the "master" tower, while the inner or "slave" towers could be kept in alignment by controlling the water valves at each. By the spring of 1952, he had built a complete five-tower pivot that could irrigate 40 acres. Zybach filed for a patent on his new irrigation system on June 27, 1949. And the U.S. Patent Office granted him Patent No. 2,604,359 on July 22, 1952, for the revolutionary but unheralded invention.

Zybach's first concept was that this sprinkler would be used to irrigate shorter-growing, higher-value crops such as potatoes, sugar beets, melons and alfalfa. It wasn't until later, after he and A.E. Trowbridge formed a partnership, that he raised the pipeline to 6 feet high. By raising the towers, he had made a system that could water taller growing crops such as corn.

As is often the case with inventors, Zybach lacked the capital to exploit his invention. However, while living in the Columbus area he had come to know A. E. Trowbridge, an automobile dealer and entrepreneur. (Zybach was raised on his parents' farm near there, and his father had a farm shop and served as the neighborhood blacksmith. This gave Frank his first experience working with metals.) Trowbridge had helped finance some of Zybach's earlier inventions and, while none of them struck "paydirt," he had great faith in Zybach's inventive genius. During the summer of 1952, Zybach invited Trowbridge to Colorado to see his newly patented irrigation system.

IANR

Frank Zybach

Trowbridge obviously thought the invention had potential because the two men formed a partnership that fall to build center-pivot systems in a rented machine shop in Columbus. Needing capital to begin, Zybach sold 49 percent of his patent rights to Trowbridge for $25,000 and moved his family to Columbus in the fall of 1952. The two men hired a few workers to build the new systems and in 1953 and 1954 began producing them for sale with the higher pipeline.

During the first two years, the fledging firm built only 19 systems. They had great difficulty selling the new concept because it was so different from any other method of irrigation. Farmers with irrigation experience, accustomed to gravity irrigation, were reluctant to buy an untried system. Even if they could convince a few to try it, few, if any, agricultural lenders would loan the money to buy a system with no track record of success.

In addition, Zybach and Trowbridge, who died in December of 1980, also at 86, didn't mesh well in a business struggling to introduce a new technology to an unreceptive public. It became clear that a deep conflict existed regarding how the new venture should be managed. Zybach, ever the inventor, found little challenge in building the same design over and over, so he was constantly making "improvements" as each system was being made. Trowbridge, on the other hand, a hard-headed businessman, realized that it would be impossible to make a profit if each system differed from the one before it!

Out of mutual frustration, they decided to sell the manufacturing and sales rights to Robert B. Daugherty, president of a small, short-line farm equipment company, Valley Manufacturing Company (becoming Valmont Industries, Inc. in 1966) in Valley, Nebraska. The agreement took effect in September of 1954. Valley took over exclusive rights under a 5-percent royalty agreement, a deal that turned out to be very good to all concerned. In a 1970 interview, both men said they had each received more than $1 million in royalties before the original patent expired in 1969.

At the time, Valley Manufacturing Company was making farm equipment such as grain and ear-corn elevators, bale-loaders, hay rakes and other small items. Daugherty decided it was time to diversify, even though the firm then had a contract with Sears, Roebuck and Company to build farm equipment under another name. Since the contract could be cancelled on short notice—even though it accounted for 75 percent of the firm's output—he threw it over for center pivots.

The early years were ones of development, modification and improvements leading to greater reliability and flexibility in the the systems, according to John Chapman, an agricultural engineer and director of Valmont's research and development division. Despite tremendous labor-saving advantages and the capacity to irrigate on hilly terrain or sandy soils (on which uniform water distribution by gravity systems is impossible), acceptance of the new method came slowly.

In 1955, Valley built only seven of the systems. By 1960, the annual production had reached only 50 systems a year. During this time, the company encountered the usual resistance to a new technology, especially since it wasn't well understood by those with the widest experience with irrigation—either knowledgeable farmers or irrigation experts. Undoubtedly, some of Valley's executives (and even the firm's bankers), wondered many times if this venture was a mistake. But Daugherty's perseverance and faith in the future of center-pivot irrigation eventually paid off, even though several years passed before his firm saw sizable profits from it.

During these early years, Valley made important improvements in the design. They developed a variable-speed drive so different amounts of water could be applied per revolution. They installed end guns to irrigate extra acres at the end of the system and an automatic shut-off if the water pressure was lost or the towers got out of alignment. They also discovered that some fertilizer and pesticide solutions could be distributed with the water with far greater accuracy and at far less cost than with any other method. And in the late 1960s, Valmont began building hot-dipped galvanized irrigation systems far more resistant to corrosion than the earlier painted steel systems. Today most such systems are galvanized.

Bill Curry

A. E. Trowbridge

Success and Expanded Application

As more people left farms during the 1950s, 1960s and 1970s, farm labor became scare and higher priced. Even more importantly, some far-sighted farmers and landowners recognized the potential for center-pivot irrigation on land with sandy soils and/or hilly terrain, land not irrigable with gravity systems.

In the mid-1950s, this kind of land, with good groundwater supplies, sold in north-central Nebraska for as little as $40 to $50 an acre. In an area north of Atkinson in Holt County, section after section of nearly level land with sandy soils was available. Groundwater supplies and quality generally were good and located about 50 to 125 feet below the soil surface. These soils, predominantly sandy loam or loamy fine sand, overlaid coarser sands and gravel that were excellent water-bearing deposits.

The first center-pivot system installed in Holt County in 1955 was near Page, Nebraska, on a farm owned by Mr. and Mrs. C.J. Getz. Bill Scheinost, the tenant operator, assembled the system. In 1957, Trowbridge and his son-in-law Bill Curry, both living in Columbus, bought land north of Atkinson and put in center pivots to grow corn and alfalfa, the first multiple system installation under single ownership. They installed more systems north of Atkinson but later separated their farming interests there. Their farm manager, from the fertile Platte Valley, was an expert corn grower. With the new center pivots and good cultural and soil-fertility practices, he achieved yields equal to or better than those from choice gravity-irrigated lands in the Platte Valley.

Their success drew increasing interest from farmers and landowners, not only in Holt County but in surrounding counties and states. Gradually, more and more center-pivot systems were installed on lands not suitable for gravity irrigation. Eventually, by the late 1970s and early 1980s, farmers began to install them on choice, gravity-irrigated lands. By then, many realized they could save 80 to 90 percent of the labor needed with gated-pipe while saving a third to half of the water needed with gravity irrigation.

Successful applications were still being discovered when, in the fall of 1966, the first center-pivot system ever located at an agricultural experiment station was provided by Valmont Industries under a grant to the North Platte Agricultural Experiment Station. It was an eight-tower, cable-supported water-drive system that was used for a study of irrigated pasture for beef cattle. The studies were conducted by Bert Somerhalder, agricultural engineer; Donald Clanton, professor of beef cattle nutrition; and John Vallentine, assistant professor of range and forage science. Both cow-calf production and growing weaned calves were studied, with as much as 1,000 pounds of beef per acre achieved during the weaned-calf study.

The 52-acre circle was divided into five pastures with cross fencing, and a series of various gates allowed the towers to pass from one pasture to another. Four of the pastures were seeded to a mix of brome grass and alfalfa while the fifth was planted to a warm-season grass mixture. After two years, the pasture with warm-season grasses was converted to a brome grass-alfalfa mixture, since the latter produced considerably more forage under irrigated conditions. The cattle were rotated from one pasture to another on about a 5-8 day rotation to allow for quick use of the forage, followed by a regrowth phase. This project drew attention and visitors from several states and even some foreign countries. Several farmers and ranchers installed center-pivots on their pastures as a result.

Since most of the center-pivot systems then on farms were for corn production for grain and I was selecting a topic for my Ph.D. dissertation, I decided to study the economics of corn production using center-pivot systems in the nine-county Southwest Nebraska Cropping District. This was the first graduate thesis written about center-pivot irrigation. I then selected 19 cooperating farmers with a total of 54 center-pivot systems on their farms, and divided them into three size groups: those with one to two systems; those with three to five systems and those with six or more systems. The results of this research showed the following costs:

Group	Total Annual Costs/Ac.	Ave. Cost Per Bushel
I (1-2 Pivots)	$110.01	$.952
II (3-5 Pivots)	$100.88	$.855
III (6 or more Pivots)	$ 88.77	$.733

This research indicates the economies of size involved in center-pivot irrigation. While the annual fixed costs of the three groups were very similar, the variable costs per acre declined as the size of operation increased.

The Center-pivot Companies

In an article titled, "Center-Pivot Irrigation," in the June 1976 issue of *Scientific American,* William E. Splinter, formerly chair of the UNL Department of Agricultural Engineering, said: "Center-pivot irrigation (is) perhaps is the most significant mechanical innovation since the replacement of draft animals by the tractor."

With that appraisal in mind, it is easier to understand why the first competitors to Valley's center-pivot systems didn't wait until the patent expired in 1969 to begin manufacturing their systems. A number of patent infringement suits against the early competitors resulted before the original patent expired. Most of these cases were decided in Valley's favor or were settled out of court.

Among the various competitors, Leo Sokol of Columbus was the first. He built a similarly designed center-pivot system. After a short time of very limited production and the threat of a patent infringement suit, he quit operations.

The first major longer term competitor was Layne and Bowler at LaPuente, California, which in 1959 introduced a center-pivot system marketed as "Rain-Cat." This really was a modification of a center-pivot system that had been developed in Australia and sold there under the name "Grasslands." It featured electric-drive motors instead of the water-drive system used by the Valley systems; it also featured "caterpillar"-style tracks at each tower. Instead of the cable-support system to hold the pipeline, it used an undertruss of bracing to support it between the towers. Later on, the track system was changed to rubber wheels with electric motors. Due to a limited market for center-pivot systems in California, a few years later this division was sold to the A.O. Smith Corporation. Two years later, it was sold to Gordon Johnson, who moved the manufacturing operations to Greeley, Colorado. The firm quit operating during the farm crisis of the early 1980s, when demand for pivots declined sharply.

The next competitor was Olson Brothers, headed by Ted and Carroll Olson, at Atkinson, Nebraska. In 1966 they began making a steel-wheeled system us-

"Circles of green" dot the plains in this aerial view of center-pivot irrigated corn north of Atkinson, Holt County. The first extensive use of center-pivot technology occurred near here. The first center-pivot system in Nebraska was installed near Page, Nebraska (Holt County), on the C.J. Getz farm.

Les Sheffield, UNL, via Nebraska Farmer

Les Sheffield, UNL

Center pivots irrigating pasture near Chase County in southwestern Nebraska

ing a cable support system very similar to that used by Valley, except they used an oil-hydraulic system instead of a water-drive to propel it. Later on, the Olson brothers also built center-pivot systems propelled by electric motors at each tower. The Olson brothers eventually separated, and at present each has his own small firm that makes a limited number of center-pivot systems near Atkinson.

Also in 1966, the EnResCo (Engineering Research Corporation) company of Colorado Springs began making a water-drive center-pivot system sold as ''HiGroMatic.'' It was a cable-supported system with rubber tires propelled by a water spinner connected to a gearbox in each tower and then by chains to the wheels. In 1975, Butler Manufacturing Company of Kansas City bought the rights to make and sell the ''HiGroMatic'' systems, keeping the same plant location. After a few years, with the demand for pivots low due to low grain prices, the plant was closed.

A firm in the Texas Panhandle by the name of Gifford-Hill, under a license agreement with Reinke Manufacturing, began making an electric-drive, undertruss center-pivot using roller chains connected to each wheel; it sold under the name ''Gifford-Hill 360.'' Later models offered an electric motor connected to a gearbox at each wheel. This firm became a major manufacturer in the industry, and in 1970 built a plant at Colby, Kansas. Due to the extremely competitive and cyclical demand in the center-pivot industry, quite a few firms closed operations in the early 1980s, including Gifford-Hill.

Also in 1966, Reinke Manufacturing Company, founded by Richard Reinke in the early 1950s, began making an electric-drive center-pivot system that featured several innovations. Marketed as ''Electrogator,'' it was the first reversible electric-drive system with rubber wheels. Reinke also was the first to develop and patent the ''bow-string'' undertruss system now used by most center-pivot makers, as well as the first to use an electric motor in the center of each tower connected by driveshafts to worm-gear drives at each wheel.

Reinke also was the first to develop an internal collector ring to transfer the electric power from the pivot point to each tower. This eliminated the need for wrapping and unwrapping an electrical cord at the pivot each time a system made a revolution. All totaled, Richard Reinke invented and patented more than 30 innovations for center-pivot systems. He was the first to develop a system made of aluminum alloy called the ''Alumigator,'' and he also was the first to develop short, one-to-two-tower ''Kwik-Tow'' systems that could be used on smaller acreages.

In the late 1960s, Leo Dowd, an oil jobber and automobile dealer who had lived at Schuyler, Nebraska, developed a cable-supported, oil-hydraulic-drive center-pivot system that used sled-like skids at each tower instead of wheels.

Center-pivot irrigation allowed for irrigating land with hilly terrain and sandy soils that previously was not irrigable, as shown here. In some instances, this capability allowed land to be broken out of sod that should not have been cropped, but many of these fields have been reseeded to grass.

Valmont Incorporated

Les Sheffield, UNL

A high-pressure, water-drive center-pivot system in Chase County, Nebraska

Dowd built a plant at O'Neill, Nebraska, where he sold his system as "Dowd Walking Water." In the mid-1970s he sold the rights to his system to the Farmhand Corporation at Minneapolis, which closed the division after a few years.

In 1968 Paul Geis, an automobile dealer at York, Nebraska, also began making a center-pivot system. Marketed as "Kroy" ("York" spelled backwards), it featured a cable-supported, steel-wheeled pivot with an air compressor at the pivot and used air-activated cylinders at each tower to move the system. In 1971 the firm offered a new undertruss design pivot. He later sold the rights to the system to a well driller at Sidney, Nebraska, but the firm quit business in the late 1970s. However, Kroy Industries still makes aluminum and polyvinyl chloride (PVC) tubing widely used in the irrigation industry.

In the late 1960s, Gus Thieszen, a well driller at Henderson, Nebraska, began building a center-pivot system under the name "Oasis," but he stopped making the systems after a few years. There were many individuals and small firms in Nebraska, as well as other states, that entered the center-pivot industry in the late 1960s and 1970s with their own concepts and ideas, but few were able to succeed in establishing a regional or national dealer network or in finding a market for their systems for more than a few years. As a result, a lot of "orphan" systems still exist, but the five major firms in Nebraska offer kits to convert the older systems to more modern technology.

While many of its competitors were building the more widely-accepted electric-drive, undertruss-supported systems, Valmont continued to build only a water-drive, cable-supported system until 1969. Early that year, Valmont bought a small competing firm at Grant, Nebraska, that was building an electric-drive, undertruss system marketed as "Walking 7." A farmer-rancher by the name of Hoyt Hendricks started the firm with two partners, Leslie Potts, a Grant doctor, and Walter Jones, a contractor who owned the large steel building in Grant where the systems were built. After buying the firm, Valmont operated the plant at Grant only a few months before moving its operations to Valley. They completely redesigned the system and now offer only an electric-drive, undertruss-supported center-pivot system, although they continued to make the water-drive systems for many years and at one time also made an oil-hydraulic-drive system.

One of the firms that had a humble beginning, but later grew very rapidly to become the second-leading firm in the industry was started by Paul Zimmerer and his two sons, Art and Bernie. After Paul began making tow-line systems in 1958, he was joined by Art and Bernie in 1963. In 1965 they formed Lindsay Manufacturing Company. In 1968, Art and Bernie began working on a new product line, an electric-drive center-pivot system with a number of new features. They were the first to use a "uni-knuckle" joint at each tower (others used a ball-joint at each tower). This allowed for maneuvering over rough terrain; they were also the first to develop an external collector ring at the pivot—in contrast to Reinke's internal ring—to transfer power to propel the system.

By the fall of 1969, the small firm had built and sold six center-pivot systems under the trade name "Zimmatic." Production and sales expanded rapidly each year as new facilities were added. Art concentrated on engineering and production, Bernie worked on developing a multi-state dealership, and Paul managed retail sales. By 1973, the firm had expanded such that it needed more capital to keep meeting the demand for their systems. In January of 1974, the Zimmerers sold their firm to DeKalb AgResearch at DeKalb, Illinois. Art and Bernie continued to manage the firm as a wholly-owned subsidiary and were on the board of DeKalb AgResearch. However, Art was killed in an auto accident on Aug. 16, 1978.

Lindsay Manufacturing continued to operate as a division within DeKalb AgResearch until Oct. 12, 1988, when Gary Parker, president of the division, arranged to separate the firm from DeKalb through a public over-the-counter stock offering. Today, Lindsay is the nation's export leader in center-pivot systems and is pressing Valmont for the "number one" position in the industry. Under Parker's leadership, the firm diversified its manufacturing operations and, among other things, builds equipment under contract for firms such as Deere and Company and Caterpillar.

T-L Irrigation at Hastings, who had been making different types of irrigation equipment, including gated-pipe, skid-tow systems and other farm equipment, entered the center-pivot industry in 1969. T-L Irrigation, with LeRoy Thom as founder and president, offered a system different from most others; it uses oil-hydraulic motors at each tower to propel it. In the 1970s, the T-L system was modified by using a planetary gear drive instead of a worm-gear drive with gearboxes at each wheel. This allows the system to move continuously, instead of the micro-switches used at each tower on electric-drive pivots that cause towers to start and stop frequently to keep the system in alignment.

In 1969, the Lockwood Corporation of Gering, Nebraska, bought a small firm at Dalhart, Texas, that had been making a center-pivot sold as "Hydro-Cycle." Founded by T.J. Lockwood of Gering in 1935 to make potato-farming equipment, the firm has had a solid reputation in the potato-equipment industry but has diversified into other equipment lines over the years. After buying the rights for the "Hydro-Cycle" system, the systems were called "Lockwood" and built at the Dalhart plant until operations moved to Gering in 1970. Later, they completely redesigned the system and now are one of the five major producers.

All in all, more than 80 individuals and companies have tried making and selling center-pivots. But most never lasted more than a few years because of the cyclical nature of the demand and the difficulties of maintaining a strong sales and service network. Those that have survived generally have diversified manufacturing lines that offset periods of slack demand for irrigation systems. Today only about seven companies in the country still make the systems year around. The top five firms in the industry that have survived all are located in Nebraska. They are: Valmont Industries, Inc., Valley: trade name "Valley"; Lindsay Manufacturing Co., Lindsay: trade name "Zimmatic"; Reinke Manufacturing Company, Deshler: trade name "Electrogator"; T-L Irrigation Company, Hastings: trade name "T-L Irrigation"; Lockwood Corporation, Gering: trade name "Lockwood."

Growth and Benefits

From 1970 to the early 1990s, by far the fastest growing method of sprinkler irrigation in the United States has been the center-pivot system. In Nebraska, an estimated 90 percent of the growth of irrigation from 1972 to 1982 was due to center-pivots. Their success and wide acceptance have been because of the labor and water savings and the ability to apply water to land with sandy soils and uneven terrain. With proper maintenance, the systems are highly automated and dependable. The irrigator can leave them to operate with very little supervision.

As a result of these benefits, many landowners have installed multiple systems over large areas. It is not uncommon for one person who can service and operate them to look after 10 to 15 systems or more. Since the standard pivot can irrigate about 130 acres of each quarter-section (160 acres), one person can supervise irrigation on 1,200 to 2,000 acres or more. Where no buildings or other obstructions interfere, systems up to a half-mile long can be installed, but these aren't as efficient as the quarter-section systems, so the demand for them has been very limited.

With gated-pipe irrigation, one person can irrigate about 320 to 480 acres effectively with many more hours of labor per acre. If gated-pipe irrigation is to be effective, row length shouldn't exceed a quarter mile, while most gravity-irrigated fields today are a half-mile long. Also, the irrigator needs to walk the rows regularly and adjust gate settings to regulate the water flow for uniformity of water applied. Over-irrigation in the first half of the run and under-irrigation in the last half is a serious problem with gravity irrigation. Without a re-use system, most gravity-irrigation averages 40-60 percent efficiency in uniform water application. By contrast, a center pivot with an electric pump is virtually a "push button" system, although the operator needs to exercise care and good judgment about when to use it. Good irrigation practices and a basic understanding of soils and water requirements for a given crop are vital to achieving the best yields.

Contrary to popular perception, a typical center-pivot system installed on land with suitable soils, using an efficient pumping plant and power unit, can achieve about 80-90 percent efficiency in water applied to the root zone. Many studies by UNL agricultural engineers and in other states have shown that center-pivot irrigation is one of the most efficient ways to apply water with a minimum lost to evaporation. Only drip irrigation is more efficient, but the cost per acre for this method is far more.

A number of other Nebraska firms not mentioned in this chapter have been involved at one time or another in making and selling center-pivots, and at least a couple of firms in Nebraska still make a few systems on demand, although they have no extended dealership. Two other firms, the Pierce Corporation at Eugene, Oregon, and the Tumac Corporation at Grand Junction, Colorado, still make and sell center-pivots but lack the national dealer network and size of operation typical of the largest five Nebraska firms.

Based on conversations with officials of the five largest Nebraska firms, it is estimated that they have about 90 percent of the nationwide and export market for center pivots. In 1990, one of these executives estimated that more than 100,000 center-pivot systems presently operate in the United States. In addition, they have been exported to more than 50 countries around the world.

During a visit to Brazil in 1988, I was surprised to find that 18 firms there made and sold center-pivots. The top five firms, most of them affiliated with U.S. firms, accounted for about 90 percent of the market there while the others were more minor players. At that time, an estimated 3,500 center pivots operated in Brazil; the demand for them varied widely from year to year, depending on prices for irrigated farm commodities and government subsidies. One major irrigation project I visited there had 45 pivots. This venture had joint financing from the government of Brazil and private Japanese interests.

Based on satellite imagery provided to UNL by the National Aeronautics and Space Administration (NASA) under a series of grants to study the growth of center-pivot irrigation from 1972-1988, it was possible to monitor the center-pivot boom for that period. James Drew, Donald Edwards, William Splinter and I, all of UNL, obtained the first of such grants, one of the first of any sort given by NASA after the launching of ERST-1 (later called "Landsat"), the first of NASA's land-observation satellites. Based on that data, the growth of center-pivot irrigation in Nebraska is shown in the table below.

Corner Arms

The first center pivot with an arm at the end of the pipeline to extend irrigation into field corners was developed by David Seckler at Fort Collins, Colorado, and David Parot, an engineer from Massachusetts. They obtained a patent

Growth of Center-Pivot Irrigation in Nebraska, 1972-1988

Year	No. of Pivots Observed	Increase from Previous Year	Percentage over Previous Year
1972	2,725	NA	NA
1973	3,876	1,151	42.2
1974	6,237	2,361	60.9
1975	8,903	2,666	42.7
1976	12,394	3,491	39.2
1977	14,651	2,257	18.2
1978	16,457	1,806	12.3
1979	17,393	936	5.7
1980	18,785	1,392	8.0
1981	21,261	2,476	13.2
1982	22,820	1,559	7.3
1983	24,410	1,590	7.0
1984	25,216	806	3.3
1985	26,208	992	3.9
1986	27,287	1,074	4.1
1987*	27,014	-268	-0.1
1988	27,617	603	2.2

*First year of USDA Payment-In-Kind Program that diverted acres from corn and feed grain production

Les Sheffield, UNL

A linear- (lateral-) move system in Fresno, California

on the corner-pivot system on April 9, 1974, and began to market it. In 1975 they sold the manufacturing and sales rights to Valmont Industries. The first corner-pivot system worked off a cam at the pivot point that provided guidance for the arm to extend and retract as it moved into the corner and back. In 1976 Valmont developed a more advanced corner-pivot that uses a buried low-voltage cable that the corner arm could follow into corners; it retracts by use of a signal sent to an antenna on the arm's tower.

Most of the major makers of center-pivot systems now offer a corner-pivot system. Most of the systems use a buried low-voltage line and antenna system. This allows the corner arm to retract to clear obstructions (roads, telephone or power lines, etc.). The corner-pivots irrigate another 20-24 acres over a quarter-section pivot. These became popular with farmers who had higher-value land and wanted to irrigate the 5-6 acres in each corner. However, in the Midwest, where feed grains such as corn and sorghum dominate, the federal farm program has allowed non-irrigated corners to be classified as "set-aside" acres. This provision has sharply cut demand for corner-pivot systems in the Great Plains.

Lateral-move Sprinkler Systems

The first lateral-move sprinkler system was invented by Bob Stafford of Eugene, Oregon, but he was never successful in marketing his product. One of the early lateral-move sprinkler systems was produced by R. M. Wade and Company at Portland, Oregon, in 1970 and sold as "Square-Matic." The linear-move sprinklers are similar to center-pivots but they move laterally across a rectangular field rather than in a circle. Some systems are a half-mile wide. Lateral-move systems have one main advantage over center pivots: they apply water uniformly from one end of the pipeline to the other. By contrast, a center pivot applies more water at the outer end of the pipeline. The area under the pipeline between the outer towers is far greater than that between the inner towers. Water applied declines as one moves inward. To compensate for the much smaller acreage covered toward the center, either the number of sprinklers or the size of the sprinkler hole decreases.

The principal disadvantage of lateral-move sprinklers is that when the system moves from one end of the field to the other, it must be stopped for several hours to allow the last water applied to soak into the soil or run back "dry" to the original position. Later, the system was changed to drag a high pressure hose instead of pumping water out of a ditch. This works well if the land is uneven.

The first "Square-Matic" lateral-move system developed by R. M. Wade and Company, mentioned above, was an automatic lateral-drive system powered by an air-cooled engine and moved by a winch and cable. A 660-foot hose that was 5 inches in diameter supplied the water, and the winch pulled the hose. The machine's application rate could be adjusted according to the soil type, the infiltration rate and the type of crop. The unit had a system of towers that suspended the pipeline with a cable support system. Tri-Matic, a firm in Texas, also developed a lateral-move sprinkler irrigation system; one of these was installed

Mobile Irrigator Planting System Would Do Nearly Everything

William M. Lyle is leading a study at Texas A and M University that could revolutionize the way crops are produced. Lyle, an agricultural engineer, and others are developing a mobile-irrigator planting system (MIPS) that would perform every function necessary to produce a crop. Basically a lateral-move irrigation system fitted with a row-crop planting device, the MIPS system could eventually replace the expensive tractors and multitude of implements used today. A lateral-move system is one which moves perpendicular to the rows directly across a field, rather than in a circle like a center-pivot.

The MIPS system rose out of a 1986 multifunction irrigation system (MFIS) study to enhance production through improved control of chemical spray. In that study, Lyle and others achieved continuous uniform movement of the lateral-move irrigation system, which provided excellent chemical-application results. The researchers then realized such uniform motion made the planting of seed possible. A seed-planting device was built and installed on the MFIS system, turning it into a MIPS system.

The seed-planting attachment forces water through tubes spaced at proper row-crop intervals. The forceful stream of water excavates a seed trench. One by one, seeds drop out of the water tube and onto a planting shoe, which places the seed into the furrow and then covers it with soil. Throughout the growing season, the irrigation system is used to apply water, pesticide and liquid fertilizer to the field. It can be used to fertilize the field before planting as well. The system cannot yet harvest the crop, but research will turn to that problem in the near future.

Initial research has shown that plant stands and yields using MIPS are comparable to those planted conventionally. The pre-plant and growing-season applications of fertilizer greatly increased yields. Corn, cotton and grain sorghum have all been planted and cultivated with the system.

—Brad Rundquist, weekend and copy editor,* Ft. Dodge (Iowa) Messenger, *and formerly CSD editorial assistant

on a farm north of Wahoo in the mid-1970s.

Lindsay Manufacturing Company introduced one of the first lateral-move sprinkler systems in June of 1977 when a prototype was installed on the Mike Mielack farm near Columbus, Nebraska. After successful operation there, they sold and installed their first commercial lateral-move system in October of 1977 to the Rosedale Ranch, owned by Tenneco in California. Their second lateral-move system was sold to the Heidrick Farm at Woodland near San Francisco. Their third lateral-move sprinkler system had a half-mile line on either side of the supply ditch; it was installed on the Glenn Kluck farm near Richland, Nebraska.

Valmont Industries, Inc. was next to introduce a linear-move system in November of 1977 at a demonstration in Arizona. Valmont later bought the rights to a very sophisticated lateral-move system that could connect and disconnect from pipe risers at regular intervals. The cost of these latter systems and risers has limited demand for them. Only about 20 of these systems have been sold in the United States.

Because of the equipment cost, a lateral-move system needs to have a long run (generally one mile or so) to spread the cost over more acres. In California, where many high-value crops are raised and land prices are much higher than the Great Plains, linear-move systems are more popular than center pivots. Center pivots still "reign supreme" in the Great Plains because of their much lower initial investment. Most of the major center-pivot manufacturers, including the top five in Nebraska, now offer lateral-move systems.

Surge Irrigation Systems

Surge irrigation was developed at Utah State University by Jack Keller and associates in the Department of Agricultural Engineering. This system is based on the principle of intermittent, rather than continuous, flow down the furrow. Surge systems have a computer programmed to control flow through a valve that opens or closes gates on one side or the other of a pipe. Most surge valves operate by solar power.

For a surge system to work effectively, a main pipeline must sit at the head of the field. A short connecting pipe is attached from the main water supply to gated pipe parallel to the main pipeline. The valve with the controller sits at the center of two lines of gated pipe with gates pre-set at the proper opening. In some cases, the gates are opened wider to allow greater flow than is normal without a surge system. After the water runs through the gates on one side of the valve for about 30 to 60 minutes, the controller switches flow to the other side for a similar period.

Since water running down the furrow after the first surge doesn't soak in as much, it advances more rapidly down the furrow than with continuous flow. The controller can be programmed for different periods. Toward the end of the time needed for water to reach the end of the rows, the controller may be set for shorter cycles of flows from one side of the valve to the other. Surge irrigation, if properly operated, achieves more uniform distribution of water and can save up to 30 percent of the water normally used for a gated-pipe system.

Surge systems have gained popularity among farmers with gated pipe, especially in the early 1980s in the Texas Panhandle. In addition, research by C. Dean Yonts and Dean E. Eisenhauer, of the UNL Department of Biological Systems Engineering, found only minor yield variations between surge and continuous-flow gated-pipe irrigation. Their paper reported test results from studies of surge systems in Buffalo, Clay, Scotts Bluff and Box Butte counties. A few different companies make surge controllers, and the number of farmers gaining experience with the method is increasing.

Cablegation Systems

"Cablegation" systems were developed in the late 1970s at the USDA Agricultural Research Service's Snake River Conservation and Research Center in Kimberly, Idaho. This unique method of water distribution was first introduced in Nebraska in the Imperial area by SCS. Other such demonstration units were installed at a few other locations, including a farm in Kearney County in the early 1980s.

With cablegation, water flow into the furrow is heavy at the start to force water down the row rapidly and then tapers off to provide only enough water for a good soaking without excessive runoff. With a cablegation system, the main pipeline is installed at the head of the field. The gated pipe is rotated upwards at about a 30-degree angle from the top of the pipe. All the gates are opened, but water is discharged only through a set number behind a moving "plug" that is slightly smaller than the diameter of the pipe. The pressure of the flowing water pushes the plug through the pipeline, and it is restrained by a cable hooked to a braking wheel at the upper end of the pipeline. Cablegation gets its name from this cable action.

SCS irrigation specialists in Nebraska say that a 75-80 percent water-application efficiency is possible with cablegation properly installed and operated, compared with 45-60 percent efficiency with conventional gated pipe. Installation of a re-use pit can increase the efficiency of a gated-pipe system to 75-85 percent. Few farmers in Nebraska or the Great Plains use these systems.

Solid-set Sprinklers with Risers

Solid-set sprinklers with risers have been used more widely in states like California with orchards, nurseries and specialty fruits and vegetables than in the Midwest where lower value field crops predominate. The mainlines and laterals generally are buried underground, usually below the frost line. Risers are located at suitable intervals, depending on the crop's height. Solid-set sprinklers with risers have been used for research at various locations by UNL since they

Buried solid-set sprinklers on irrigated pasture near North Platte

Les Sheffield, UNL

Pat McGrane, SCS

Various emitters used in drip irrigation of shelterbelts in Morrill County. These emit 1-2 gallons of water per hour.

lend themselves to smaller-plot irrigation research. While there have been a few installations of solid-set systems by farmers, these have been mostly experimental. Byron Sadle, owner/operator of a ranch, farm and feedlot at Paxton, Nebraska, installed a solid-set system in the late 1960s for irrigated pasture. He later abandoned it because of leaks that occurred in the buried pipeline and risers. At the time, the cost for such an installation was about $1,000 per acre, considerably more than some of the alternative systems, such as center-pivots.

One of the problems with solid-set sprinkler systems for field crops is that each riser needs a marker or protective device so that tractors and field equipment don't damage them during field operations. Very few farmers want to invest three to four times the cost of center pivots, for example, unless the crops produced have considerably greater value than the field crops grown in Nebraska.

Drip Irrigation Systems

Drip or trickle irrigation began commercially in the United States in the 1960s. But the first experiments with it occurred in Germany in 1860, according to *Sprinkler Irrigation,* by The Irrigation Association. Studies in Colorado in 1913 showed it was too expensive for practical use. But in 1920, an important breakthrough occurred in Germany when perforated pipe was introduced. Following World War II, the development of certain plastics made drip irrigation economical for high-value crops, as are grown in California, Arizona, Texas and Florida.

Israel, which used successful drip irrigation methods earlier and more widely than any other country, first reported research with drip irrigation in 1963. One reason the Israelis developed drip irrigation was the higher salt content of much of their water. By applying much less water but putting it right near the plant, they minimized the salt build-up of other methods. By 1992, thousands of acres were drip irrigated in California, Arizona, Florida, Texas and Hawaii. Israel, Australia, New Zealand, South Africa and several other countries also have extensive drip-irrigated acreages, especially where high-value crops are grown. In the last several years, this method has been used for ornamental plantings near buildings or in parks and other locations.

Drip irrigation delivers small amounts of water at frequent intervals as near the roots as possible, either on the surface or with buried lines. The tendency, however, has been toward surface installations, which can be checked and serviced readily. The main problems with this method are plugging of the emitter by salts in the water or soil and line breaks causing nonuniform water distribution.

Kansas State University is studying buried drip irrigation for field crops at the Garden City and Colby research and extension centers. While the costs for such installations at present are too high, these studies will help determine the relative efficiencies of water use of this method on field crops in the Great Plains.

A paper presented at the 1975 Drip Irrigation Conference and Trade Show on ''Water Scheduling with Drip Irrigation,'' by F.K. Aljibury and associates supplied the following information:

Approximately 150 acres of drip irrigation were installed in California in 1970, another 1,500 acres in 1971, 20,000 acres in 1972, 45,000 acres in 1973, 60,000 acres in 1974 and finally about 75,000 acres in 1975. Drip irrigation had its start in this country in southern California avocado and citrus orchards. Those installations were made primarily to help alleviate high water costs of over $35 per acre-foot and where excessive salinity were problems. Water costs in areas of the west side of the San Joaquin Valley in California are a major factor in the development of vast drip irrigation fields. Irrigation labor requirements and material costs are given serious attention by many westside farmers.

Les Sheffield, UNL

Drip irrigation for grape vines near Fresno, California

By the early 1990s, drip irrigation for row crops or alfalfa was not yet economical in Nebraska. In November of

1984 at the Paloma Ranch near Gila Bend, Arizona, 10,000 acres of roll-up drip irrigation lines were installed for irrigating cotton and wheat. That installation cost more than $27 million, or $2,700 per acre, not counting the wells and pumps. The system was abandoned three years later because costs were prohibitive.

Recent Technology Changes for Sprinkler Systems

Low-pressure Sprinklers or Spray Nozzles

"Gooseneck" drop tubes for low-pressure spray-nozzle packages were developed by Marion Miller at Colorado Springs, Colorado, to distribute water below the main pipeline so as to prevent icing during cold weather. They also helped cut evaporation. But after the energy crisis of 1973-1974, the low-pressure sprinkler packages or the even lower pressure spray-nozzle packages were in demand because they helped cut energy costs. Water-drive center pivots required about 80 to 85 psi at the pivot point. Electric-drive or oil-hydraulic drive pivots need about 50 to 60 psi with low-angle, impact-style sprinklers. Since each psi at the pivot equals lifting that water another 2.31 feet, when the price of diesel fuel rose to 36 cents a gallon in 1974 (and $1.25 a gallon by 1980), nozzles that cut pressure to 30 or 40 psi became quite attractive.

Les Sheffield, UNL

Center pivots with spray nozzles on the Merlyn Carlson Ranch near Lodgepole, Nebraska

Les Sheffield, UNL

A linear-move irrigation system with drop tubes near Yuma, Arizona

The principal disadvantage to these packages is that the wetted area under them is about a third to a fifth of that of higher-pressure sprinklers. So, unless the gallons pumped are cut sharply (which means pulling the pump out and changing the number of bowls or impellers), the system may apply more water per acre than the soil can absorb readily, resulting in greater runoff. Another disadvantage is that enough pressure to operate end guns effectively isn't available unless a booster pump is added at the last tower. This costs $1,800 or more per pivot and usually adds repair and maintenance costs as well. The limits to adopting the lower pressure sprinklers or spray-nozzle packages are type of soils (sandy compared with silt or clay-loam), the infiltration rate and the type of terrain. The finer textured the soils are and the more uneven the terrain, the greater the runoff.

Using low-pressure packages with all but very sandy soils can lead to severe runoff problems, which defeats the purpose of irrigating—*to get water into the root zone as uniformly as possible, where it is readily available to the crop*. This problem is acute with center-pivot systems on fine-textured soils, where the application rate at the outer end is the highest. Uneven distribution of water results in more runoff, especially on land with undulating terrain, and can cause lower yields from a low-pressure system, compared to a high-pressure system.

LEPA Sprinkler Systems

William Lyle, an agricultural engineer at the Texas A and M Research and Extension Center at Lubbock, began experimenting in the late 1970s with "drop tubes" on sprinklers that release water nearer the soil surface. He also tried drop tubes with trailing plastic sleeves to avoid spraying water into the air. Lyle was concerned about sharply declining groundwater tables in the Texas High Plains. He reasoned that the way to conserve groundwater was to decrease the water evaporating from small airborne droplets or off the plant. This method also reduced wind drift.

Lyle's research led to the development of his "low-energy, precise-application," or LEPA, system. His studies prompted some sprinkler companies to make LEPA nozzles for center-pivot or linear-move systems. However, each nozzle must be set by hand at present. LEPA nozzles operate in four ways: bubble mode—an umbrella-shaped water pattern with the drop tubes down between the rows; horizontal spray mode—water dispersed horizontally in a circular pattern; low-angle spray mode—water dispersed downward at about a 30-degree angle; upward spray mode—water dispersed upward at about a 45-degree angle (ideal for applying insecticide through the irrigation system, as it reaches the underside of the leaves).

Water is lost from rotating, impact-style sprinklers as very small airborne droplets evaporate (especially during hot, windy days with low humidity) and as water evaporates from plant and soil surfaces. Since LEPA nozzles drop into the crop canopy (sometimes 1 to 2 feet above the ground), evaporation of small droplets is minimized and far less plant surface is wetted. When the crop is planted in circular rows (as is common in the Texas Panhandle), drop tubes and LEPA nozzles can be placed between rows to improve distribution and curb evaporation and wind drift with winds more than 10 miles per hour.

However, unless the volume of water pumped through the system is cut considerably by changes to the pump, the LEPA nozzles generally result in a water-application rate that is very intense. This can cause serious runoff, especially on fine-textured soils and uneven terrain. Special tillage techniques, such as "diker-dammer" cultivators that make small basins at short intervals, have helped some make better use of LEPA systems.

Some companies have developed other techniques to minimize evaporation losses by using drop tubes with "wobbler" nozzles and spinners with low-presure nozzles to better spread water. While each of these systems helps cut evaporation relative to rotating sprinklers, they generally cost more to install because they need more drop tubes and nozzles compared with the sprinklers required with the overhead packages. But the biggest drawback to low-pressure packages is they can create a much less uniform water distribution, with resulting runoff, especially on fine-textured soils with uneven terrain. The sharp reduction in the wetted area under the sprinkler is critical in many cases, compared with the overhead rotating sprinklers.

Both the sellers of irrigation systems with low-pressure packages, as well as the buyers, need to know the water-intake capability of the soils and the terrain to be irrigated before deciding which sprinkler or low-pressure package to buy. It is expensive to switch systems. In some cases, irrigators may have to change the number and type of bowls or impellers in the pump to maintain pumping efficiency if they change the water-application package.

Soil Conservation Service

A soil probe is a cheap, easy way to determine available moisture

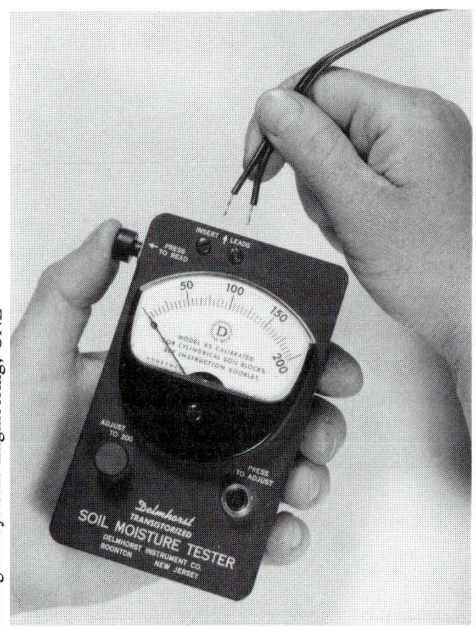

Biological Systems Engineering, UNL

A soil moisture meter

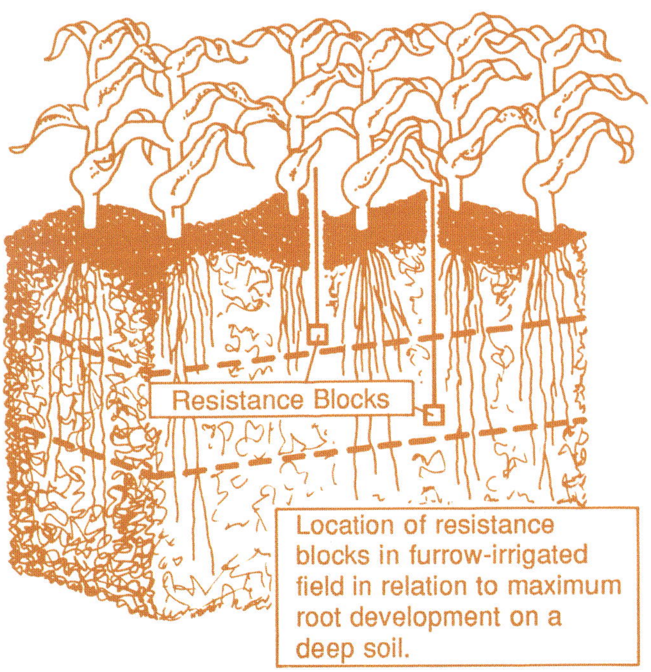

Soil Conservation Service

A drawing of resistance blocks to monitor soil moisture

Soil-Moisture Monitoring

One of the most critical management decisions an irrigator must make is when to irrigate and how much water to apply. Under-irrigation at critical times can result in a sharp drop in yields. Over-irrigation can result in deep-percolation water losses or a water-logged soil, leaving inadequate oxygen for crop roots. Also, applying too much water at the wrong times is not only wasteful of water but of the energy that provides the water and causes needless wear on equipment, besides using excess labor.

The actual water requirement of a given crop varies widely during the growing season. After germination and emergence of a crop like corn, the water requirement is low until the plant reaches about 3 feet. From then until the reproductive period (tasseling, pollination, ear and seed formation), the water requirement increases rapidly. Studies have shown that during the reproductive period, on hot days when the temperatures exceed 90 degree F and are accompanied by low humidity and winds more than 20 miles an hour, the daily evapotranspiration rate can reach as much as 0.4 of an inch of water a day. After the seed begins to mature, when the kernel is fully developed, the crop-water requirement for corn decreases, and more water will not increase yield.

Consequently, irrigators can limit water applied if they know how much moisture is at varying depths of the root zone. For centuries, irrigators used a spade to check soil moisture. With improvements in technology, farmers have used soil probes to sample deeper with less labor, and many learned to use the "feel" method of estimating the moisture in the soil.

More recently, various soil-moisture measuring devices have been developed to estimate available soil moisture. Among these were tensiometers (rods inserted to various depths) and gypsum electrical-resistance blocks, inserted with a soil probe at varying depths at regular intervals to be read with meters. Such devices take most of the guesswork out of the timing and amount of irrigation. Allowances also need to be made for unexpected rainfall during the irrigation season, so most of the precipitation percolates into the soil profile.

During the last 15-20 years, a number of consulting firms have offered services such as soil sampling with fertilizer recommendations, soil-moisture analysis with irrigation scheduling recommendations and insect monitoring to determine when it would pay to apply insecticides. It has been estimated that irrigation farmers in Nebraska have contracted with agricultural consulting firms for as much as 15 to 20 percent of the state's corn acreage. Fees charged by these firms vary according to the amount and frequency of services provided, but can range from $5 to $10 per acre. By saving one or two irrigations during the growing season or with a modest yield increase, the cost of such services can be paid for easily.

Remote Control of Center-pivot Systems

Several center-pivot firms offer remote control and monitoring of their systems, especially for multiple pivot installations. For example, Valmont offers the Computer-Assisted Management System, or "CAMS." The control box at the pivot is connected by radio to a remote location with a computer to control, monitor and record the operations of one or more pivots. Tied into CAMS is an alarm that goes off if the system shuts down so the irrigator can check on why it quit. This minimizes down-time during the irrigation season, critical during the period of highest water requirement.

Lindsay offers a similar system sold as the "Automated Irrigation Management System," or AIMS. The computer can be programmed for operation and monitoring of the system and keeps a record of its operation. The added cost is about $2,000 per pivot, and multiple installations can be hooked up to a remote office by radio. Some of the other major makers of automated irrigation systems offer similar systems.

A firm in Denver, Remote Devices International (RDI), headed by Gerald Abts, offers a very complete electronic monitoring and control system for multiple installations of pivot systems. Not only can a system be started or stopped from a remote location, but also the travel direction, the speed of operation and timing of fertilizer or pesticide injection can be programmed remotely. Included is an automatic alarm system to alert the irrigator if a system has quit. One large farm operator who grows potatoes and other crops near Alliance has installed this system for 34 center-pivots.

These kinds of systems could be very useful in REA districts trying to control the peak electrical demand during the summer. Such systems could allow REA to remotely shut off and start up center pivots assigned a lower rate through the irrigator's participation in an "electric-load" control program. By temporarily shutting off some of these systems when peak demand is rising, REA can reduce penalty fees based on highest peak demand during the non-irrigation season. (Those who allow REA to better control peak-demand power get a lower rate for their irrigation systems.) These systems allow irrigators to prioritize pivots for temporary shutdown.

Better Hybrids and Use of Ag Chemicals

Just after World War II, Nebraska had 873,960 irrigated acres, and the average irrigated corn yield in Nebraska was below 50 bushels an acre. Today, farmers of irrigated corn often get average yields of more than 200 bushels an acre. Much credit for this progress rests with the many studies conducted by agricultural experiment stations in various states on use of agricultural chemicals and with the breeding of new high-yielding hybrids by U.S. seed companies. Erect-leaf hybrids, for example, make more efficient use of sunlight, and new hybrids dry down faster, allowing farmers to begin harvesting earlier. The newer hybrids also drop fewer ears and have improved disease resistance.

Coupled with the development of vastly improved hybrid corn has been much greater use of commercial fertilizers, especially nitrogen, but also mixed fertilizers with essential trace elements where needed. These have caused dramatic increases in irrigated corn yields. Whereas, immediately after World War II, most fertilizer was packaged dry in bags, the development and widespread use of anhydrous ammonia (a liquid nitrogen compound), which provides the lowest cost per pound of available nitrogen, has contributed greatly to better yields and profits. Also, the development and use of liquid fertilizer has eliminated the labor required with dry bagged fertilizers. Today, most farmers apply liquid starter fertilizers at planting by using plastic saddle tanks, and thereby, use less by banding it near the seed. Both the agricultural experiment stations (in cooperation with USDA) and the agricultural chemical industry have made great progress in developing and testing various chemicals for improved crop production.

Weed control in row crops like corn also has undergone vast changes since World War II. The development of herbicides (like 2,4-D) and later, more sophisticated selective herbicides have allowed for far more effective weed control than was possible with cultivation alone. Instead of using as many as three cultivations to control weeds in corn, many farmers can use a single or no cultivation after planting by using wide-spectrum herbicides at planting or with post-emergence treatment. The field operations eliminated not only save fuel, labor and wear on equipment, but each cultivation saved means less soil moisture lost.

Since corn is the principal field crop that responds most readily to irrigation with properly timed water and fertilizer applications, as irrigated acres increased rapidly in Nebraska, the trend has been toward monocultural farming—corn after corn after corn. There are many irrigated fields in Nebraska where nothing but corn has been produced for more than 30 years. This trend has had some adverse effects, encouraging insect and disease problems. Corn rootworms, corn borers, red spider mites and other insects have built up large populations where only irrigated corn is grown. The widespread continued use of a single-spectrum insecticide to control a particular insect results in a selection process that increases resistant insect populations, an unfortunate side effect.

Changes in Irrigation and Some Problems

On-farm methods of delivering water to crops have changed more in the last 50 years than in all of the previous history of irrigation. The highly automated systems have eliminated much of the drudgery of applying water to crops that was commonplace for centuries. But not just the labor of irrigation has changed. The highly automated irrigation systems now apply far less water (saving as much as one-third to one-half of the water applied by gravity irrigation, while achieving equal or better yields). They also apply water more uniformly than was possible for centuries.

The net result not only saves a precious resource, but also improves yields with higher and more uniform quality of crops. While the irrigation farmer may be the primary beneficiary of improvements in irrigation technology, the consumer also realizes considerable benefits.

Never before have consumers in this country been offered such a wide array of high quality, nutritious foods. This includes products made from dryland and irrigated crops or meat produced from livestock fed those crops. Consumers also benefit from a more dependable food supply at lower cost than would be the case without modern irrigation technology. The USDA estimates that average consumers in the United States spend no more than 10 percent of their disposable income on food products. Improvements during the last 50 years leading to labor-savings and improved water-use efficiency have been truly marvelous developments, with few adverse impacts on society or the environment.

However, some problems result from almost any sort of development, as changes nearly always bring economic, sociological and environmental side effects. There is little doubt, for instance, that mechanized irrigation equipment favors the larger farm owners or operators, since multiple systems are feasible with resulting economies of scale. From this standpoint, the widespread development of multiple center pivots on larger holdings no doubt has helped hasten the demise of some of the smaller farms.

In addition, individual land owners and outside investors made mistakes by installing numerous pivots on land that was poorly suited to irrigation, especially during the inflationary 1970s. This was particularly true in the Nebraska Sand Hills, where large supplies of groundwater make such development tempting. Not only were some of the center pivots installed on choppy dune sands (after smoothing dunes with earth-moving equipment), but also many pivots were installed on wet meadows where the water table is very near the surface, especially in the spring, resulting in poor drainage and poor yields. In 1971, I saw one center pivot in the Sand Hills under which had been planted various crops five different times; in September less than 30 percent ground cover existed and sand was blowing. Few native Sand Hillers made these mistakes, as they were aware of the fragile soils and shallow groundwater.

During that same period, federal tax laws allowed for investment tax credits and rapid depreciation of investments (including irrigation equipment); also, interest on borrowed money for investment and operating expenses were deductible expenses. The 1986 Federal Tax Act eliminated most of these incentives for unwise investments. Also, the poor yields achieved on sites unsuitable for irrigation resulted in severe financial losses since the fixed and variable costs far exceeded the income generated—even with the tax incentives. Many outside owner/operators with no farming experience invested in center-pivot installations packaged as tax shelters. Most of them ended up with far greater losses than what the tax shelters saved them.

Most who made these errors either lost the land and the irrigation equipment through foreclosure or took advantage of the Conservation Reserve Program, which provided guaranteed government payments on an annual basis for 10 years. In the latter cases, taxpayers ended up the losers, not only from the tax incentives for poor investments but from subsidies provided through the Conservation Reserve on land that never should have been broken out of grass in the first place.

One legitimate concern with increased use of agricultural chemicals under irrigation is the risk these chemicals or their residues pose if they percolate to the groundwater or run off to nearby surface water. At present, especially in areas in the fertile Platte Valley, nitrate-nitrogen levels in the groundwater exceed the federal maximum of 10 parts per million. In such cases, for health reasons, many farmers and people in small towns and cities are using bottled water.

It behooves every irrigator to use the utmost care in choosing agricultural chemicals and to keep the rate of application as low as possible while still achieving good yields. After all, the irrigator and his family, as well as any farm livestock, depend in most cases on the quality of groundwater available in the domestic and stock wells nearby. It is in the irrigators' best interests to use all available management skills to minimize the risk of pollution of surface water and groundwater.

Sources: *Beyond the Hundredth Meridian,* Wallace Stegner, 1982, University of Nebraska Press, Lincoln and London; "Cablegation" (booklet), USDA Soil Conservation Service, 1984; "Center-Pivot Irrigation," William E. Splinter, *Scientific American,* June 1976; "Development of Irrigation Well Drilling in Nebraska," Vance Anderson, (retired vice president, Western Land Roller Co.), 1981, Western Land Roller Co., Hastings, Nebraska; "An Early History of Center-Pivot Irrigation," Ralph Jones, undated, unpublished, Lockwood Corp., Gering, Nebraska; "Economic Analysis of the Costs and Returns from the Production of Corn Using Center-Pivot Irrigation Systems, Southwest Nebraska, 1970," Leslie F. Sheffield, 1971, unpublished Ph.D. dissertation, Department of Agricultural Economics, University of Nebraska-Lincoln; "The Economic Impact of Irrigated Agriculture," Leslie F. Sheffield, 1985, The Irrigation Association Educational Foundation, Arlington, Virginia, Sept., 1985; *Land of the Underground Rain,* Donald E. Green, 1970, University of Texas Press, Austin; *Irrigation Handbook and Directory,* Irrigation Data, Inc., 1972, Lincoln, Nebraska; *Irrigation Pioneers: A History of the Tri-County Project To 1935,* Gene E. Hamaker, 1964, Warp Publishing Co., Minden, Nebraska; "Nebraska Agricultural Statistics Historical Record, 1866-1954", and annual statistical reports for 1947-1950, Nebraska Agricultural Statistics Service, Lincoln, Nebraska; "The Pathfinder Dam on the North Platte River" (booklet), U.S. Bureau of Reclamation, undated; *Sprinkler Irrigation,* Claude H. Pair, Walter W. Hinz, Crawford Reid, Kenneth R. Frost, 1975, 4th ed., The Irrigation Association, Arlington, Virginia; Various papers and articles by William Lyle, Texas A and M Agricultural Research and Extension Center, Lubbock, Texas; "Volume-Gun Sprinklers," John W. Addink, presentation at 1974 Nebraska Irrigation Short Course, Lincoln, Nebraska; "Water Scheduling with Drip Irrigation," Falih Aljibury and others, 1975, Drip Irrigation Conference proceedings, University of California-Davis.

Interviews with: Gerald L. Abts, president, Remote Devices International (RDI) Denver, Colorado; John Chapman, agricultural engineer, Valmont Industries, Inc.; Robert B. Daugherty, chairman, Valmont Industries, Inc., Omaha; Ralph Jones, Dean Edwards, Charles Johannsen, Lockwood Corp., Gering, Nebraska; William Lyle, agricultural engineer, Texas A and M University Agricultural Research and Extension Center, Lubbock, Texas; Charles Meis, vice president-engineering, and Robert Snoozy, vice president-marketing, Lindsay Manufacturing Co., Lindsay, Nebraska; Marion Miller, Marion Miller and Associates, Colorado Springs, Colorado; Richard Reinke, president, Reinke Manufacturing Co., Deshler, Nebraska; Bill Scheinost, O'Neill, Nebraska; LeRoy Thom, president, T-L Irrigation Company, Hastings; Frank Zybach, A. E. Trowbridge, on July 30, 1970 (both deceased in 1980).

A Brief History of Early Well-drilling

by Vincent H. Dreeszen
Hydrogelogist Emeritus
and Formerly Director, CSD
and Leslie F. Sheffield
Coordinator of Outreach Programs, Water Center
and Formerly Extension Irrigation Economics Specialist, UNL

The source of water for the first settlers in Nebraska was from springs and streams. But it was not until better drilling equipment and pumps allowed people to tap large groundwater supplies that living on the High Plains became more profitable and practical for farm families. Wells were first dug by hand along river valleys, particularly the Platte, where the water table was the shallowest. Digging these first wells was fraught with danger. The sides were apt to cave in because the wells were often dug through unstable soils. Later, so-called "spudder" drilling rigs were developed. They punched a relatively small hole by raising and dropping a heavy steel bit into the ground. However, the rigs were unsuited to drilling larger wells because larger bits required much larger, heavier and more expensive rigs, which were less mobile.

One particular type of well that was of great interest to the scientists and settlers alike was the "artesian" well. (Artesian wells are wells where water rises higher in a water-level monitoring device than the existing water table due to natural pressure by a confining layer in the aquifer. They sometimes flow at the surface.) These wells are common in Nebraska. However, flowing wells occur almost exclusively along stream valleys and terraces. The interest stemmed from the convenience of a steady supply of water and the scarcity of reliable pumps and sources of power for pumping. Flowing wells have been drilled in every county in Nebraska, some several hundred feet deep and others only a few tens of feet in depth. Some, as at Lincoln, spawned a health spa industry in the late 1800s. Others were used as a source of power for milling while most were the source of water for domestic and stock use.

The artesian-well boom in the Dakotas in the 1880s and 1890s bred a new generation of well drillers. As discussed by Donald Green in *Land of the Underground Rain*, Peter Norbeck, who later became a South Dakota governor and senator, made a fortune drilling small-bore artesian wells (1 to 3 inches in diameter). Norbeck greatly improved the "jetting" method of drilling wells, which lowered the cost of such wells. ("Jetting" means using water under pressure through the drill bit to help break loose and remove the soil particles as a well was being drilled.)

Sources of water for flowing wells include: sands and gravel of Pleistocene and late Tertiary age in most groundwater regions of the state (*see map of underground water regions and hydrostratigraphic table in "Introduction"*); the Chadron sands in groundwater regions 6, 7 and 13 (and only rarely in groundwater region 12); sandstones of the Dakota Group in groundwater regions 10, 11 and 13; and Paleozoic limestones, dolomites or sandstones in groundwater regions 10 and 11. Two well-known "artesian basins" were at Cook in Johnson County and Beaver Crossing in Seward County. The early interest in "artesian water" is evidenced by the enactment of laws by the state legislature in 1897 to prohibit waste from flowing wells and to provide for penalties if wasting occurs.

Most of the flowing wells were drilled during the late 1800s to the 1920s, and many have been abandoned due to reduction in flow as the pressure dropped or the well failed. However, there are many flowing wells drilled in recent years in groundwater region 1.

Methods of constructing wells in Nebraska have changed over the years for several reasons. Cost of obtaining water is one factor. A closely related factor is that of the development and availability of manufactured machinery to "dig" the hole in the ground and advancements in drilling technology. The types of materials to case and screen wells have also evolved as new technology and materials became available.

Most private wells and many municipal and industrial wells prior to 1900 were dug either by hand or, in the case of large-capacity wells, with digging machinery. Early settlement concentrated in the major stream valleys, where the depth to water was shallow. Private wells usually 3 to 6 feet in diameter were dug by the landowner with help or by a local "well digger." Commonly these wells were only a few tens of feet in depth, although as settlement moved across the land, some wells in the uplands were hand-dug to depths more than 300 feet. Because of caving problems, once the water table was reached, the wells were usually dug only a few feet below the water level and so were drought-prone.

The dug well was lined with available local materials. Limestone was commonly used in southeastern Nebraska and nearby areas. Brick was also used, and where limestone and brick were not readily available, wood cribbing was substituted. Large 10 to 20 foot (or more)

C. Layton and his mother on Cliff Table putting down a well, 1889. These wells were from 350 to 400 feet deep, according to the caption from the Nebraska State Historical Society. Without a well, water had to be hand carried many miles from a river or stored in a small reservoir.

Nebraska State Historical Society

Conservation and Survey

Harry Burleigh (left) pumps water into one of two nearby mud pits. Miller (first name unknown) stands beside the rotary-hydraulic Dempster drill rig. In the mud pit, clay is mixed with water to form a drilling mud that is pumped down the drill stem. This solution hardens and acts as a casing to prevent cave-ins. Cuttings settle out in the second pit and are laid on the table for closer inspection.

diameter wells were dug as the supply for railroad engine boilers and other industrial and municipal wells. Wells often were left open at the top or were covered with loose planking. They were easily polluted by surface wash, wind-blown material or by small animals and insects.

Mahlon E. Layne, another driller from South Dakota, developed an approach that made it possible to drill non-artesian wells of large diameter, the kind later suited to the "pit-less" pumps he developed in 1903. Layne began drilling in 1882 near Hurley, South Dakoka. In 1886, he began drilling larger wells—18 to 36 inches in diameter—using a large auger turned by a horse. In 1887, he developed a unique well screen placed at the bottom of the well casing. As word of his success spread, he drilled large-diameter wells for farmers, towns and railroads in South Dakota, Nebraska, Minnesota, Wisconsin and Iowa. He also built improved rigs powered by as many as six horses but converted to steam power in 1896. Steam turned the auger faster and more efficiently. Similar rotary well-drilling rigs had been developed by oil-field drillers in Louisiana and Texas.

Water was obtained from the private wells by rope and bucket, by shallow-well suction pumps or by cylinder pumps. Suction pumps such as the pitcher pump or centrifugal pumps depended upon atmospheric pressure to force water from the source of supply to the pump. Practical lifts for a pitcher pump in Nebraska are about 22 to 25 feet and somewhat less for a centrifugal pump. For the large-capacity wells, the centrifugal pump was used and often was placed on a platform in the well. Homemade windmills were a source of power for many domestic and stock wells and even to irrigate small plots of land. Windmills may have been manufactured commercially as early as the late 1870s in Nebraska.

The fate of many of the dug wells is uncertain. Some have been reported as open or covered holes on farmsteads. Large diameter wells probably were dug in most cities, villages and at many railroad stops. Many of the wells probably have been filled with debris and soil as they were abandoned or started to collapse. Possibly due to settling of the fill some sites may be marked by depressions on the ground.

In Nebraska very few wells were "hand dug" after 1930, but a number of private dug wells are still in use and many of these have been reconstructed to meet sanitary standards. Geologic conditions were most favorable for construction of dug wells in areas of shallow depth to bedrock or glacial till. The largest number of wells dug by hand were in groundwater region 11 and to a lesser extent in groundwater regions 9, 10, 12 and 13.

Other early types of wells that were constructed include the bored, driven, jetted and cable-tool (percussion) drilled well. The method of boring or augering wells moved west with the settlers. Very few wells of this type were constructed after the late 1940s. Wells were bored in soft unconsolidated materials by means of an auger turned by hand, horse and later gasoline power. The size of the hole bored ranged in diameter from a few inches to as many as 48 inches. The large-diameter augered well became in effect a replacement for the old dug well. Silt, clay and sand were removed from the hole as the bit was pulled and cleaned. Sand below the water table was removed by a bailer.

Bored or augered wells were constructed in many parts of the state. The wells were cased with various materials including concrete or clay tile, and may or may not have been screened depending upon the casing used. A common type of casing used in Nebraska was clay tile. One end of the tile was oversized to fit on the next joint. The joints were not water tight and water moved into the well through the joints and the bottom of the well, which was not plugged. Some wells were cased with screwed or welded joints and were screened for large capacity wells, but for most domestic and stock wells, a drop pipe with a well screen was installed. Probably because of cost, at times thin-walled stove-pipe casing was used. This casing, too, was not tight and tended to rust very quickly. Water was pumped (if the well was shallow) by a suction pump, or if the water level was deep, by means of a plunger or cylinder (tubular well). Windmills were often the power source for stock and domestic wells.

A driven well was and still is a method of constructing a shallow small diameter well. The "drive point," a perforated pipe pointed at one end, is driven into the ground. Pipe is added as the well is deepened. The pipe constitutes the casing. The type of pump used depends upon the diameter of the pipe. Pitcher pumps commonly are used.

The jetting method of well construction does not require drilling equipment but does require a pump and a water source. The equipment for jetting consists of an inner tube, the drilling or jetting tube, and an outer tube, which is the well casing. A pump with suitable hose attachments supplies continuous water pressure during drilling. Casing is usually sunk as drilling proceeds. The method of jetting wells is best suited to soft sandy formations and small capacity wells.

Other than the horse-powered drilling equipment used for boring wells, the earliest drilling equipment used in the state was the cable-tool drilling rig powered by steam engine. Private, industrial and municipal wells were drilled by cable-tool rigs. These early rigs (also called churn drills or spudders) had wooden framed masts and other wooden parts. Drillers and manufacturers adapted the equipment for jetting or a combination of jetting and percussion and converted from wood parts to iron and steel. Once the gasoline-powered engine and pumps became available, hydraulic rotary methods followed.

The size of the cable-tool holes drilled ranged in diameter from a few inches to several feet. Cuttings were removed with a bailer, and iron or steel pipe screwed, welded or pinned was driven as the drilling progressed. In wells more than 100 to 200 feet in depth, the diameter of the hole and casing was reduced with depth. Manufactured screen was installed in loose sandy formations, but the hole often was not cased or screened in the water-yielding section of consolidated rock such as limestone, dolomite, sandstone or siltstone. Areas where early wells in bedrock were left open include: groundwater region 11, wells in Permian or Pennsylvanian limestone, sandstone or siltstone or Dakota sandstone; region 10, Paleozoic limestone, dolomites or sandstone largely in the Omaha area, or Dakota sandstone and Niobrara chalk in Knox and Cedar counties; and in regions 6, 7, and 12, bedrock of Tertiary age (often called "red rock" in region 12). Since the late 1940s screen usually has been used in the bedrock wells.

The early cable tool rigs operating in Nebraska were not truck mounted but were moved and set up at each site. Later portable cable-tool rigs were the dominant rig for drilling wells in the late 1800s and early 1900s. Cable-tool rigs gradually were replaced by hydraulic rotary equipment, and few cable-tool rigs were in use in Nebraska by the late 1940s.

Nebraska was well adapted to use of hydraulic rotary equipment because much of the state has extensive deposits of unconsolidated clays, silts, sands and gravels or loosely consolidated sandstones such as the Dakota Formation (in eastern Nebraska) and the Ogallala Group (in central and western Nebraska). Hydraulic rotary equipment was trailer- or truck-mounted, and the power source was the internal combustion engine.

The portable hydraulic rotary drilling rig became the dominant drilling rig in Nebraska for drilling domestic and stock wells and for drilling test holes to determine the siting of large capacity wells. Auger rigs are still commonly used for foundation investigation.

The hydraulic rotary method of well drilling apparently originated in the oil fields of Louisiana as early as the 1890s. A bit is rotated mechanically on a string of pipe, and water with drilling fluid additives is pumped down the pipe, out of holes in the bit and up through the hole to the surface, where it is recirculated. The drilling fluid carries the cuttings from the hole. The sides of the drilled hole are kept from collapsing or sloughing by the pressure of the column of drilling fluid maintained in the hole. The hole remains open to permit the placement of casing and screen.

The principal advantage of the hydraulic rotary rig over other types of rigs drilling 4- to 12-inch wells is speed. Completion of a well drilled by cable tool often took one to several weeks; the well drilled with hydraulic rotary equipment could be completed in one or two days. Compressed air added to the fluid stream has facilitated the drilling of large diameter and deep wells. Another advantage of the hydraulic rotary drilled hole is that the uncased hole can be maintained in an open-hole condition to permit geophysical logging.

Somewhat primitive hydraulic rotary equipment, made either by adapting a cable-tool rig or by manufacturing one, began to be used in Nebraska in the early 1900s. The equipment probably was used to drill mostly 3- to 6-inch holes, primarily for domestic and stock use. Galvanized metal and iron pipe (also "stove pipe") casing was used. Slotted pipe or manufactured screen or sand points were screwed or welded to the casing.

With the added capability of heavier machinery and larger "mud" pumps, large diameter holes could be drilled with the hydraulic rotary rig. In the late 1940s a practice began of drilling holes 2- to 14-inches larger than the casing to be installed. The space between the screen and casing and the walls of the drilled hole was filled through the water-bearing section and often to or near the surface. Wells constructed in this manner could take advantage of more than one water-bearing interval. The "gravel pack," "gravel envelope" or "formation stabilizer" served, too, as an extension of the screen vertically and laterally. If the graded "gravel pack" is carefully selected and placed, the tendency for a well to pump sand can be reduced.

Two potential problems have emerged because of this practice. One is that if an inadequate grout or no grout is placed in the upper 10 or more feet of the hole, an avenue is created for contaminants to move from the surface downward. A second problem occurs, particularly in the eastern part of the state (*groundwater regions 10 and 11*), if the gravel pack is brought up the hole into a "perched water" zone or an upper formation that is contaminated, water from the upper zone, although small in quantity, can trickle downward.

A major change in water-well construction came with the availability of various types of plastic tubing. Because of the light weight and relatively low cost compared to black and galvanized pipe, drillers in the mid-1960s began installing pipe not intended or acceptable for potable water wells. Plastic pipe manufactured for use as sewer and drain pipe was used as well casing for stock and perhaps some domestic wells.

As the industry became aware of its potential, plastic casing and screen were manufactured from polyvinyl chloride (PVC). Standards for plastic slowly evolved during the 1970s, and plastic now is the dominant material used as well casing for domestic, stock, many industrial and municipal wells and some irrigation wells. Manufactured and shop slotted plastic pipe are used as screens or, if more open area is needed, machined or "wire wrapped" slotted screens are installed. Lengths of casing and screen are either screwed or cemented to provide water-tight joints.

The development of the "pitless adapter" in the early 1960s, and later the pitless unit, were also major technological advancements in supplying domestic water needs. They attach directly to the casing and provide water-tight subsurface connections for suction lines

Miller (first name unknown), Harry Burleigh and A.L. Lugn (left to right) lower the well screen and casing into a hole drilled by the old Dempster rig.

Conservation and Survey

or pump discharge without the use of a pit or well house. Some wells were constructed in basements. The pitless unit, if properly installed, provides a sanitary connection from the well to the pressure tank and eliminated the problem of frozen pipes in the winter. Undoubtedly some of the earlier types of installation are being used for water supplies.

The average depth of wells has gradually increased through the years as drilling and pump technology evolved, as electricity became available to the rural user, as the demands for larger capacity wells increased and as knowledge about the presence and characteristics of deep-lying aquifers increased. Most domestic and stock wells were dug or drilled only a few feet below the water table or a few feet into the aquifer prior to World War II. Now many wells for even stock and domestic purposes are drilled through the water-bearing section. The deep wells have the advantage of being less susceptible to contamination and to interference from other wells as water levels lower during the summer months. Because of the latter, well drillers and pump installers now usually set the pump well below anticipated summer pumping water levels in the area.

Pumps, too, evolved from the days of the bucket lowered into the well. Pitcher, cylinder, centrifugal and, later, vertical centrifugal, turbine, jet and submersible pumps all have been used.

Except for some use of the pitcher pump today, and the cylinder pump for stock wells, most small capacity wells are equipped with submersible pumps (pump and motor are lowered into the well). Large capacity wells of several hundred gallons per minute are equipped with turbine pumps or in some cases with submersible pumps.

Large capacity wells were hand dug in the late 1800s and early 1900s. Later this type of well was constructed with the cable tool rig or an adaptation of the cable tool rig to include sand bucketing.

An outer well casing (blind casing) was sunk or driven as the well was deepened. Later a smaller casing, some type of screen and gravel was placed as the blind casing was removed. Black or galvanized pipe, usually welded, was used for the permanent casing. Manufactured screen was used for some municipal and industrial wells, but slots were cut by saw or torch into the casing to act as screen in irrigation wells. Some of the earliest irrigation wells were cased with wood, and water entered the well between slats. Because sand or gravel would come into the well from an open bottom, the bottom of the well was sealed inside the casing or screen by some kind of plug, usually concrete. Wells ranged in diameter size from 10 to 24 inches.

Although some large-capacity wells were drilled with the hydraulic rotary rig, the size of hole and hence pump was generally restricted to a diameter of 6 to 10 inches by the capacity of the drilling rig. The drilling of large capacity wells was time consuming and relatively expensive until the development of the reverse hydraulic rotary rig.

Around the end of the 1930s and the beginning of the 1940s, the reverse-rotary hydraulic drilling rig was developed in two places at about the same time, and it became a force that revolutionized the industry. While the details are sketchy, John B. Alfs, a well driller from Shickley, Nebraska, and Meder Smith, of Alliance, both pioneered this technique. Whether they knew of the other's work is not clear, but the new technology made it possible to drill deeper wells in far less time. Today, a crew can set up a modern reverse-rotary rig, drill a 36-inch well as deep as 500 feet, install 18-inch steel casing, put gravel packing around the outside of the casing, test-pump the well and be gone in 24 to 36 hours.

Drilling is accomplished in reverse rotary by maintaining a hole full of water from a large sump; the water moves down the hole and is pumped up through the drill stem along with the cuttings. Large capacity pumps and large diameter drill stems (to accommodate boulders of several inches in diameter) are required.

Irrigation-well drilling quickly moved from the river valleys to the uplands with the successes of reverse rotary drilling. Vance Anderson, in *Development of Irrigation Well Drilling in Nebraska*, states that in the mid-1950s, "Winter-Weiss of Denver improved their reverse rotary rig by putting air on it. This was done by running a pipe along the side of the drill stem and then running it into the drill stem a few feet above the bit. The air coming up the drill stem proved to be a big help in bringing sand and gravel and cuttings up through the pipe. This made it easier to drill deeper . . . you could drill as fast at 500 to 600 feet as you could down to the first 200. And so by 1957, most of the drillers had either added air to their old rigs or bought new Winter-Weiss rigs with air attachment in order to compete. To drill a 400-foot well, to install the casing and gravel pack in one day was not uncommon."

The reverse hydraulic rotary rig, ideally suited for the drilling and construction of large capacity wells in Nebraska, revolutionized the industry. Nearly all irrigation wells and most industrial and municipal wells are drilled with the reverse rotary method. Contractors, mostly established firms from Nebraska, drilled an average of about 500 irrigation wells per year in the 13-year period 1940 to 1952. Along with drought conditions of the mid-1950s came an influx of new drillers, mostly without previous experience. About 13,000 irrigation wells were drilled during the next 5 years. Many wells were commonly "gravel-packed" to the surface. Thickness of gravel pack ranged from about 6 to 9 inches. Spot-welded galvanized steel or concrete casing was used. Large slots were provided in relatively long sections of pipe as screen. The "gravel pack" may or may not have been selected to match the formation and the screen slot size. Sand pumping was common and, particularly in groundwater region 6, it was the practice to add gravel pack as it subsided. Some wells were not drilled plumb, causing pump column problems; other wells collapsed. Many of the wells drilled during this period and for the next five years or more were not drilled deep enough to take advantage of the aquifer. Very likely many of the irrigation wells drilled prior to the late 1960s have been replaced or are little used or unused. Well registration with the Nebraska Department of Water Resources, required in 1957 for municipal, irrigation and industrial wells, provides information about location, year drilled, status of well and some information about construction.

In the early 1940s, Henningson Consulting Engineers of Omaha contacted the Conservation and Survey Division for assistance in helping Nebraska City solve its water-quality problems. The city pumped water from the Missouri River, and drilling was needed on a Missouri River island. Here, a barge moves drilling equipment and crew to the island.

Profiles:

Ralph O. Canaday: Central's Chief Counsel and, Later, Manager

Ralph Canaday devoted most of his career to Nebraska and the public power and irrigation concept. Born in Minden, Nebraska, in 1891, Canaday graduated from the University of Nebraska College of Law in 1915 and practiced law in Bridgeport until 1931, when he moved to Hastings.

Canaday was one of the originators of the Central Nebraska Public Power and Irrigation District, which he served until his retirement in 1959. He became interested in the project early in life as his father, Joseph S. Canaday, was an early supporter. Canaday served as attorney for the promotional organization prior to the district's organization and became its chief counsel and secretary when it was chartered in 1933.

He continued to serve in these capacities until his retirement. In addition he served as manager of the project from 1950 to 1957, and as general manager of the Hydro and Irrigaton Division from 1957 to 1959.

Upon his retirement, Canaday received a plaque which read, in part, "To Ralph O. Canaday in recognition of his service to the people of the State of Nebraska in the planning, organization and management of the Central Nebraska Public Power and Irrigation District, which he served with devotion, determination and high legal and managerial ability and integrity during its first 25 years."

The Canaday Steam Plant south of Lexington was built at the request of Nebraska rural electric organizations and the Nebraska Public Power System. The plant was named in recognition of Canaday's service to the district and his effective efforts to get the plant financed and built.

—Karen Stork, CSD administrative assistant

George E. Johnson: Designing Irrigation and Public Power Facilities

Born in Wymore, Nebraska, in 1885, George Johnson devoted his life to public service. He had many interests, but probably foremost was trying to improve life in Nebraska through the development of public power and irrigation systems.

At the age of 21, Johnson received degrees in both electrical and civil engineering and began his career designing and constructing electrical power plants and distribution systems for towns in Missouri, Kansas and Nebraska. He served as state engineer for Nebraska from 1915 to 1923 and laid out a system of state highways. From 1919 to 1923 he was a member and secretary of the State Capitol Commission, which oversaw the building of the State Capitol Building.

In 1926 he learned to fly. He bought and operated the aviation school where Charles Lindbergh learned to fly and became the first Nebraska businessman to fly his own airplane. He was also a pioneer in the early days of radio and in using Nebraska farm products to make industrial alcohol.

Still, much of Johnson's energy and drive went into the promotion and development of the Tri-County irrigation project and the Nebraska Public Power System. He helped promote the Tri-County project from 1915 until it was approved in 1935. In *Public Power in Nebraska*, Robert E. Firth noted: ". . . No man is more important in the history of public power development in Nebraska than George E. Johnson . . . as Chief Engineer and General Manager of the project, he supervised its complete design and construction, including Kingsley Dam, the system of canals and reservoirs, and the three hydroelectric plants which generate the district's energy." Johnson was appointed Chief Engineer and General Manager of the Tri-County Project on Sept. 1, 1935, and served until he left to work in Argentina in 1946.

In 1955 when Johnson was more than 70 years old, he was put in charge of designing and constructing the Canaday Steam Plant near Lexington, as well as training the staff and putting the plant in operation.

Johnson's views were expressed in a formal report to the government of Argentina upon the completion of his work there: "What happens to the land, the soil, the water and the minerals within the earth determines what happens to its people. It is upon these resources that men and nations must build. These are the foundations upon which our hopes and dreams for a future of prosperity and security are based."

—Karen Stork, CSD administrative assistant

Outlet works under construction, Kingsley Dam, Oct. 19, 1937

Department of Water Resources

Otto H. Liebers: Spearheaded the Salt Valley Watershed Project

Called the "father of the Salt Valley Watershed," Otto H. Liebers was born near Minden in Kearney County, Nebraska, in 1887. He attended the University of Nebraska College of Agriculture and graduated at the top of his class in 1913. That same year he became Nebraska's first county agricultural agent, serving in Gage County. That was the beginning of Liebers' life-long involvement in Nebraska agriculture and soil and water conservation. In 1961 he was cited by the Nebraska Hall of Agricultural Achievement for "outstanding contributions to agriculture in the Cornhusker State."

In 1919, Liebers became organizer and manager of the Nebraska Dairy Development Society, and with others was instrumental in bringing purebred dairy cattle into the state. He served as an honorary director of the Nebraska State Guernsey Breeders Association and as director of the American Guernsey Cattle Club. Until 1967, Liebers was president of Skyline Farms Dairy of Lincoln and was associated with the dairy from its inception in 1926.

Liebers served as state senator from the 18th District for four terms from 1951 to 1959 and was a member of the budget committee and chairman of the tax study committee. In 1953 he introduced the Watershed District Enabling Act (LB 99) to the legislature. This act was approved on May 1, 1953. Called the Watershed Act of Nebraska, this was the first significant watershed legislation passed in Nebraska. The Salt Valley Watershed was one of those formed under this legislation.

His leadership in forming the Salt Valley Watershed, the flood control, conservation and recreation project of many small reservoirs that surround Lincoln, is usually cited as his most outstanding achievement. Liebers was named president of the Watershed District in 1960. He resigned as president in 1968 and was elected to the office of president emeritus.

In addition to his local leadership, Liebers also chaired the Governor's Advisory Council on Watershed and Flood Control and took an active interest in federal watershed legislation in the mid-1950s. Among his other activities, Liebers was director of the U.S. Chamber of Commerce and director and member of the executive board of the Ford Foundation's Resources for the Future, headquartered in Washington, D.C.

—Karen Stork, CSD administrative assistant

Nebraska Farmer

So-called "Lake Street Flood" in Lincoln, probably during the late 1930s or early 1940s

Economics and Finance

by Steve Schafer

When the Rains Fail

The Dust Bowl era of the 1930s combined the Great Depression with years of drought to drain the vitality of the entire Great Plains. By 1937, one out of every six farm families in Nebraska depended on public assistance. The value of crop production during the 1930s dropped to less than half that of the previous decade. The number of farms shrank from a peak of 134,000 in 1935 to 121,000 in 1940. Total population dropped 4.5 percent during the decade. Excluding births and deaths, more than 12 percent of the people deserted the state during this time.

Both drought and economic depression contributed to the misery. The Depression caused crop prices to tumble because of the nationwide contraction in the money supply and reduced demand by families of unemployed workers. Corn went from 77 cents a bushel in 1928 to 28 cents a bushel in 1932. When prices climbed to 87 cents a bushel in 1934 and $1.16 a bushel in 1936, the drought left farmers with nothing to sell. Average corn yields in those two years fell to single digits—3.2 and 3.5 bushels per acre. Anything close to normal rains would have produced about 24 bushels of corn to the acre. But together, drought and depression robbed farmers of their wealth and livelihood, eliminating bank credit and bringing foreclosures.

However, not all farmers watched their crops burn up during the drought of 1931 to 1941. In the U.S. Bureau of Reclamation's North Platte Project, tens of thousands acres of cropland in Nebraska stayed moist with water from the North Platte River. Snowmelt in the Rocky Mountains, stored in reservoirs before wetting the project's fields, kept the local economies alive. In Scotts Bluff County, only one out of 25 farm families depended on government assistance, less than one-quarter of the state average.

Plenty and Paucity: Water History and Economics

Like a river cutting through a channel, water has formed the state's economy and like the rock walls of a canyon, economic and financial factors have in turn affected the course of water use. In the past, low and unpredictable rainfall stunted the productivity of Nebraska's rich soils, but vast irrigated crops can now grow without rain. The economic impact of that irrigation is a major story in the history of water use in Nebraska. Another story worth telling is how economic constraints have guided development of both surface water and groundwater.

Irrigation dominates the discussion because about three-quarters of water use in Nebraska goes for crop production. To be sure, water has had economic implications beyond this single use. The Missouri River made Omaha an early transportation center and gave birth to many other river towns that have grown and receded in importance. The economic signature of the Missouri River is still visible as a water highway, as a source of coolant for power plants, until recently as paymaster for commercial fishermen, as a source of recreation for boaters, anglers, hunters and bird watchers and as a natural boundary that separates local economies in eastern Nebraska from their neighbors in South Dakota, Iowa and Missouri. Mainstream dams on the Missouri provide cheap electrical power to the whole region.

The access of Nebraska's municipalities to plentiful supplies of high quality groundwater also has had a definite economic dimension. Water for cities and industries keeps businesses alive and flourishing and, consequently, pocketbooks full. It also keeps households and their inhabitants clean and urban lawns green. As supplies shrink and quality degrades, the cost of providing alternatives will challenge many local budgets, perhaps writing the death warrants for some towns. Larger cities react like a mother bear protecting her cubs whenever their sources of water are threatened. Grand Island, for example, has a long and continuing history of challenging the water

A dust cloud rolls over a western Kansas town, 1935.

Conservation and Survey

A color-infrared aerial photo of center-pivots on cropland near Imperial, Chase County, 1973. This is a drought-prone area with limited average annual precipitation that depends on irrigation for economic stability.

Les Sheffield, UNL

Nebraska State Historical Society

One of the earliest economic benefits from the state's water resources was the "water highway" traveled by riverboats moving goods up and down the Missouri River. Here, the "Elisha Woods" passes under Omaha Bridge near Omaha. Photo undated.

rights of projects that would divert Platte River flows away from its well fields. Lincoln wants the state legislature to give it a legal divining rod that it can use to keep its Platte River aquifers saturated. The stakes are high because local economies are potentially at risk.

Recreation is another use of water with economic implications. Ogallala would be different—likely smaller—without the summer migration of water worshipers to Lake McConaughy from across the state and from eastern Colorado and Denver. The expanding mud flats of the shrinking Harlan County Reservoir recently proved the importance of recreation-seeking tourists to that area's economy. All across the state, water-based recreation adds to the quality of life of local residents, even if there is no precise way to measure the actual value of its economic contribution.

These are the positive aspects of water. But when too much rain comes, streams show their destructive face. And this destruction can cost communities and regions millions of dollars in damages. Major floods on the Missouri and Republican rivers finally ushered in the massive federal Pick-Sloan program of dam building that brought the region large and continuing benefits. Even so, every year, the slot machine of the clouds dumps its treasure on another "winner." But unlike pocketfuls of quarters, rain in over-abundance is nearly always destructive. Some years back DeWitt "collected" twice in near succession. Then Howells had a turn; then Crawford. No one should minimize the economic wreckage of such floods; they waste the slowly accumulated wealth of townspeople, wash out crops and impair future growth. Flood insurance and federal disaster assistance may try to act like a financial cavalry riding in to the rescue, but those dollars always stop shy of their mark.

All of these other aspects of water are important, but none compares with irrigation for its statewide and far-reaching effects. Irrigation has transformed production agriculture, the historical engine of the state's economy.

The Economic Impact of Irrigation

Farmers irrigate for only one reason: it pays. It pays in greater production per acre. It pays in assuring a crop in dry years. It pays in planting higher value crops than otherwise could be cultivated and it pays in higher land values.

After a century of development, irrigation in Nebraska has grown to major dimensions in its importance to agriculture. Of 56,000 farms, 22,000 use irrigation. Of more than 18 million acres of harvested cropland in 1990, 38 percent used supplemental water and accounted for more than half (57 percent) of the total value of crop production.

If irrigation pays, corn supplies most of the money. Of all irrigated acres in 1990, corn took up 74 percent of the land and contributed 76 percent to the total value of irrigated crop production. Soybeans (at 7 percent), alfalfa (6 percent), dry edible beans (6 percent) and sugar beets (3 percent) provide nearly all the rest of the value of irrigated crop production, but none comes close to King Corn.

Nebraska's 734 million bushels of irrigated corn in 1990 swamped its 200 million bushels of dryland corn and made it the third biggest corn-producing state in the nation behind Iowa (1.5 billion bushels) and Illinois (1.3 billion bushels).

Nebraska is also third in the nation in production of dry edible beans and sixth in production of sugar beets. Both depend entirely on irrigation and contribute mightily to crop values in the Panhandle and especially Scotts Bluff County.

These numbers illustrate irrigation's popularity with Nebraska farmers. Many fields and many crops depend on it. Corn growers, especially, use irrigation extensively. Or, more accurately, the opportunity for extensive irrigation has turned Nebraska farmers into corn growers, especially.

Direct Economic Impacts of Irrigation

The direct economic impact of irrigation is measured by the productivity that it brings and the value of that production compared with what might be produced without irrigation. "Irrigated agriculture" in the economic sense is a package of inputs, including supplemental water, labor, chemicals, equipment, land and management, that makes increased crop production possible. In 1988, the Bureau of Business Research

Farmers irrigate because it pays. In this undated photo from the Tri-County area, Charles G. Gallace (right) of Hastings shows off a well for gravity irrigation.

Nebraska State Historical Society

at the University of Nebraska-Lincoln published a study that compared 1985 irrigated-crop output with hypothetical dryland substitute crops. The total direct effect for the entire state was $980 million. That number was 30 percent of total crop value in 1985 and 13 percent of total farm marketings of crops and livestock.

Higher production accounts for part of this gain. Yields on irrigated land far outstrip dryland yields, even in years of normal precipitation. In the period 1984 to 1986, when precipitation statewide was better than typical, yields on dryland corn averaged 93 bushels to the acre. Irrigated corn typically yielded 139 bushels per acre, half again as much as dryland. For the same period, irrigated sorghum bested dryland yields by 26 percent, and irrigated soybeans outperformed non-irrigated fields by 24 percent.

The 1988 study also found a significant shift in cropping patterns under a scenario of no irrigation. Corn would drop from 42 percent of all dryland and irrigated acres to just 28 percent. Acres devoted to sorghum, hay, wheat and soybeans would offset corn's fall from grace, increasing by roughly 9, 3, 2 and 1 percent, respectively.

Nebraska State Historical Society

Irrigated sugar beets at the University of Nebraska "studio" (experiment farm) in Scotts Bluff County, 1928. The state ranks sixth in the nation in sugar beets.

Indirect Economic Impacts of Irrigation

Besides the direct economic impact of more crops, irrigation has stimulated or made possible considerable other economic activity. Accounting for these other "indirect" economic impacts involves tabulating all the production, services and sales associated with a particular enterprise. It is a far bigger category than the narrow pigeonhole of "direct" economic impacts, which measure just the "added value of production." In contrast, indirect economic impacts trace the ripple effect of money as the income from an activity filters through the economy.

It measures people's interdependence as much as anything. A hermit enjoying a life of bare subsistence and complete independence has no "indirect" economic impacts to speak of. At the other extreme, in the days of the old West, for example, a thirsty cowboy just off the trail drive with a month's pay in his pocket would have had great potential for indirect economic impacts. The cowboy's direct economic impact was the value of helping transport a herd of cattle. Much of this value was spent on the cowboy's wages and other costs of the drive. The indirect economic impacts are the additional rounds of spending and respending that occur with the money from the trail drive, including the economic commotion the cowboy causes at the saloon and other places.

A color-infrared aerial photo of Jeffrey Reservoir, its canal and the Platte Valley. The economic benefits of irrigation have been particularly evident along the Platte.

Les Sheffield, UNL

Irrigated agriculture is more like the example of the cowboy in terms of generating indirect economic impacts. A few years back, one observer, speaking of farm foreclosures and low net income, summarized agriculture's ability to distribute its earnings: "I suspect one of these times we're going to have to quit serving up farmers on the platter as the sacrificial lamb in order for a whole host of other people to make money." Irrigation is enormously expensive and the furious pace of spending associated with it trickles widely through the state's economy.

Even before the crop is harvested, an irrigation farmer's obligations to others makes for a long list. Just to get access to water means drilling a well and setting a pump, or paying the costs for project water. Then there's the expense for pipe or pivot, or the labor costs for tending open ditches. To produce high yields takes proportionately larger amounts of fertilizers, pesticides and seed. A 1984 study found that 80 percent of agricultural chemicals sold in Nebraska go on irrigated land. At the very front of the line are the tax collector and the lender—

who demands interest on loans for land, equipment and ballooning operations.

Irrigation's much-loved companion has been the economic gain it brings to surrounding communities. Business grows and bank deposits multiply as the money from added crop production circulates from person to person and from business to business. In 1975, the mayor of Ainsworth described the blessings bestowed by center-pivot irrigation in his area: "Seven new churches, a new hospital and an average of one new business in the community each year [have been added] since irrigation began here."

Fancy economic models attempt to measure the total economic commotion associated with increased crop production. The 1988 Bureau of Business Research (BBR) report mentioned above found that each dollar of irrigated crops generated a total economic activity of $2.21 within the community and state. Compared with this, dryland crops had a "multiplier" (the term for quantifying indirect economic impacts) of $2.07 per dollar of production. The higher multiplier (because of more inputs) and the higher yields associated with irrigation explain irrigation's large economic ripples. After estimating the hypothetical size of the economy with no irrigation, the BBR report found that irrigation added $2.3 billion to the state's economy. This was 9 percent of Nebraska's gross state product.

Irrigation's benefits also extend to nearby livestock fed on crops grown with irrigation, due to the benefits from low transportation and storage costs and from integrated crop and livestock operations.

Nebraska Farmer

Other Economic Effects

Irrigation's economic coattails go beyond the effect of a bigger harvest and all the spending it brings. There are several other major types of economic impacts. One is the investment by firms that have expanded to supply the larger demands of irrigated crops. Agricultural chemical producers and distributors, manufacturers of grain-storage facilities and other farm buildings, well drillers, farm-equipment makers, even distributors of moisture-monitoring equipment and computer dealers, are examples. To the extent those businesses have expanded and the investments occurred within the state, this further stimulates Nebraska's economy. The center-pivot business, born and raised by Nebraska's irrigated agriculture, has developed broader markets and products that make it a spinoff industry, no longer dependent solely on crops in this state, but still important to the economy.

Another type of impact has been the growth of industries dependent upon irrigation's bountiful harvests. In recent years, Nebraska has often been first in the nation in the number of cattle marketed, second in the nation in the number of cattle on feed and fifth in the nation in the number of hogs. The livestock industry, which pumps $6 billion of added production into the state's economy, owes part of its size to a plentiful and assured supply of grain with low transportation costs. The size of the livestock industry in turn contributes to Nebraska's attractiveness to food-processing industries.

Yet another impact has been the creation of tremendous wealth for landowners. Gravity-irrigated cropland sells at 88 percent above the cost of dryland cropland with irrigation potential. Center-pivot-irrigated cropland, not counting the cost of the pivot, sells for 35 percent more than dryland cropland with irrigation potential. Compared to cropland with no irrigation potential, the prices are 154 percent and 82 percent more for gravity-irrigated and pivot-irrigated cropland, respectively. Irrigation gives new emphasis to the adage that farmers live poor and die rich. Property taxes, which march in step with land values, help make the first part of the adage true, providing an important benefit to local government from irrigation.

Besides bringing these positive effects, irrigation shares at least some responsibility for a few problems. Groundwater quality has been a victim of the growing reliance on fertilizers and pesticides that come with irrigation. Monoculture—planting corn (or any single crop) season after season—can affect soil fertility, increase disease and insect problems and change the economic characteristics of the farm unit. Irrigation also reinforces the trend toward large farms, which has had a major impact on life in rural Nebraska.

What Happens When the Rains Fail?

This review of the economic impact of irrigation so far assumes that dryland cropping remains a viable alternative. During the occasional year of poor rainfall, that assumption shrinks with the nonirrigated yields. During periods of extended drought, the assumption cracks apart like parched topsoil.

The stability of irrigated harvests year in and year out is one of the greatest strengths of Nebraska's agricultural economy and of everyone who feeds on it. Nebraska bankers in irrigated regions never worry about a failed crop. They can loan out virtually every penny, with no reserves against widespread disaster. Drought, especially if it hits growers in Iowa and Illinois, becomes good news in Nebraska, even if it hits here, as prices rise for the state's 5 million acres of guaranteed 145-bushel-per-acre corn.

Putting numbers to this aspect of irrigation's advantage, two economists at the University of Nebraska-Lincoln recently simulated the effects of drought with and without irrigation. It showed the impact of a moderate drought with precipitation 20 percent below normal and a severe drought with precipitation 40 percent below normal. In their analysis, a moderate drought increased the total economic impact of irrigation by 41 percent, and a severe drought increased it by 78 percent. Numerous assumptions that might vary from one economist to another go into producing comparisons like these, but the trend is obvious: the drier the weather regionally, the more important irrigation becomes.

Agriculture's Albatross

Irrigation's happy story sours a little in the face of some other economic trends: the number of farms has been dropping since the Depression; and by 1980, it had fallen under the count of a century before, when the pioneers were settling the state. The number of people leaving farms and rural areas grew during the 1980s; during that decade, 409 of the state's 535 towns lost population, more than twice the count for the 1970s. On-farm employment has dropped 45 percent in the same period. The average price of corn for the 5-year period ending in 1990 was 10 percent less than the comparable figure a decade ago; after wringing out inflation, the true reduction is close to 50 percent. From 1980 through 1987, government payments grew from just 1 percent to a sizeable 14 percent of total farm cash income; for the decade, government payments averaged more than half of net farm income.

Despite the important contribution of irrigation, the agricultural and rural sectors of Nebraska's economy show signs of weakness. Low commodity prices are forcing out the marginal producers even while irrigation, good management and low debt allow others to expand and still turn a healthy profit. As the "pie" of total agricultural income is redivided into fewer slices to support fewer farms, the economies of scale bring consolidation in related businesses as well. The shakeout ultimately means fewer jobs in rural Nebraska.

The deep agricultural recession of the mid-1980s scared the business community in rural areas and throughout the state. Severely low commodity prices and a stressed agricultural sector demonstrated the economic linkages between town and farm, and made many businesses decide to hitch their wagon to a different horse. The goal since then has been to diversify. In the 1988 report, "New Seeds for Nebraska," the authors criticized the state's heavy reliance on a vulnerable agricultural sector that faced diminished demand for basic commodities in domestic and international markets. The report also warned of an uncoupling of the economies of large cities from rural areas. Economic-development strategies written for Omaha acknowledge agriculture's importance (it's responsible for 25 percent of the city's employment) but mostly point away from agriculture in seeking future growth. Emphasizing the already frayed ropes tying rural and urban economies, the analysis found that financially squeezed farmers did not matter as much to Omaha's economy as national demand for its large food-processing industry.

The Economic History of Irrigation

Irrigation now nourishes about 8 million acres of crops in Nebraska. At the start of this century, the best guess is that about 150,000 acres of land had irrigation. The history of this growth is the story of two very different but overlapping developments. One is the tale of surface-water projects. It largely involves collective action to make surface waters of the state available to more farmers. The other is the record of groundwater development. It reflects the collective outcome of individual decisions of thousands of farmers to tap the groundwater supplies underlying their lands.

Les Sheffield, UNL

While surface water was developed first for irrigation, after about the mid-1950s, groundwater pumping came on fast to far outstrip acres irrigated by surface water. This high pressure center-pivot system operates in Lincoln County.

One important difference between surface-water and groundwater irrigation is magnitude. Nearly 87 percent—6.9 million acres—of all irrigation in Nebraska uses groundwater supplies. Of the remaining 1 million or more acres, 56 percent—619,000 acres—belong to federally funded irrigation projects and depend on water held behind numerous dams in Nebraska and Wyoming. The balance, some 429,000 acres, represents farms immediately adjacent to waterways, privately funded ditch and canal projects and self-financed irrigation districts.

These numbers, however, cheat the true importance of the federally funded projects. Return flows (water diverted from a river but with the unconsumed portion returned downstream plus groundwater seepage from lakes and canals and from irrigation water applied that percolates to the water table and then provides baseflow to streams) provide a true "trickle-down" benefit to hundreds of thousands more acres. More than 34 private irrigation districts in the North Platte River valley rely on return flows of the Bureau of Reclamation's North Platte Project to irrigate 171,580 more acres. Groundwater recharge from surface-water projects reduces pumping costs and ensures a continued supply of groundwater for large areas. The major surface-water projects have increased groundwater levels by 10 to 50 feet for nearly 1.5 million acres, particularly just south of the North Platte and Platte rivers east of Lake McCounaghy. Also, the North Loup Project (Calamus Reservoir) is not reflected in the totals. When complete, it will irrigate up to 53,000 acres. Still, even giving stored water projects their due, irrigation in Nebraska today is mostly a groundwater phenomenon.

Another vital difference is timing. Surface-water development predates significant groundwater use by up to a century. Surface-water irrigation's firstborn status and its many years as an only child gave it a crucial advantage in the state's legal household. (By 1941, the land under surface-water irrigation had reached about 80 percent of the present total. Much the younger sibling, groundwater irrigation was still in its infancy at the start of World War II, with about 180,000 acres recorded.) Water law, especially the first-in-time, first-in-right appropriation system, reflects the unique needs of surface-water irrigators. Groundwater users generally dislike being caught up in this system. So reconciling the growing rivalry over who owns and controls the "family inheritance" will not come easily.

Yet another key difference is how these laws regulate surface water and groundwater use and affect water conservation. Surface-water irrigators first have to ask the state's permission to use the water (to obtain a water right) and then must use the water or lose the right. Fear of losing one's right can impede water-conservation efforts. Groundwater users, in comparison, only have to register with the state their decision to use water under their land and, except in a groundwater control area, can generally use as much as they want. Pumping costs (and some groundwater control regulations) encourage conservation, but unlimited access by all landowners to a shared groundwater resource can lead to significant drawdowns in supply.

Other important differences between surface-water projects and groundwater development include technological breakthroughs, organization and financing. All of these differences become important when examining the dual economic histories of surface-water and groundwater irrigation.

The Economic History of Surface-water Irrigation

Who Owns the Surface Water?

The legal system governing a resource becomes a critical economic factor affecting its consumption. A good understanding of surface-water law and its history is important to knowing the economic history of surface-water irrigation development.

The first recorded attempts at diverting streamflow for irrigation date from before the Civil War and the federal Homestead Act. By 1860, farmers had constructed 4 miles of canals near North Platte. By 1889, small systems like this supplied water to about 9,000 acres. Steady growth continued with more rapid expansions during dry periods.

In the late 19th Century, the Nebraska legislature passed several laws designed to smooth the way for irrigation promoters. In 1877, the legislature granted irrigation companies the right of eminent domain for canals, dams and reservoirs. A law in 1889 recognized the right of anyone to acquire a vested interest in water by applying it to a beneficial use, such as irrigation. In 1895, the legislature completely revamped the state's irrigation laws, creating a formal registry of water rights, giving priority to those with the oldest water rights (first-in-time means first-in-right) and establishing the state Board of Irrigation (now the Department of Water Resources) to enforce the law. This 1895 act, then commonly called "the Akers law," after irrigation pioneer and politician William Akers, still applies today.

Before these changes, only a person owning land adjacent to a stream had a right to use its water. Even that right was limited by the "reasonable" needs of downstream neighbors. This system of "riparian rights," so called because "riparian" means related to the banks of a river, was inherited from England. Under it, landlocked property owners had no legal access to water.

The 1895 act made surface water a public commodity. While still protecting any riparian rights existing prior to it, the act in essence divorced ownership of water from adjacent lands, gave it to the state and then allowed any and all comers to make permanent use of it on a first-come, first-served basis. With this law, irrigation companies could serve non-riparian tracts of land and protect their investments from any future upstream and downstream developments.

This new property right making water a commodity in itself came with certain key restrictions, like easements to a title. One such restriction is the legal interest of any remaining riparian rights. Another is the system of priorities that gives "bumping" rights to the water used by others according to the date of the water right and a list of constitutional preferences regarding use (domestic first, then livestock, then agricultural, and last, industrial). Yet another restriction is the use-it-or-lose-it rule. The water right vanishes if the owner fails to use the water for more than three consecutive years, except in any of a number of special circumstances. Finally, the new water right attaches to a specific piece of land. Buy or sell that land, and the water right remains with it. Until 1983, water rights detached from the land could not be sold and reverted to the state. LB 23, passed in 1983, now permits the sale of water rights separate from the land. Such sales must be in the same river basin for the same use and subject to approval by the director of the state Department of Water Resources.

The 1895 law unleashed the economic potential of water for new lands. But it carefully channeled that potential within the system of preferences and guaranteed that water would remain free to the user by outlawing transfers.

Early Frustrations

Conditions did not turn rosy for private water developers after the legal victory in 1895. The new water law was necessary but not enough by itself to ensure success. Visionary projects died in various stages of development from many different causes all having a common symptom: financial anemia. Some fell to the recurring hard times of their farmer-customers, too poor to pay the costs of preparing their land for irrigation or the price of delivering the water. Other projects went broke when rainclouds would return after a drought and wash away the economic advantage of irrigation and sometimes the irrigation works themselves. For example, the South Side Irrigation and Land Company, organized in 1894, built a 41-mile ditch to divert North Platte River water south of the South Platte River. It irrigated land intermittently from 1896 to 1903, but in 1903 the wooden flume (channel) across the normally dry South Platte washed out. Subsequent hard times caused farmers to lose money and then, interest in the project.

Ample rainfall and economic hard times in the decade following the dry years of 1893-1894 defeated several more projects. Joel Hull of Minden worked tirelessly to promote a 54-mile canal from the Gosper County line to Minden but could never raise the money. The Mirage Flats project, finally built in 1946, had its start in 1895, when a farmers orga-

A "ditch gang" in 1913, perhaps for the Frenchman Valley Irrigation District

Nebraska State Historical Society

Nebraska State Historical Society

An undated photo of the early days of irrigation in the Tri-County area

nization built canals. Low flows on the Niobrara River during irrigation season, fires that destroyed its wood flumes and above-normal precipitation led to the project's failure. From 1895 to 1910, 26 canal and ditch companies in the Platte River basin, alone, abandoned their water rights.

Financing of these privately owned systems came from various sources. Often, farmers would build the canals in exchange for part ownership and a right to use the water. Other efforts involved raising capital by selling shares of stock. A private investor from Minneapolis built the Cozad Canal in 1895-1896 and sold it in 1901. The Dawson County Irrigation Canal Company borrowed $110,000 from eastern capitalists to build its system in 1894, but the debt load caused the company to close.

The Reclamation Act of 1902

The pioneers who came to Nebraska fulfilled a national policy to settle the West. The Homestead Act of 1862, the seductive promises of railroad companies and the glorious visions of promotors enticed many. After arriving with few resources and no experience with growing crops in a climate of untimely or sparse rains, they lived on the margin. "Sowing in hope, cultivating in doubt and harvesting in despair" became a well-known adage. Many failed.

The federal government eventually realized that the goal of populating the West with small landowners was in danger. The unique arid and semi-arid conditions beyond the Missouri River demanded unique policies. During the 1900 election, both political parties lauded reclamation as the cornerstone of a new policy for western lands. After President McKinley's assassination shortly after taking office in 1901, when Theodore Roosevelt became president, he took up the reclamation banner. In a speech to Congress, he called for a federal reclamation service with a mission of "homemaker" that would justify federal help:

> The reclamation and settlement of the arid lands will enrich every portion of the country, just as the settlement of the Ohio and the Mississippi valleys brought prosperity to the Atlantic States. . . . Successful homemaking is but another name for upbuilding of the nation.

The next year, Congress passed the Reclamation Act of 1902 to provide national support for western irrigation development. Sponsors designed the bill so that public lands would serve as leverage for developing the West. The act created a Reclamation Fund to finance construction of irrigation projects in the 16 "public-land" states through interest-free loans. Revenues would come from the public lands, sale of publicly owned resources in those states and loan repayments by water users. One important feature of the bill, designed to silence critics and serve the "homemaking" (settlement) goal of reclamation, was the 160-acre limitation on the size of any estate using reclamation water. Congress wanted primarily to develop the West while still serving the ideal of wide dispersion in land ownership. By administrative fiat, the Bureau of Reclamation later interpreted this to mean 160 acres for each spouse. In 1982, Congress finally bent to strong western demands and increased the limit to 960 acres.

In 1903, Congress authorized five reclamation projects, including the North Platte Project in Wyoming and Nebraska. This project now irrigates 164,000 acres in Nebraska's Panhandle. In 1911, Congress passed the Warren Act, which expanded the potential of reclamation by allowing the sale of surplus water from federal projects to nonfederal lands and the construction of canals to serve such lands. In Nebraska, six irrigation districts serving 96,000 acres receive water from Wyoming's Pathfinder Reservoir under this provision.

The stream of revenues from public lands quickly ran dry as the homesteading era ended and sales of public lands dropped off. In 1921, Congress augmented these revenues with proceeds of federal mineral leases, potassium royalties and water-power licenses. Congress also began making general-fund appropriations to supplement the Reclamation Fund as early as 1908. Eventually, direct general-fund appropriations would surpass the Reclamation Fund in importance but not during the early 1900s.

Authors of the Reclamation Act envisioned the Reclamation Fund as a type of self-perpetuating financing machine, fueled by revenues from public lands, oiled by loan repayments and, thereby, creating more no-interest loans for irrigation. Reality tarnished the vision. The fuel and oil turned out to be low grade, lacking in fiscal octane and lubricant. The original act called for a 10-year repayment period, but settlers could not afford the bill. Costs of preparing land for irrigation and planting the first crop exhausted their resources, and annual profits from the harvest barely covered current expenses; nothing was left over to pay on the irrigation contract. In 1914, Congress extended the repayment period to 20 years. But this still proved unworkable, and several projects failed, especially during the agricultural depression after World War I and the Great Depression. For a while in the 1920s, Congress permitted contracts that assessed 5 percent of the gross value of the crop for repayment. Some of these contracts will take up to 104 years to repay.

Most of the contracts in the North Platte Project use the crop-production plan of repayment. Congress authorized a general moratorium on repayments in 1921 and then year by year during the Great Depression. Finally, the Recla-

What's in a Name? Bi-County, Tri-County or Quad County?

Little about the history of the Tri-County project is simple, least of all the name. Even Gene Hamaker's authoritative book, *Irrigation Pioneers: A History of the Tri-County Project to 1935*, was at a loss to identify precisely which counties the name referred to at all times. Hamaker relates: "As first used back in the 1880s, the term probably referred to Gosper, Phelps and Kearney counties. The 'Tri-County Supplemental Water Association' organized at Holdrege in 1913 was a bit indefinite as to the territory included in the 'Tri-County' description. Apparently Gosper was the third county generally included but some old timers will insist that even at that early date Adams County was the third county referred to."

Later, after supporters advocated irrigation of more than 100,000 acres in Adams County, "Tri-County" generally referred to Adams, Phelps and Kearney counties. Central Nebraska Public Power and Irrigation District organized in 1933 and included four counties: Adams, Phelps, Gosper and Kearney.

In 1936, when a Nebraska Supreme Court decision eliminated transbasin diversion to Adams County, plans were for irrigation of two counties. However, by 1954, two laterals had been extended to Gosper County and the "Tri-County" name was again appropriate. Originally named the Central Nebraska Public Power District, now the project is known as the Central Nebraska Public Power and Irrigation District, by its acronym, CNPPID, or simply as "Central."

—Steve Gaul, water resources planner, Nebraska Natural Resources Commission

mation Project Act of 1939 created a 10-year development period with no repayment of construction costs, followed by 40 years of payments on a graduated scale. (In addition to construction costs, contract holders also pay all operating costs of irrigation projects.)

Ultimately, far more important to financing irrigation development than moratoriums or extended payment terms was the Reclamation Project Townsite Act of 1906, which authorized the Reclamation Service to build hydroelectric power plants, sell the energy and use the proceeds to offset repayments from irrigators. Later, the Pick-Sloan program of flood-control, irrigation and navigation projects would include hydropower as a key source of revenue to help offset construction costs. Because Congress established the whole Pick-Sloan program as a single "project," revenues from all hydropower facilities are available to pay the costs of any authorized irrigation project. By 1966, two-thirds of all repayment costs came from hydropower revenues. Twenty years later, hydropower revenues would cover 91 percent of the estimated repayment costs of the North Loup Project in Nebraska.

Drought, Depression, Public Works and Public Power

The drought and Great Depression prompted a second surge of water development in Nebraska. When finished in 1938 to 1941, five federally funded projects irrigated a total of 285,000 acres of cropland. Today, this still represents one-quarter of all surface-water irrigation and more than 40 percent of all irrigation projects in the state. The construction and financing of these Depression-era projects had one other economic effect. Of necessity, they turned Nebraska into a public-power monopoly.

The nation's original reclamation program stalled for lack of revenues to pay the huge construction costs of projects. With the Depression in full swing, President Franklin Roosevelt's Public Works Administration (PWA) took up the fiscal slack. In November of 1933, the PWA approved the Sutherland project sponsored by the Platte Valley Public Power and Irrigation District. This project now irrigates about 100,000 acres. Final costs were $2.9 million in grants and $9.1 million in loans.

In September of 1935, Roosevelt approved the Tri-County project, organized now as the Central Nebraska Public Power and Irrigation District (CNPPID or "Central"), which currently irrigates about 130,000 acres. Final costs were $19 million in grants and $24 million in loans. The PWA also funded the North Loup and Middle Loup projects during this period. North Loup received $655,000 in grants and loans of $916,000, and Middle Loup received about $500,000 in grants and $728,000 in loans. Together, these two projects irrigate about 55,000 acres.

Other federal entities participated in the effort to create jobs and opportunity in Nebraska. The Reconstruction Finance Corporation provided $10 million in grants and loans to the Loup project near Columbus, which was exclusively a hydropower operation. The Rural Electrification Administration (REA) was active in funding transmission lines and rural connections.

A major weakness of the original Reclamation program was lack of revenues to repay construction loans. Crops, by themselves, could not support the debt burden of building dams and extensive irrigation works. Learning from those experiences, project sponsors in the 1930s sought to harness the paying potential of hydroelectric power to their irrigation plans. Both the Sutherland and Tri-County projects included major hydropower production and the Loup project was exclusively hydropower. The large generating capacities of these three projects killed the economic justification for adding power facilities anywhere else. Although the North Loup and Middle Loup projects also sought power generation, the PWA refused to fund those components. Stripped of hydropower revenues, the North Loup and Middle Loup projects lost their financial feasibility.

Economic jealousies plagued all the projects and were often played out in the courtroom. The North Loup and Middle Loup projects challenged Loup Public Power District's (LPPD) water right because it might interfere with their irrigation plans. North Platte irrigation interests worried that the inter-basin diversion features of Tri-County would hurt the state's lawsuit against Wyoming. Tri-County opposed Sutherland and Sutherland opposed Tri-County as both competed for limited water and PWA funding.

The City of Grand Island, fearful for its future if its Platte River well fields went dry, joined others in challenging the water rights of Sutherland and Tri-County and opposed legislation enabling their creation. The resulting lawsuit against Tri-County ended in a partial victory when the Nebraska Supreme Court ruled in June of 1936 that water cannot be diverted to another watershed (Osterman vs. Central Nebraska Public Power and Irrigation District, later overturned in Little Blue NRD vs. Lower Platte North NRD, 1980). This forced Tri-County to eliminate half of its irrigation territory, drop two reservoirs planned for Plum Creek and increase generating capacity to keep the project financially feasible. Private power companies also sued to prevent federal funding of electrical generating facilities and won a temporary injunction against Tri-County in 1936. The U.S. Supreme Court's ruling in January of 1938 that the government could aid public power projects finally cleared the way for Tri-County and 51 other power-related projects with PWA allotments to proceed.

The Public Power Monopoly

Economic survival motivated the private power companies' opposition to the hydropower districts. Hydroelectrical power would be cheap, and the generating facilities of the private companies were often old and always inefficient by comparison. With demand for electricity stagnant even without the Depression, the private power companies foresaw a future of destructive competition. At the time, total electrical consumption in Nebraska was about 600 million kilowatt-hours per year. The combined capacity of the hydropower districts was about 500 million kilowatt-hours per year.

Economic survival motivated the hydropower districts, too. Their projects depended on selling electricity, but the private power companies dominated the retail markets. As early as 1933, PWA staff became worried about the lack of demand for the power these projects would generate. By 1940, PWA officials were deeply concerned about the financial viability of the three projects. Sutherland's only major power customer since completion in 1938 had been Tri-County for construction of Kingsley Dam, and Sutherland's bonds were in default. Efforts to market its power to others or to buy the retail markets of private companies had failed. Tri-County and the LPPD faced similar marketing problems and certain financial problems to follow.

PWA refinanced all three hydropower districts in 1940, replacing the original 4 percent interest rate with a variable rate that began at 1/4 percent the first year and increased by small increments to 4 percent in 15 years and then to 5 percent near maturity of the bonds. Total interest would equal a straight 4 percent. (In 1949, Congress approved another refinancing with a fixed, 2 1/2 percent interest rate.) In the 1940 refinancing, the PWA also lengthened the repayment period from 30 years to 64.

As a condition of this refinancing, the federal government insisted on a joint operating agreement that would pool the facilities and revenues of all three districts in order to achieve the most efficient production schedule. The districts did this by creating the Nebraska Public Power System in 1940. During the winter, LPPD assumed most of the generating load while the other two stored water. During summer, when the Loup River was lower, Platte Valley and Tri-County assumed most of the load. The cooperative agreement divided the pooled revenues remaining after paying all power-generating costs according to the indebtedness of each district, rather than the district actually generating the power.

During this time the hydropower districts made several efforts to buy one or more private companies as an outlet for their power production. These efforts failed because the federal government refused to give the first claim on assets in case of default to private bond holders. This closed the door to the only source of private financing for such purchases.

To bypass this barrier, backers of LPPD proposed creation of a new public power district. With no prior indebtedness, it could issue revenue bonds, buy the private companies of Nebraska and purchase power from the hydropower districts for resale. The state approved the formation of Consumers Public Power District on Aug. 5, 1939. Although there was no legal connection between the Loup and Consumers districts, the same men served on the boards and as officers of each. Seven days later, Consumers signed a lease purchase contract with the private holding company that owned the utility in Columbus. Within the next 18 months, Consumers issued $47,150,000 in bonds to purchase 14 utilities representing all the private power companies in the state, except Omaha. Financing costs were $2.9 million, working capital was $3.4 million and the purchase price of the properties was $40.7 million.

Certain political and economic factors helped Consumers during its whirlwind buying spree. In 1935, Congress had enacted an anti-monopoly measure that dictated a breakup of large utility holding companies. Even though the Nebraska power companies opposed public power, they were all subsidiaries of several eastern holding companies that were seeking buyers.

Another factor was the groundwork that LPPD had already laid. Agents representing the district, who were highly regarded by the private utilities, had worked since 1937 to appraise the properties and begin negotiations for purchase. The new Consumers district, with the same key people as the Loup district, was able to conclude these efforts. Its willingness and ability to pay reasonable prices for the properties ensured rapid

Farmers Home Administration Funds Water and Waste Disposal Systems

The involvement of the U.S. Department of Agriculture in providing funds for rural water and waste disposal systems has evolved and expanded over time. A predecessor of the Farmers Home Administration (FmHA), the Resettlement Administration (1935-1936) was born of the farm depression of the 1930s. President Franklin D. Roosevelt established the agency by executive order in 1935. Its first mission was to help impoverished farm families work their way off government relief rolls with the aid of loans, grants and counseling. The agency became the Farm Security Administration (1937-1947) and later the Farmers Home Administration (FmHA).

In 1937 Congress enacted a Water Facilities Act to provide loans for water systems of individuals and associations in the 17 western states (including Nebraska) affected by drought and inadequate water supplies. In the late 1950s Congress authorized use of FmHA credit to benefit both "farm and other (non-farm) rural people," and extended the 17-state western water-facilities program to the entire country. The program was expanded in 1961 to authorize rural towns of up to 2,500 in population to come into the service areas of the FmHA-financed water systems, and in 1965 the water-facilities loan program was converted to a loan and grant program for water and waste-disposal systems, and communities of up to 5,000 became eligible. Later, in 1972, Congress authorized FmHA to make loans for community facilities of unlimited variety with the service jurisdiction including rural towns up to 10,000.

Loans and grants made through FmHA supplement the amount of credit and capital available from commercial lenders in rural areas. In most programs, the agency makes loans to qualified applicants who can find no other sources of financing available on terms or conditions they can meet. Many other federal agencies, too numerous to explore thoroughly, have also cooperated or contributed to the funding and building of water delivery and disposal systems.

—Vincent H. Dreeszen, hydrogeologist emeritus and former director, CSD

Nebraska Farmer

Byron Bunn and his stock going home 6 miles east of Blair, Washington County, after the Missouri River flooded in 1952

progress. This in turn was due to an assured supply of cheap hydropower. Even with indebtedness exceeding the purchase value of its physical assets, Consumers was able to meet its bond payments, pay the Nebraska Public Power System an attractive rate and still cut the retail price of electricity in 1941.

The fate of public power in Nebraska is still closely linked to those original hydropower districts. In 1956, Central (Tri-County) built the Canaday steam plant with a $17 million REA loan to prevent an impending power shortage in the state. Today, the Nebraska Public Power District carries on the responsibilities of the Nebraska Public Power System, Consumers and the Platte Valley district. It continues to operate all of the power facilities of Central. Those revenues essentially pay all the costs associated with Lake McConaughy and allow Central to focus on irrigation.

Pick-Sloan Funding

The job-creation efforts of the PWA ended as the Depression years turned into World War II. Although Nebraska had benefitted greatly, far out of proportion to its share of the nation's population, vast regions of the state and other midwestern states remained vulnerable to drought and hard times. The outmigration of population that began during the Depression continued as rural economies fell further behind the income levels and opportunities in the cities. Serious floods also plagued the region intermittently. In 1935, an 8-foot wall of water rushed through the Republican River valley, killing 110 people and destroying $8.7 million of property, the equivalent of about $100 million today. During the spring of 1943, five major floods in the Missouri River basin killed six people. Damage from South Dakota to Missouri was extensive.

In response to these problems, development of the Missouri River basin held a high priority on the national agenda as the Depression faded and World War II erupted. The National Farmers Union and Senator James Murray of Montana eagerly promoted creating a "Missouri Valley Authority" modeled after Senator George Norris' success in creating a Tennessee Valley Authority. Norris, a progressive Republican from the McCook area, had won the support of the president, but resistance to a strong, independent authority was stiff, especially by the governors of all 10 states in the region.

To derail the push for a regional entity, the governors urged that the Bureau of Reclamation and Corps of Engineers prepare water-development plans for the basin. The competing agencies responded with separate, dissimilar plans that Congress virtually stapled together and passed in the 1944 Flood Control Act. Called the Pick-Sloan Plan after its authors, then Brigadier General (usually known as "Colonel" due to his rank while promoting the plan) Lewis A. Pick of the Corps and William Glenn Sloan, assistant commissioner of the U.S. Bureau of Reclamation, it sought to maximize and integrate the Corps' priorities of flood control and navigation and the Bureau's priorities of irrigation and power production. When signing the bill, a disappointed President Roosevelt warned that his approval was "not to be interpreted as jeopardizing in any way the creation of a Missouri Valley Authority, which should receive the early consideration of the next Congress." Congress ignored the advice, and the Pick-Sloan Plan, with an early $350 million in authorizations, became the focus of regional development efforts.

Funding under the Pick-Sloan program is a hybrid of direct appropriations by the federal government and repayments by some project beneficiaries. Power users must repay their part of construction costs, including interest. Flood-control beneficiaries have a free ride, with their share of the tab picked up entirely by the federal taxpayer. The same is true of navigation. Irrigators must pay back their costs but on "affordable" terms. The repayment period is long, interest is forgiven and hydropower revenues eventually offset project costs beyond the officially determined "affordable" levy against irrigation. In addition, a part of the hydropower (with construction costs repayable with no interest) is reserved for irrigation pumping. This has subsidized deep-well irrigation development in Nebraska. Because major parts of the original Pick-Sloan irrigation plan were never completed, especially in South Dakota, large amounts of interest-free hydropower have also benefitted industry and urban users.

Pick-Sloan has served Nebraska well. It has tamed the flooding "instincts" of the Republican and Missouri rivers and turned the immense energy of the latter into electrical power. A deep channel provides navigation on the Missouri River south from Sioux City, and numerous lakes offer recreation to millions of citizens. The Bureau's irrigation works, built under authority of Pick-Sloan, now water about 190,000 acres in the southwestern, south, central and northwestern parts of the state. The flood-control and irrigation reservoirs serving these lands include Enders, Swanson, Hugh Butler, Strunk, Harlan County, Box Butte, Merritt and Sherman. The Calamus and Davis Creek reservoirs, part of the North Loup Project, will irrigate 53,000 acres when all of the canals are in place.

These are the achievements of Pick-Sloan. Critics point to several perceived failures. The Midwest's tendency to lag economically behind the rest of the nation leads to speculation about whether a strong regional entity could have accomplished more. William E. Warne in *The Bureau of Reclamation*, has said that, compared to the TVA, "Missouri River

Basin development has suffered from less flexible, inconstant governance. . . . [It] has not been comprehensive enough to halt the downward spiral of the economy of the Great Plains, which remains a critically depressed region'' [brackets added]. Those words were published in 1973. In Nebraska, major remnants of Pick-Sloan left on the drawing board include the Mid-State project (rejected by voters in 1975, but largely reborn in various proposals to develop more of the Platte River for irrigation) and the O'Neill Unit (with the controversial Norden Dam on the Niobrara River). In South Dakota, few of the promised irrigation projects have materialized.

Randy Epperson, Soil Conservation Service

Flooding of the Elkhorn River and Logan Creek near the mouth of Logan Creek, east of Hooper in Dodge County, June 18, 1990

The End of Large-Scale Water Development History

Pick-Sloan represents the end of an era. From the Reclamation Act of 1902, continuing with the PWA projects of the Depression and including passage of the Flood Control Act of 1944, the federal government provided funding for irrigation as a tool for the economic development of a region lagging behind the rest of the nation.

Economic development of the West and Great Plains through irrigation and huge public works no longer claims a spot on the national agenda. A gang of political competitors have muscled aside large water projects for irrigation. Growing federal budget problems, a national economy facing an insecure future, farm surpluses and environmental worries have been at the front of the fray. Critics cry foul over federal participation in water projects so that beneficiaries can qualify for more federal price supports on corn already in surplus. Less noticeable, economic calculations have tripped up the justification for publicly financed irrigation.

The first Reclamation projects qualified for funding after demonstrating only their engineering prowess. In 1924, after 20 years of too many financial failures, Congress imposed the first economic requirements on new projects. Prior to funding, a project would have to demonstrate adequate water supply and productivity of land to back up claims of financial viability.

Fifteen years later, the Reclamation Project Act of 1939 added the stipulation that cost-benefit ratios of a project had to exceed 1 to 1. This meant that Congress viewed irrigation projects partially as an investment and expected to offset their costs by future economic gains to the region and nation. Henceforth, each project had to make economic sense. All of the Pick-Sloan projects built or under construction today have met these higher standards.

Beginning in 1968, the federal government raised the ante yet again by demanding higher discount rates in the economic analyses. The discount rate is used to recompute all benefits and costs for the life of the project into a single set of numbers within a single year. It shrinks the size of future dollars into one current number called ''present values.'' This gives a fairer comparison of their relative values than simply adding up total benefits and costs. Even without inflation, a promise of a future dollar is worth less than a dollar in hand. Both the chance of a broken promise and the opportunities of other investments for the dollar in hand erode the worth today of a promised buck.

Until 1968, the discount rate in federal projects was 3.25 percent. It rose to 4.625 percent that year and then to 5.375 percent in 1971. It now stands at 8.5 percent. Such a simple and mundane adjustment dramatically alters the economic prognosis of water-development proposals. For example, an irrigation ''sweepstakes'' may promise total benefits of $1 million paid out over 50 years. The true economic worth today of that promise is $491,000, calculated at 3.25 percent, or only $231,000 calculated at 8.5 percent. Higher discount rates and long periods of calculation cause tremendous shrinkage in economic values. In a cost-benefit analysis for an irrigation project, benefits will shrink far more than construction costs because the latter are more immediate, rather than extending over many years. Neither the North Loup Project (Calamus Reservoir) nor the O'Neill Unit (Norden Dam) would have made the ''steeper grade''; neither would have survived using the current discount rate. The federal authorization for each used the old rate of 3.25 percent. So far, no irrigation project in Nebraska has proven economically feasible at the current discount rate, especially when the analysis also uses relatively low commodity prices for determining benefits. Until prices improve or the rules become more favorable or both, gaining federal authorization of irrigation projects presents a challenge, even if the funds and political support were there.

What counts as benefits in economic evaluations is also crucial. ''Multipliers,'' which, as mentioned earlier, measure the ripple effect of an economic stimulus in a community or region, do not apply in a cost-benefit analysis because they are considered a double counting of true benefits. Also important is the geographic point of view for tallying costs and benefits. For example, the increased value of production is a benefit to the local area but may represent added costs to the nation at large due to more price support payments and perhaps lower crop plantings in less productive, dryland states. Another example is construction costs. Construction represents a cost to the nation but is a benefit to the recipient and local or regional businesses. Without the reclamation theme of using irrigation as part of a conscious national policy of stimulating regional development, large irrigation projects lose much of their

The Sporhase Ruling: Water as a Commodity of Interstate Commerce

Joy Sporhase wasn't trying to make legal history; he was just trying to irrigate the land around his well. The problem was that the Colorado state line was located 55 feet from that well. In 1972 he purchased 640 acres of land at a farm auction near Lamar, Nebraska. Sporhase wasn't dim-witted. He had purchased the land at a good price and likely realized it was worth a good deal more money if he irrigated.

In 1976 the Nebraska Department of Water Resources warned him that proceedings would start against him and his partner, son-in-law Delbert Moss, unless they stopped pumping water and using it in Colorado without a Nebraska permit. However, when he applied to drill a well in Colorado, he was turned down because it was determined that the aquifer was already overused in that area. When interstate irrigation from the Nebraska well continued, the Nebraska attorney general brought a successful suit, successful, that is, until it reached the U.S. Supreme Court.

When the U.S. Supreme Court ruled on the case in 1982, it was a landmark opinion. Nebraska statutes had permitted transfers under conditions where the withdrawal is reasonable and not contrary to the conservation and use of groundwater. However, Nebraska's laws also required that the other state grant reciprocal rights to withdrawals. The Nebraska law was overturned because it was found that the reciprocity provision "operates as an explicit barrier to commerce" and was therefore in violation of the commerce clause of the constitution. The ruling did not, however, limit the state's right to conserve and preserve its water resources.

The Sporhase ruling probably had as many national ramifications as it did for Nebraska. It specifically made water an article of commerce under the constitution. This means states will find it difficult to treat water users from other states differently than those in their own state. The far-reaching implications of this ruling are still being determined.

—Steve Gaul, water resources planner, Nebraska Natural Resouces Commission

economic justification.

The Carter and Reagan administrations also introduced cost-share requirements for new projects. The Bureau of Reclamation now looks for a local sponsor to provide half the cost of investigating a potential project. Cost sharing for construction costs vary from project to project but would likely be at least 25 percent.

Taken all together, the erosion of the reclamation theme, growing political challenges to major surface-water projects and changes in the economic rules for project approval have marked the end of an era.

The Economic History of Groundwater Irrigation

The economic keys to surface-water development have been public financing of major irrigation works and a legal system designed to protect early irrigation investments against new water users. The economic history of groundwater irrigation has followed an entirely different scenario. Advances in technology within a legal setting that until recently imposed few barriers to groundwater withdrawals by individual landowners have been the key factors.

The History of Development

The history of groundwater development dates back to the first windmill-watered garden, but significant irrigation by pumping had to await technological advances. Progress was slow and late. By 1941, only about 1,500 wells were installed. By 1953, the year well-drilling began to take off, only about 8,000 wells were registered. At last count, irrigation wells numbered more than 76,000, an average of close to 1,800 new wells per year over the last four decades. Sometime in the early 1950s, the number of acres irrigated by groundwater overtook the number of acres irrigated by surface waters. The latest estimate is that more than 85 percent of all irrigated land in

Recent efforts in irrigation gained momentum in upland areas after the development of center-pivot technology and the droughts of the 1950s and 1970s. This rye field northeast of Imperial, Chase County, started blowing in February of 1975, and a 24-hour windstorm in March caused a total loss.

G. Alexander, Soil Conservation Service, from Resource Notes, CSD, v. I, no. 3

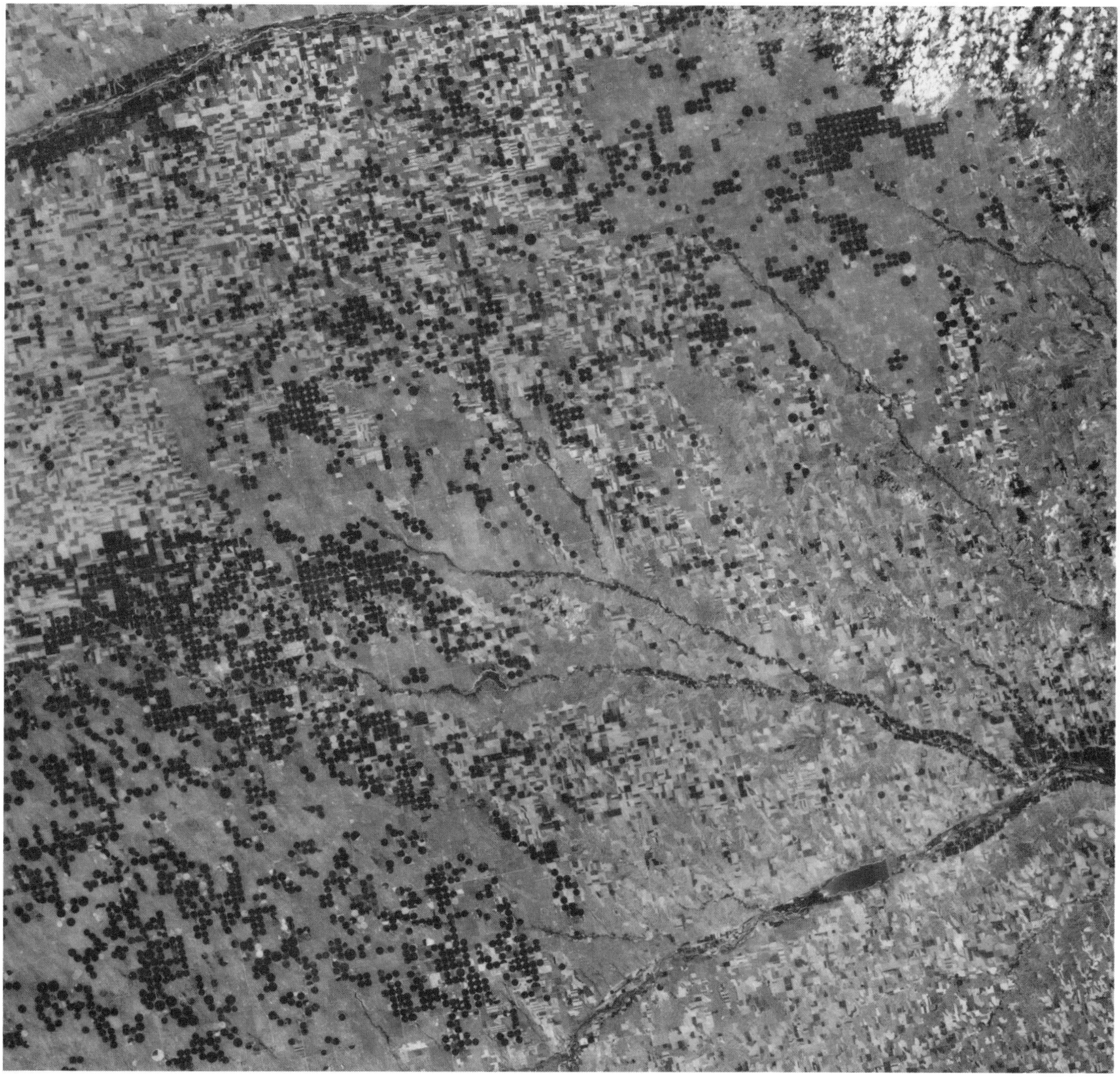

A Landsat satellite photo shows the distinctive pattern of center-pivot development dotting the land in southwestern Nebraska between the South Platte (upper left) and Republican (lower left to middle right) rivers, including all or parts of Dundy, Chase, Perkins, Hitchcock, Hayes and Lincoln counties.

the state uses groundwater supplies.

Groundwater development from the 1940s and into the 1950s was gradual—usually less than 500 new wells in any year. But the severe drought from 1953 through 1957 brought virtually non-stop drilling activity. In those five years, some 13,000 new wells increased groundwater development by 144 percent. There was a comparative lull until 1964. Then a few dry years ushered in a long, unbroken period of massive expansion of groundwater irrigation. For almost two solid decades, the number of irrigation wells increased by an average of more than 2,000 wells annually, with three peak years of more than double that during the dry spell of 1975-1977. This was an era marked by strong grain prices, escalating land values and adaptation of center-pivot technology. In 1983, the irrigation boom faded as a deep agricultural recession took hold. The number of new wells drilled each year dropped from about 2,200 in 1982 to 700 in 1983 and fell to a 50-year low of about 200 in 1987. Economic recovery and drought at the end of the 1980s have now restored the pace of irrigation development somewhat.

Who Owns the Groundwater?: How Water Law Affects Development

The state owns the water in the ground, but as a practical matter, whoever pumps it out for a beneficial use on the overlying land acquires a right to its continued use. Although shy of a property right in the water itself, the law allows landowners to use groundwater as if they owned it, so long as that use does not cause waste or injury to other users.

For the purposes of legal niceties, groundwater in Nebraska is public property, but this wasn't spelled out until a state Supreme Court decision in 1981 (State of Nebraska vs. Sporhase). Before then, beginning with a decision in 1933, case law governing groundwater established what is termed the "Nebraska Rule of Reasonable Use." This rule allows a

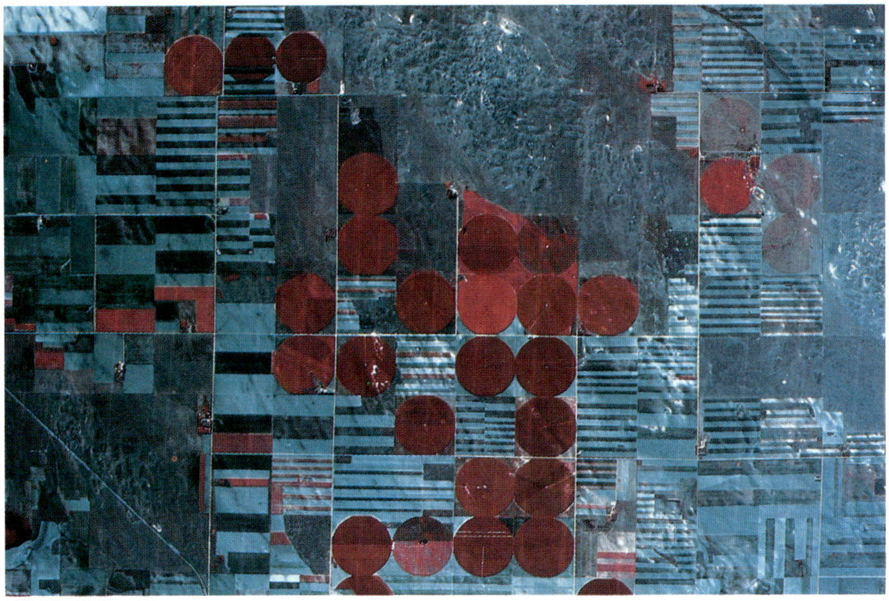

Les Sheffield, UNL

A color-infrared photo of western Nebraska shows center pivots among land that otherwise often must be cropped by alternating fallow strips to conserve water and fight wind erosion.

landowner to use underground waters if such use is beneficial, reasonable and does not injure others with similar rights to the waters. The rule also requires all owners to share any natural shortage of groundwater supply. The Nebraska Supreme Court in 1941 emphasized the proprietary interest that landowners have in the use of waters underlying their lands. In that decision, the Loup Public Power District had to pay damages when its power canal drained a private landowner's subirrigated field. Later, in 1978, the Nebraska Supreme Court required a groundwater irrigator to pay the cost of replacing a neighbor's domestic well, because irrigation impaired the neighbor's use of water.

In the Sporhase case, the state Supreme Court emphasized several points regarding ownership and regulation of groundwater. In no uncertain terms, it declared that groundwater is publicly owned, subject to state regulation and restricted to use just on overlying lands unless otherwise authorized by the legislature. In the court's words, ''Water is a unique commodity subject to state regulation to assure that it is available to everyone in the state in relation to their need, rather than their ability to pay for it.'' This confirmed the legislature's power to regulate groundwater use and transfers. Landowners, subject to state regulations, may use the water underlying their lands but have ''no property right in the water itself.''

When the Nebraska Supreme Court found that no private property right in groundwater existed, the legislature quickly reacted. In 1982, the Unicameral passed LB 375, which entitled every landowner to the ''reasonable and beneficial use of the groundwater underlying his or her land.'' This guarantee is subject to the state's groundwater statutes and the ''correlative rights of other landowners when the groundwater supply is insufficient for all users.'' (Correlative rights mean the mutual legal interest of all landowners using the same limited supply of groundwater.)

Nebraska groundwater law gave legal nourishment to rapid expansion in groundwater irrigation. For the most part, each landowner could act independently of others with full confidence of continued access to whatever groundwater there was. Even in situations of known groundwater declines, the basic law permitted and encouraged maximum groundwater development, because the impact of each additional well is shared by all. Massive irrigation development, where the dots representing wells on a small map blur into whole regions of smudges, would have been cumbersome unto impossible under an appropriative (first-in-time is first-in-right) system, similar to what governs surface water. Both the cumbersome process of applying for a water right and the insecurity of the water supply for new irrigators would have acted as brakes on groundwater development under such a system.

For the same reason, Nebraska's equal-opportunity-for-all groundwater law has contributed to some serious problems of groundwater depletions. Each additional well draining an overtapped aquifer marginally reduces the supply of water available to all, but the full quantity of water from the newest well goes exclusively to its owner. The economics of Nebraska's groundwater law (Rule of Reasonable Use) allows the owner to capture the full benefit of the new well while shifting most all the costs (due to decline of water supply) to others.

Over the years, the legislature has tried to come up with statutory fixes to the problem of overwithdrawals. In 1957, it mandated well registrations and minimum spacing requirements. In 1975, it passed the Groundwater Management and Protection Act, which allows natural resources districts (NRDs) to regulate pumping for irrigation. Except when these regulations prohibit new wells, which protects established economic interests against newcomers, the development incentive of groundwater law remains intact.

The Economic Impact of Technology on Groundwater Development

Technological advancements have made pump irrigation feasible and economical. At the turn of the century, the internal combustion engine revolutionized machinery but could only feebly lift water out of the ground with the centrifugal pumps available. Still, it was a start, and by about 1910 irrigation began to spread to valley lands with water-saturated sand and gravel at shallow depths of 20 to 50 feet. Application of the turbine pump to irrigation in the 1930s permitted much heavier lifts and far deeper wells. Adoption of new drilling techniques during this period cut well-installation costs by slashing the time for drilling deep wells from more than a week to a couple of days. This freed irrigation from being confined to the shallow aquifers adjacent to streams.

Although development would be gradual, as farmers and their bankers approached groundwater pumping with varying degrees of skepticism, these technological advances brought irrigation potential to millions of acres of prime Nebraska farmland. The drought of the 1950s, after more than a decade of watching the new technology, prompted thousands of farmers into action. Except for a brief lull in new well drilling from 1958 to 1962 and again during the agricultural recession of the 1980s, this was the start of an irrigation boom that continues today.

In the 1970s, center pivots accelerated the irrigation boom by bringing irrigation potential to millions of acres of farmland unsuited to gravity irrigation. Coming during a period of high, infla-

(Continued on p. 129)

The '70s Boom Meant Well Installations Skyrocketed; the '80s Bust Saw a Well-drilling Downturn

In the summer of 1973, when the Soviet Union began buying large amounts of U.S. wheat and feed grains following disastrous grain harvests at home, the prices of wheat and feed grains rose dramatically. Corn prices at country elevators in Nebraska that had gone as low as $1.27 per bushel in early winter of 1973 reached an average of $2.50 per bushel by August of 1973. Even larger purchases of feed grains occurred into the summer of 1974, and average corn prices for Nebraska farmers reached $3.35 per bushel in August and October of 1974.

In some cases, when the "basis" (difference between corn futures prices in Chicago and the price quoted at a local elevator) was very narrow, some farmers were able to contract with their local elevators for future delivery of corn at prices in excess of $3.50 per bushel. Also, in the summer of 1973, the price of No. 2 diesel fuel averaged $.16 a gallon, anhydrous ammonia prices were about $80 a ton and farm-equipment prices hadn't increased substantially. So, since the prices of some of the key inputs hadn't increased rapidly, as occurred after the 1973-1974 oil crisis triggered by the OPEC oil-producing nations, farmers with irrigation and high crop yields achieved substantial profits during the years of low input costs coupled with high corn prices.

During the 1970s, real interest rates (nominal rate minus the rate of inflation) for several years ranged from 1 to 4 percent, and even were negative a couple of years, and banks and other agricultural lenders were moving aggressively to increase their farm-loan portfolios. Since farmland values were increasing from 10-25 percent a year during the boom years of the mid-1970s, farmers who owned or had good equity in their farmland found their assets were increasing in relation to their liabilities—thereby showing a healthy balance sheet—but it was paper wealth since they weren't selling out at the time. During this expansionary period of agricultural lending, most of the lenders were more than willing to make loans on the basis of a farmer's equity or the "net worth" shown on an individual farmer's balance sheets.

In the mid-1970s, many farmers were able to obtain 75 to 100 percent loans from agricultural lenders for the purchase of additional quarter-sections of land plus the initial costs to drill an irrigation well and install a pump, motor and center-pivot system. In some cases, lenders were meeting each other calling on farmers they considered to be good credit risks. The result of high grain prices coupled with rapidly escalating land values and easily available credit at low real rates of interest offered too much temptation for many farmers.

Unfortunately, the days of reckoning for the "go-go" years of agricultural expansion and easy credit arrived during the farm crisis years from 1982-1987. The values for irrigated land declined by nearly 50 percent during this period with disastrous results for both borrowers and lenders with loans based on balance sheet lending. As land values declined drastically, the "paper net worth" values for farmers with large debts evaporated, and many found themselves with negative net worth and very limited or no borrowing power with their lenders.

At the same time, the real rate of interest had risen substantially. Many farm borrowers either lost land and equipment through foreclosure or were forced to sell all or part of their land and machinery. Some very good farmers who had expanded to take in offspring or their offspring's inlaws, with otherwise good management ability, found themselves having to eliminate as much debt as possible by whatever means available. In many cases the agricultural lenders "wrote off" part of the debts that were uncollectable. During this period, agricultural lenders returned to the sound principles of making loans on the basis of "cash flows" that could be generated by borrowers from any loans and abandoned "balance sheet" lending.

At the same time, exports of farm commodities declined as competition from other exporting nations increased, especially from the European Economic Community. This added to the problem through declining prices for most grains as supplies exceeded demand. The 1985 Food Security Act (farm bill), which took effect in 1986, helped farmers to recover gradually, but it wasn't until 1987-1988 that the agricultural economy began a slow recovery period.

In recent years, as government farm program payments have declined steadily, there is renewed concern by many involved in production agriculture, since prices for grains are again under severe supply-side pressures and below the true costs of production for most farmers as this is being written. In mid-October of 1992, the prices quoted for No. 2 yellow corn at most country elevators in Nebraska ranged from $1.80 to $1.95 per bushel.

The table below shows the correlation between the prices for corn received by Nebraska farmers, the number of irrigation wells drilled and the increase in irrigated acres in Nebraska from 1969-1991. As can be noted in this table, in 1970, there were slightly less than 4 million acres under irrigation in Nebraska, with a total of 38,593 registered irrigation wells as of Dec. 31, 1970. In 1973 when the first big grain sales to the Soviet Union occurred, the number of new irrigated acres increased by 305,000 acres. From 1974 through 1981, the state's irrigated acreage was increasing from 250,000 acres to as much as 500,000 acres (1976 and 1977) each year, and by 1981 the total acres that could be irrigated reached 7.5 million acres. In 1976, the number of newly registered irrigation wells increased by 5,907, a truly astounding figure.

During the farm crisis years of 1982-1987, the number of new irrigation wells declined to well under 1,000 per year to as low as 126 new wells (1986), and the number of new irrigated acres dwindled to the point where no net new irrigated acres were reported. During the first year of the Payment-in-Kind (PIK) farm program, when farmers could idle both non-irrigated and irrigated acres and still receive up to 92 percent of their farm-program payments, there were many thousands of acres of irrigated land signed up for this program.

Note that the average price farmers received for corn in 1986 was down to $1.52 per bushel, far below the true costs of production, and in 1987 the average price was still below most farmers' cost of production at $1.96 per bushel. As a result of these years of very difficult economic times for agriculture, both farmers and agricultural lenders were extremely cautious about expanding irrigation in Nebraska.

Weighted Average Corn Prices and the Number of Irrigation Wells Drilled and Increase in Irrigated Acres in Nebraska, 1969-1991

Year	Ave. Price Farmers Rec'd for Corn $/Bu.*	No. of Total Irr. Wells**	Irrigation Wells Increase***	Changes in Irrigated Acres: Total Irr. Acres***	Changes in Irrigated Acres: Irrigated Ac. Increase**
1969	$1.09	36,792	—	3,783,000	—
1970	$1.25	38,593	1,801	3,995,000	212,000
1971	$1.11	40,800	2,207	4,200,000	205,000
1972	$1.53	42,899	2,099	4,478,000	287,000
1973	$2.41	45,035	2,136	4,783,000	305,000
1974	$2.92	48,254	3,219	5,050,000	267,000
1975	$2.48	52,938	4,684	5,400,000	350,000
1976	$2.06	58,845	5,907	5,900,000	500,000
1977	$1.97	62,790	3,945	6,400,000	500,000
1978	$2.22	64,246	1,456	6,700,000	300,000
1979	$2.48	66,247	2,001	6,950,000	250,000
1980	$3.08	68,071	1,824	7,200,000	250,000
1981	$2.47	70,307	2,236	7,500,000	300,000
1982	$2.82	71,001	694	7,600,000	100,000
1983	$3.13	71,488	487	7,700,000	100,000
1984	$2.58	71,981	493	7,800,000	100,000
1985	$2.22	72,210	229	7,900,000	100,000
1986	$1.52	72,336	126	7,900,000	zero
1987	$1.96	72,467	131	7,900,000	zero
1988	$2.48	72,991	524	7,900,000	zero
1989	$2.30	74,000	1,009	8,000,000	100,000
1990	$2.40	74,847	847	8,000,000	zero
1991	NA	75,363	616	NA	NA

*Data from E.C. 92-883-C, "Crop & Livestock Prices for Nebraska Producers," Compiled by Allen C. Wellman, UNL Co-

(continued on next page)

operative Extension Service, June, 1992; Weighted Average Prices weighted by monthly sales within marketing year, Sept. 1 through Aug. 31.

**Registered Irrigation Wells from Nebraska Agricultural Statistics Service Annual Reports and from Nebraska Department of Water Resources Reports on Registered Irrigation Wells and Data Bank at the Nebraska Natural Resources Commission.

***Data from Nebraska Agricultural Statistics Service Annual Reports Lincoln, Nebraska, 1969-1991

—Leslie F. Sheffield, coordinator of outreach programs, Water Center, and formerly irrigation extension economics specialist, UNL

What Happens When It Costs Too Much to Pump: The Socio-economic Effects of the Conversion from Irrigated to Dryland Farming in the Great Plains

Humans have long been aware of the benefits to food production that come from irrigated agriculture. Evidence suggests that as early as 5000 B.C. the Egyptians were using water from the Nile to irrigate crops alongside the life-giving river. Early irrigation developments also occurred in China, Phoenicia, Peru, Mexico and what is now the southwestern United States.

Humanity's ingenuity led to widespread experimentation with techniques for storing and transporting water. Truly fascinating practices evolved from the simplest of procedures through increasingly intricate systems of dams, reservoirs, canals and a variety of ingenious application strategies.

Parallel to this development, the world has witnessed an almost exponential growth in the demand for water. Not only is this valuable natural resource needed to enhance food supplies and personal amenities for a burgeoning world population, but demands for industry, river navigation, electric power generation, recreation and the preservation of wildlife areas have become vigorous competitors for both surface water and groundwater.

Researchers have documented well how the emergence of irrigation in an area has a favorable socio-economic influence on community life. However, now they must turn to what happens when irrigation waters are depleted. The minimal data at hand have revealed some probable consequences.

First, farms will probably grow larger, presumably through the combination of parcels no longer considered economically feasible. Second, as farms grow larger, rural population is likely to decrease. This exacerbates the downward trend in rural residents that long has troubled rural communities, particularly in maintaining viable social and economic institutions. Further, the expanded acreage for dryland farming may cause a decline in the overall profitability of farming.

Where irrigated land must revert to dryland farming, a resultant drop in the property tax base of an area will probably also follow, unless manufacturing or some other type of income-generating activity can be introduced. The property tax base of a community, including both farm and village assets, is crucial to maintaining the community's infrastructure and public services, to say nothing of indirect influences on the ability of churches and other nongovernmental institutions to function.

We are at the threshold of an era where the adequacy of our water supplies is threatened by ever-increasing demands. Essential watchwords for the future, in rural as well as urban areas, include water conservation and intelligent, fair-minded water-management policies, coupled with further sociological research.

—Otto Hoiberg, UNL emeritus professor of sociology

Managing (or Avoiding) the Transition to Dryland Farming Depends on the Right Questions and the Right Information

Two factors determine a farmer's choice to farm with or without supplemental water: a ready supply and the cost of using the water, which, along with other costs, must be low enough to ensure a profit. When groundwater levels decline or energy and other operating costs climb too high, farmers often convert from irrigated to dryland farming.

In October of 1985, a symposium in Denver explored why that transition occurs and discussed ways to make it pay off. Called "Causes and Consequences of the Transition to Dryland Agriculture," it was sponsored by the Great Plains Agricultural Council and the University of Nebraska-Lincoln (UNL) Water Center.

The conversion to dryland agriculture is neither inevitable nor final, according to Harry Mapp, professor of agricultural economics at Oklahoma State University. Such a conversion is linked to the economically viable life of the High Plains Aquifer, which is determined by how long it will be profitable to produce with irrigation compared with dryland farming.

Irrigation scheduling and precision-application systems can greatly reduce the cost of irrigation and extend the life of the supply, explained Mapp, a risk-management and farm financial-management expert. The use of these kinds of innovative management practices is a key to avoiding complete conversion to dryland farming.

In addition, certain expenditures might be reduced that have an impact on the profitability of irrigated farming compared with dryland agriculture. Such expenses include seeding and fertilizer rates; amounts of herbicide and insecticide and price of those inputs; the cost of buying and operating machinery; and crop prices. Other factors include the soil type and availability of water, as well as soil and water management. With so many factors to consider, it's difficult to estimate how long irrigation from the High Plains Aquifer will be feasible.

If commodity prices rise and water is available, Mapp said, farmers will be able to adjust to energy-cost increases and continue to irrigate. But if commodity prices remain constant, irrigated acres will decrease.

Speakers at other sessions of the symposium stressed the importance of researching all possible scenarios for Great Plains agriculture. Joseph R. Johnston, director for the Southern Plains Area of the U.S.

(continued on next page)

Department of Agriculture Agricultural Experiment Station in College Station, Texas, said, "It's crucial to know what the production potential is for all classes of land."

Leo E. Lucas, then dean of the Cooperative Extension Service at UNL, said, "we have to lay out the consequences and the alternatives of a host of different scenarios in an educational program." Economics are the basis for the scenarios, he said, and records must therefore be kept to create those scenarios.

R. Wayne Shawcroft, irrigation agronomist with Colorado State University Cooperative Extension, said the farmer needs to be presented with a number of options. "We need to work with farm-management specialists to develop economically viable alternative cropping systems," he said.

"Going back to strictly dryland farming is not all a bad situation, and this may be the farmer's best alternative," Shawcroft added.

The primary purpose of the symposium was to develop recommendations to the Great Plains Agricultural Council for local, state and federal policies that would ease the transition from irrigated to dryland farming. These included:

—Program development: accelerate programs for the revegetation of abandoned land and for weather-modification studies on hail suppression and for groundwater recharge.

—Research priorities: determine which research should be conducted first to best address both short-term and long-term problems.

—Water law, policy and planning: encourage local, state and federal agencies to examine existing water laws, policies and plans and propose those that would enhance more efficient use of water.

—Institutional arrangements: ask the council to examine the structure of local, state and federal agencies working with farmers and eliminate cross-purpose regulations that might encourage improper land use and marginal irrigation operations.

—Extension programs: combine all systems of extension to develop alternative management schemes, including risk management.

—Brad Rundquist, weekend and copy editor,* Ft. Dodge (Iowa) Messenger, *and formerly CSD editorial assistant; adapted from a special issue on the symposium, published by the UNL Water Center, written and edited by Pat Larsen.

(Continued from p. 126)
tion-driven land values and high crop prices, some boosterism set in. Marginal lands came under the plow and established the limits of irrigation when land and crop prices fell. In 12 Sand Hills counties, 50,000 out of more than 215,000 acres developed for irrigation proved to be a bust. In most cases, low crop prices idled the pivots, but poor site selection was the culprit at times. High water tables some places caused saturated soils, marshes and ponded water. In other cases, poor management led to wind erosion with loss or burial of topsoil, mired pivots and damaged crops.

Advances in irrigation technology unleashed the potential of groundwater and allowed farmers to turn millions of acres of dryland into irrigated cropland. This doubled and often more than doubled the productive potential of each acre, with a guarantee of a crop in dry years thrown in as a bonus.

The economic advantages of recent irrigation technology doesn't end there. Compared to using ditches, pumping water from the ground is more efficient in several categories. For one, the water is there virtually on demand. Surface-water irrigation depends on the stream being full at the right time, and, assuming it is, the water comes all season long, whether it is needed or not. In theory, a project may supply each farmer-customer with enough moisture to grow corn on 160 acres, and in a good year, with enough snow and rain at the right times, this is possible. In practice, a project has to keep water flowing in the canals constantly. (This is why return flows for downstream users are so high.) To be sure of having enough water during critical growing periods, farmers must either cut their plantings far below the 160-acre allotments, supplement project water with groundwater or gamble on rains.

Another efficiency of pump irrigation is in the conservation of water. The open canals of most surface-water projects typically lose half or more of the water to evaporation and seepage before it even reaches the farm. On-farm losses are almost as bad wherever open ditches distribute the water in the fields. (This explains the massive groundwater mounding near these projects.)

Center-pivot technology offers all the advantages of other methods of water distribution and more. It can be sparing in the use of water, through low-pressure sprinker or drip systems or carefully timed applications of moisture. It is also sparing in the use of scarce labor. One exhausted person can keep up with about 400 acres of gravity-irrigated crops during the irrigation season. The same person could manage twice that acreage of pivot-irrigated crops and sleep better at nights too. By applying chemicals through the pivot, the farmer can avoid the time and cost of several trips across the fields. These advantages are bringing a conversion of gravity irrigation to center pivots, despite the higher capital costs of pivots.

Financing Pump Irrigation

Coming up with the cash to water dryland acres is a private matter between individual farmers and their creditors, thank you.

This has been one of the great strengths of pump irrigation and one of the major reasons groundwater irrigation quickly outpaced surface-water projects once technology allowed it to compete. Free from the whims of public financing, free from government dictating acreage limitations (other than those that come with federal farm programs), free from political debate about growing more crops that are in surplus, free from worries about snail darters or other endangered or threatened species, farmers had only their bankers to convince that irrigation was smart. And if one banker was doubting, as often happened early on, others were more alert to irrigation's pay-back potential.

The close-to-the-chest nature of financing for pump irrigation has left few public records of who provided the money or what factors influenced the flow of capital to irrigators. By category, however, the "who" included local bankers, national insurance companies and the Federal Land Bank. In the beginning, banks surrounded by acres irrigated by surface-water projects were quicker to foresee the safety of a loan for pumps and pipe. Financiers in places like Bertrand and Hastings, who had watched Tri-County-watered harvests for a decade or more, found it easier to commit their institutions' deposits to private irrigation schemes. (Perhaps this partly explains why irrigation is so extensive in these areas.) Terms of the loans helped ease bankers' worries about repayment. Typically repaid within five years, with the equipment or land as security, not much could go wrong, especially with a harvest guaranteed against dry spells.

In the early days of pump irrigation, the makers and sellers of pumping equipment often played an important role in helping farmers secure loans. One example is the Western Land Roller Company in Hastings and its relationship with the First National Bank in Hastings. In a presentation to the 1972 Nebraska Water Conference, the president of the bank described a 15-year period beginning in 1956. During that period, the bank aggressively accepted contracts from the entire trade area of Western Land Roller, with total loans for wells and pumps exceeding $7.6 million.

Early on, the factors influencing loans focused on the collateral for the loan and management abilities of the borrower, assuming of course that a dependable water supply existed. With gravity irrigation, the need for collateral sometimes posed a barrier to financing the costs of leveling the land. Cost for grading would sometimes far exceed the price tag for the well and irrigation pipe. In cases where an indebted farmer could not refinance an existing mortgage on the land, irrigation would have to wait until another day. Everything else being equal, the perceived management abilities of a loan applicant would dictate the lender's response. In the words of one banker, "A poor dryland farmer, if he can get water to his crops, will be a poor irrigation farmer."

Once the technology became familiar, the importance of collateral gave center-pivot loans an advantage. Little, if any, land-preparation costs, security in the system itself and the impact of irrigation on land values combined to make center pivots an attractive addition to any loan portfolio.

Federal tax treatment of agriculture has also had an impact on private financing of irrigation. Full deductibility of irrigation expenses such as land leveling, depreciation schedules for equipment and cost-accounting techniques allowed farmers to shift some costs to the U.S. Treasury. Until the 1986 tax reform bill, farmers could take advantage of the investment tax credit as well. All of these favorable tax provisions helped to reduce the federal income taxes otherwise owed by irrigators, which encouraged more development.

Since the collapse of overly inflated land values in the 1980s, however, lenders have added cash flow as a critical test for any agricultural loan, including for irrigation. Many have computer programs to calculate the farmer's ability to repay a loan under different scenarios. Burned by their experiences during the agricultural recession of the last decade, lenders no longer rely on paper wealth alone to justify a major irrigation investment.

Speculation Fever

Agriculture faced a major financial restructuring during the recession of the 1980s. One study found that by 1984, 20 percent of all Nebraska farmers held 60 percent of all farm debt and faced moderate or severe financial problems. Another statistic looms large: financial lenders had committed 30 percent of their loan volume to farmers with debt-to-asset ratios of more than 70 percent. This was the group most heavily caught up in the debt-leveraged land boom.

During the entire decade prior to the recession, borrowing to expand farming operations was a gambler's dream. Land values marched steadily upward with inflation and often did better than that. (In Nebraska, they quadrupled from 1970 to 1981! Even after subtracting inflation, they doubled during this period.) The true cost of borrowed money, after subtracting for inflation, was cheap. From 1970 through 1981, the average real interest rate charged by the Federal Land Bank exceeded 4 percent for only eight

A weir, perhaps in the Frenchman Valley Irrigation District (F.V.I.D.), 1913

Nebraska State Historical Society

quarters. It was 1 percent or less during 14 quarters. And commodity prices were good to excellent. Within this economic environment, a well-leveraged farm with debts equal to 70 percent of assets was no more speculative than the traditional 20 percent down payment on a home mortgage. It seemed to make good financial sense.

The causes of the financial squeeze on farmers after 1980 were basic: high interest rates, declining inflation and too much production of all types. For a rare period, cattle, hogs and grain were all unprofitable at the same time. Sharply falling land prices, dropping 25 percent from 1981 to 1984, destroyed the inflated balance sheets that had served as the main justification for too many land and equipment loans.

Irrigation did not instigate these problems, but it became mixed up with them, especially in some parts of the state. The pace of center-pivot developments in the Sand Hills on fragile soils prompted much concern. Cheap land, ample water and a favorable income tax code created a vision of extensive conversion of rangeland to crops. One university agricultural economist remarked, "We've seen ranchers become farmers almost overnight."

A 1976 study by the Center for Rural Affairs in Walthill documented the apparently disproportionate involvement of outside investors in the irrigation boom in two Sand Hills counties and in irrigating marginal, sandy soils. A few dramatic examples of partnerships and corporations buying huge tracts of land and immediately installing a horde of pivots lent credibility to these warnings.

The pace of irrigation development in the Sand Hills became linked to concerns about rising land prices generally and the tax advantages of wealthy investors. High land prices would defeat the beginning farmer. A major shift of land ownership to absentee investors would disrupt the traditional social structure of rural areas. Mismanagement of soil and water by unknowing and uncaring "tax farmers" would degrade the state's natural resources for future generations. These were the fears that Sand Hills irrigation reinforced.

Some economic and financial facts supported these concerns. A beginning farmer could not meet the combined debt load of over-priced land and the expensive irrigation equipment needed to make it competitive. Also, prior to 1987, the federal tax code gave important tax concessions that helped some agricultural producers. Cash accounting rules, depreciation schedules, lower rates on capital gains and the investment tax credit benefitted everyone with agricultural income to some extent but gave greater benefits to those with higher incomes and higher income-tax rates. In addition, the ability to apply losses from agriculture against other sources of income turned farmland into a potential tax haven for wealthy city folk.

Nebraskans responded by amending the state constitution in 1982 to restrict corporate ownership of farm and ranch land. Since then, other changes have eroded the economic foundation for speculation. The Tax Reform Act of 1986 dropped the investment tax credit, eliminated agriculture as a tax shelter, reduced tax rates and restructured the treatment of capital gains. The collapse of land prices restored a sense of reality and discipline to farm financing, and lower commodity prices have broken the investment fever. In addition, the conservation provisions of the 1985 and 1990 farm bills imposed a variety of restrictions that discourage irrigation development on erodible soils.

Unless many of these factors change again, groundwater irrigation will not likely return to the fast pace of development during the 1970s and early 1980s. The institutions that finance land sales and irrigation development now focus on the cash-flow potential of a farm, the farmer's management ability and the quality of the land as the true collateral against default. Gradual expansion of irrigation to the remaining acres of good cropland that offer good access to water will probably continue, as will the conversion of gravity-irrigated systems to center-pivot technology. But the boom is off, and the Sand Hills appear to be safe from massive irrigation development.

Sources: "Analysis of Tax Incentives for Intensive Irrigation Development in the Nebraska Sandhills," Maurice E. Baker, 1983, Nebraska Water Resources Center, Institute of Agriculture and Natural Resources, University of Nebraska-Lincoln; "Availability and Use of Water in Nebraska, 1975," June 1979, Ray Bentall and F. Butler Shaffer, Nebraska Water Survey Paper No. 48, Conservation and Survey Division, Institute of Agriculture and Natural Resources, University of Nebraska-Lincoln; *The Bureau of Reclamation*, William E. Warne, 1973, Praeger Publishers; "Draft Report on the Sandhills Area Study," Nebraska Natural Resources Commission, State Water Planning and Review Process, March 1992; "Economic Impact of Irrigated Agriculture," Leslie F. Sheffield, 1985, The Irrigation Association Educational Foundation, Arlington, Virginia; "The Economic Impact of Irrigated Agriculture on the Economy of Nebraska," Theodore W. Roesler, F. Charles Lamphear and M. David Beveridge, 1968, No. 4, UNL Bureau of Business Research; "The Economic Importance of Irrigation in Nebraska," Charles F. Lamphear and Merlin W. Erickson, February 1990, Bureau of Business Research, University of Nebraska-Lincoln; "A Financial Profile of Indebted Farm Operators in Nebraska—1984," Bruce B. Johnson, April 1984, Department of Agricultural Economics, Institute of Agriculture and Natural Resources, University of Nebraska-Lincoln; "Impact Analysis of Irrigated Agriculture on Nebraska's Economy 1967-1970," Bureau of Business Research, September 1974, Nebraska Economic and Business Report No. 8; "The Importance of Irrigated Agriculture to Nebraska's Economy," Charles F. Lamphear and Merlin W. Erickson, in Business in Nebraska, v. 44, no. 531, December 1988, Bureau of Business Research, University of Nebraska-Lincoln; "The Law of the Missouri," John P. Guhin, in South Dakota Law Review, vol. 30, p. 350-487, Spring 1985; "1990-91 Nebraska Agricultural Statistics," Nebraska Department of Agriculture, October 1991; "Nebraska Agricultural Statistics Historical Record, 1866-1954," State-Federal Division of Agricultural Statistics, U.S. Department of Agriculture, undated; "Nebraska Farm Real Estate Market Developments," Bruce B. Johnson and Ronald J. Hanson, various from annual series, Department of Agricultural Economics, Institute of Agriculture and Natural Resources, University of Nebraska-Lincoln; "Nebraska Farm Real Estate Market Developments in 1983-84," Bruce B. Johnson and Ronald J. Hanson, July 1984, Department of Agricultural Economics Report No. 139, Department of Agricultural Economics, Institute of Agriculture and Natural Resources, University of Nebraska-Lincoln; "Nebraska Statistical Handbook 1990/1991," Nebraska Department of Economic Development; "Nebraska Water Resources and Irrigation Development Seminar for the 70's," Leslie F. Sheffield, ed., Proceedings, March 2-3, 1972, Water Center, Institute of Agriculture and Natural Resources, University of Nebraska-Lincoln; "Property Rights in Groundwater," Nebraska Natural Resources Commission, January 1983, Report No. 4, in Policy Issue Study on Selected Water Rights Issues, State Water Planning and Review Process; *Public Power in Nebraska—A Report on State Ownership*, Robert E. Firth, 1962, University of Nebraska Press, Lincoln and London; "Reclamation Project Data," Bureau of Reclamation, 1948, U.S. Government Printing Office, Washington, D.C.; "Report on the Framework Study," Nebraska Soil and Water Conservation Commission, May 1971; "Report on the Framework Study—Appendix C: Land and Water Resources Problems and Needs," Nebraska Soil and Water Conservation Commission, September 1971; "Report on the Framework Study—Appendix D: Survey of Nebraska Water Law," Nebraska Soil and Water Conservation Commission, June 1971; "Report on the Platte River Basin," Missouri River Basin Commission, June 1976, Nebraska Level B Study; "Surface Water Use in Nebraska's Platte River Valley," Michael Jess and Stanley M. Christensen, 1980, Nebraska Department of Water Resources; *Water in Nebraska-Use, Politics, Policies*, James Aucoin, 1984, University of Nebraska Press, Lincoln and London; "Wheels of Fortune—A Report on the Impact of Center Pivot Irrigation on the Ownership of Land in Nebraska," Bert Evans and others, 1976, Center for Rural Affairs, Walthill, Nebraska.

Circles of Green on the Plains: Frank Zybach, Valmont and the Center-pivot Revolution

by Leslie F. Sheffield
Coordinator of Outreach Programs, Water Center
and Formerly Extension Irrigation Economics Specialist, UNL
and Brad Rundquist
Weekend and Copy Editor, Ft. Dodge Messenger
Formerly Editorial Assistant, CSD

Few in this world leave a mark to rival that made by Frank Zybach. His can be seen from space.

"Traveling in orbit around the earth at an altitude of some 270 miles, the Skylab astronauts rotated their spacecraft to establish a bearing on one of their principal check points: a cluster of several hundred round green spots, each a half a mile in diameter, arrayed in an orderly pattern on the earth's surface below them," wrote William R. Splinter, former head of the agricultural engineering department and now dean of graduate studies at the University of Nebraska-Lincoln, in the June 1976 edition of *Scientific American*.

Center-pivot irrigation in the Nebraska Panhandle

Stan Hass, NU

Those round green spots were fields watered with Zybach's invention—the center-pivot irrigation system. Center pivots now dot most of the Great Plains and parts of other arid nations. In Nebraska alone there were an estimated 26,741 systems in 1988.

Zybach, who had but a seventh-grade education, designed and built the first system in the summer of 1948. His goal was to eliminate the tremendous amount of labor needed to operate conventional sprinkler irrigation systems.

The invention of labor-saving devices was nothing new to the Oregon native. Zybach, born in 1894, moved with his family to a farm near Columbus when he was 3 months old. Many childhood hours were spent watching his father at work, providing blacksmith services to neighboring farmers. He quickly adopted a passion for metalwork and learned from his father all he could.

Instead of grumbling about the tedious task of walking behind the horse-drawn harrow for hours on end, the 13-year-old Zybach did something about it. His first invention was a cart with a seat and two swivel wheels he could lock to move in a straight line. The front end of the device attached to the harrow and Zybach rode in relative comfort across the field.

On Oct. 8, 1920, at age 26, Zybach filed for his first patent. His invention was an automatic tractor guide for steel-wheeled tractors, very few of which existed in those days of horse or humanpower. Using the device, the farmer was able to leave the plow after digging the first furrow around the outside of the field. The guide followed on the inside of the first furrow. It continued to move inward a furrow at a time until the tractor reached the center of the field. Zybach created an emergency kill-switch in case the tractor jumped the furrow or overheated. Shortly after receiving his patent, he sold it to the Chase Plow Company in Lincoln. Chase manufactured the device for several years, but the concept never quite caught on.

The design and construction of an automatic automobile transmission was Zybach's project in the 1930s. He successfully built and nearly sold his invention to the Chrysler Corporation before patent lawyers discovered that one component of his invention resembled an already patented device. Unable to build the transmission without the component, Zybach dropped the project.

By 1947, Zybach had moved to work as a tenant farmer on a dryland wheat farm near

A satellite photo of center pivots in Holt County from the early 1970s

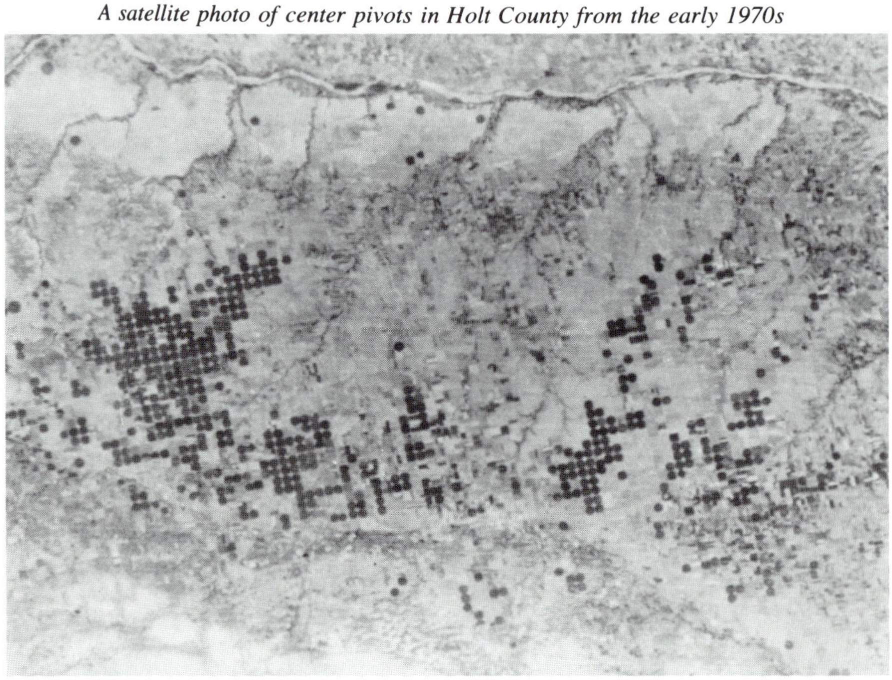

Les Sheffield, UNL

Strasburg, Colorado. He and a friend attended a sprinkler-irrigation demonstration that summer on a nearby farm. During the demonstration, Zybach noticed how the workers had to set up the sprinkler lines by hand, wading through thick mud to disconnect the pipe and move it one piece at a time to the next setting. Convinced there was a better way, Zybach began to develop his idea of an automatic, self-propelled sprinkler irrigation system. He went to work in his garage almost immediately.

By the next summer, he had completed a small center-pivot system propelled by the water flowing through it. The system had two sprinkler towers. Unable to find a device to keep the towers in alignment as the system moved, Zybach developed his own—a system of cables, pulleys and automatic water-pressure adjustments. He applied for a patent on July 27, 1949, for the "Zybach Self-Propelled Sprinkling Apparatus," and was granted Patent No. 2,604,359 on July 22, 1952.

In the spring of 1952, Zybach had built a full-scale, five-towered center-pivot system to irrigate 40 acres. The device was installed that year on the farm of Ernest Engelbrecht near Strasburg, Colorado, and was used until the early 1970s to grow alfalfa.

Zybach lacked the financial ability to mass produce center-pivots and so contacted his long-time Columbus friend A. E. Trowbridge, a car dealer and entrepreneur. Trowbridge agreed to finance production of the systems in a rented machine shop. Zybach sold 49 percent of the patent rights to Trowbridge for $50,000 and moved back to Columbus to go to work.

During the first two years of the partnership, only about 19 units were built. The new concept wasn't easy to sell. The systems looked strange, didn't operate like any of the proven irrigation methods and hadn't really been put to the test. Most farmers were, therefore, reluctant to gamble. Those who were willing to take a chance were often turned away at the bank. Most bankers were unwilling to finance a new-fangled device. Many told the partners that their system would never work.

Another problem was the partnership itself. The men were great friends but possessed starkly different philosophies about how the business should be run. The inventor, Zybach, was forever making changes to this or that part of the system and no two he built were alike. (It was at this time Zybach altered his design to accommodate taller crops like corn. The pipeline of his earlier system had been suspended only 2 or 3 feet above the ground.) The businessman, Trowbridge, quickly saw it would be impossible to make a profit if each system differed from the one before it.

So, during the summer of 1954, Trowbridge and Zybach decided to sell their manufacturing rights. The partners sold them to Robert B. Daugherty of Valley Manufacturing Company in Valley, Nebraska—now called Valmont Industries, Inc.—in exchange

Les Sheffield, UNL

A.E. Trowbridge and Frank Zybach

for 5 percent royalty rights for every unit sold until the patent expired in 1969. Distribution rights remained with Zybach and Trowbridge until the late 1950s. Valley had until then manufactured mainly short-line farm machinery such as bale loaders and grain and ear corn elevators for farm use.

Daugherty took a chance when he purchased manufacturing and sales rights to the relatively unknown and untested "center-pivot" irrigation system in 1954. The system was at the time unheralded and unpopular with most farmers and bankers. But he had seen a pivot in action and was convinced the system had merit.

The first system was installed in Holt County near Page in 1955. The next year, Trowbridge and his son-in-law purchased land north of Atkinson and put in several center pivots to produce corn and alfalfa, marking the first multiple installation under the same ownership. Soon, the Trowbridge farm was producing corn yields equalling or exceeding those of prime irrigated lands in the Platte River Valley.

In the mid- and late-1950s, the system underwent many changes by Valley's engineers to improve its reliability and flexibility. They developed a variable-speed drive so different amounts of water could be applied per revolution. They also added end guns with automatic shut-off and safety switches to shut down the system when pressure was lost or the system lost alignment. Valmont workers also discovered that some liquid fertilizer could be injected into the system and distributed with great accuracy.

The company only built 7 systems in 1954 and by 1960, annual production had reached only 50 a year. But Daugherty and his colleagues were persistent and interest finally caught on by the early 1960s as more and more people left farms and moved to the cities, increasing the demand for labor-saving agricultural devices. Interest was also spurred when word got around about the economic potential of converting relatively low-cost sandy or uneven lands into valuable irrigated lands through the use of center-pivots.

Center-pivot irrigation was first used on a widespread basis in Holt County in the late 1960s. There, farmers began to realize how well the systems worked to produce corn on shallow, sandy soils. Center-pivot irrigation proved to be the salvation of people farming in such an area near Atkinson, known locally as "poverty ridge." Using center-pivots, corn yields on the ridge were quickly boosted to levels comparable with those on choice irrigated lands in other parts of the state.

The real center-pivot boom came in 1973. That year, corn and wheat prices rose dramatically as the Soviet Union began to buy up large quantities of American grain. The 1972 center-pivot inventory, published by the Conservation and Survey Division of the University of Nebraska-Lincoln, recorded 2,735 systems in Nebraska. By 1975, the number had risen to 8,517. The latest inventory, conducted in 1988, showed 26,741 center-pivot systems in the state.

Zybach received the first Pioneer Irrigation Award given by the state Water Conference Council and the University of Nebraska-Lincoln in 1973. In 1974 he was presented with the Sprinkler Irrigation Association's Industry Achievement Award. He died on Aug. 19, 1980, in Columbus at age 86. By 1969, Zybach and Trowbridge had both made more than $1 million on the 5 percent royalty deal with Valmont.

Valmont Irrigation, a division of Valmont Industries, Inc., was the first to manufacture extensions to water the corners of fields, the first to produce and market a linear-move system which moves across the length of a field instead of in a circle, and the first to market a linear-move system that automatically detaches from and connects to groundwater pipe risers as it moves.

The oldest and largest of all center-pivot manufacturers, Valmont was the first company to sell manufacturing rights to the Soviet Union and the first to export their product to foreign countries. In 1977 they became the first to sell center-pivot systems to China. The company has sold units in all 50 states and 70 foreign nations. Valmont's irrigation-related net sales for 1991 were $128,210,000. That year the company, which trades its common stock on the NASDAQ market, employed 4,320 people.

All told, more than 75 companies have manufactured and marketed center pivots. Most were in business only a few years. Only seven companies survived, and five are in Nebraska. Besides Valmont, they are: Reinke Manufacturing Company of Deshler, Lindsay Manufacturing Company of Lindsay, T-L Irrigation Company of Hastings and the Lockwood Corporation of Gering.

It is estimated that the Nebraska firms together have about a 90 percent share of the domestic and export markerts.

In the 1976 *Scientific American* article, Splinter also said, "What is being observed (by the Skylab astronauts) is perhaps the most significant mechanical innovation in agriculture since the replacement of draft animals by the tractor."

Profiles:

Four Men Named Kingsley: Fundamental to the Tri-County Legacy

Four Kingsleys have played a prominent part in the development that resulted from the Central Nebraska Public Power and Irrigation District.

F. R. Kingsley, a Minden banker, was active in the plan sponsored by Joel Hull in the 1880s to use Platte River water for hydroelectric power and industrial development in the Minden and Holdrege area and eventually became president of the Nebraska Canal and Improvement Company when it was incorporated in 1889. Through the years he was an enthusiastic supporter and promoter of the development.

George P. Kingsley, brother and banking partner of F. R. Kingsley, joined Charles W. McConaughy in about 1908 in championing the use of Platte River spring and fall flows for supplemental irrigation in Phelps and Kearney counties and became vice-president of the Tri-County Supplemental Water Association on its organization at Holdrege in 1913. In 1922 when the Central Nebraska Supplemental Water Association was formed at Hastings, he was elected vice-president. Upon his death on October 29, 1929, a few hours after he had driven back to Minden from Kearney with Joseph S. Canaday after a water conservation meeting, the area newspapers published the following eulogy delivered at his funeral, "His death was a grievous loss and a severe blow to the cause of irrigation—a businessman of sound judgment, unswervable determination, and broad vision, he was a commanding figure in this region. His contribution of time, ability, and money to the cause of conservation proclaimed him the leader of that movement which has for its objective the conservation and utilization of Nebraska's waters." It was in his honor that "George P. Kingsley Dam" was named.

Fred R. Kingsley, a son of the above named F. R. Kingsley, was active in promotion of the project and became a member of the original board of directors of the Central Nebraska Public Power and Irrigation District with its chartering on July 24, 1933, serving until 1941. He was treasurer of the district from 1937 to 1941.

Donaldson W. Kingsley, a son of George P. Kingsley, followed in his father's footsteps in his enthusiastic support of the project and became Central's first president in 1933, serving during the turbulent eight years it took to get financing and build the $45 million development. At the time of the dedication of Kingsley Dam in 1941, *The Hastings Tribune* paid tribute to "All who had undertaken an important and irreplaceable part in making the dream come true," and, naming Don W. Kingsley of Hastings, said, "The Dam and the water it holds, which one day will . . . bring happiness and prosperity to thousands of people in South Central Nebraska, is his monument too." "Without Don Kingsley the work of the others would have been in vain," it said.

—Adapted and reprinted from* Irrigation Pioneers: A History of the Tri-County Project to 1935, *by Gene E. Hamaker

Charles W. McConaughy: Saw Flood Flows as the Key to Irrigation

Charles Washington McConaughy was born in Champaign, Illinois, in 1859. He came to Holdrege, Nebraska, in 1892 and worked as a clerk and salesman. In 1893, he entered the grain business. He also served as mayor of Holdrege for eight years. These business and civic activities aroused his interest in irrigation development.

While serving his fifth term as mayor, McConaughy proposed a plan to supplement annual rainfall by soaking the subsoil with water from the Platte River during its flood stages in the spring and fall. His inspiration for this idea came from observing a field of wheat which "in the good spots was as high as my waist," but which elsewhere was stunted and headed out before maturity. From this observation, he conceived his idea of supplemental water and devoted many years to pursuing his dream.

McConaughy strongly promoted his supplemental water theory and spoke at many public meetings during 1913. Finally, the Tri-County Supplemental Water Association was organized in November 1913, and Charles W. McConaughy was elected president. This was the first of three organizations to be created prior to the success of the Tri-County project.

The Central Nebraska Supplemental Water Association was established in 1922, and again McConaughy was elected president. He also served as director of the Central Nebraska Public Power District (established on July 1, 1933) and its successor, The Central Nebraska Public Power and Irrigation District (established in November of 1933). McConaughy resigned as director of the district in December of 1935 in protest against the substitution of the Keystone (Kingsley) dam and reservoir (now Lake McConaughy) for two reservoirs on Plum Creek.

Early in his struggle to promote his ideas, McConaughy remarked, "I have been called a dreamer. I am proud of the name. I cannot think of one thing accomplished by knockers." He lived to see much of his dream realized but passed away on April 13, 1941, a few months before the formal dedication of the Tri-County project.

—Karen Stork, CSD administrative assistant

Gerald Gentleman: Instigator for Central Nebraska's Power, Irrigation

One of the pioneers of the public power and irrigation programs in Nebraska that had their inception in the early 1930s, Gerald Gentleman was a native of Platte Center, Nebraska. He moved to North Platte in 1925, managing a department store and later serving as a deputy collector for the Internal Revenue Service.

In 1933, Gentleman joined with many North Platte area citizens to petition for the creation of the former Platte Valley Public Power and Irrigation District. He served as its secretary and general manager from 1934 to 1965.

During those 30 years, Gentleman played an active role in the formation and statewide expansion of two former organizations, the Nebraska Public Power System (NPPS) and Consumers Public Power District. He served as secretary, vice-chairman and chairman of the board of managers of NPPS from 1940 until 1965.

Journal-Star Printing Company

Gerald Gentleman

Gentleman also helped the district buy private utilities in Nebraska to provide a market for western Nebraska's hydroelectric plants. To solidify operation of the Platte District properties, he assisted in establishing the Dawson County Irrigation Division at Lexington in 1941, the Gothenburg Division in 1943 and the McCook Division in 1943.

Gentleman continued to promote irrigation from Ogallala to Kearney and was instrumental in organizing the Greater Platte Valley Irrigation Development Association. As a consultant from 1966 to 1970, his knowledge and financial background proved invaluable in solving the many intricate and difficult problems encountered in accomplishing the merger of Platte, Consumers and NPPS properties into the new Nebraska Public Power District (NPPD). The NPPD power generating station near Sutherland Reservoir was named for Gerald Gentleman in recognition of his many contributions to irrigation development in Nebraska.

—Karen Stork, CSD administrative assistant

Extension and Education

by Kathy Sullivan

Water

The rain is plenteous but, by God's decree
Only a third is meant for you and me;

Two-thirds are taken by the growing things
Or vanish Heavenward on vapour's wings;

Nor does it mathematically fall
With social equity on one and all;

The population's habit is to grow
In every region where the water's low:

Nature is blamed for failings that are
Man's,
And well-run rivers have to change their
plans.

—Sir Alan Herbert

Whether in response to the social inequity of rainfall or the failings of human beings, educational programs on water have played a much-needed role in the scheme of water use in Nebraska.

And while education on water issues has come from many sources, one of the driving forces has been the University of Nebraska (NU)—by 1969 called the University of Nebraska-Lincoln (UNL)—and its outreach programs, particularly those of the Cooperative Extension Service (CES).

The unique three-fold focus of the CES made it a natural for disseminating information on various water issues:

—Its tie to the knowledge and research base of the land grant university;

—Its network of field staff that reaches into every county in the state; and

—Its cooperative partnerships with other agencies and with the grass-roots clientele.

Many organizations in the state have been involved in water education (some of which are profiled at the end of this chapter). However, extension's fundamental and early role in this area makes its outreach efforts regarding the development, conservation and preservation of water resources an especially important topic.

In tracing extension's role in this education, water seems an appropriate analogy. What began as a trickle became a stream of ideas, programs and knowledge that flowed across the state, rivers of information on such major efforts as irrigation development, water-use efficiency and quality and habitat protection and preservation. Periodically, efforts were diverted to critical events and issues such as floods, droughts and energy crises. But whatever the issues, extension programs have been reliable sources of information aimed at improving the quality of life of the state's people, and in this case, helping them make more efficient and wise use of their natural resources.

The Early Years

It is not surprising that early outreach efforts by extension on water issues focused on the potential for irrigation. Lewis Hicks was more visionary than he probably ever imagined when he penned remarks on ''Irrigation in Nebraska'' for the first bulletin put out by the NU Agricultural Experiment Station. Published in 1887, Hicks comments were as follows:

> I am convinced that the wealth and resources of western Nebraska will be largely increased and its development greatly quickened by an intelligent and persevering use of the pure and copious streams of water flowing over it. . . . The time will come when these lands will be too valuable to neglect any means of increasing their productiveness, and then suggestions regarding irrigation and fertilizers will receive the attention they deserve.

A demonstration of various kinds and diameters of siphon tubes. Jack Steele, extension agricultural engineer (barely visible behind bareheaded man left of ditch) explains the importance of well-made and good-sized field ditches. Width of the ditch is shown by the spade in front of the check dam.

Biological Systems Engineering, UNL

State Irrigation School: A Dream that, Once Fulfilled, Faded

The Nebraska State Irrigation School had been in the works for more than four years when it finally opened its doors in 1921. The doors were closed less than three years later.

Nevertheless, the school, proudly proclaimed at the time to be the only one of its type in the state, was a bold venture led by those concerned with the future of the young technology and with the future of the young men who would direct it.

Thomas H. McCarthy, the school's superintendent, said during the 1921 convention of the Nebraska State Irrigation Association that the trend of rising prices of water for irrigation would lead to the desire among farmers to use water less wastefully. It was his goal, and the school's duty, he said, to provide education to irrigators wishing to practice economy.

Impetus for the school was provided by water users in the North Platte Valley, who, at the 1916 Nebraska State Irrigation Association meeting, adopted a resolution calling on the legislature "to appropriate a sum sufficient for and to establish a school of agriculture, irrigation and mechanic arts in Scotts Bluff County."

Early the following year, the legislature approved such a school, but made no appropriation for it. It was to be a branch of the state university in Lincoln.

During the two years it took before the school received any state money, university officials and North Platte water users decided that the school would offer instruction in the engineering phase of irrigation, rather than in purely agricultural aspects. As such, direction of the school was given to O. V. P. Stout, then dean of the University of Nebraska College of Engineering.

In 1919, the legislature appropriated $60,000 for the new school. Ten acres of land along the east bank of Winter's Creek in the western suburbs of Scottsbluff were donated to the state by the city, on which now sits the Cleavland Athletic Field. An additional 80-acre plot adjacent to the south was purchased by the school, with the help of $2,000 donated by the citizens of Scottsbluff.

A year before the school's opening, Stout said, "The interest that has been shown, the disposition to help, and the fact that unprecedented numbers of boys and young men all over the country are now, under post-war conditions, seeking technical instruction of various kinds, suggests the possibility or even probability that the Scotts Bluff Irrigation School may within a short time after its opening, be called upon to care for students in much greater numbers than were in the minds of those who first advocated its establishment."

When the doors finally opened in January 1921, only 23 students, primarily young men from surrounding irrigated farms, enrolled.

McCarthy defended the low enrollment, saying it was understandable because the school offered only one course—elementary irrigation engineering—while other schools invariably have several courses to offer. Partial blame was also laid on tough financial times and the fact that the school catered only to the interests of young men, instead of to both men and women.

The two-year course was on a sub-collegiate level and the only entrance requirement was that a student had completed the eighth grade or was 18 years old and had experience with irrigation. There was no tuition and a fee of $5 covered all charges.

Instruction was given to provide theoretical and practical experience in plane surveying, measurement of land, measurement of water, construction of ditches and irrigation structures on the farm, grading of land, methods of applying water, drainage and the administrative and legal principles pertaining to irrigation.

Adequate crops were grown to facilitate engineering operations, but most of the agricultural demonstrations were done at the Scotts Bluff experiment sub-station, located near Mitchell. The experiment station was under the dual supervision of the U.S. Department of Agriculture and the University of Nebraska.

In addition to instruction in strictly irrigation-related disciplines, students were taught to write proper business letters, build with concrete and wood, make mechanical drawings and produce simple maps. They were also required to study arithmetic, algebra, geometry and trigonometry.

Despite extensive efforts to advertise the irrigation school, which in some cases included personal visits by faculty to farms in the region, only 14 were enrolled in 1922.

Citing low enrollment, the founders of the school closed down the Nebraska State Irrigation School late in 1923.

—Brad Rundquist, weekend and copy editor,* Ft. Dodge (Iowa) Messenger, *and formerly CSD editorial assistant

History shows that it took some time for Hicks' vision to bear fruit. In 1895 O.V.P. Stout, a civil engineer, was hired by the experiment station to conduct research in irrigation. He also published a bulletin on the "Water Supply in Nebraska."

In 1914 Ivan D. Wood was employed as the first full-time extension agricultural engineer. He, too, had a vision of what irrigation could do for the state's agricultural industry. In a Model T Ford, he covered miles and miles of Nebraska's muddy and dusty roads, educating and running demonstrations on irrigation and other topics.

In 1917, the second experiment station bulletin, "Pump Irrigation in Nebraska," was published by E. E. Brackett and O. W. Sjogren, visionaries both because when they reported, only six irrigation wells were documented in the state.

While pump irrigation was developing at this time, surface-water irrigation was the predecessor and bellwether in irrigation development. Education was a two-way street in those early years as new surface irrigators and extension staff learned together in developing the technology, equipment and expertise needed to irrigate.

The Soil Conservation and Domestic Allotment Act of 1936 stimulated growth in the application of engineering science and technology in improving land and in the use of water resources. Interest in irrigation began to grow rapidly. In 1930 about 800 irrigation wells were pumping in the state; this grew to 3,000 by 1940.

During this period changes came rapidly in mechanization, rural electrification, farmstead improvement, soil conservation, irrigation and drainage.

But it was the post-World War II era and its technology that brought interest in irrigation to the forefront. The drought of the 1930s had been devastating to farmers. Coupled with the Depression, which had "drained" them financially, physically and emotionally, the message was clear—they could not survive another such drought. Therefore, technology perfected during the war was well received.

Vince Dreeszen, then groundwater geologist with the Conservation and Survey Division (CSD), later its director and now professor emeritus, recalls that test drilling for irrigation wells had, up to that time, been rather slow and arduous, limited mostly to the Nemaha, Republican and Platte River valleys. The consensus had been that one could not afford to lift water more than 50 feet.

Speeding research and development, added emphasis was received from a wartime National Priority on Food. Irrigation was part of that priority, so test drilling continued and expanded.

Then, Dreeszen recalls, Eugene C. Reed came "out of the oil patch" to work for Conservation and Survey. It was his wartime experience, particularly drilling for oil in Venezuela, that gave a scientific basis to the well-drilling experiments in Nebraska. The result was a "grid pattern" of test drilling across the state that revealed the geologic framework and characteristics of the groundwater reservoirs. Because this information needed to be transferred to well drillers, irrigators and the general public, CSD and the CES cooperated in educating people on its application.

Although this mix of new technology and growing interest made for strides in irrigation use, most needed was the management expertise to enable a producer to change from dryland to irrigated farming—without going broke.

Irrigation Development Farms

The desire to help farmers successfully and economically make the transition from dryland to irrigated farming spawned what became the Irrigation Development Farms.

These farms developed out of a cooperative venture among several federal and state agencies, specifically, the U.S. Bureau of Reclamation, USDA's Soil Conservation Service and Agricultural Research Service, as well as the university's CES and Agricultural Experiment Station.

Biological Systems Engineering, UNL

An early demonstration of gated pipe. Car dates photo from the mid- to late 1940s

Development farms had been used successfully in other states to demonstrate new irrigation technology, but Nebraska was the first to use owner-operated farms as demonstration sites. The first Nebraska development farm was started in 1950 in the Upper Republican River valley. Owned by Mac Anderson of Arapahoe, it was known as the Frenchman-Cambridge Irrigation Development Farm.

A gathering of the McCook town band and "boosters" at a field day at the N. Platte experiment station, "about 1916," says photo notation

IANR

IANR

A demonstration of one of the first hydraulic scrapers, used for making small dams and terraces, probably about the mid- to late 1940s.

The aim of the development farms was to educate Nebraskans to become irrigation farmers and not just part-time irrigators. Efforts centered not just on practices for irrigating crops but on designing the total crop and/or livestock program to produce a maximum return.

A team approach was pursued all the way. Technicians and specialists worked directly with and alongside the farmer and his family from the planning through the implementation stage; the technical staff provided the technology and new information and the farmers provided the demonstration site and labor while receiving hands-on experience.

Several extension staff have over the years devoted a considerable amount of time to working with the farms. Jack Steele, retired professor of ag engineering, was one of the first. He headed up the first development-farm technical committee, which selected the farms and developed work plans with the operators. Paul Fischbach assumed responsibility for the farms in 1952 and later turned it over to Bob Mulliner in south-central Nebraska and John Decker in the southwestern part of the state. These extension irrigationists worked closely with the development farm owners but also drew heavily on the expertise of other university staff for assistance in studying and analyzing what was being done on the farms.

The list of areas studied and analyzed was extensive. It included the value of adequate and timely irrigations, good crop rotations, land grading, crop selection and yields, irrigation and consumptive use and fertilizer, insecticide and herbicide use. All information was thoroughly analyzed for cost of production and net returns from irrigated crops.

The development farms proved to be effective in promoting good irrigation-management practices. Local media kept the citizens of the area informed of activity at the farms, and area farmers paid regular visits to check progress and learn about techniques in use. Twice a year the farms would host field days that would bring in hundreds to see the latest irrigation technology and hear about the projects. These field days often would lend a social and entertaining aspect to the hard work of irrigating. One of the social activities that got a lot of attention and laughs was the "siphon-tube-setting" race. Youngsters and adults, male and female, participated in the competition to see who could set the tubes the fastest.

All in all, from 1950 to 1965, there were nine different development farms. Farmers whose acreages were used were:

—-Thomas Hays, Edison, Frenchman/Cambridge Irrigation District;

—Willard Stenson, Hardy, Bostwick Irrigation District;

—O.H. Scheuneman, Franklin, Level Benches, Bostwick Irrigation District, Franklin Division;

—F.D. Graff, Ainsworth, Ainsworth Irrigation District;

—Carl L. Rogers, Sargent, Sargent Irrigation District;

—Ed Lauer, McCook, Meeker-Driftwood Irrigation District;

—Paul Gorecki, Farwell, Farwell Irrigation District;

—Raymond Janssen, Juniata, Deep Well and Pump Irrigation Farm.

After the droughts of the 1950s, interest in irrigation blossomed, and extension saw a growing need to get more research-based information to farmers on better irrigation practices.

In the early years of irrigation, farmers knew little about the "behind-the-scenes" workings of this new technology. Irrigation was new and interesting, but not much attention was given to its economics and good management. Wells were cheap. Often a system could be paid for in a year. And with the use of fertilizers, producers were irrigating for maximum yields.

Ed Janike, then director of extension, knew there was a need to focus on better management techniques for irrigators. He rallied extension specialists such as Deon Axthelm to develop and take to producers programs on how to get a good well, how groundwater moves and what happens in an irrigation well. Some demonstrations even included putting a camera down a well to check casing integrity, screen plugging and evidence of corrosion and encrustation.

In addition, it was becoming apparent that groundwater was not an inexhaustible resource, and figures from the CSD were confirming this. So the division and the extension service teamed up again to take this information to the people. At times, however, they were met with an unwelcome response.

It was not an easy pill to swallow as

farmers and producers were told that, in part, their irrigation practices were responsible for dropping water tables. In one such meeting in western Nebraska, the audience was told the water table had dropped 50 feet in some places. The local newspaper editor was present and accused the university of "trying to ruin the local economy with that ridiculous message."

However, persistence, testing and research eventually produced facts that had to be reckoned with, and farmers recognized the need to conserve.

Partnership between extension and CSD encouraged that change in attitude. Axthelm remembers: "Educating the public on groundwater supplies, occurrence, use and waste was a big task. Messages were not readily accepted by local users. Had it not been for the [U.S.] Geologic Survey surface structure maps and the groundwater occurrence maps of the [Conservation and Survey] Division and their assistance in water-level measuring and gauging, educating the irrigators would have been slower and vastly more expensive for landowners."

Biological Systems Engineering, UNL

A demonstration of a sprinkler system. Cars date from the early 1950s.

In some cases, the initiative to change and improve practices came from producer groups. A classic situation was the Benedict Project. A group of farmers from the Benedict area came to the local groundwater management association and extension to request a program on metering and irrigation scheduling. Tom Harrington of Benedict, Nebraska, hosted a meeting in his home for more than 30 farmers who wanted to discuss what could be done to improve irrigation practices in their area.

With the help of state extension specialists and the local extension agent, a program was outlined identifying a watershed area, measuring all the irrigation runoff that occurred and asking the cooperating farmers to submit part of their irrigated land to scheduling. Gauges were used to measure rainfall, meters measured the water pumped from the wells

Paul Fischbach (back to camera), extension irrigation engineer, discusses crop drying with unheated air at the Frenchman-Cambridge Irrigation Development Farm near Edison at the farm's second annual field day, Sept. 23, 1954.

IANR

Biological Systems Engineering, UNL

A sprinkler irrigation demonstration in Dundy County, April 16, 1952

and the stipulation was that irrigation practices recommended by the university be followed.

The results were enlightening, particularly to the farmers. J. D. Hirschfeld, one of those cooperating who still farms in the area, said, "Without a doubt we all saved money without hurting our yields. The most water we used in that season was 12-14 inches, and most of us had been using a lot more than that." Hirschfeld believes the study led to many of those cooperating farmers and others in the area improving their irrigation practices. Axthelm remembers one farmer being particularly pleased with the results, announcing that he had saved enough money in the study to buy himself a new pickup truck.

It was not always that easy, however, to bring about change among producers. One county groundwater association wanted to enforce rules to prevent runoff of groundwater. A producer in the area refused to comply, believing the association did not have the power, much less right to regulate him. The case went to court and the association won, providing background for a landmark ruling in water rights and management.

The policy-making role, with water or any other issue, has never been one extension has played. Extension has, however, been a significant player in helping interpret regulations to those affected, as well as facilitating responses from clientele when regulations were being formed.

To this extent, extension has made available water-law expertise through David Aiken, who since 1977 has been an extension water and agricultural law specialist. His primary role has been helping Nebraskans understand laws and regulations pertaining to water. He works regularly with water-related groups and organizations, as well as with individuals. His "*Nebraska Water Law Update*" newsletter, published from 1980 to 1986, kept readers informed of current happenings in water legislation.

Water Research and Extension Programming

The use of Nebraska's abundant water supply has occupied the minds and efforts of many people, including a long list of researchers and scientists at the University of Nebraska's College of Agriculture, which became the Institute of Agriculture and Natural Resources (IANR) at UNL. These are another group of "irrigation pioneers" who have studied and probed a myriad of water-related subjects.

Their efforts, particularly with respect to irrigation, have had significant impact on Nebraska's water resources. Their research has brought about changes in attitudes, practices and thoughts about water. Studies have spawned inventions, improved water-use technology and resulted in literally millions of dollars saved and earned for the state's economy. Through the years NU researchers have studied virtually every aspect of water in nearly every location in the state.

Cattle feeders on the Nebraska Public Power District tour look at the Paxton cut. Cars date from mid- to late 1960s.

Harold Stevens family collection

IANR

Farmers examine a "Parshall flume" for channeling ditch water, named after its inventor, date unknown.

The study of water resources at the University of Nebraska (NU) began early in the institution's history. In 1895 Oscar Van Pelt Stout, professor of civil engineering, published a bulletin on the "*Water Supply in Nebraska*." Even earlier, in 1888, the first bulletin published by the NU Agricultural Experiment Station, written by Lewis Hicks, was "Irrigation in Nebraska."

Nearly every department in the NU College of Agriculture, later the Institute of Agriculture and Natural Resources devoted research time and effort to water issues. In 1937 the Department of Rural Economics (later Agricultural Economics) initiated the first of a long series of studies on the economics of irrigation and focused on the benefits of irrigation from Kingsley Reservoir. In addition, a survey of pump-irrigated farms in Buffalo County was reported in 1944, and in 1948, agricultural economist A. W. Epp evaluated alternative systems of farming for the Tri-County irrigation area.

Work by agricultural economists has also yielded information on irrigation profitability, including the first analysis of the economics of center-pivot irrigation.

But irrigation development was perhaps where university research had its greatest impact, helping the state become a giant in irrigated agriculture. It is estimated that the net economic impact on Nebraska from irrigated crop production is now about $2.3 billion annually.

The Agricultural Engineering department (now Biological Systems Engineering) has a long history of developing irrigation technology. Agricultural engineer Ivan Wood was instrumental in developing and popularizing the siphon tube. In the early 1950s, agricultural engineers studied many deep-well pumping plants in the state to obtain data needed for planning and developing design standards. The results of these studies ultimately helped many pump irrigators improve their practices. And Paul Fischbach helped develop the automated gated pipe, computer scheduling of irrigation and the use of photovoltaics (solar cells) to pump water for irrigators. In addition, Thomas Thompson developed a scheduling program for irrigation by computer. In yet another project, engineers at UNL and the Massachusetts Institute of Technology installed an automatic gated pipe system powered by photovoltaic electricity to pump water for irrigators. Ag engineer George Meyer developed the first automated computer-controlled, remote irrigation system encompassing plant-growth models and the determination of evapotranspiration using an automated weather station.

Ivan D. Wood

Ivan Wood family collection

Researchers in the Department of Agronomy provided valuable information to producers on the timing and placement of nitrogen fertilizer in relation to irrigation water management. R. A. Olson and other agronomists made major strides in bringing greater consistency and integrity to soil-testing recommendations. Their research on soil fertility and crop nutrition had a major impact on recommendations aimed at reducing the amount of fertilizers required for crops and ultimately their potential role in groundwater pollution.

The five NU agricultural experiment stations and their satellites were sites for many NU researchers to test new technology and relate it specifically to the varying topography of the state.

A facility for irrigation research was put into partial operation in 1951 at the North Platte Experiment Station under the direction of B.R. Somerhalder. Also in the 1950s, field facilities at the Scottsbluff Experiment Station were used to test the effectiveness of various installations and designs such as underground concrete pipelines and bordered level irrigated benches.

In 1964, Paul Fischbach and Paul Schleusener wrote the specifications for the first irrigation well at the NU Field Laboratory at Mead. Land grading was done on a 10-acre plot to do research on surface irrigation. Along with this effort, a variety of different types of pipeline and a reuse system to catch runoff were tested. UNL was an early leader in promoting irrigation reuse pits (tailwater recovery systems) to recycle runoff water from surface-water irrigation.

A back-flow prevention valve system developed at UNL by Fischbach was crucial to making chemigation (the application of fertilizer and pesticides through irrigation water) a safe technology. This valve prevents chemicals from flowing back into a well during chemigation and now is required by law in Nebraska and several other states. A testing procedure to monitor equipment and evaluate the risk of contamination also is the result of UNL research.

An unidentified farmer (from left), Jack Steele, extension agricultural engineer, Mark Weldon, extension soil specialist, and J. C. "Chet" Swinbank, extension agronomist, examine a centrifugal pump. Cars date from late 1940s to early 1950s.

Biological Systems Engineering, UNL

UNL scientists have developed methods to select the most profitable crops to irrigate and to determine how many acres to irrigate when water supplies are short. Also, numerous extension publications, newsletters and reports served as a way of disseminating this research information to the people of Nebraska.

Through their studies, NU (and UNL) researchers have been able to offer objective analysis and comparisons of changing technology. F.B. Hamilton and J.F. Schrunk started comparative studies of the sprinkler and surface-water systems of irrigation and subsequently organized the annual irrigation short course that has attracted the interest of many farmers and dealers through the years. UNL ag economists have for years studied the economic aspects of irrigation, including the first analysis in the nation of the economics of center-pivot irrigation by Dr. Les Sheffield *(see "Technology," p. 96)*.

Many deep-well pumping plants in Nebraska were studied in the early 1950s by Schrunk, Schleusener and J.J. Sulek in an effort to plan design standards for these units. At that time standards for artificially-packed irrigation wells were developed and published as a cooperative effort of the Department of Agricultural Engineering, the UNL Conservation and Survey Divsion and the Nebraska Well Drillers Association.

Another means of informing clientele on the latest research technology in irrigation was the annual irrigation shortcourse. The course was started in 1952 and was designed primarily for dealers and manufacturers of irrigation equipment. Over time it became more producer oriented. Each year the auditorium at the Nebraska Center for Continuing Education was filled with those eager to hear of new irrigation developments and research. Topics covered in the course included all aspects of irrigation, but the main emphasis was efficient crop production under irrigation.

In 1967, under the leadership of then coordinator Paul Fischbach, a notebook of proceedings from the shortcourse became an annual publication. These reports, cataloged and put in notebook form, were made available to all county agents, libraries and other sources as a ready reference on the latest technology in irrigation. The last irrigation shortcourse was held in 1990.

From the time of that original work on pumping-plant efficiency and safety of the unit, NU and UNL researchers have studied and looked for ways to improve the efficiency and safety of the pumping plant. A demonstration project on irrigation pumping-plant efficiency developed by UNL during the energy crisis of the 1970s helped farmers cut their energy bills dramatically by supplying testing of the pumping plant and servicing the pump with the necessary adjustments or replacing the unit.

While NU water-research projects have keyed in on production and efficiency, the focus has also been on maintaining the quality of the water supply for domestic use. As early as the 1950s, researchers at the university were aware of increasing concentrations of nitrates in the groundwater, due in large part to excessive application of nitrogen fertilizer. Management practices developed and promoted by NU over the years have stressed the application of optimum amounts of nitrogen and irrigation water and, as a result, have been successfully aimed at minimizing nitrate leaching below the root zone.

In 1984 a Presidential Initiative identified water quality as a key national concern. This has continued to spur more research and demonstration work in proper timing and application of nitrogen, especially with irrigation. In many instances, the research is taken directly to the field in cooperation with producers. Projects like the Management Systems Evaluation Area (MSEA) site near Shelton are aimed at helping farmers evaluate irrigation and fertilizer practices with the aim of lowering inputs without sacrificing yields.

In another development, as part of a national program to better manage water research and set up water research centers in every state, the Nebraska Water Resources Research Institute, later known as the Water Resources Center and then the Water Center, was established in November of 1964. It has served as a clearinghouse, depository and coordinator of funds for water research and programming. Its purpose was to bring together water researchers, users and funding sources and integrate and coordinate federal, state and local efforts. All water-related research at IANR and funding for such projects is funneled through the center.

One of the first grants from the center helped establish a climatology research site at the UNL field laboratory at Mead. Subsequent grants helped establish the Center for Agricultural Meteorology and Climatology, which later became the Department of Ag Meteorology.

Through the years the center has looked at many issues, including the legal aspects of water use, providing information to policy makers as they ponder questions about ownership, use and transfer of the state's water resources. It has also analyzed the impact of increased livestock production on the water supply. Studies funded through the center have examined such issues as how much beef feedlot manure can be applied on the land without causing water or soil pollution. Information such as this is used by cattle feeders, environmental quality personnel and engineers in designing feedlot waste-management facilities.

Other center-funded projects have studied the quality of runoff produced from urban areas and developed models for treatment and management of urban stormwater flows. A number of center-funded projects have looked at herbicide mobility—how fast certain herbicides travel through the soil. Findings such as these are being used by irrigators to choose and apply herbicides in a manner that is least hazardous to groundwater quality.

Still other center projects have sought to find ways to reduce water losses and soil erosion on cropland. Conservation and tillage systems designed by university researchers have resulted in a 90-percent reduction in soil losses and a 50-percent reduction in water lost. In addition, both flood-hazard reduction and pollution control have resulted from these efforts. Other center-funded research has looked at eutrophication—the excessive growth of unwanted plants in shallow waters of lakes due to fertilizer runoff.

In the area of training and education, the center has helped develop a major in water resources in the natural-resources curriculum of the College of Agricultural Sciences and Natural Resources.

Peaks and Valleys

As irrigation steadily increased over the years, the early 1970s saw another big push for irrigation development. Money was available from lenders. Farm support subsidies were good, and the cost of energy was low. It was also then that D.B. ''Woody'' Varner, NU president, became an eloquent spokesman for water development.

In a talk prepared for a water resources seminar at the Nebraska Center in 1972, Varner said that ''Nebraska's greatest 'secret weapon' is its water resources, including both our surface and underground waters. We have enough water in this state to increase by more than four times the number of acres that we now have under irrigation.''

He cited records on Nebraska's rivers and streams showing that ''about 1 million acre-feet of water enters the state, but between 7 and 8 million acre-feet leaves the state each year. Much of this water, Varner said, ''is not utilized nearly as effectively as it could be for irrigation, recreation and fish and wildlife habitats, if it were impounded through upstream watershed land treatment and in reservoirs as it flows through the state.''

Varner predicted that if Nebraska's irrigated acreage were to double between 1970 and 1980—from 4 to 8 million acres—the increased economic activity in the state would amount to $4 billion. His enthusiastic and motivating message that day translated into a special emphasis identified by Varner and the Board of Regents that would lend dollar support to water resources and irrigation development in the form of new and expanded programs at the university. The Irrigation Development Program was established and Leslie Sheffield was appointed coordinator of the special effort. CES joined this effort by adding three new extension irrigation specialists to the staff at the North Platte, South-Central and Scottsbluff stations.

Also out of that first seminar in 1972 emerged the annual Nebraska Water Conference, coordinated by Sheffield and planned by the Water Conference Committee, representing both public agencies and private interests. In addition to the conferences, the annual Nebraska Irrigation Tours, led by Sheffield, have taken thousands to irrigated farms and water development projects within and outside of the state and even on several international excursions. On one such tour Nebraskans studied irrigation in China.

Varner was not far off the mark in suggesting that Nebraska's irrigated acreage could double in the 1970s. By 1982, approximately 7.8 million acres of land were under irrigation. But not everyone agreed with the aims of the Irrigation Development Program.

Environmental groups and interests raised concerns about the effects the increased focus on irrigation might have at the expense of streamflows and fish and wildlife habitats. Organized opposition emerged to Bureau of Reclamation irrigation projects, such as the Norden Dam on the Niobrara River.

Another concern expressed in the late 1970s and early 1980s concerned the rapid pace and location of irrigation development—especially by absentee owners—on the fragile soils of the Sand Hills. This concern proved valid as many of the extensive irrigation development projects in the Sand Hills have either failed or brought about severe financial problems.

Through all this the extension staff continued its conservation and preservation message, cautioning against overdevelopment, particularly with respect to fragile lands.

The Irrigation Development Program was phased out in 1977. But an even more pressing and irritating complication arose that gave needed emphasis to the conservation message. The oil embargo and subsequent energy crisis of the 1970s saw fuel prices, including those for irrigation, go sky high. From 1974 to 1978, energy prices increased 300 percent, from 15 cents to 45 cents a gallon for diesel fuel and from 57 cents to $1.71 per thousand cubic feet of natural gas. Electricity went from 1.5 to 4.5 cents per kilowatt-hour. Efficiency finally started to mean something.

Extension staff rallied to the cause, producing programs, publications and other forms of information to help farm-

Nebraska Farmer

Demonstration of multi-level samplers, used to get water samples from various depths, at the Management Systems Evaluation Area (MSEA) site near Shelton

Water-education Efforts Reach Back to Early SCS Demonstration Projects

Conservation education by the U.S. Soil Conservation Service, or any of its educational activity about natural resources, has a long history whose origin is tough to pinpoint. However, certainly these activities increased after the Great Depression and Dust Bowl era of the 1930s. (And since soil loss by water has been a primary concern of SCS, water-related education has been fundamental to all its educational efforts since nearly its inception.)

For example, it was on Aug. 25, 1933, that Secretary of the Interior Harold Ickes earmarked $5 million of funds to start the Soil Erosion Service (now the Soil Conservation Service and part of the U.S. Department of Agriculture.)

Much of the educational work of the Soil Erosion Service and other federal agencies at that time was aimed at adult education. While there were some youth activities, restoring, healing and protecting the land from erosion received much of the focus. Demonstration projects, some through the Civilian Conservation Corps, were started across the country as educational "beacons" and "pilot projects." Three of the first 40 of these projects were in Nebraska near Syracuse, Douglas and Albion.

All totaled, 16 of the 23 Nebraska CCC camps were under the management of the Soil Conservation Service. The work by the more than 30,000 men employed in these camps fell into three categories: the demonstration of practical methods of soil conservation on farms; actual work on private lands in cooperation with landowners; and the development and improvement of erosion-control techniques through research.

The roots of today's conservation education efforts for adults still pretty much grow from these methods; the technology to transfer that information has improved through mass communication methods, and private business has taken over the construction of the actual practices, but the basic methods remain the same. Today, local demonstrations, field plots, tours of a neighbor's conservation practices are still practical means of adult education used by SCS and many agencies.

Conservation education for "future adults" and "future owners of the land" also has a long history. The 1945 annual report of the chief of the Soil Conservation Service to the Secretary of Agriculture says, "the chief problems in education are a lack of suitable teaching materials and the fact [that] few teachers know enough about conservation to integrate it with their courses." Today there are improvements, but some of these same problems seem to hold true.

While organized efforts in the 1940s like 4-H Clubs and Future Farmers of America gained a large part of the rural youth audience, greater outreach was needed. Public schools, churches and other institutions were contacted and gave more support to soil and water conservation programs.

Today it is included to some degree in the kindergarten-through-12th-grade educational curriculum. Local natural resources districts, state and federal agencies, and private organizations, all with an interest in natural resources and a desire to see more education, have become involved in sponsoring materials and workshops for teachers and/or students.

Some of these activities include: Ag in the Classroom—aimed at helping students understand the food and fiber system and the role of agriculture; the Children's Groundwater Festival—to educate about groundwater quality, quantity and uses; Project Wild and Project Learning Tree—these discuss wildlife activities and uses of the forest as window into the natural world; Stop, Look and Learn About Our Natural World—this includes educational activity that fits natural resource education into science, social studies, language arts and math, as do many other programs and materials.

The interest in natural resources education in Nebraska has grown recently to result in these organizations and individual interests uniting to form a Nebraska Environment Education Association. The mission of this association will be to unite these individual efforts and serve as a center for teachers, students and organizations interested in the natural resources.

—Pat McGrane, public affairs specialist, Nebraska state office, SCS

---

ers save dollars by becoming more efficient water users. Publications like one entitled "Energy Use in Nebraska" showed how significant savings could be realized by installing re-use systems for irrigation runoff, automated surface-irrigation systems, increased pumping-plant efficiency and operating electric irrigation pumps off the daily peak-use rate of electricity. Ironically, extension staff had "preached" for years on all of these subjects, but the "teachable moment" had arrived, and, as in many cases, when one's pocketbook is affected, people finally take notice.

One major conservation effort developed by NU extension staff was irrigation scheduling. With a half-million dollar grant from the U.S. Bureau of Reclamation, the Cooperative Extension Service, together with the Department of Agricultural Engineering, trained county extension agents, irrigators and consultants how to determine the most effective times to irrigate using the least amount of water. ASCS cooperated in this effort by providing cost-sharing funds to irrigators for purchase of water meters and monitoring equipment.

The cost savings brought about by the massive effort on scheduling was nothing short of amazing. It was estimated that in 1978 with the more than 1 million acres scheduled, more than 18.5 million gallons of diesel fuel—the equivalent of nearly $8 million—were saved. Additional savings of more than $5 million were realized on these lands from the lowered fertilizer applications.

Another program aimed at improving irrigation efficiency was the Pump Unit Management Program, or PUMP. Implemented in 1980 with funding from the U.S. Department of Energy and the Nebraska Energy Office, it focused on making the actual pumping plant more efficient.

A key element of this project was the concept of demonstration sites on cooperating farms. Public demonstrations were held at these sites, with technicians showing farmers the procedures for testing the pumping plant, simple adjustment procedures after noting inefficiencies and a general discussion of the results. From 1980 through 1982, more than 4,000 irrigators heard this message, and pumping sites in nearly every county in the state were tested. It was estimated that the equivalent of more than 350,000 gallons of diesel fuel was saved from just 1,500 pumping plants.

Focus on Water Quality and Protection

Emphasis on protection and preservation of water resources has always been a cornerstone of the extension message. As early as the mid-1950s, NU researchers expressed concern over groundwater contamination, particularly as it related to nitrate levels. Over time, university extension programming has encouraged producers to adopt management practices that minimized nitrate leaching into the groundwater. So called best management practices (BMPs) such as soil and irrigation-water sampling, adjusting nitrogen-application rates according to

(Continued on p. 146)

Water Education and Public Involvement in Protecting Our Water Resources

The spectacle of stinking ooze and rotting fish carcasses coursing down channels where bubbling brooks and grand rivers once flowed jolted the American public into a realization that our water could not endlessly endure abuse. Public outcry over the destruction of rivers, streams and lakes culminated in passage of the federal Clean Water Act of 1972. This historic legislation provided for the comprehensive planning, technological development, financial assistance and meaningful enforcement needed to achieve "swimmable and fishable" conditions in all surface waters of the United States. While the original target dates set by the Clean Water Act for achieving specified water-quality standards were not successfully met, the processes put in motion by the act, along with constant public pressure, continue to drive us toward restoration and protection of our water resources.

Initial efforts to protect the quality of surface waters were primarily directed at point-source pollution. This type of pollution originates from fixed and identifiable sources such as municipal wastewater facilities and industrial plants. Since these pollutants originate from fixed sources, technological solutions can be economically applied to prevent them from entering our surface waters. Because these sources typically contribute identifiable and traceable pollutants to receiving waters, it is relatively easy to monitor them for compliance with water-quality regulations. So a strategy of command and compliance was adapted to address the problem of point-source pollution. The impersonal nature of the offending institutions also encouraged public support for their regulation.

Subsequent amendments to the Clean Water Act placed greater emphasis on the problems of nonpoint-source pollution of surface waters, as well as groundwater supplies. (Nonpoint-source pollution is the accumulation of contaminants from broad areas or diffuse point sources into a receiving body of water.) These pollutants include sediment, nutrients, pesticides, bacteria and petroleum products carried in agricultural and urban runoff. Groundwater supplies are especially susceptible to contamination by chemicals dissolved in water that percolates through the soil and reaches the underlying water table. Generally, technological solutions to nonpoint-source pollution are not possible or are not economically feasible. Since each of us makes incremental and untraceable contributions to nonpoint-source pollution as we pursue our daily activities, command and compliance is not an efficient strategy to regulate this problem. We are faced instead with the task of modifying our own attitudes and behaviors to prevent nonpoint-source pollution.

Point-source pollution has been relatively low in Nebraska due to the absence of numerous large urban areas and a fairly small and relatively benign industrial base. Intensive agricultural development, on the other hand, has caused a significant level of nonpoint-source pollution of the state's surface water and groundwater. Sedimentation of rivers, streams, and lakes resulting from unchecked erosion of croplands, roadways and construction sites is the primary cause of water-quality degradation. Eutrophication of surface waters, especially lakes, from excessive nutrient runoff is also a problem. Degradation of surface-water quality from pesticide runoff or animal waste runoff tends to be a transient or localized problem. The problems of surface-water degradation are being addressed by encouraging the use of best management practices by farmers, ranchers, road maintenance personnel and developers.

Nitrate is the chief nonpoint-source contaminant of groundwater in Nebraska. The heavy application of nitrogen fertilizers combined with intensive irrigation is thought to be the primary cause of this problem. This is especially true for regions along the Platte River where the soils are porous and the water table is shallow. Other areas of the state no doubt will develop groundwater problems in the more distant future as nitrates continue to move through denser soils toward deeper water tables. Pesticide contamination of groundwater tends to occur at isolated locations. Many of these incidents can be attributed to spills or runoff into improperly maintained or abandoned wells. Back-siphoning from improperly designed chemigation and chemical mixing facilities have also contributed to this problem. Educational efforts on proper budgeting of nutrient and water inputs and on safer and more efficient application techniques are being directed at producers to prevent further nitrate contamination of groundwater supplies. Both regulation and educational efforts are being used to prevent direct contamination of groundwater through spills and back-siphoning into wells.

Nonpoint-source pollution is being addressed in Nebraska through four separate, but interactive, state and federal programs made possible directly or indirectly by the Clean Water Act. These programs are the Nonpoint Source Pollution Management Program, the Clean Lakes Program, the Wellhead Protection Program and the Special Protection Area Program. Each of these programs has provisions for direct input and participation by citizens. And the Department of Environmental Quality (DEQ), along with other agencies such as the natural resources districts, helps manage much of the participation in such programs.

The Nonpoint Source Management Program, authorized under section 319 of the Clean Water Act, directs the state to develop multi-agency efforts to reduce nonpoint-source pollution on a small watershed basis. These projects pool the resources and talents of the local, state and federal water quality agencies to implement the adoption of best management practices by producers in the watershed. Each project contains an information and education component to assure public awareness of the project and to facilitate information and technology transfer. Watershed projects are usually developed and managed by the local natural resources district. Citizens advisory committees provide those affected by the project a chance to express their concerns and to help develop pollution-control strategies that are compatible with the needs and desires of the local community.

A statewide information and education effort is also provided for in the Nonpoint Source Management Program. This effort is intended to increase awareness and knowledge about nonpoint-source pollution by the general public. Seminars, workshops, publications and promotional materials and programs are components of this effort. The ultimate goal of this effort is to create an informed Nebraska citizenry that is motivated to take individual and collective action to prevent nonpoint-source pollution of Nebraska's water resources.

A unique feature of the Nebraska Nonpoint Source Management Program is the Information and Education Mini-grant Program for Nonpoint Source Pollution Reduction. This program provides funds to local agencies, schools and private organizations to create materials and programs that will promote awareness and knowledge about nonpoint-source pollution. The objective of this program is to recruit and harness the talents of a wide range of individuals to assist in the effort to educate Nebraskans about the causes of nonpoint-source pollution and possible solutions to this problem.

The Clean Lakes Program, authorized under section 314 of the Clean Water Act, provides resources to assess water-quality problems in publicly owned lakes and for restoration of those lakes impaired by pollution. Nebraska lakes are primarily degraded by nonpoint-source pollution. Sediment and nutrient input are the chief problems due to inadequate control of erosion and poor nutrient management in the upper watershed. Control measures such as construction of sediment traps at the inflow and restoration measures such as dredging are made possible by funds from the Clean

(continued on next page)

Lakes Program. As with the Nonpoint Source Management Program, Clean Lakes Projects are managed by the local natural resources district with input from a local citizens advisory committee. The high value placed on lake resources by the public has made this a popular and successful program in Nebraska.

Public awareness of lake-pollution problems in Nebraska has been enhanced by creation of the Volunteer Lake Monitoring Network. This network consists of private citizens who volunteer to periodically monitor water quality in Nebraska's publicly owned lakes. Water clarity readings are collected monthly by the volunteers and submitted to DEQ for analysis. The volunteers receive an annual report that describes the trophic state (nutrient level) of the monitored lakes and a general description of their state of degradation. The report also includes educational materials on techniques to properly manage lakes and protect them from degradation by pollution. The Volunteer Lake Monitoring Network provides a mutually beneficial partnership between DEQ and the volunteers. The department receives valuable information that would otherwise not be collected on the condition of public lakes while the volunteers receive accurate knowledge about proper lake management to share with other interested citizens.

—Elbert Traylor, DEQ nonpoint-source pollution information and education coordinator

(Continued from p. 144)
nitrogen contributed from the soil and irrigation water, and irrigation scheduling to avoid excessive deep percolation of water were some of the concepts stressed.

In 1961 a team of extension specialists secured a grant from the state Department of Health to study water quality throughout Nebraska and to look especially at nitrogen in the water. The result was the "Nebraska Water Quality Survey." County extension agents helped in the assessment and sampling, which included samples from 5 percent of all irrigation wells and some irrigation canals. Laboratory analysis included testing for nitrates and other elements. The study identified areas of the state with high concentrations of nitrates, which were then sampled a year later to note changes. This water-quality study provided a benchmark for tracking the quality of Nebraska water for years to come.

Other water-quality issues in extension programs have focused on efforts to stop leaching of pesticides and other chemicals into the groundwater, septic tank construction and maintenance and livestock-waste management. There have also been several specially funded projects that have provided assistance to parts of the state experiencing critical water quality problems.

One of the largest was the Burlington Northern Foundation project at the South Central Research and Extension Center near Clay Center. This $1 million, 5-year research demonstration project begun in 1984 looked specifically at chemigation. According to some estimates at the time, more than 26,000 center pivots were operating in Nebraska, of which 40 percent were used to apply chemicals. Risk of contamination of the groundwater or of surface water is high in this situation unless proper anti-pollution safety devices are used. The study also looked at cleaning up a contaminated aquifer should a backflow occur. The research emphasized the need for prevention, showing that backflow events are difficult and sometimes impossible to correct. The study also showed that chemigation was an effective way to apply pesticides, and when used properly, did not contribute significantly to the leaching of chemicals into the groundwater.

* * *

Another project centered on an area of the state that has long been plagued with groundwater contamination problems. When the Hall County project began in 1978, nitrate-nitrogen levels in part of the project area were as high as 30 parts per million (ppm), well beyond the 10-ppm federal limit for safe drinking water. And rates were increasing at the rate of 1 ppm per year. Investigators believed the high concentration was due in part to excess fertilizer use by corn growers. Extension, in cooperation with other agencies, developed nitrogen- and irrigation-management techniques to address this problem. Through these educational programs, nitrogen applications were reduced nearly 80 lbs. per acre while maintaining corn yields, and, with irrigation scheduling, an average of 1.7 inches of water was saved.

* * *

The Maple Creek project, which ran from 1978 to 1982 in Colfax, Platte and Stanton counties, was aimed at controlling nonpoint pollution. Educational programs and demonstration projects promoted such practices as conservation tillage, construction of terraces and installation of sediment- and erosion-control devices.

* * *

The Long Pine Project, started in 1981 in Brown and Rock counties, and the Logan Creek Project in the Pender area in 1985, sought to pinpoint specific water-quality problems in those areas and developed educational programs to solve them.

* * *

In 1988, a series of "Presidential Initiatives" were identified by the extension service and the USDA that focused on areas of national concern and the need for programming in those areas. One of the areas identified was water quality.

Nebraska followed the federal mandate and developed a major program known as "Enhancing Water Quality." It focused on three main areas: providing a safe and adequate domestic water supply; managing the use, storage and disposal of pesticides; and managing nitrates to prevent them from reaching the groundwater. The program thrusts took an interdisciplinary approach, using the expertise of various specialists who all participated in the problem-solving on an issue.

The outreach issues have been many and varied. County extension newsletters provide water-quality information about pesticides. A video feature on the popular Nebraska Educational Television program "Backyard Farmer" discussed the proper use of pesticides. Hundreds of extension programs across the state looked at such issues as crop protection clinics, chemical horticultural clinics and chemigation training. Extension continues to provide the educational training for commercial and private pesticide applicators. The Sprayer Calibration Clinics were an example of another program in this area. In Butler County alone over 250,000 acres of cropland were treated by the calibrated sprayers. A 5 percent improvement in accuracy resulted in an estimated savings of $125,000 in farm chemicals and reduced the risk of groundwater contamination.

In the area of nitrogen management, the interdisciplinary team continues to focus on BMPs (as explained above). Field demonstrations on farmer-owned land continue to be the most successful method of bringing about changes in producer practices. These on-farm demonstrations account for over 300,000 crop acres annually. These are in addition to special demonstration projects co-sponsored by CES with the Soil Conservation Service and other agencies.

Extension publications, fliers, exhibits, and displays at county fairs all contribute to what has become an all-out effort to help farmers adopt BMPs to reduce the amount of nitrogen applied to the soil. These water quality efforts and others have made major contributions to improving water quality, helping minimize groundwater contamination, conserving water resources and reducing soil-robbing runoff.

Field Staff: Close to the Source

Extension field staff, both county agents and district specialists, have played a key role in educating Nebraskans on water issues. Working in the field and "close to the source," they have provided a valuable link between the technical information of the research environment and the ability to apply that information. And their proximity to the clientele has made it possible for them to identify user needs and concerns.

Nearly every extension field person has at one time or another devoted some time in his or her educational programs to water. Their newsletters, newspaper columns, meetings and other programs have offered advice on everything from how to conserve water at home to timely scheduling of irrigating crops. In times of crisis, such as when floodwaters ravage an area, extension agents have been in the front lines providing timely advice on flood cleanup and care and use of contaminated items. A few of these people are profiled at the end of this chapter.

Domestic Water Use

Not all extension programming regarding water has been agricultural in nature. In its mission to serve Nebraska families, extension has been concerned with assuring that a safe and adequate domestic water supply be available to homes, farmsteads and communities in Nebraska.

In addition, over the years when flood problems plagued Nebraskans, extension agents provided assistance and information on dealing with flood-contaminated food and water supplies. And during the energy crisis of the 1970s, a major effort on home energy conservation was begun. Extension consumer-education specialist Janet Wilson developed a program called "Winning Ways with Water" that was carried to more than 28,000 home extension club members and other community organizations. Using extension publications and mass media campaigns, extension staff promoted a variety of practices for conserving water in the home.

In 1988, the Extension Domestic Water Quality Team began organizing an interdisciplinary program that included not only extension staff but cooperators from many other agencies such as the state Department of Health, the natural resource districts, the NU Medical Center, the Water Quality Association and the Nebraska Groundwater Foundation. This coordinated effort encouraged and promoted the sampling of wells and private sources of drinking water. Materials and instructions for sending in test samples were available at county extension offices. The state Department of Health reported a dramatic increase in domestic water samples sent in for testing as a result of this effort.

Other areas of focus on this team effort included informative programs on home water-treatment systems, nitrates in drinking water, plugging abandoned wells and disposal of hazardous wastes. Several agents and specialists are also working with clientele to bring about a reduction in the amount of water and fertilizer used in home landscaping.

The Department of Textiles, Clothing and Design has for several years conducted studies on the laundering of pesticide-contaminated clothing. Dr. Joan Laughlin and her staff have published more than 400 bulletins, papers and articles on the subject. It has become a nationally known research emphasis. Information from these studies has proven valuable to farm families and particularly operators who work with agrichemicals. Guidelines produced through this research have been disseminated through the extension network and are used as part of the training required of pesticide applicators.

"The Medicine Show" and Other Youth Programming

While extension programming regarding water has been directed at many different audiences and generally at the welfare of all people, it may have been most influential when focused on those who will ultimately be responsible for the future—the young.

And one tireless promoter of soil and water conservation among youth was H. H. "Hi" Gilman. For nearly 20 years this extension conservationist took his "Medicine Show" across the state preaching to young children the benefits of soil and water conservation. At the time of his retirement in 1972, he had presented that show to more than 70,000 young people.

The first Medicine Show was presented in Madison County. Gilman and his wife, Museta, who at the time was county superintendent of schools, decided that young people should learn about soil and water conservation. So they organized a program for seventh and eighth graders in the county, presented it in a Norfolk theater and took the group to a farm. Gilman was then a conservationist with the Madison County Soil and Water Conservation District. It was the first of many shows he was to present.

The Medicine Show was a "pots and pans" approach to teaching conservation, and in 30 minutes it kept youngsters' eyes glued to the activity. Gilman told the audience that the "medicine the Medicine Man sold in pioneer days wasn't worth a hoot, but the medicine I'm selling—soil conservation—is." And then, with cake pans of soil, plugs of grass, water sprinklers and other paraphrenalia, he demonstrated how water erosion occurs, its effects on the soil and methods to curtail it. It was a birdseye view of the importance of soil and water conservation using examples and language the children could understand.

He gave the show in rodeo arenas, theaters, court houses, auditoriums, schools, city parks and open fields. In 1971 participation reached an all-time high in one year, with presentations at 51 locations to more than 7,000 persons.

Prior to his Medicine Show, Gilman was with the USDA Soil Conservation Service for 19 years. He joined the university's agronomy department as an extension conservationist in 1956 and continued in that role until his retirement on June 30, 1973. Oddly enough, this popular conservation position was not refilled after Gilman's retirement. However, extension programming in water use and conservation aimed specifically at youth continues in a variety of ways.

The Groundwater Foundation's annual Children's Groundwater Festival has served as a catalyst for many such activities. Extension agents from across the state participate in that event and bring youth to it. They have also developed similar programs in their counties. In 1991 more than 2,000 fourth-through-sixth graders learned about the value of water through these county water festivals. The programs have been so successful, plans are to expand them in coming years.

In addition, the 4-H school enrichment project, ''Water Riches,'' was used with at least 8,500 participants in recent years to help educate youth about groundwater.

Conclusion

The Cooperative Extension Service has reached out to Nebraskans in many ways with educational efforts on water. Its goal has been to help educate Nebraskans on how best they can use, protect and preserve one of their most precious natural resources. As a goal for the future, this will no doubt remain the mission of extension programming. But as we already see, the diversity of interests among water users, the increased demands for use of this resource and the recognition that it may be in peril will require informed decision-making and wise choices as never before. But certainly, extension will continue to be a reliable, objective source of information for people to make those informed choices on how the future of water use in Nebraska will evolve.

Sources: *College of Agriculture of the University of Nebraska: The First Century*, Elvin F. Frolik and Ralston J. Graham, 1987, The Board of Regents of the University of Nebraska, Lincoln; Various progress reports and annual reports from various years of the Cooperative Extension Service; Various annual reports of the irrigation development farms of the University of Nebraska; Various IANR Farm and Ranch (later Farm, Home and Ranch) Quarterlies.

Interviews with: retired and former extension service staff: Deon Axthelm; Paul Fischbach; Harry Hecht; Jack Steele; Harold Stevens.

Biological Systems Engineering, UNL

Taking it easy between speakers at an irrigation demonstration farm field day, site and date unknown

Biological Systems Engineering, UNL

A demonstration of sprinkler irrigation, solid-set or tow-line type, in west-central Nebraska, probably put on by the then North Platte Experiment Station. Cars date from the early 1950s.

Young and old participate in a soil-judging contest in Pawnee County, ''about 1954,'' according to notation on the photo.

IANR

Nebraska State Irrigation Association: Educating on Technology, Policy, History Over 100 Years

by Brad Rundquist
Weekend and Copy Editor Ft. Dodge Messenger
Formerly Editorial Assistant, CSD

The beginning of the Nebraska State Irrigation Association and the irrigation movement were "like the beginnings of streams themselves—little rills here and there, almost unnoticed, and well nigh failing to persist until they were able to unite into a recognizable stream."

That's how O. J. Ferguson, then dean of the University of Nebraska College of Engineering, began his report on the history of the Nebraska State Irrigation Association. Presenting the report at the association's 1933 meeting in North Platte in Ferguson's absence was E. E. Brackett, an NU engineering professor.

There were indeed a few "rills" of irrigation activity in Nebraska as early as the 1860s, but irrigation was generally regarded as undesirable and unnecessary. Several ditches and wells pumped by windmills cropped up in the next thirty years but still, irrigation was not a common practice. In 1889, for example, there were only 9,026 irrigated acres in the state, according to the U.S. Census Bureau. All that changed when farmers watched their crops dry up in 1890. Interest in irrigation by surface water, and later groundwater, swelled to feverish levels on occasion, spawning the organization of local irrigation associations, county and state irrigation meetings, and, finally, a state irrigation association. By 1900, as many as 150,000 acres were being irrigated; the stream had indeed begun to flow.

Interest was stirred by a series of six editorials appearing in the *Omaha Bee* in January and February 1891. The articles, written by William E. Smythe, prescribed irrigation as the cure for the state's drought woes. Many people listened, and Smythe was propelled to the forefront of irrigation leadership. Smythe's work helped bring delegates from 36 counties to Lincoln for the first statewide irrigation convention, held Feb. 11-12, 1891. Both days were spent drafting legislation to promote irrigation development and to create a systematic appropriation system.

The proposed legislation was sent to the Nebraska State Legislature, where it was passed around in one form or another for several years, finding little support. At that time, irrigation was not a popular notion. To admit additional water was needed for crop production was to admit the state was not suited to agriculture. The draft was modified over the years, and finally was passed into law in 1895. That law still provides the framework for today's state water-law system.

The state convention and other regional meetings made Smythe a recognized irrigation expert and led to the formation of the Nebraska State Irrigation Association. Soon, Smythe was leading national and international irrigation meetings all around the country.

Back in Nebraska, Ivan A. Fort, a North Platte irrigation promoter, began working to pull together a state irrigation association early in 1893, operating in western Nebraska under the slogan "Irrigate or Emigrate." At a North Platte irrigation meeting held Dec. 19, 1893, called for by the local Business Men's Association, delegates voted to organize a Nebraska irrigation association. Temporary chairman J. G. P. Hildebrand presided.

The *Independent Era*, a North Platte newspaper, reported that "When I. A. Fort first began to agitate the question of holding a convention in this city for the purpose of forming a state association to devise ways and means for a system of irrigation for western Nebraska, many of our citizens were prompt to declare in slang parlance he had 'wheels in his head.' Since the meeting, however, the scoffers are the repentant sinners who have remained to pray that the fullest fruition may crown the effort."

At the meeting, procedures were drawn up and bylaws were adopted. The logical choice for the position, Fort was quickly nominated and elected association president by the 117 delegates in attendance. Hildebrand was made secretary.

On the first day delegates dealt with organizational activities and the adoption of resolutions. One such resolution was to support a resolution made at an interstate irrigation meeting held the previous September in Kansas. That resolution stated that Congress should appropriate money to investigate the feasibility of using the "underflow" for irrigation and to determine the extent to which reservoirs can be built to retain storm waters. Delegates to the first Nebraska association meeting also agreed to cooperate with the State Board of Agriculture to gather all available information on state irrigation experiments and to distribute it to interested parties; to urge farmers to plant trees; to endorse the creation of the post of state engineer; to call on the state to create and maintain agricultural experiment stations in western Nebraska; and to seek funds for the publication of the association's proceedings.

Testimony concerning the successes of irrigation farmers filled the second day. The first meeting of the irrigation association was declared a great success, and newspapers around the state ran lengthy stories detailing the event.

The following year, delegates to the Nebraska State Irrigation Association met in Kearney on December 18 and 19. At that meeting, Fort handed out 16-page booklets filled with information on the amount of water available to irrigators, the amount of land that could be irrigated, various types of pumps and windmills, how to build canals, the location of existing canals and the proper treatment of land. Fort also emphasized the need for irrigation law in his booklet.

More than 200 delegates heard letters of support from William F. "Buffalo Bill" Cody, William E. Smythe and J. Sterling Morton. It was resolved to support a legislative measure to develop an irrigation commission to protect water appropriation rights. Again the meeting was a major Nebraska event, and the proceedings filled the front page of many papers.

A. G. Wolfenbarger, a Lincoln irrigation lawyer, was elected secretary for the following meeting. Fort remained as president.

The 1895 meeting was held in Sidney on

An advertisement for the 1896 Nebraska Irrigation Annual

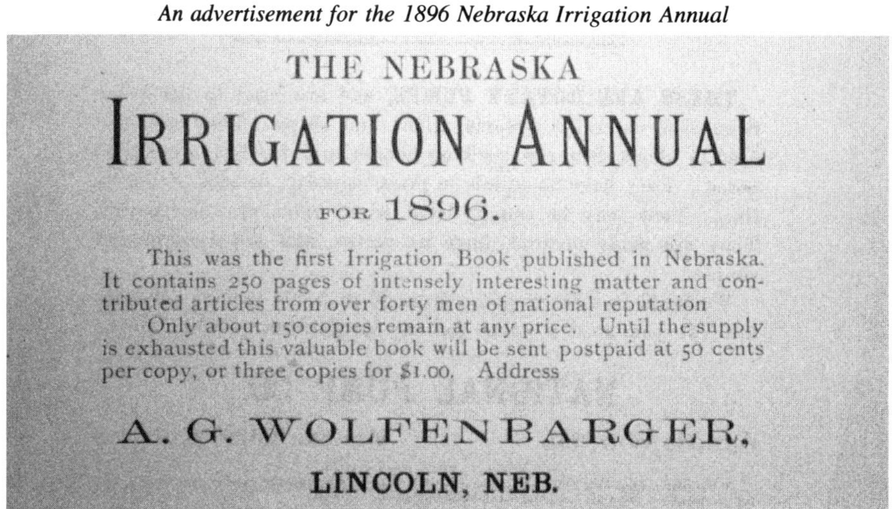

V.H. Dreeszen, CSD

December 18 and 19. Wolfenbarger lectured delegates about the need to publish the proceedings. He told them that no money had yet been donated to the association, and that he had mailed newspaper articles detailing the first two meetings at his own expense, as no proceedings had been published.

William R. Akers, secretary of the newly formed State Board of Irrigation, addressed the delegates later in the meeting. He outlined the water appropriation bill passed earlier that year by the legislature. Akers, who as a state senator pushed for the bill's passage, told the group he could not accept all the credit. He explained the new law was the work of many people and thanked them for their support. Akers is still called the father of the modern irrigation law, and the law itself is commonly referred to as the "Akers Law."

Wolfenbarger was elected president for the following year, J. L. McIntosh became secretary and Fort was made state and national irrigation lecturer.

During the meeting, the nearly 250 delegates agreed that the association should sponsor an irrigation fair. They appointed a fair committee, with William L. Park of North Platte as president, to organize such an event. The Great Nebraska Irrigation Fair opened in North Platte on Oct. 9, 1896. It was a great success, but too costly for the association to sponsor a second time.

Following the 1895 meeting, Wolfenbarger collected and edited "The Nebraska Irrigation Annual, 1896." The 234-page book contains a stenographic account of the third annual convention. At the back of the booklet is a section containing pictures and biographies of prominent irrigation enthusiasts. Forty-eight men and one woman are included. Akers later said the book was "the first work of its kind published in the United States." Cost at the time was $1 for a clothbound edition and 25 cents for the paperbound version.

The fourth annual meeting was held November 20 and 21 in Lexington. Much of the discussion at that meeting centered on the success of the Nebraska irrigation fair.

The association was incorporated, and its name was changed from the Nebraska State Irrigation Association to the Nebraska Irrigation Association.

University of Nebraska geology professor Erwin H. Barbour presented a paper on windmill and pump irrigation and told of the tremendous potential for using the state's underground water as a water supply. Barbour described in detail available techniques and equipment for pump irrigation, and showed pictures of the "merry-go-round," "go-devil," "jumbo" and "baby jumbo" mills. He stressed the importance of operating the pump year-round and storing the water in a reservoir, so it would be available when needed.

Wolfenbarger again published meeting proceedings in two editions of the "Irrigation Annual for 1897." The first contained 151 pages of stenographic records, as did the second. The second edition, however, contained an additional 51-page collection of pictures and biographies of people interested in irrigation, including three women.

Wolfenbarger writes in a preface to the book, "The day is near at hand when the most economic and universal conservation of the water supply will enable the people of the arid and semi-arid regions of our great country to rival the whole world in character, comfort and earthly blessings of our rural homes. . . . Trusting that this little volume will help disseminate the needed knowledge on this important question, and stimulate thought and investigation, it is consigned to those who read."

The number of delegates was small, and the meeting received little attention from the press. One explanation may be that the irrigation framework for the state had been set by that time, and the state and national governments, rather than the irrigation association, led the irrigation movement. The movement was shifting from private leadership toward bureaucracy. As a result, the irrigation association never again held as much prominence as it once did.

Wolfenbarger was elected president for 1897, and Joseph Oberfelder was made secretary. The 1897 meeting was held at North Platte. No proceedings were published and information about the gathering is scarce. Akers was elected secretary for 1898 and Wolfenbarger remained president.

Meetings may have taken place between 1898 and 1902, but it is doubtful they were held on a regular basis. Wolfenbarger said before the 1902 meeting of the Nebraska Irrigation Association, "We assemble in annual convention and conference after a lapse of more than the regular term, and it has been deemed a matter of great importance that the association, which was the pioneer of practical irrigation in its broadest and most comprehensive view, should again take up educational and promotive work looking to the material, economic and general welfare of our great commonwealth."

Wolfenbarger served as president to the 1903 and 1904 meetings, both held in Lincoln. The organization probably ceased to exist between 1904 and 1909. Ferguson says as much in his history.

When the association began again in 1910, meeting officials called it the "first annual meeting of the Nebraska State Irrigation Association," as did state newspapers. H. H. Andrews, speaking at the 1920 meeting, said, "The organization of this association was decided upon at the session of the National Irrigation Association at Pueblo, by the Nebraska Delegates." That Colorado meeting was held in 1910 and Andrews was a Nebraska delegate.

Between 1910 and 1918, association meetings were held at Bridgeport on an annual basis. No meeting was held during the war year of 1917, as many members devoted most of their spare time to the war effort.

Some irregularity appears in the numbering of the association meetings. The 1920 meeting, held in Gering, was called the ninth annual, when it was actually the tenth, throwing the numbering system off by a year until 1929. In 1929, the system was changed to recognize the pre-1910 meetings. The new system, which called the 1929 meeting the 37th annual, took into account that no meeting was held in 1917 but did not recognize that no meetings were held between 1905 and 1909 and that meetings weren't held regularly between 1898 and 1902. Assuming for the sake of simplicity that annual meetings were held between 1898 and 1902, and taking into consideration that two meetings were held in 1921, the 1929 meeting should have been called the 32nd annual. No meetings were held in 1952 and 1975 but two meetings were held in 1954 and 1977. Therefore, the 1993 meeting is actually the 96th annual, instead of the 100th. Nevertheless, 1993 is the 100th anniversary of the birth of the irrigation association.

When R. H. Willis, head of the Nebraska Bureau of Irrigation, Water Power and Drainage, was elected secretary in 1919, more detailed records were kept. Willis knew the importance of historical records and didn't want discussion at the irrigation meetings to pass forgotten. Proceedings were again published from 1920 to 1976. To help with publication costs, the Nebraska State Legislature in 1919 appropriated $1,000 to the association. Smaller appropriations were made on an irregular basis until the mid-1930s, when payments stopped. After that, the association relied on annual dues from member irrigation districts. Irrigation association proceedings were last published for the 1977 meeting. Rising publication costs were to blame.

Willis served as secretary until 1949, and as secretary emeritus until his death in 1954. In 1950, Dan S. Jones, who took over at the Bureau of Irrigation, Water Power and Drainage when Willis retired that year, became association secretary. The association has had only three secretaries since Jones stepped down in 1957. Stan A. Matzke served from 1958 to 1963, Don P. Long from 1964 to 1989 and Norma Sitzman from 1990 to 1993.

The irrigation association met jointly with the Nebraska Reclamation Association from 1960 to 1970. It has met with the Nebraska Water Resources Association (NWRA) since 1971. For a few years in the late 1970s, the Nebraska Association of Resource Districts met with NWRA and the irrigation association.

In 1992, the association was again established as a non-profit corporation under the name Nebraska State Irrigation Association. The 1896 incorporation, under the name Nebraska Irrigation Association, was apparently voided, either when the organization ceased to hold regular meetings in the late-1800s and early-1900s or during one of several changes in the state's laws of incorporation.

Over the years, meeting discussion has focused on new irrigation technology, policy, regulation, history, and methods. The group has been addressed by state senators, U.S. senators and representatives, state governors and prominent irrigation scientists and promoters from around the country.

No matter what changes have affected the course of irrigation development in Nebraska, the aim of the Nebraska State Irrigation Association to further irrigation knowledge has remained.

The 1896 words of state senator and irrigation enthusiast J. S. Hoagland illustrate the importance of the association: "This organization deserves the earnest commendation of the people of Nebraska for the great progress it has made along the line of education in the art and science of irrigation. Much of such progress may be attributed to the wisdom displayed in the selection of association officers, who have devoted time and talent to the subject of irrigation, and to the creation of an interest in our annual meetings.

"The history of our association shows a grand work; a work that benefits not only those who are actively engaged in irrigation farming, but the preacher, the doctor, the lawyer, the merchant, the wage earner, all receive some benefit by and through an increased prosperity which comes to our citizens through an increase in the yield of food products by irrigation enterprise."

Meetings of the Nebraska State Irrigation Association 1893-1993

Year	Date	Location	President	Secretary
1893	Dec. 19-20	North Platte	I.A. Fort, elect	J.G.P. Hildebrand
1894	Dec. 18-19	Kearney	I.A. Fort	J.G.P. Hildebrand
1895	Dec. 18-19	Sidney	I.A. Fort	A.G. Wolfenbarger
1896	Nov. 20-21	Lexington	A.G. Wolfenbarger	J.L. McIntosh
1897	?	North Platte	A.G. Wolfenbarger	Joseph Oberfelder
1898	?	?	A.G. Wolfenbarger	W.R. Akers
1899	?	?	A.G. Wolfenbarger?	?
1900	?	?	A.G. Wolfenbarger?	?
1901	?	?	A.G. Wolfenbarger?	?
1902	?	?	A.G. Wolfenbarger	?
1903	Jan. 23	Lincoln	A.G. Wolfenbarger	Stevens
1904	Jan. 21	Lincoln	A.G. Wolfenbarger	H.O. Smith
1905	NO MEETING			
1906	NO MEETING			
1907	NO MEETING			
1908	NO MEETING			
1909	NO MEETING			
1910	Nov. 15-16	Bridgeport	T.C. Egleston	W.L. Minor
1911	Nov. 14-15	Bridgeport	T.C. Egleston	W.L. Minor
1912	Oct. 22-24	Bridgeport	J.J. Halligan	J.E. LeBlanc
1913	Oct. 28-30	Bridgeport	J.G. Beeler	J.E. LeBlanc
1914	Dec. 2-4	Bridgeport	O.W. Gardner	J.E. LeBlanc
1915	Dec. 7-9	Bridgeport	J.T. Whitehead	J.E. LeBlanc
1916	Dec. 6-8	Bridgeport	W.M. Barbour	J.E. LeBlanc
1917	NO MEETING — WORLD WAR I			
1918	Dec. 18-20	Bridgeport	J.G. Woodman	J.E. LeBlanc
1919	Dec. 17-19	Gering	H.H. Andrews	R.H. Willis
1920	Jan. 28-30	Gering	A.N. Mathers	R.H. Willis
1921	Jan. 18-20	Gering	A.N. Mathers	R.H. Willis
1921	Dec. 7-9	Bridgeport	Math Schumacher	R.H. Willis
1922	Dec. 6-8	Bridgeport	F.F. Everett	R.H. Willis
1923	Aug. 21-23	Bridgeport	A.B. Wood	R.H. Willis
1924	Dec. 10-12	Bridgeport	R.S. Butterfield	R.H. Willis
1925	Dec. 9-11	Bridgeport	F.A. Dutton	R.H. Willis
1926	Dec. 8-10	North Platte	C.B. Manuel	R.H. Willis
1927	Dec. 13-15	Scottsbluff	B.J. Seger	R.H. Willis
1928	Dec. 6-7	Kearney	P.M. Whitehead	R.H. Willis
1929	Dec. 10-11	Lexington	R.O. Canaday	R.H. Willis
1930	Dec. 8-9	North Platte	R.O. Chambers	R.H. Willis
1931	Dec. 3-4	Bridgeport	H. Yensen	R.H. Willis
1932	Dec. 12-13	Scottsbluff	H.F. Parsons	R.H. Willis
1933	Dec. 11-12	North Platte	H.L. Williams	R.H. Willis
1934	Dec. 10-11	North Platte	C.M. Kearney	R.H. Willis
1935	Dec. 9-10	Scottsbluff	C.E. Allen	R.H. Willis
1936	Dec. 7-8	North Platte	A.F. Buechler	R.H. Willis
1937	Dec. 9-10	North Platte	T.W. Parry	R.H. Willis
1938	Dec. 8-9	Scottsbluff	C.T. Young	R.H. Willis
1939	Dec. 7-8	North Platte	J.W. Shoup	R.H. Willis
1940	Dec. 9-10	McCook	Leo Daniels	R.H. Willis
1941	Dec. 8-9	Kearney	B.J. Hilsabeck	R.H. Willis
1942	Dec. 7-8	North Platte	Frank Robinson	R.H. Willis
1943	Dec. 9-10	North Platte	C.F. Gund	R.H. Willis

Year	Date	Location	President	Secretary
1944	Dec. 7-8	North Platte	P.N. Foster	R.H. Willis
1945	Dec. 6-7	North Platte	H.L. Sams	R.H. Willis
1946	Dec. 5-6	Holdrege	A.A. Kjar	R.H. Willis
1947	Dec. 5-6	North Platte	L.C. Thomas	R.H. Willis
1948	Dec. 9-10	Scottsbluff	E.O. Daggett	R.H. Willis
1949	Dec. 8-9	Grand Island	G.E. Condra	R.H. Willis
1950	Dec. 7-8	Hastings	Cletus Nelson	D.S. Jones-R.H. Willis
1951	Dec. 6-7	Ogallala	C.G. Klingman	D.S. Jones-R.H. Willis
1952	NO MEETING			
1953	Feb. 19-20	Kearney	Lester Danielson	D.S. Jones-R.H. Willis
1954	Feb. 4-5	North Platte	Max Junkin	D.S. Jones-R.H. Willis
1954	Dec. 9-10	North Platte	H.P. Sutton	D.S. Jones
1955	Dec. 8-9	Scottsbluff	G.H. Storm	D.S. Jones
1956	Oct. 25-26	Holdrege	G.A. Munn	D.S. Jones
1957	Oct. 17-18	Grand Island	G.H. Robbins	D.S. Jones
1958	Oct. 23-24	Hastings	G.V. Brownfield	S.A. Matzke
1959	Dec. 10-11	Columbus	A.A. Kjar	S.A. Matzke
1960	Dec. 8-9	Sidney	A.L. Hoyt	S.A. Matzke
1961	Nov. 30-Dec. 1	Norfolk	J.C. Adams	S.A. Matzke
1962	Nov. 28-30	North Platte	J.C. Adams	S.A. Matzke
1963	Nov. 13-15	Kearney	A.C. Johnson	S.A. Matzke
1964	Dec. 9-11	McCook	Glen Beerline	D.P. Long
1965	Dec. 1-3	O'Neill	J.W. Boyd	D.P. Long
1966	Nov.30-Dec. 2	Holdrege	S.L. Bowman	D.P. Long
1967	Oct. 4-6	Lincoln	L.H. Whitehead	D.P. Long
1968	Dec. 4-6	North Platte	Ted Johnson	D.P. Long
1969	Dec. 3-5	Ogallala	A.C. Splattstoesser	D.P. Long
1970	Dec. 9-11	Grand Island	W.W. Lyons	D.P. Long
1971	Oct. 20-21	Omaha	D.B. McOstrich	D.P. Long
1972	Nov. 27-29	Kearney	James Pringle	D.P. Long
1973	Nov. 26-28	Scottsbluff	Ford Jacobsen	D.P. Long
1974	Nov. 20-22	York	R.D. Dirmeyer	D.P. Long
1975	NO MEETING			
1976	Jan. 12-14	Lincoln	Harold Fintus	D.P. Long
1977	Jan. 17-18	Kearney	Alex Heimbouch	D.P. Long
1977	Dec. 5-6	North Platte	Henry Lange	D.P. Long
1978	Dec. 4-5	North Platte	Henry Lange	D.P. Long
1979	Dec. 3-4	North Platte	Jack Thompson	D.P. Long
1980	Dec. 8-9	North Platte	Vernon Laverack	D.P. Long
1981	Dec. 14-15	North Platte	Vernon Laverack	D.P. Long
1982	Nov. 29-30	Kearney	Vernon Pearson	D.P. Long
1983	Dec. 5-6	North Platte	Harlin Welch	D.P. Long
1984	Dec. 6-7	Scottsbluff	Paul Pritts	D.P. Long
1985	Dec. 5-6	Kearney	Keith Davis	D.P. Long
1986	Nov. 20-21	North Platte	Frank Dragoun	D.P. Long
1987	Dec. 10-11	Kearney	Gary Westphal	D.P. Long
1988	Nov. 14-15	Grand Island	Paul Pritts	D.P. Long
1989	Dec. 4-5	Kearney	Harlin Welch	D.P. Long
1990	Dec. 3-4	North Platte	Tom Knutson	Norma Sitzman
1991	Dec. 2-3	Grand Island	Doyle Lavene	Norma Sitzman
1992	Dec. 7-8	Kearney	Doyle Lavene	Norma Sitzman
1993	March	North Platte		Norma Sitzman

Profiles:

Harold Stevens: Water Tours Showed Irrigators 'Where the Water Came From'

In his 42 years of helping Dawson County residents through a myriad of educational programs, issues and concerns, retired county extension agent Harold Stevens considers some of his greatest accomplishments to be in the area of water awareness.

A philosophy that guided Stevens was that "most people will help themselves if they are shown the way." Showing them about their water resources was exactly what Stevens did through his water tours. He believed that it was crucial for everyone, rural and city dweller alike, to know where their water comes from and to develop an appreciation for it.

His "5-0-5" tours became popular events in Dawson County, so named because participants left at 5 a.m. and returned at 5 p.m. In between, participants took a 300-mile tour of key water sites such as Kingsley Dam, Lake Ogallala, the Gerald Gentleman Station at Sutherland, Central Nebraska Public Power and Irrigation District's Diversion Dam on the Platte River at North Platte and many canals along the Platte. Co-sponsors of these tours were the Nebraska Public Power District and Central Nebraska Public Power and Irrigation District. From 1960 to 1987, Stevens hosted 44 water tours, logging about 13,000 miles.

Harold Stevens family collection

Harold Stevens

A similar tour he hosted several times was the VIP-Top Brass North Platte River Tour. The four-day affair was geared toward policy makers such as state senators and key irrigation people. In addition to in-state stops, this group traveled to Colorado and Wyoming to observe the major dams developed off the North Platte River.

On both the 5-0-5 and VIP tours, Stevens took photos at each stop. These became the basis for water presentations he gave to community organizations in Dawson County to continue spreading his message about "where your water comes from."

Stevens believes his biggest educational accomplishment in water programming was in the early 1980s when he arranged for the Dawson County ditch irrigators to bring their water rights up to date. A public meeting masterminded by Stevens in 1981 brought state water specialists and regulators together with local water users. The results of this meeting led to the organization of the Central Platte Irrigators Association, which arranged for all surface irrigators in Dawson County to bring their water rights up to date, and significantly influenced state legislation in 1983 on LB 198 regarding groundwater recharge.

Stevens retired from the extension service in 1987. He and his wife, Virginia, continue to reside in Dawson County, and Harold remains active in many agricultural issues, including water.

—Kathy Sullivan, free-lance writer and formerly extension communications specialist

Paul Fischbach: 'Mr. Irrigation' Pioneered New Techniques and Spread the Word

The research and programming provided by extension specialists, both at the state and district level, have added immeasurably to the wealth of knowledge amassed and passed on to the people of this state with respect to water issues.

Each have had their own message, but all have shared the same goal—to provide educational information to the people of Nebraska to help them make informed decisions and wise choices about water use.

None tells this story more appropriately

Paul Fischbach (center, in suit) discusses the benefits of irrigation with an unidentified farm family, probably during the 1950s.

IANR

than "Mr. Irrigation," Paul Fischbach, who for more than 30 years devoted time and effort to helping Nebraskans do a more efficient and effective job of irrigating their crops.

Fischbach's roots are in irrigation. He remembers as a boy on the family farm near Orleans helping his father pump water out of their first irrigation well dug in 1926. Fischbach recalls the day when he climbed down into the well for the daily task of oiling the pump bearings, which ran the full length of the well. Unfortunately, prior to doing this, the young Fischbach forgot to turn off the engine that drove the pumping plant. As he began his descent, the drive shaft caught his pant leg, quickly removed his well-worn overalls and threw him down the remaining length of the 30-foot well. Fortunately, Fischbach lived to tell that story but learned a valuable lesson in pumping-plant safety.

Fortunately also for Nebraska, Fischbach went on to study agricultural engineering at the University of Nebraska. He joined the extension service in 1952 as a district extension engineer working specifically with the irrigation development farms. In 1956 he moved to Lincoln and began his career both as an extension irrigation specialist and researcher.

His schedule, like that of many extension specialists in those early years, was tiring to say the least, exhausting at worst. Typically, Fischbach gave three talks a day in three different counties, once being so tired he fell asleep talking.

The message Fischbach passed on to irrigators was often the result of his own research on various types of irrigation equipment and techniques. In 1964 Fischbach and fellow ag engineer Paul Schlessner wrote specifications for the first irrigation well at the IANR field lab near Mead, Nebraska. Working with superintendent Warren Sahs, they developed four different 10-acre fields to do surface-irrigation research. Working with manufacturers, Fischbach obtained various types of irrigation pipe made of a variety of materials. Installed in 1965, these pipelines were still in operation in 1991. Also installed was a re-use (tailwater) system to catch runoff.

Fischbach was one of the first to experiment with the application of nitrogen through the irrigation system. Pioneering in this field meant having to build his own equipment and finding willing cooperators. One such farmer was Bob Wolf of Albion, who helped Fischbach test some of this early technology.

Later Fischbach designed and installed three center-pivot sprinkler systems and one solid-set sprinkler system to research the amount and frequency of irrigation water. In 1977 he designed an automatic gated-pipe system powered by a 25-kilowatt solar photovoltaic system to irrigate 80 acres at the Mead field lab.

Fischbach's work on irrigation scheduling and pumping-plant testing saved Nebraska farmers millions of dollars. He has been recognized both statewide and nationally for his efforts in leading Nebraska in the development of one of the strongest irrigation extension programs in the United States.

Among his many accomplishments, kudos and awards, Fischbach received the Agricultural Achievement Award by the Knights of Aksarben, which honored him for "his imaginative leadership both as an inventor and a driving force in developing irrigation in Nebraska."

—Kathy Sullivan, free-lance writer and formerly extension communications specialist

Harry Hecht: A Leader in Groundwater Conservation

Harry Hecht, extension agent in York County from 1956 to 1975, was widely recognized as a leader in the conservation and wise use of water resources. He was years ahead of his time in perceiving the potential problems of heavy demands on groundwater through increased irrigation, domestic and industrial uses.

York County has been one of the most intensely irrigated counties in the nation. In 1974, more than 1,800 irrigation wells pumped an average of 935 gallons per minute while in operation. They operated from 600 to 800 hours per season on approximately 160,000 acres.

In 1957, Hecht organized the York County Irrigation Association, and in 1958 began systematic measuring of the static water level in 68 irrigation wells. This laid the foundation for a subsequent successful campaign to convince irrigators of the need to form a Ground Water Conservation District. He tailored an educational program, involving many meetings and heavy use of mass media, to explain what the Ground Water Conservation District would provide.

An important piece of information that influenced the successful outcome of the petition drive and subsequent favorable vote was co-authored by Hecht and Deon Axthelm. This bulletin, "York County's Ground Water," cited scientific data and knowledge about York County's water levels. It used data from the Conservation and Survey Division, U.S. Geological Survey and USDA Soil Conservation Service, as well as data from an automatic recording well that Hecht had installed at the York County Fairgrounds. The graph on the front cover of the publication clearly showed that the water level at the fairgrounds well had declined 5 feet from 1958 to 1965.

Hecht also promoted the use of "catch-pits" (re-use pits) to recirculate runoff water from irrigation systems. Irrigation interests in York County, with assistance from Hecht, conducted a study of costs of irrigation re-use systems in 1970. Hecht, who received the USDA's Superior Service Award in 1974, now lives in Dickinson, North Dakota.

—Kathy Sullivan, free-lance writer and formerly extension communications specialist

Profiles of Organizations (Alphabetically):

Groundwater Foundation Educates to Save our Precious Resource

As Susan Seacrest, founder and president of the Nebraska Groundwater Foundation, stood behind the podium before the Groundwater Education in Michigan conference on May 10, 1992, in Chicago, her thoughts raced back to nearly a decade before when she first envisioned an organization to educate citizens about Nebraska's most valuable resource—groundwater.

It was the early 1980s when declining groundwater tables and accelerating nitrate levels threatened Nebraska's groundwater. This troubled Mrs. Seacrest, a professional educator. Consequently, she, and volunteers, began immediately to organize the Nebraska Groundwater Foundation (NGF). In a letter dated September 20, 1985, Seacrest announced the formation of the NGF.

She wrote: "This reality represents the culmination of a personal dream—a dream of an informed and motivated citizenry caring about and for our groundwater resource." And she immediately began to inform and educate Nebraskans and others about groundwater's vital role in everyone's lives. This was spearheaded by the Nov. 5, 1985, Nebraska Groundwater Symposium on "Groundwater for the Future."

Her goal was to emphasize an educational focus, stressing the absence of political and economic interests. She committed the NGF to a purely educational mission, saying, "The Nebraska Groundwater Foundation does not represent any single point of view on this very complex subject, but hopes instead to bring together for the close examination of the public, many minds and many perspectives."

Now annual fall Nebraska Groundwater Foundation symposiums, begun in 1985, bring nationally recognized experts to Lincoln to share expertise with usually about 300 participants. In that first symposium, participants examined the impact of climate on our lives and water supply and created their own climate response plans for a rapidly changing weather outlook.

Keynoter Rita Schmidt Sudman, California water educator, spoke on "Winds of Change: Climate and Groundwater Supply," At that first symposium, it was affirmed that the foundation's mission was "dedicated to educating the public about the conservation and management of groundwater."

Another symposium that attracted key water professionals and lay persons was the fall 1988 NGF Symposium, "The Platte River Valley and Groundwater: A Mystery in Three Acts." Speakers included Christine Olsenius, then of the Freshwater Foundation, of St. Paul, Minnesota, and Jay Lehr, then executive director of the National Water Well Association.

Special participants included Marlene Rasmussen and her 5th and 6th grade students of Litchfield, Nebraska, Public Schools. This prompted Seacrest and volunteers to establish in March 1989 the annual Children's Groundwater Festival held each spring at the Grand Island Central Community College.

This festival annually attracts more than 3,000 4th, 5th and 6th graders from across the state who learn about groundwater in a festive context without minimizing serious issues relating to the management of this vital resource.

In addition to the fall symposiums and the spring Children's Groundwater Festival, an award-winning quarterly, The Aquifer, is published by the foundation with a circulation of 2,500. It contains local, state, and national groundwater information from different points of view.

And during spring, Nebraska Groundwater Week is observed with a special awards banquet to honor outstanding contributors to the conservation and preservation of groundwater in Nebraska.

A special bonanza for the Nebraska Groundwater Foundation during the first Groundwater Week came from Lona and Jack Shafer, of Wood River. They presented the foundation with a 10-acre parcel of land at the northeast corner of the Wood River I-80 interchange. Funding is being sought to build a $2 million visitors' and educational center planned "to be the nation's first building to raise public awareness of groundwater."

The Nebraska Groundwater Foundation is now nationally recognized. While at the May 1992 Groundwater Education Summit in Chicago, Seacrest conducted Children's Groundwater Festival workshop sessions for about 200 persons. She shared the successful pattern, easily duplicated by other states.

The NGF is also heading up the newly created National Groundwater Education Consortium which met for the first time in September 1991. Seacrest sits on national committees that include the National Drinking Water Advisory Council that makes recommendations to the U.S. Environmental Protection Agency. The Environmental Protection Agency also invited Seacrest to join its Management Advisory Group on water issues in 1991.

The Nebraska Groundwater Foundation has received national awards such as the Outstanding Environmental Achievement Award, presented by the Soil and Water Conservation Society of American in 1991, as well as being recognized by "Renew America" in Washington, D.C., and the Friends of the United Nations Environmental Program in 1990.

State honors and awards include: an education award presented by the Central Platte Natural Resources District; the J.L. Higgins award from the Nebraska Department of Environmental Control in 1989; Seacrest has been named "Soil and Water Conservation Steward" by former Governor Kay Orr, in 1990 and "Water Conservationist of the Year" by the Nebraska Wildlife Federation in 1987.

The most recent award was from the Nebraska Water Conference Council in March 1992 when the NGF received the council's Progress Award for "its outstanding educational programs relating to preserving and protecting both the quantity and the quality of Nebraska's vital groundwater resources for present and future generations."

—Pat Larsen, commmunications specialist, Water Center, UNL

Charles Flowerday, CSD

A multitude of water-related subjects are presented at the annual Children's Groundwater Festival, some more seriously, others light-heartedly. Here, someone dressed as a raindrop and another as Sesame Street's "Grover" entertain children at the 1992 festival in Grand Island.

Susan Seacrest (left), president of the Nebraska Groundwater Foundation, and Nebraska Governor Ben Nelson make a presentation to three children from Mexico and their sponsors at the 1992 Children's Groundwater Festival in Grand Island. This was the first time the festival attracted an international audience.

Charles Flowerday, CSD

The Nebraska Water Conference Council: A Forum on Water Resources

Shortly after D. B. ''Woody'' Varner began his duties as the president of the University of Nebraska (NU) system in 1970, he participated in a tour of the Central Nebraska Public Power and Irrigation District facilities, including Lake McConaughy and the distribution system that provides water to about 115,000 acres in south-central Nebraska. As a result of that tour, it became obvious to him that the state's surface water and groundwater consituted its most valuable natural resources—and the key to its economic future.

In speeches Varner made around the state, he referred often to the tremendous potential offered by Nebraska's ''gold''—its water. In 1971, he and the NU Board of Regents requested more than $1 million for a ''Regents Discretionary Fund.'' Included in that package was a request for $266,480 for a Water Resources Development Program for the University of Nebraska.

While the Nebraska legislature and then Governor J. J.Exon didn't fund the entire package, they did provide $150,000 for a Water Resources Development Program. Of that amount, $30,000 was allocated to the Nebraska Water Resources Research Institute (later to become the Water Center) to start a research project on groundwater recharge in an area in Hamilton, York, Seward, Clay and Fillmore counties—an area with some of the most concentrated irrigation in the state.

A total of $80,000 was also allocated to establish an ''Irrigation Development Program'' in the University of Nebraska-Lincoln (UNL) College of Agriculture that helped hire four new faculty to work on irrigation development. I was fortunate enough to be selected as the coordinator for this new program in Aug. 1, 1971. Three other positions also were established: an extension irrigation engineer in the UNL Department of Agricultural Engineering, John Addink; an extension irrigation engineer at the Panhandle Research and Extension Center at Scottsbluff, David Fonken; and an extension irrigation engineer at the West Central Research and Extension Center at North Platte, Darrell Watts.

One of the first assignments that Varner gave me as coordinator for the new Irrigation Development Program was to organize a two-day ''Water Resources and Irrigation Development Seminar for the 1970s.'' He also tabbed me for chairman of a 39-member committee that he had appointed to accomplish this task. The result was a very successful two-day seminar held on March 2-3, 1972, at the Nebraska Center for Continuing Education.

In his opening statement at this conference, Varner stated: ''I am sure that I need not point out to you again that we know now that we could double the number of acres of land under irrigation in this state during this decade. We have the topography, we have the water in place, and it can be done with the resources we have. Now, I say this not thoughtlessly, because I am well aware, as are you, that our water resources are expendable, that we must be concerned with wise utilization, with sound conservation practices amd always mindful of ecological concerns. If we can accomplish this, it would mean for the state of Nebraska that we would have added a $4-billion industry in this state.''

At the close of the conference, Varner proposed the formation of a group, eventually called the Nebraska Water Conference Council, with the following challenge: ''I therefore propose to you that we form ourselves into what I will suggest we call the Nebraska Water Conference, that is an on-going assembly of interested people and agencies and institutions concerned about the wise development, utilization and management and conservation of this enormous resource, which is so crucial to the wellbeing of the state of Nebraska; (and) that we, therefore, being an ongoing group known as the Nebraska Water Conference, meet at least once a year in this kind of seminar situation; but, perhaps more importantly, to have emerge out of this a continuing concern with the broad base of representation for the whole range of problems that face all of us and the business of developing to the maximum the resources of this state, not to replace anything, not to supplant anything, but to reinforce and to work with and to bring together into a common group the various interests we share in this great concern in this resource in this state.''

Varner then announced the appointment of 25 people to serve on the Nebraska Water Conference Committee (now Council) to meet and think through the things that might be done, but particularly to focus attention on a continuing effort on the part of this kind of broad-based representation. He appointed me as his representative and as secretary for this group, and Vance Anderson of Hastings was elected at the group's first meeting to serve as the chairman.

In the fall of 1973, the group was expanded to include 50 representatives of the public and private sectors, all with involvement and interest in water resources and irrigation. In 1974, the first of what became the annual ''Nebraska Irrigation Tour'' was held and traveled to the Blue River Basin in York and Hamilton counties and to Phelps County to tour irrigation facilities. In 1979, the group was expanded to 100 members with even wider representation of the many agencies, organizations and individuals interested in the state's water resources. At that time, the name was changed to the ''Nebraska Water Conference Council'' (NWCC).

The principal activities for the group were identified and listed in a charter adopted on Oct. 6, 1979, as follows: 1) a fall one-day meeting to identify current and anticipated water problems and issues and to arrange for presentations relating to these topics; 2) an annual two-day water resources conference held in cooperation with the UNL Institute of Agriculture and Natural Resources during the winter or early spring to provide a forum for presentation of pertinent information and discussion relating to Nebraska's water and land resources; 3) an annual tour during the summer to view water resources facilities and associated management practices, and/or water problem areas, to facilitate exchange of information on water needs and issues in various regions; and 4) other meetings and public events as may be necessary to carry out the objectives and activities indentified in this charter.

One of the annual Nebraska Irrigation Tours put on by the Nebraska Water Conference Council looks at bank stabilization along the Elkhorn River.

Biological Systems Engineering, UNL

Except for the 1990 conference, which was held in Kearney, all of the 21 annual Nebraska Water Conferences prior to 1993 have been held in Lincoln. Another exception has been the 1993 conference, tied to the centennial observance of the founding of the Nebraska State Irrigation Association and held at North Platte. Besides the traditional sponsors (the NWCC, the UNL Insitute of Agriculture and Natural Resources, the University of Nebraska Water Center), additional sponsors for the 1993 conference were the Nebraska Groundwater Foundation and the Nebraska State Irrigation Association. Attendance at the annual conferences has ranged from 130 to as many as 390 people.

—Les Sheffield, coordinator for outreach programs, UNL Water Center, and formerly UNL extension irrigation economics specialist

Stewards of the Platte: Educating About the Platte River Ecosystem

Stewards of the Platte, based in Grand Island, is a non-profit, non-political organization that seeks to promote awareness and provide education about the Platte River and its related ecosystems.

The group was formed in 1990 in response to the lack of educational opportunities available to the general public. It strives to represent all interests of Platte River use.

On-site seminars are the main teaching tool. Throughout the year, the group sponsors field trips to different parts of the river. During the 1992 spring waterfowl migration, Stewards of the Platte led more than 1,500 adults and 2,500 elementary school children on trips.

The group, which has members from 36 states, also publishes a quarterly newsletter that it distributes to members and all Nebraska elementary schools as a teaching aid. Distributed with the newsletters are one-page flyers containing information on Nebraska wetlands, geared toward school-age children. The ''wetlands notebook'' flyers, as they are called, are published with the help of a grant from the U.S. Fish and Wildlife Service. A long-term goal is to construct a visitor center near the river.

In addition, Stewards of the Platte participates in Wings Over the Platte, a three-day celebration held in Grand Island during the height of the spring migration, and the Nebraska Children's Groundwater Festival, sponsored by the Nebraska Groundwater Foundation.

—Brad Rundquist, weekend and copy editor, **Ft. Dodge (Iowa) Messenger,** *and formerly CSD editorial assistant*

Profiles of Other Water-Education Efforts (Alphabetically):

Below is a partial listing of agencies, organizations and programs that play a role in educating the public about the use, conservation and importance of Nebraska's water resources to its people and wildlife. Much of this information is compiled from the *Nebraska Environmental Education and Information Resources Directory,* available from the Nebraska Natural Resources Commission, P.O. Box 94876, Lincoln, NE 68509-4876, (402) 471-2081.

Central Nebraska Public Power and Irrigation District: Pioneering The Platte is a program designed to teach children in the 4th through 6th grades about the role of the Central Nebraska Public Power and Irrigation District on the Platte River in the areas of irrigation, hydropower, wildlife and recreation.

Environmental Education Institute: This is a two-week environmental education course held annually at Chadron State College for teachers of kindergarten through 12th grade in all subject areas. The course addresses water and water-quality issues as part of the curriculum.

Nebraska Game and Parks Commission: The game and parks commission cooperates with federal, state and local agencies to promote water education, primarily directed at school-age children. The commission is a sponsor of Project WILD Aquatic, a nationwide environmental education program for teachers of kindergarten through high-school-age young people. The program is designed to increase awareness and knowledge of aquatic ecosystems and to encourage responsibility for their protection. Since the program was launched in Nebraska in 1987, it has reached about 6,000 teachers across the state. A companion program, the Nebraska Aquatic Education Program, supplements Project WILD Aquatic with texts and audio-visual materials designed to bring the aquatic world into the classroom.

Children also learn about the central role of water in the environment through the Volunteer Fishing Instructor Program, jointly sponsored by the commission and the Ak-sar-ben Aquarium in Gretna. In addition to learning how To fish, children are taught basic environmental ethics and the importance of water quality to maintaining a healthy habitat for fish.

Whenever possible, the commission is an active supporter of water-education initiatives spearheaded by sister agencies. For example, the commission is a partner in efforts by the Lower Platte North Natural Resources District to introduce Riverwatch, a worldwide environmental education program developed at Michigan State University in the early 1980s. Once the program is in place, children in the Lower Platte North NRD will have a hands-on opportunity to learn about the Platte River ecosystem by sampling water and observing its wildlife. They will then share this information with children in other states and countries through a computer network.

Nebraska's Natural Resource Districts: The state's 23 natural resource districts (NRDs) are involved in a wide range of educational and regulatory activities related to water quality and quantity. Most NRDs have education programs that include school visits, demonstrations and camps for teachers, with the purpose of explaining proper resource management. On the regulatory side, rules to reduce and prevent groundwater contamination in vulnerable areas—through irrigation scheduling and water and soil sampling, for example—have generated education and outreach programs to bring farmers into compliance at the local NRD level.

Project Learning Tree: This is an interdisciplinary environmental education program for kindergarten through 12th grade used in the United States and abroad. In Nebraska the lead sponsors are the Nebraska Forest Service and Cooperative Extension Service at the University of Nebraska.

Southeast Nebraska Environmental Education Seminar: The seminar is a two-week course for teachers of kindergarten through 12th grade held annually at Camp Jefferson near Fairbury. The purpose of the course is to give teachers basic training in environmental education.

Stop, Look and Learn About Our Natural World: A Nebraska Natural Resource Elementary Education Guide: This is the title of a three-volume series of lesson plans and activities for use in the elementary grades. The volumes are available from the Nebraska Natural Resources commission.

Water Riches: This is a water education curriculum directed toward 4th and 5th grade students. Sponsored by the 4-H Youth Department of the University of Nebraska Extension Service, the course content focuses on water quality and conservation, hydrologic cycles and aquifers.

Environmental Quality and Changes in Ecosystems

by Sally J. Herrin
and Charles A. Flowerday

The history of the ecosystems of this state is a history of diversity and succession. The geo-political plot we know as Nebraska has been, at various places and times, an ocean floor, a desert, dotted with swamps and bogs, a savanna parkland, a woodland and an ice mass, as well as the more familiar grassy plain laced with streams and rich in wet meadows and wetlands. In the eastern half of the state, where rainfall averages hover around 25-35 inches a year, these wetlands are mostly near rivers or are runoff-fed. Where present in the far drier west, such aquatic riches are due largely to the abundance of groundwater. Through climatic variations and corresponding changes in vegetation, Nebraska was home to an impressive array of wildlife, including many dinosaurs, the great land mammals of the ice ages, and later, enormous herds of grass-eaters, particularly bison, that grazed over hundreds of grasses and the carnivores who fed on them, as well as to a wealth of less dramatic but no less important species such as freshwater mollusks (clams and snails, for example) and aquatic insects. And, of course, it has been a staging ground for great migrations of cranes, ducks and geese, eagles and many other birds for some millions of years.

By comparison, human beings are relative newcomers to the state, probably arriving within the last 10,000 to 15,000 years. Like the other older inhabitants, the plants and other animals, the first people in Nebraska survived largely through adaptation. Prior to the introduction of the horse, Native Americans likely contributed to the extinction of mammoths, set occasional prairie fires and cut timber in isolated spots in river bottoms, but their activity probably changed the landscape little more than, say, the activities of beavers.

Beavers are fascinating mammals and the subject of a great deal of folklore and popular literature. At times, they can seem like small, furry humanoids, with their purposeful building of dams and lodges, yet it would distort the picture to give them overly human qualities. For example, some have claimed that beavers are so clever they can fell trees in any direction they want. The truth is that trees often fall toward water because of a natural tendency to grow leaning toward open areas above streams where the light shines through.

But as a means of discovery, let's briefly examine the impact of beavers on the environment. Beavers alter their environment in obvious, even dramatic ways, and offer an interesting comparison to humans. They create ponds or wet meadows by damming shallow streams with barriers of sticks and logs they mortar with mud. These small reservoirs protect their lodges, or the dens they sometimes dig in stream banks, from predators, provide waterways for the transportation of building materials and food and create a swimming area below the ice in winter.

An example of the native tallgrass prairie at the Lillian Annette Rowe Sanctuary near Gibbon in Buffalo County. The sanctuary is managed by the National Audubon Society to provide habitat for migratory birds and other wildlife.

Jon Farrar, Nebraska Game and Parks Commission

Many trees can't survive in the waterlogged soils in areas flooded by beavers; these die and are replaced by species adapted to saturated ground—especially the willows, cottonwoods and alders, the bark of which is a staple for the beavers. So, the damming of streams protects the beavers' homes and creates a high local water table that favors their preferred food species. A rough analogy can be made with human-made impoundments, which are designed partly to control flooding and to provide irrigation water for farming—that is, to protect human homes and other interests and to favor our food production. But the differences between the two species with respect to dams are greater than their similarities.

Probably the most striking difference between human and beaver impoundments is scale. Beavers work shallow water, and their dams create wet meadows, which are habitat for other wildlife such as muskrats, mink, waterbirds and shorebirds, as well as fish varieties that favor still water. Humans, on the other hand, construct much larger dams, and in the process, some new wildlife habitat is created. For example, cold water releases below Kingsley Dam are perfect for trout, which otherwise would never

Mike Anderson, Soil Conservation Service

A view looking away from the Harlan County Reservoir. This area provides habitat for fish and other wildlife such as bald eagles. While dams eliminate some habitat for some wildlife, they also add habitat for certain species.

Rocky Hoffmann, Nebr. G & P Comm.

A beaver along the North Platte River near Hershey, Lincoln County

be found in the Platte, and cottonwoods on sandbar islands provide more habitat for deer and wild turkey.

However, much of the water we impound returns too late or in quantities and ways not useful to the same wildlife that benefited from the previous riverine wetlands and the broad channels of shallow water and sandbars scoured by peak spring flows. These species benefiting from such habitat would include migratory birds and their food, mollusks and aquatic insects.

What beavers can do rather elegantly on a small scale using renewable energy sources (their own food), humans require enormous amounts of non-renewable energy (petroleum fuels) to achieve on a large scale. In addition, prime farmland is often inundated, at great social and personal cost to the people involved. In fact, large impoundment projects are so costly that they almost never are accomplished without large amounts of federal funds. And, of course, no structure lasts forever. When the useful life of our dams is done, it's not simple or easy to decide what to do with the silted-over reservoir. Dredging is expensive, and cropping is difficult, if not impossible.

A beaver dam on the North Fork of Birdwood Creek, McPherson County. Even small human impoundments dwarf most beaver dams, which, when the beavers are done with them, disintegrate relatively rapidly.

Charles Flowerday, CSD

Beaver dams silt up too, and when this happens, they move on to a new site and start again. Without maintenance, abandoned beaver dams disintegrate and drain, usually leaving the land no worse than before and in some cases better.

Beavers are interesting for another reason also. They are historically significant. Beyond a few isolated Spanish explorers, the first whites to spend any real time in Nebraska, less than 25 years after Shakespeare's death, were French fur trappers. A fashion for beaver hats evolved. The foundation was laid for one of the first industries that convinced Europeans how profitable the western half of this continent could prove. As a result, beavers were harvested so heavily that their numbers seriously declined. Fortunately, recent efforts to restock beavers in Nebraska have been extremely encouraging, an example of how human ingenuity, working toward mitigating its own effects, can redress some injuries done to an ecosystem, something no animal can achieve. In the Nemaha River basin, in fact, beavers are so plentiful today that some farmers are beset with flooded fields and trappers once again are doing business.

The earliest indigenous people of Nebraska were primarily hunters and gatherers, some of whom later farmed on a very small scale. There is some evidence to suggest they burned the prairies, probably to stampede bison, and so, inadvertently managed the grassland. But by the late 19th Century, the activity of large numbers of European immigrants began to alter the landscape of Nebraska much more significantly by breaking the sod for widespread cropping and by introducing domestic cattle to succeed native bison.

In the 20th Century, changes in Nebraska's ecosystems caused by human

Rocky Hoffmann, Nebraska Game and Parks Commission

A beaver dam on a channel of the South Platte River near the city of North Platte, Lincoln County

activity have accelerated dramatically. While diversion of river water for irrigation, particularly along the Platte, and expanding crop acreage by draining wetlands have brought prosperity to many farmers and rural communities, reduced streamflows and diminished habitat have produced stress for migratory bird populations and aquatic animals.

In addition to concern about instream flow for wildlife, other changes wrought by humans in the previous ecosystem involve diminished groundwater quantity—reduced water stored underground in the pore spaces between the "grains" of the earth materials. (In addition, movement of water through canals has contributed to water-level rises in some areas.)

Pumping of groundwater for irrigation has caused water-level declines great enough that groundwater control areas, a means of regulating irrigation pumping, have been established in the Upper Republican Natural Resources District (NRD), the Upper Big Blue NRD and part of the Little Blue NRD.

In addition, many people have become concerned that the residues of nitrogen fertilizer and pesticides, the environmental by-products of intensive farming, are threatening water quality. Not largely visible generally, these effects are easy to see where fertilizer-laden runoff enters standing water. This makes the water over-rich in nutrients and fosters an overgrowth of vegetation damaging to much animal life, a process called *eutrophication*. And perhaps most importantly, while not the most glamorous of environmental problems, soil erosion, especially by water, may be the most difficult long-term effect of intensive agriculture to contain, in large part because of its incremental and nearly invisible nature.

So, after more than a century of environmental change resulting from the lifestyles of European settlers in Nebraska, the environment may be said to be reasserting itself. That is, changes created by human activity have created in effect a new environment, which is acting upon the life of the people, and the people, in turn, are responding to these changes and challenges.

It would require more than a chapter or even a whole book to completely describe the workings of Nebraska's natural systems, the changes that have occurred in this century and the attempts to undo or mitigate some of the losses due to these changes. While the history of management of environmental problems is presented in the accompanying chapter (*see "Conservation Management and Environmental Protection"*), a great deal must be passed over here, but not necessarily because any one ecosystem or species is more or less important than any other. In fact, biologists now understand that to study just one species or individual in nature is fairly misleading and not very useful.

Nature is not a collection of individuals, a box of dusty butterflies on pins. It is, on the contrary, a system of systems, plants and animals, large and small, sturdy and fragile, and the water, air, soil and sunlight, ceaselessly interacting together. That *everything is connected* is a statement too true to be trite. So a change in any part of the system has repercussions for the whole, spreading like rings from a stone dropped into water.

(Continued on p. 162)

Woody vegetation encroaching on sandbars of the Platte River has provided more habitat for wild turkey than existed before damming of the river stopped scouring floods.

Soil Conservation Service

The Occurrence of Nitrate Contamination in Nebraska

Groundwater is the source of drinking water for almost all of Nebraska's 330,000 rural households and 84 percent of the public water supplies. Groundwater quality, like quantity, is vital to the state's economy. The widespread use of large quantities of nitrogen fertilizer and pesticides, which have the potential to contaminant groundwater, and the risk of harmful health effects from ingestion of these contaminants in drinking water cause nitrate and pesticides to be the contaminants of greatest concern to Nebraskans. Atrazine, by far the most heavily used pesticide in Nebraska, also is the most frequently detected pesticide. The concentrations, however, are generally very low. In a recent statewide assessment of pesticide contamination in ground water, less than 1 percent of the 2,260 wells sampled for atrazine exceeded the maximum contaminant level (MCL) of 3 parts per billion (ppb). In contrast, nitrate concentrations exceeded the MCL of 10 parts per million nitrate as nitrogen (10 ppm NO_3-N) in about 20 percent of the 5,826 sampled wells, causing nitrate to be the most frequently detected contaminant in Nebraska's groundwater. Maximum contaminant levels are enforceable drinking-water standards established by the U.S. Environmental Protection Agency to protect human health. Although water quality in private wells is not regulated, MCLs are used as criteria to evaluate their status.

Methemoglobinemia is the primary adverse health effect that can result from the ingestion of nitrate-contaminated groundwater, and it is this noncancerous acute toxicity that is the basis for establishing an MCL of 10 ppm NO_3-N. The MCL protects infants under 6 months of age from methemoglobinemia, a condition in which nitrate in the infant's stomach is converted to nitrite, which inhibits the oxygen-carrying ability of the blood. The tissues become starved for oxygen and asphyxiation and death can result. Other adverse health effects linked to the ingestion of nitrate-contaminated drinking water have not been proven.

A recent compilation of nitrate data from 5,826 domestic, irrigation, public supply, stock and monitoring wells sampled throughout Nebraska between 1984 and 1988 indicated that the nitrate concentrations in 20.8 percent of the wells exceeded 10 ppm NO_3-N. The locations of these wells are shown in the accompanying map on the state's nitrate-nitrogen in groundwater. About 51 percent of the contaminated wells were in areas where the groundwater is highly vulnerable to contamination. Many of these wells were concentrated in the highly vulnerable areas in the Central Platte Valley, both north of the Platte River from Kearney in Buffalo County to Columbus in Platte County and south of the river in northern Phelps and Kearney counties and in Holt County north of the Elkhorn River. These areas are characterized by fence-row to fence-row irrigated corn monoculture on well to excessively well-drained soils and a water table within 50 feet of the surface. Irrigated crop production brings with it increased amounts of nitrogen fertilizer, a potential groundwater contaminant, and the vehicle, water, to transport the contaminant to the groundwater.

Irrigated small grain crops require almost twice as much nitrogen fertilizer as dryland crops and irrigated corn requires more nitrogen fertilizer than other irrigated crops grown in Nebraska. Nebraska's application of approximately 755,000 tons of nitrogen fertilizers in 1987 ranked the state third nationally in commercial fertilizer usage. Water in excess of crop needs can move nitrate below the crop root zone where it is no longer available to the crop and through the unsaturated layer to the water table. Attenuation of the nitrate occurs in this zone, the intermediate vadose zone, and is affected by both the composition and thickness of the sediments. The thicker the intermediate vadose zone, or the greater the depth to the water table, the more likely the contaminant will be attenuated. The higher the clay and silt content of the unsaturated sediments, the less likely the pollution potential.

Periodic reassessments and monitoring since the areas of nonpoint nitrate contamination in the Central Platte Valley and Holt County were delineated in the mid- to late

(continued on next page)

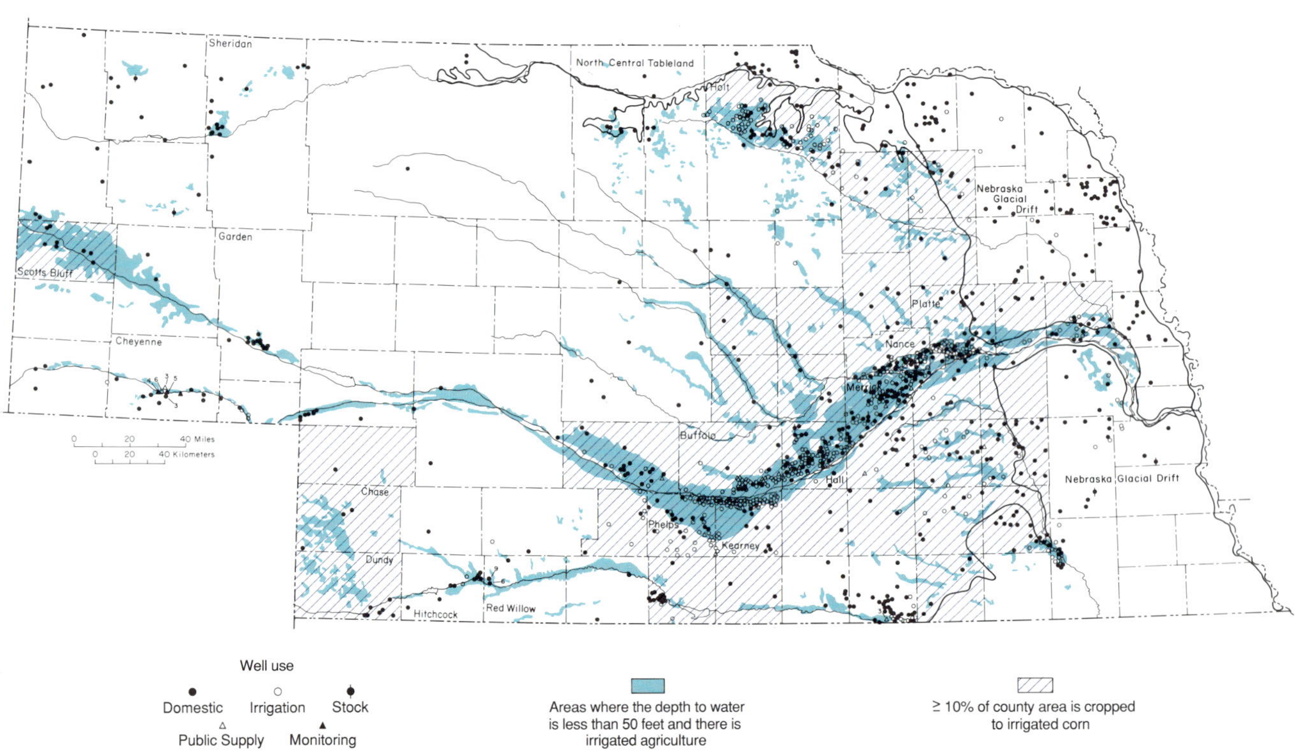

Occurrence of Nitrate Contamination in Nebraska

1970s show that the problem is dynamic and that both the areal extent of the contamination and the nitrate concentrations are increasing. In eastern Buffalo, Hall and Merrick counties, where more than 500,000 acres are underlain by nitrate-contaminated groundwater, the nitrate is from a nonpoint (dispersed) source, and that source is leachate from commercial fertilizer applied to cropland. Between 1974 and 1985, the extent of the contaminated area more than doubled while the concentrations increased at rates of 0.4 to 1 ppm NO_3-N per year. Concentrations in eastern Buffalo and western Hall counties and in much of Merrick County exceed 20 ppm NO_3-N. The contamination extends into Nance and Platte counties as far north as the Loup River and as far east as Columbus. During the 1980s average annual increases of 1 ppm NO_3-N were the rule for irrigation wells south of the Platte River in northern Kearney County. Concentrations above 20 ppm are encountered frequently. In central Phelps County, where the soils are finer textured and the water table is deeper (about 40-50 feet), the incidence of contamination was less than in northeastern Phelps County. Concentrations in both domestic and irrigation wells in this area have increased since the late 1970s. Although the soil type and principal crop are the same in the area highly vulnerable to groundwater contamination that extends into Kearney County, the frequency of contaminated wells was dramatically less.

A majority of the irrigation wells sampled in the highly vulnerable area of Holt County north of the Elkhorn River have nitrate levels above 10 ppm NO_3-N, and many have concentrations in excess of 20 ppm. Concentrations of 30 to 40 ppm NO_3-N are not uncommon in the irrigation wells north of Atkinson. Between 1976 and 1990 nitrate concentrations in these areas of nonpoint contamination increased at average annual rates of 0.4 to 0.7 ppm NO_3-N. The areal extent of the contamination also increased. A minimum of 140,000 acres are underlain by groundwater with nitrate concentrations in excess of the MCL. The source of this nitrate is leachate from commercial fertilizer.

There are numerous small areas where the groundwater is highly vulnerable to contamination. Many of these areas are in narrow river and creek valleys and, while the county as a whole may not be conducive to irrigated corn production, these areas can be irrigated and cropped almost exclusively to corn. The Sidney area along Lodgepole Creek in Cheyenne County and the Oshkosh area on the North Platte River in Garden County are two relatively small areas of nonpoint nitrate contamination. The source of the contamination in the Oshkosh area is leachates from commercial fertilizer applications, while in the Sidney area the source of most of the nitrate is manure applied to cornfields. Other suspected areas of contamination are near Mitchell in Scotts Bluff County, Mirage Flats in west-central Sheridan County and in the Republican River valley between Culbertson in Hitchcock County and McCook in Red Willow County. Nitrate-nitrogen concentrations in many of the domestic wells in these areas range from 10 to 20 ppm.

In Dundy and Chase counties, there are fairly extensive areas cropped to corn and highly vulnerable to groundwater contamination. Nitrate-nitrogen concentrations in the few irrigation wells that were sampled generally were below 5 ppm. The principal aquifer is the Ogallala sandstone, which is less transmissive than the sand and gravel aquifers that are the primary source of shallow groundwater. Because there is significantly more drawdown around pumping irrigation wells in this sandstone aquifer and the effects of pumping are more extensive than in sand and gravel aquifers in the Central Platte Valley and northern Holt County, domestic wells are developed at the bottom of the aquifer to ensure they will not be affected by the 20 to 30-foot water level declines that occur during the irrigation season. Consequently, there are few, if any, existing wells that can be sampled to define water quality at the top of the aquifer. Monitoring wells are needed in these areas to determine the extent of the nitrate contamination.

In southern Phelps County and in southwestern Kearney County, there are a considerable number of irrigation wells with nitrate concentrations that exceed the MCL and concentrations in many other wells are approaching the MCL. This is an irrigated, corn-producing area with depths to groundwater in excess of 50 and 100 feet. It is the only irrigated corn-producing area with more than 50 feet to groundwater where contaminated irrigation wells were encountered fairly frequently. The cause of these elevated nitrate concentrations has not been investigated. It is known, however, that there is the potential for nitrate contamination in excessively fertilized and irrigated areas underlain by fine-textured sediments.

Associations between the areal distribution of groundwater nitrate concentrations, land use and general hydrogeology also are useful in evaluating the potential for point-source contamination, which originates at discrete locations. Potential point sources of nitrate contamination are barnyards, corrals and abandoned feedlots. Because domestic and stock wells often are located near these potential point sources, they are more susceptible to point-source contamination than are irrigation wells. Point-source contaminated wells can be characterized by very high nitrate concentrations. One percent of the 5,826 wells used to determine the distribution of nitrate concentration in the state had nitrate concentrations greater than 50 ppm NO_3-N. Fifty-two of the 61 samples were from domestic wells. Most of the domestic wells were in areas that are not highly vulnerable to contamination. These areas include the North Central Tableland and the Nebraska Glacial Drift regions. About 26 percent of the domestic wells sampled in the two regions had nitrate concentrations between 10 and 50 ppm NO_3-N.

Because the bedrock outcrops or lies close to the surface in the tableland and glacial till overlies the bedrock in the glacial drift region, both areas have a very limited supply of groundwater. If a well can be developed in the tableland, it usually is in the sands and gravels immediately above the bedrock where there is little groundwater for dilution of contaminants. In the glacial drift region most wells are developed in perched aquifers of small areal extent and consequently the yields of water are small. There are, however, deposits of sand and gravel along stream valleys and in buried paleovalleys that are a source of groundwater. The excessively high nitrate concentrations found in the two regions, their occurrence in areas that are not highly vulnerable to nonpoint nitrate contamination and the large variations in the distribution of the concentrations leads to the conclusion that nitrate in most of these domestic wells originates from a point source of contamination. The conclusions from a study in the glacial drift of southeastern Nebraska can be extrapolated to both areas. In that study the incidence of nitrate contamination associated with poor siting of wells near potential sources of nitrate contamination, namely, intermittently used and abandoned barnyards was augmented by the large number of poorly constructed wells. While some of the nitrate from the dissolution of manure in the barnyards does infiltrate through the unsaturated zone to the groundwater, the preponderance of wells with open-jointed casing increases the number of potential entry paths for a contaminant and short circuits the "normal" infiltration route of a contaminant. These findings clearly illustrate that leachates from irrigated agriculture are not responsible for all the nitrate contamination in Nebraska. Probably as these older wells are sealed and replaced, the incidence of point-source contamination in the north-central tableland and glacial drift regions will decline.

Additional information on the occurrence of nitrate as well as pesticides can be found in "Occurrence of Pesticides and Nitrate in Nebraska's Ground Water," available through the University of Nebraska-Lincoln Water Center.

—Mary Exner Spalding, CSD research hydrochemist

(Continued from p. 159)

In coming to terms with this connectedness, we have begun more and more this century to understand the importance of the ecosystem as the fundamental unit of study, survival and value in the natural world. Although definitions will vary and new and better ones will replace the less useful, the USDA 1991 Yearbook of Agriculture, *Agriculture and the Environment,* offers the following: "Ecosystems can be visualized as dynamic segments of the environment, involving the biology of plants, animals and microorganisms; the chemistry and physics of moisture, heat, wind and solar radiation; and the geology and hydrology of surface materials in a landscape."

Nebraska State Historical Society

Paper birch along the South Branch of the Niobrara River in Cherry County. This photograph was taken by G. E. Condra, the dynamic first director of the Conservation and Survey Division during the early years of his tenure.

In general, we have come to understand them as places where successions and communities of plants and animals interact in characteristic and loosely predictable ways, evolving a balance in populations over time, some over hundreds and thousands of years. What is crucial to understanding human-wrought changes in ecosystems is the relative speed with which humans can create changes, some larger, some smaller, in these systems relative to much longer periods of time needed for much of the biological community to adjust. In addition, we know we can also make changes to which some species cannot adjust.

While agriculture has brought prosperity to European immigrants and their descendants in Nebraska, for native plant and animal species—and for indigenous peoples—this new lifeway introduced by a new race has been hard on other species, contributing to the extinction of a few species of animals, the extirpation of some plants and animals and a reduction in numbers of many plant and animal species that once were quite common. (Extirpation is the driving out of a species from a given region.) From the point of view of biodiversity, in purely technical terms, it has been a "catastrophe," not unlike the glaciers.

Lest such terminology gets anyone overly worked up, it should be noted that farming on the Great Plains, particularly irrigated, row-crop agriculture, has been a kind of miracle, a technological, as well as a humanistic, miracle. It has fostered a comfortable way of life for many more people than ever was possible before modern research and technology were applied to "making the desert bloom." However, that miracle has had a cost. It has involved the expenditure of ecological, as well as economic, capital. While the blessings and benefits of this development have been chronicled in most of the other parts of this book, this chapter is largely about counting the cost of that "miracle." There is little to be gained from throwing stones at the people of the past. That is not what is intended here. But in the interest of telling the whole story, one must try to come to terms with what has been lost, as well as what has been gained.

As European settlers came into this land, they created a completely new culture. In doing so, they mostly did what was expedient, what seemed most necessary at the time. In settling the frontier, only a minority, if any, felt compelled to ask about the fortunes of native plants, animals or peoples as they built this new culture. If such questions were asked from time to time, no doubt the prevailing sentiment was that the end justified the means, that building a life here that could support many more people was most important, more important that worrying about what the ecosystem had to offer before they began "improving" it.

Only more recently have we begun to ask: what may we, or others, be giving up through this development? In effect, we know what we have built, and we feel more secure in this environment because of it. That in part may now allow us to ask what we have razed in the process.

In addition, in light of the efforts of people who have asked the hard ecological questions, who have advocated for the "welfare" of this land and its native species, it is appropriate here to document some of the history of the conservation programs and processes that have attempted to mitigate some of the losses we are about to recount. As with any other history, it would impossible to list every agency, group or individual ever involved in environmental protection, mitigation and preservation. But so as not to seem to overwhelm anyone with a litany of wrongs done, in the accompanying chapter on conservation management and environmental protection, we have sought to present some of the strategies, plans and policy decisions that have evolved for establishing or re-establishing ecological "health and wealth."

Plowing the Prairie

Twenty thousand years ago, much of the land just east of us in Iowa was covered with hundreds, even thousands of feet of ice. The last of many ice advances and retreats between about 2-3 million and 20,000 years ago (some of which caught the eastern edge of the state), this glaciation began to retreat about 18,000 years ago. Even today, in isolated locations, we can find relicts of that time, plants that are typical of a much cooler climate. The paper birch and aspen that grow along the Niobrara River are examples, as are the little orchid known as twayblade and the bogbean, both found in certain wetlands of Cherry County. Along Minnechaduza Creek, north of Valentine, grow certain northern grasses that are also abundant in upper Minnesota but do not occur in the 500 intervening miles.

Unlike Kansas, for instance, Nebraska supported no state biological survey in its early years, so information on predevelopment plant and animal populations is very sketchy by modern standards. We have records dating back to 1860 that name various prairie plants, but not until the turn of the century did researchers begin to record how plant populations were organized—what grew with what and where. However, it is es-

The Natural Vegetation of Nebraska

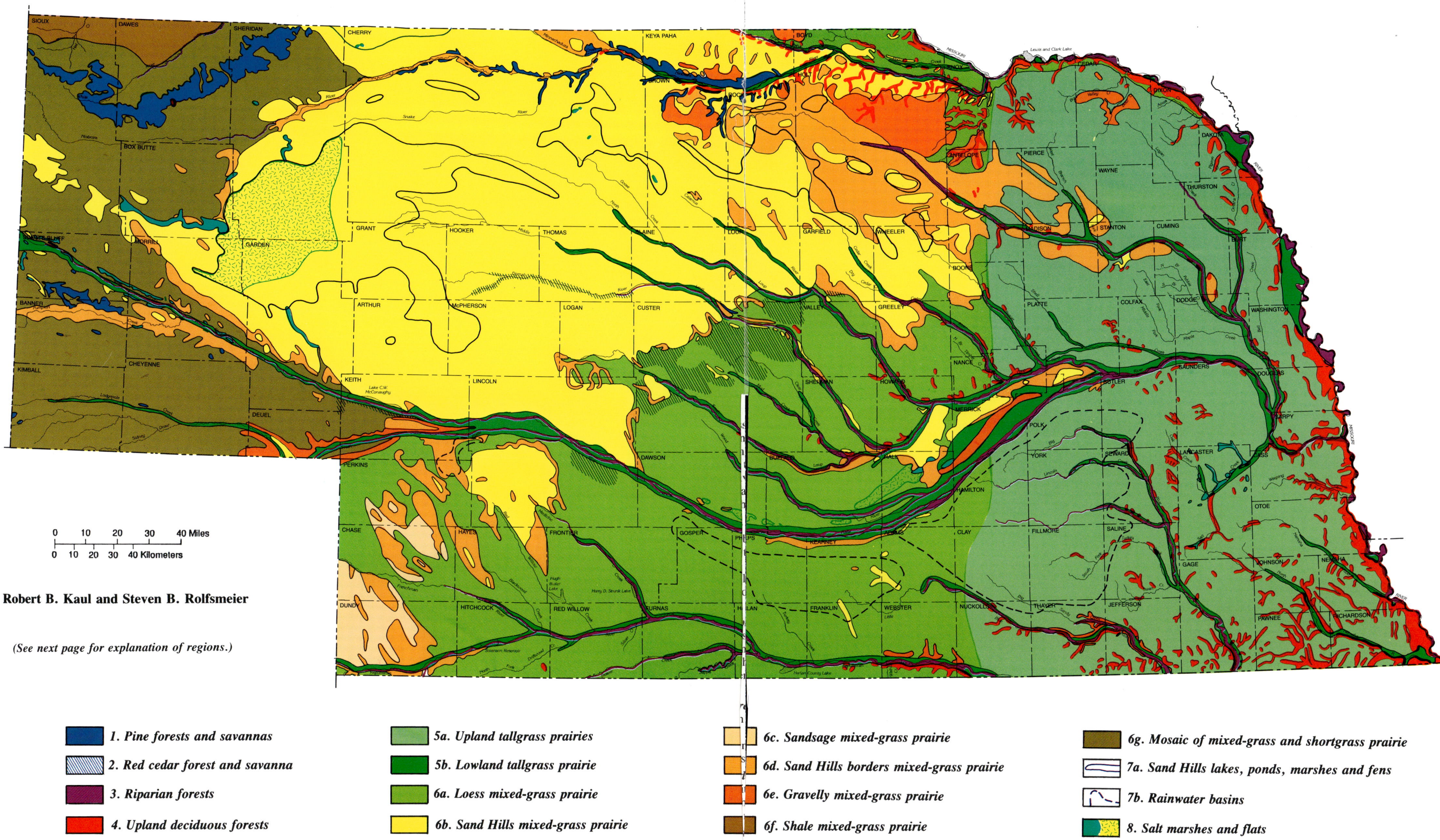

Robert B. Kaul and Steven B. Rolfsmeier

(See next page for explanation of regions.)

The Natural Vegetation of Nebraska

Robert B. Kaul and Steven B. Rolfsmeier

This map of Nebraska's native vegetation was prepared mostly from field observations of remnants of native vegetation. There are about 1,400 species of native plants in Nebraska; another 350 or so have been deliberately or accidentally introduced from elsewhere. Many native species are now rare because of loss of habitat, and some that have always been rare are even more so today. In the less-disturbed areas many of the species are probably as common today as they ever were. Most alien species are Eurasian in origin, and some are serious weeds and strong competitors with the native flora and crops.

Forests and Savannas

Evergreen Coniferous Forests and Savannas

1. Pine forests and savannas: ponderosa pine and associated deciduous trees and shrubs, with prairie species on exposed ridges and among the trees in savannas. There is disturbance by grazing and logging except on the steepest slopes. Fires are mostly suppressed but are destructive due to buildup of inflammable litter from the trees. In the Niobrara River valley the pine is invading the Sand Hills now that natural prairie fires are fewer. Some areas are protected as parks, wildlife management areas and nature preserves.

2. Red cedar forest and savanna: red cedar is common over most of the state except in the sandiest areas. It is invasive in prairies and rangelands in the absence of fire. It fills valleys and moves into unburned uplands. Box-elder and green ash are common trees in these forests, as are various prairie species from adjacent grasslands. Some naturally fire-protected places are shown on the map that have red cedar forest of large old trees. Those areas were heavily logged in early agrarian times. Suppression of fire is leading to substantial increase of red cedar in many places, and some areas logged in the 19th Century have recovered, especially those on steep slopes that protect them from cattle and fire.

Deciduous Forests

3. Riparian forests: open woodlands and dense forests on valley floors near rivers and creeks. The forests are especially dense in the Missouri, Platte, lower Loup, and lower Republican river valleys. The commonest trees are cottonwood, willow, American elm, green ash, box-elder and the introduced white mulberry. The forests are often dense with shrubs and vines, especially in the eastern half of the state. There has been a substantial 20th-Century increase in riparian woodlands in smaller valleys because prairie fires are now suppressed. Controlled and reduced flows of the Missouri, Platte, and Republican rivers now bring stability to the floodplains, resulting in dramatic increases in woody vegetation there too. There is some disturbance to the native vegetation by grazing cattle and logging, but a few areas are protected as parks and nature preserves.

4. Upland deciduous forests: the oak-hickory forest of the Midwest reaches its western limits in Nebraska and becomes impoverished in species. It is mostly confined to the slopes of valleys and is richest in species in the southeastern part of the state. American elm, green ash, and oaks (especially bur oak) are common, hickories are uncommon except in the Missouri River Valley, and the forests have a considerable understory of shrubs and perennial herbs, but vines are mostly at forest edges. In some areas aggressive alien shrubs have invaded the forests in large numbers. The forests are often disturbed by grazing and cutting. Well-preserved examples are in state parks and wildlife management areas.

Grasslands

Tallgrass Prairies

5a. Upland tallgrass prairies with luxuriant perennial sodforming and bunch-grasses on uplands in the eastern third of the state, are now mostly eliminated by agrarian activities. Some grasses grow to 7 feet tall, especially in moist sites and in wet years, and the vegetation is so dense as to obscure the ground. Forbs are numerous and conspicuous. Some low shrubs are fire-adapted and thrive in virgin prairie. Species diversity is maintained by occasional burning, which stimulates the native species but harms the aggressive introduced ones such as musk thistle, Kentucky bluegrass, and smooth brome-grass. Absence of fire allows invasion by native and introduced trees and shrubs. These prairies are about 99 percent gone, mostly plowed for agriculture or so heavily grazed as to be severely reduced in species. Many species that were once common are now rare or extremely localized, and some could disappear from the state's flora unless protected. Fragments of virgin prairie remain, particularly in areas too steep to plow, in odd corners unsuited for cropland and in railroad rights-of-way; only a few are deliberately preserved.

5b. Lowland tallgrass prairie on moist, level floors of the major river valleys and some of their tributaries, and extending far west of the upland tallgrass prairie. Vegetation is dense and often luxuriant. Many species of the upland prairie are here, and the moister sites have species that are absent or less common in adjacent upland prairies. This area is mostly converted to cropland or mowed and grazed, and there is considerable invasion by trees and shrubs in less-disturbed places. Some areas are preserved as nature sanctuaries and wildlife preserves.

Mixed-grass Prairies

These are perennial sod-grass and bunch-grass prairies between the shortgrass prairies to the west and the tallgrass prairies to the east, and not sharply demarcated from them except the Sand Hills; includes the Sand Hills and the loess hills south of them, as well as the level and dissected loess plains south of the Platte River. The vegetation is a mixture of some grasses and forbs of the tallgrass and shortgrass prairies, with the tallgrass species in the moister sites and the shortgrass species in drier places.

6a. Loess mixed-grass prairie on loess soils: vegetation is rather dense, and the ground is not usually visible among the plants. The area is almost entirely converted to cropland in the level areas and to rangeland in more dissected terrain, except for small isolated fragments, especially on steeper slopes. Rangelands are heavily planted with and invaded by smooth bromegrass, which eliminates most of the native flora. A few areas are protected in wildlife preserves.

6b. Sand Hills mixed-grass prairie on loose sand: there are tall and short rhizomatous and bunch grasses and many forbs, and the vegetation is not as dense as in the adjacent tallgrass and mixedgrass prairies; the ground is usually visible among the plants. The area is less disturbed than other prairie zones and has fewer weeds and little cropland. Some large areas are preserved.

6c. Sandsage mixed-grass prairie on sandhills: the vegetation resembles 6b but has substantial growth of sand sagebrush, which is absent from all but the extreme western part of 6b, and fewer species of forbs. The area is rather disturbed by grazing and cropland, but some areas are in good condition. Only a few areas are preserved.

6d. Sand Hills borders mixed-grass prairie has a mixture of sandhills (6b) or sandsage (6c) species with those from adjacent prairies on non-gravelly soils intermediate with the Sand Hills sands and other soils. Species restricted to pure sand or to non-sandy soils are absent.

6e. Gravelly mixed-grass prairie grows on infertile, extremely well-drained, sandy-gravelly soils, usually on slopes. The plants are rather sparse. Some Sand Hills species are here, as are some not otherwise found in sandy prairies. These areas are mostly unsuited for cropland and so are rather undisturbed. No major areas are protected.

6f. Shale mixed-grass prairie exists on clay formed from weathered shale. The vegetation is rather sparse and some areas have none. Bunch and rhizomatous grasses are abundant, as are big sagebrush, rabbitbrush and winterfat; forb species are few but common. There is some disturbance by grazing. A good example is preserved in the Oglala National Grassland in Sioux County.

6g. Mosaic of mixed-grass and shortgrass prairies with shortgrass prairie growing in the drier sites and mixed-grass prairie in slightly more mesic sites. Lowlands and gentler slopes are heavily grazed or converted to cropland and rather weedy in places. No major areas are preserved, but numerous examples exist on the steeper slopes and ridgetops.

Aquatic Vegetation

The major areas of aquatic vegetation in standing waters and wetlands are shown. Most streams, especially those on heavy soils, now have no submersed and little emergent vegetation because of siltation and pollution; their pre-agrarian vegetation is not well-recorded, but evidence from early collections suggests that it was much richer than it is now.

Lakes, Ponds, Marshes and Fens

7a. Sand Hills lakes, ponds, marshes and fens exist in interdunal valleys. Many waters are slightly alkaline, especially in the eastern Sand Hills, and are usually dense with emergent and submerged vegetation; others are strongly alkaline, especially in Garden and Sheridan counties, and their vegetation is sparser.

Accumulations of sedge-peat sustain sedge- and grass-filled fens, especially in Cherry County, and have unusual plants typical of cool northern bogs. The numerous clear, steadily flowing streams have luxuriant growth of submersed plants.

Perhaps two-thirds of the original ponds, lakes, and wetlands remain, the others having been drained or their groundwater recharge depleted for irrigation. Good examples are protected in wildlife refuges. The aquatic vegetation of the Sand Hills is well-preserved, in strong contrast to other parts of the state.

7b. Rainwater basins exist in two large areas south of the Platte River with numerous shallow permanent and intermittent ponds and marshes, some quite large, on level, rich, fine soils in tallgrass and mixed-grass prairie. Areas of small basins are in Chase, Keith, Lincoln and Perkins counties. The emergent shoreline vegetation of the more permanent basins is often dense with cattails, rushes, reeds and grasses, but submersed vegetation is sparse. The mud flats that form when the basins dry have smartweeds and other annuals, including alien weeds.

Most of the basins have been drained and plowed for cropland and only about 10 percent remain, about half of them under government protection for migrating waterfowl, whose spring migration brings millions of ducks and geese.

8. Salt marshes and flats: saline marshes, ponds and flats are usually near streams and are subject to summer drying. The vegetation is patchy and there are large expanses of bare soil that are often white with encrusted salt when dry. Until recently, these areas were considered worthless and were treated accordingly. In central and eastern Nebraska, most salt marshes are degraded or destroyed by drainage and industrial, agricultural, and residential-construction activity, and many areas are invaded by alien weeds; a few areas are preserved. In the western Sand Hills and the North Platte River valley, the salt marshes are much less disturbed.

timated that about two-thirds of the original sod was broken by the turn of this century. By the 1930s, only about 5 percent of the eastern grasslands remained intact, scattered in tiny parcels, mostly on private land.

We can grasp some of these changes mostly by looking at some of these remnant grasslands. Nine Mile Prairie, for instance, 180 acres of virgin prairie north of Lincoln, is one of the largest plots of native tallgrass prairie in the state. It is actually an old farmstead that survived because it was too steep to plow. Due to its proximity to what was then the University of Nebraska, it has been studied since 1916. Professor John Weaver, whom some regard as the father of modern plant ecology, worked there extensively. Thanks to the efforts of Marguerite Hall, the University of Nebraska Foundation and other concerned groups and individuals, this priceless tract will be protected for the future.

Soil Conservation Service

The soil-holding properties of native grasses are evident in this photo of livestock-induced wind erosion in the Sand Hills, Holt County, June 6, 1936. Where vegetation is killed by a concentration of livestock, occasional hummocks are left where the vegetation has held or the soil is more resistant to erosion.

At first glance, the prairie can seem monotonous—except for its glorious flowers—nothing but grass, all but bare of trees and shrubs. But actually, Nine Mile Prairie has about 400 species of plants, primarily grasses and forbs (nongrass herbs), nearly all of which are perennials.

Life conditions, for the plants, as for the animals of our largely semi-arid state, are severe. The soil is relatively deep and fertile—some say eastern Nebraska had 36 inches of topsoil when the sod was broken—but water is generally scarce. This scarcity is especially evident during droughty summers, when the air is hottest and driest and the prairie winds dry out and greatly stress plant tissue.

Over the ages, prairie plant communities have evolved extremely efficient root systems that penetrate deeply. Almost two-thirds of prairie species attain a depth greater than 5 feet. Many of these systems branch widely, storing food during the long dormant periods of winter and often surviving many, many years. Just as plants in many ecosystems occupy several "stories" above ground (for example, west of the 100th Meridian, it is common to find 8-inch blue grama growing under 2-foot western wheat grass), prairie communities layer underground. Because the roots of the different species don't all draw on the same soil level for water and nutrients, many species can exist together.

The Prairie Formation, the term Weaver gave to the tall-, mixed- and short-grass prairie that extended from the eastern deciduous forests to the Palouse region of eastern Oregon and Washington and the Great Valley of California and from the highlands of central Mexico into Canada, was the most extensive of all natural units of vegetation on the continent. It once covered 38 percent of the land surface of the mainland United States.

Climax prairie, evolved over thousands of years, is an extremely stable system. Weaver, publishing in January of 1944 on the "North American Prairie" in *The American Scholar*, pointed out that it "denotes a high degree of equilibrium between the vegetation and its habitat under the control of the existing climate. Large tracts of prairie, after ninety or more years of settlement, remain practically uninvaded by weeds . . . although surrounded for years by weedy fields and pastures."

Needle-and-thread grass, Cheyenne County

Glenn Borchers, Soil Conservation Service

So the most obvious change in the ecosystems of Nebraska in the 20th Century is the shift from perennial grasses to annual monocultures, especially to corn in the eastern and central parts of the state, a change that probably could not have been as extensive without irrigation. Again, first impressions may well deceive. Corn is a grass, strictly speaking, so how great a difference could there be between then and now?

Well, for one thing, it would be a rare farmer indeed who would claim that his or her corn, or milo or soybeans, was naturally uninvaded by weeds. Or, as Jan Wojcik puts it in *The Arguments of Agriculture:* "A farmer says 'soybeans,' and the field answers back 'weeds.' The farmer contends against the weeds with chemical herbicides."

Herbicides, weed killers, are a subclass of pesticides. They, along with insecticides (insect killers), nematocides

Kevin Bartels, Soil Conservation Service

Indian grass near Imperial, Chase County

(worm killers), miticides (mite killers) and algicides (algae killers), are toxic chemicals designed to eliminate crop pests. Usually designed to kill discriminantly, not broadly, these target various plant and animal invaders that, by competing for water or nutrients or by feeding upon the crop itself, result in lower yields. Sometimes, however, they can kill or inhibit the growth of non-target species.

The human health risks of pesticides are quite low at this time for people who do not farm—insignificant compared with motor vehicle accidents, for example, or accidents in the home, or even the risks of eating daily four tablespoons of peanut butter (which contains natural "pesticides" produced by the peanut plant). In some cases at relatively low concentrations, these chemicals can pose risks to internal organs, fetal development and cholinesterase activity (vitamin/enzyme interaction essential to kidney functions), as well as being potential cancer-causers (in the case of atrazine, alachlor and dieldrin, for example). But at present, the real risk is almost exclusively limited to application and handling accidents.

In general, most legal pesticides break down relatively rapidly into "daughter" products that are not known to be very harmful to human health. However, there are certain real problems associated with pesticide application, but they are mostly poorly understood by the public because the effects are extremely complex and subtle. Many pesticides are extremely toxic in any concentrated dose, so there is a degree of public fear associated with groundwater contamination as pesticides percolate into the drinking water supply.

In addition, the scientific community has just recently begun to study what could happen if pesticides are contained within the root zone. In the interests of conserving both water quantity and quality, there is currently a move to limit both water and pesticides to root zones, which in the case of annual crops constitute the upper 18 to 36 inches of the soil. While this approach should help protect groundwater, some are concerned it could ultimately prove damaging to soil quality generally, due to the inhibiting action of some chemicals on microbial life.

Put simply, virtually all chemicals, especially the longer lasting organophosphates, bond with or leave a residue on soil particles. In this process, they may alter the microenvironment of the soil; this may have a negative impact on the ability of bacteria, fungi and other tiny organisms to break down stored energy for use by roots and other living creatures. To understand what is going on in the root zone, we need to look with a long-term, but microscopic, perspective.

Wes Jackson, of the Land Institute in Salina, Kansas, writes in "Toward a Unifying Concept for Agriculture," an essay in *Agricultural Sustainability in a Changing World Order*: "If we could adjust our eyes to a power beyond that of the electron microscope, our minds would reel in a seemingly surrealistic universe of exchanging ions where water molecules dominate and where colloidal clay plates are held in position by organic thread molecules important in a larger purpose, but regarded as just another meal by innumerable microscopic invertebrates."

With more lavish rates of irrigation, as the chemicals are flushed past the several inches at the surface of the soil, microorganism populations can bounce back quickly. But because of the need to conserve both the quality and the quantity of groundwater, it's not likely that former irrigation practices will continue. And no one knows yet what the long-term effects on microbial activity in the soil could be, according to the assistant state soil scientist with the Soil Conservation Service, Clayton Lee.

Blue grama

K. Hladek, Soil Conservation Service

Traditional monocultural farming has not been without a cost to much of the lands of Nebraska. As of 1986, 62.6 percent of the land in the state was adequately treated, which means it was adequately conserving topsoil, according to the *Base Document of the Soil and Water Conservation Strategy* of the Nebraska Natural Resources Commission. This means that about 37 percent of the state's cropland, pastureland, rangeland or forestland needed better management. This same document stated that 43 percent of all Nebraska's cropland needed better erosion control (47 percent for irrigated land, 41 percent for dryland), and about 34 percent was eroding at a rate that was damaging its ability to produce. In the case of water-related concerns, it says that most water-quality problems were the result of farming practices that needed improvement, generally to reduce the amount of sediment or agricultural chemicals getting into surface water or groundwater.

Soil compaction, decline of crumb-structure (a "micro-structural" element that enhances moisture storage and nutrient exchange), decrease in soil pore space and a diminished wicking effect (pulling moisture down) are also common problems resulting from intensive, monocultural farming. To some degree, excessive tillage may be to blame. (*Management options to deal with these problems are discussed in the companion chapter in the water erosion section.*)

But there are very real differences between annuals and perennial communities that relate to curbing water erosion and water pollution. Annual monocultures, especially where rotation and leaving fields fallow have been abandoned, require large doses of nitrogen fertilizers. Nitrogen fertilizer application and nitrogen-rich runoff from feedlots contribute to groundwater and surface-water pollution. And, unlike pesticides, present concentrations in some areas may pose real health risks, primarily to infants less than six months of age in the form of "blue-baby syndrome" (methemoglobinemia), which causes problems with oxygen metabolism. Secondarily, there are some concerns about the general health effects of long-term exposure to high levels of nitrates, but evidence yet is scanty.

In addition, although generally well below the maximum contaminant levels (MCLs) set by the U.S. Environmental Protection Agency, some peaks of pesticides in surface water during periods of high runoff exceed these maximum contaminant levels. And while the vast

Dain Root, Soil Conservation Service

Dames rocket near Ansley, Custer County

majority of Nebraska's groundwater is of good quality, atrazine, the most commonly used herbicide in the state, was detected in 13.4 percent of wells sampled as of January of 1989. But it should be emphasized that less than 1 percent of these wells had contaminant levels above the MCL of 3 ppb.

Other problems exist with point-source contamination of surface water and groundwater, especially regarding cleanup of hazardous waste sites, mostly covered by the EPA's Superfund program, and with solid-waste sites, mostly landfills, which until recently have been unregulated in the state for towns under 10,000 in population. *(These problems are discussed in the chapter on "Politics and Policy," p. 222).*

There are other problems with excessive use of agricultural chemicals, either pesticides or fertilizers. Both add to production, and consumer, costs for food. In addition, between 25-35 percent of nitrogen fertilizers are lost to natural processes of nitrification and denitrification. It so happens that rhizomes release inhibitors that suppress nitrification. And while a number of perennials have rhizomes, annuals do not.

Whereas perennial communities bind and enrich the soil with extensive root systems that abide over years, the annual crops, as Wes Jackson says, "roots and above-ground parts alike, die every year, so through much of the calendar year, the mechanical grip on the soil must rely on death rather than life."

In fact, annual monocultures have a poor record of holding soil. Against wind erosion, annuals fail about 60 percent of the time. Perennial grasses, particularly in semi-arid climates, are more efficient than other plants in producing granular soil structure. The roots of perennial grasses are extremely active in forming soil aggregates. Moreover, *communities* (for example, grasses and legumes growing together) do a much better job of stabilizing soil than any single variety growing alone.

Soil Loss by Water

Even so, Dust Bowl horror stories aside, it is not the wind but the water that carries the topsoil away. One leading agronomist estimated that in eastern Nebraska, on cropped land of an 8-percent slope or greater, which comprises about one third of the eastern third of the state, the topsoil is being rapidly depleted. To some degree, this loss has been masked by the "forgiving" nature of Nebraska subsoils, which continue at this time, with heavy inputs of fertilizer, economically viable production.

As a general rule of thumb, 5 tons per acre per year is the rate at which soil loss is considered acceptable without a loss of productive capability. It is known to soil scientists as the T (for tolerance) factor. In fact, soil loss and regeneration rates are extremely variable and highly dependent on the micro-environment—the specifics of the slope, geology, climate and vegetation of a particular place. Still, we know that soil losses of 25 tons per acre per year are not uncommon in Nebraska, that losses of 100 tons per acre per year are not unknown. That means that between 5 and 20 years' accumulation of topsoil is being lost in some places every year. And the worst-case scenario, where these losses are *never, ever* addressed, could be a loss of productivity, as other civilizations besides ours have shown. These losses could be masked for decades with commercial fertilizers but eventually would be too great to overcome completely. For an example of long-term poor management, look at Israel or Greece or the valley of the Tigris and Euphrates. Examine Haiti.

Soil loss is especially tragic because, at least in Nebraska, it is very nearly invisible. In general, erosion in our state does not take the form of gullies, as it does, for example, in the South. Much of it is result of sheet or ephemeral erosion where one pass with a field cultivator can completely mask it. And with the availability of commercial fertilizers to offset and "mask" the effects of decreased productivity, soil loss produces few obvious, immediate signs of distress. Generally, the march of this vital resource to the sea is a creeping process, slow and silent, and so it does not excite the kind of outrage that diminished habitat for waterbirds has, or threats to drinking water, or storage of nuclear waste. Like global warming, it changes the ecosystem, but without yet causing immediate discomfort for the masses.

Jim Miller, Soil Conservation Service

Sheet and rill erosion depositing sediment in the lower part of the field and in the road ditch, Madison County

Therefore, what Jackson and his Land Institute are researching is an agriculture that emphasizes perennial grasses and legumes as the best possible imitation of the pre-cultivation ecosystem. Hindering adoption of this kind of cropping have been the inefficiency in harvesting and less-than-bumper-crop yields, the very reasons that specialized cropping with commercial pesticides and fertilizers holds sway right now, that, and tradition, since all new ideas have to struggle at first to get a decent stand. This would greatly reduce costs by limiting planting; by cutting pesticide use via breeding of insect-resistant varieties that are close-seeded to crowd out weeds; and by curbing pumping costs since drought-tolerance would be a prerequisite also. It may sound like the wildest of dreams to some. But it may make good economic sense if the technical and economic problems—harvesting and marketing of alternative crops, for instance—can be overcome.

Loss of Aquatic Habitat

In a hungry world that is expected to double its population in the next few decades, the potential for loss of the land's productivity is a very serious matter, but it is certainly not the whole story. Soil in its place is a pearl beyond price, but loosed by water, it is the major water pollutant in Nebraska today and is having damaging effects on some aquatic communities.

Many aquatic insects, for example, aren't silt tolerant. There are exceptions, of course. Certain midges thrive in silt bottoms, for example, and are apparently untroubled by organic pollutants, though other varieties of them will only live in waters of the highest quality. Certain caddis flies and mayflies also are doing well in Nebraska, but in general the state is losing aquatic insects.

There's a blackfly, for example, that inhabited a little stream near Denton, but it has disappeared in the last three years. This particular blackfly was first described in a location in the desert Southwest but is relatively rare in these parts. In this case, siltation was not to blame. The landowners re-routed the spring-fed stream, which wasn't but a foot wide and a few inches deep, flowing from a small perched water table.

Such drainage is common where farmers struggle with the balky, imperfect land and the need to maximize their profits. Another blackfly population disappeared when Turkey Creek near Wilbur was pumped dry in 1991. It so happens that an unrelated population of another blackfly in the same creek was eliminated too—driven out of that habitat, perhaps not permanently—when its eggs were stranded, high and dry.

It's hard to scare up fellow-feeling for insects, especially the ones that bite or, like the midges, hover irritatingly around

Commercial fertilizer's benefits were so apparent that it is little wonder the linear logic of "some is good, so more must be better" took over. More recently farmers have begun to recognize the environmental and economic limits to nitrogen application.

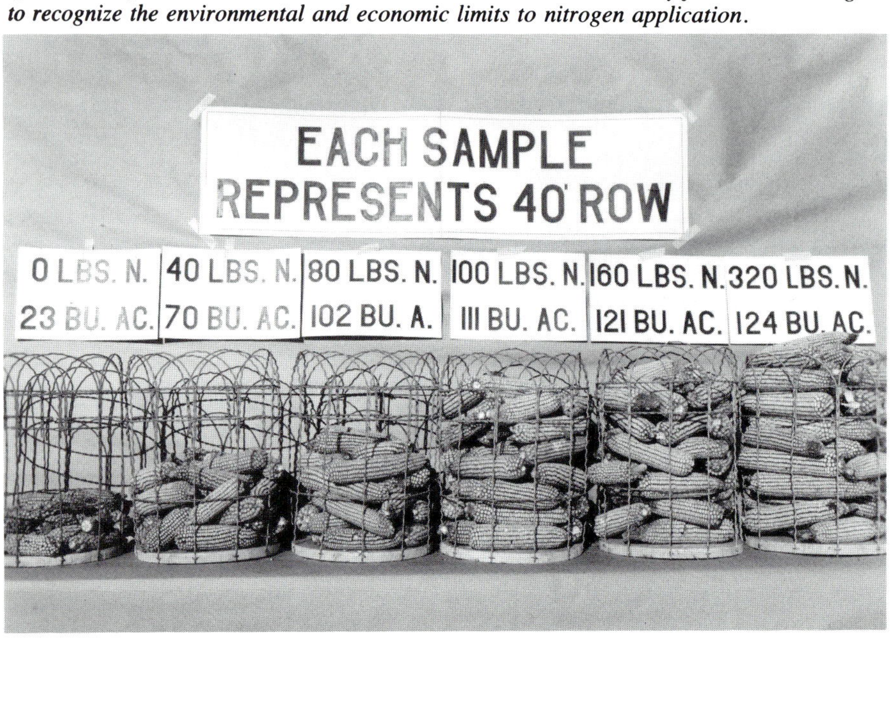

W. C. McBroom, U.S. Bureau of Reclamation

James Yates, Soil Conservation Service

An aquatic insect on grass in a subirrigated pasture, Morrill County

our faces when we hike near a woody creek in one of our favorite parks. Realistically, how much of a loss is one tiny population of blackflies, except maybe for the researcher filled with wonder at finding a Southwestern species so far from "home"?

We might speak of the loss as that of a link in a chain, but that is too linear. A better image might be a stitch in a web, for the relations of species to species run in every direction. Fish eat aquatic insects, in both adult and larval stages. Birds, especially the ducks frequenting Nebraska's lakes, reservoirs and wetlands, feed on insect larvae, but also on the fish that have eaten those insects. But the relations are even more convoluted—larvae of different species feed on each other, as well as on periphyton (sterile filaments of certain aquatic plants) and decaying algae.

The point is that the breaking down and the repeated cycling and recycling of nutrients, the ability to hold energy in a particular place through these processes, but not endlessly (since nature is not a perpetual motion machine), is performed with an elegance and a richness that human schemes can't yet replicate or approximate.

Besides siltation and draining, the other factor in the destruction of aquatic insect habitat is channelization. This activity is usually accomplished in an effort to control flooding by speeding the drainage of watersheds, with sometimes greater, sometimes lesser effects. Channelization tends in places to destabilize banks and so contributes to silt problems, but its primary effect on aquatic insect habitat is the destruction of substrate. That is, channelization robs streambeds of the debris, tree roots and other organic trash to which aquatic insects need to attach themselves in order to feed and breed.

This is a good place to address the double-edged sword of flooding and flood control. In Nebraska's water history, two examples stand out: the Pick-Sloan projects and the Nemaha Basin.

Pick-Sloan, treated in "Politics and Policy" *(see p. 212)*, has been described as a "shotgun wedding" of transportation and irrigation interests. As James Aucoin says of Missouri River development in *Water in Nebraska*, "Pick included Garrison Dam at a site rejected in an earlier study as being unsafe, and Sloan used inexperienced surveyors and inadequate soil tests. Pick used streamflow data for a period of particularly wet years to justify flood control and navigation; Sloan used flow records for a particularly dry stretch of years to show how desperately irrigation was needed." The cumulative effect, as Aucoin graphically puts it, was a series of "well-placed blows to the heart of the river."

While the Missouri in its natural state was generally faster and deeper than the Platte, its backwaters, its oxbows and chutes, its natural meandering, served a similar role as breeding, feeding and resting habitat for both resident and migratory waterfowl, as well as for shore birds, reptiles, amphibians and furbearers—beaver, mink, raccoon and muskrat. The Missouri is also one of two "core" wintering habitats for bald eagles in the lower 48 states.

Besides the destruction of these slower, shallower backwater habitats, the end to flooding of the Missouri River valley has had at least two important ecological effects. First, the journey of topsoil to the sea is accelerated in some places, both through the crumbling of banks adjacent to rapid waters and because the annual depositing of sediments by floodwaters isn't happening any more.

Secondly, the food energy cycled and recycled by plant and animal communities, the total organic carbon available on the floodplain of the Missouri River

(Continued on p. 172)

Cliff and barn swallows on Funk's Lagoon, a wetland in Phelps County

Soil Conservation Service, Nebraska office

Wetlands in the Historic Missouri Valley Brought Trappers, Explorers, Naturalists

The wild lands of the historic Missouri River had many cutoff lakes, sloughs, marshes, islands and sandbars. Wet places were created by water in river channels that meandered across the floodplain. After a spring flood, the water course could be entirely different, with a newly created oxbow lake left behind. Water overflowing the banks created woodland swamps.

Trappers taking furs knew the Missouri River region for decades before the United States made the Louisiana Purchase that bought the western territory. In 1804 the Lewis and Clark expedition ordered by President Jefferson left St. Louis in keel boats to start their "Corps of Discovery" expedition. Their notes from Nebraska include details of the nesting habits, coloration and behavior of "aquatic birds" seen on Aug. 5, 1804 while a day or two south of the Big Omaha Village.

Other explorers followed this expedition. Henry Breckenridge wrote about the many sandbars at the mouth of the Platte in May 1811. The government-sponsored Long Expedition spent the winter of 1819-1820 at the Engineer Cantonment near the Council Bluff named by Lewis and Clark. Thomas Say was the zoologist that listed nearly 100 bird species seen during their stay. His notes include the spring arrival dates of birds in the wildlife habitats of the river valley.

The German Paul Wilhelm went north to Fort Atkinson and then west to the Niobrara River in 1823. His notes refer to the "handsome tern with the black head and white band on the forehead" flying about the mouth of the Platte.

Least terns were common nesting birds on sandbars of the Missouri River and lower Platte River, as noted by Wilhelm. Many other birds were present during the breeding season along the historic Missouri River. Large cavities of cottonwood trees in forests on the floodplain were nest sites and roosts for the now extinct Carolina parakeet. Bald eagles, wood ducks and Canada geese were common along the river. Other birds, including warblers and raptors, used the diverse marshes, grassland meadows and forests of the floodplain of the river. Expanses of tallgrass prairie occurred on the uplands of the valley.

George Catlin floated down the Missouri River through Nebraska in 1833. He mentioned the "sportsman's fever was roused and satisfied; the swans, ducks, geese, pelicans" and other wildlife the target of their rifles. His stylistic drawings of Missouri Valley scenes include one of the mouth of the Platte River, showing several Indians on a hilltop overlooking both rivers.

The naturalist Prince Maximilian of Wied explored Nebraska in the spring of 1833 and 1834. The Prince sat on the boat's deck and wrote his observations in a journal. Nearby, Karl Bodmer painted river scenes. A day's entry referred to a bald eagle nesting in a large cottonwood along the river. Flocks of migrating pelicans flew overhead. At a stop near the Council Bluffs in early May 1833, the calls of whip-poor-wills joined with the songs of dancing Indians to create a musical chorus in the Missouri River wilderness, Maximilian wrote.

A decade later, the famous naturalist John James Audubon explored the Missouri River country. A delay on May 17, 1843, while in the area of the Vermillion River occurred when the boiler of the Steamboat Omega burned out. Audubon went for a hike and wrote:

> "Whilst we were on top of the high hills which we climbed this morning, and looked towards the valley beneath us, including the river, we were undetermined as to whether we saw as much land dry as land overflowed; the immense flat prairie on the east side of the river looked not unlike a lake of great expanse, and immediately beneath us the last freshet had left upwards of perhaps two or three hundred acres covered by water, with numbers of water fowl on it. From the tops of the hills we saw only a continual succession of other lakes, of the same form and nature."

Nearby in the hills, Audubon heard a lark with a new song. Specimens were collected and used later to identity the western meadowlark, Nebraska's state bird.

The Warren Expedition, searching for travel routes in the mid-1850s, noted the river would be excellent for navigation if the tree snags dangerous to steamboat navigation were removed from the river channel. This is the first hint of how the untamed Missouri River could be changed to suit human uses.

After exploration, settlement came to the Missouri Valley. Towns and farms became a part of the landscape. Birders living in the towns and cities hiked into nearby country to search for birds. The wild Missouri River continued to follow its historic cycle of high spring flows flooding channels that moved across the floodplain. Maps of the region prepared by the Missouri River Commission, a part of the Army Topographic Engineers, show the many channels and associated wetlands of the historic Missouri. During the next few decades, naturalists studied birds at wetlands at Child's Point near Bellevue, Carter Lake near Omaha and Crystal Lake near Sioux City.

A few decades ago, the Missouri River and its valley were forever changed. Workers on barges installed jetties and dikes to narrow the channel. They noted terns nesting at sandbars and swallow colonies on nearby cliffs.

A meandering channel was confined to a narrow channel. The maps of the topographic engineers are now aerial photos taken for the Army Corps of Engineers. The Missouri River now has little of its wilderness landscape. The lakes and wetlands are few. Birds occur in fewer numbers, the lack of suitable habitat forcing the large migratory flocks to rest elsewhere.

The Missouri River lands described in historic times by Thomas Say, Prince Maximilian and J.J. Audubon and others have become cities and farmland. Natural expanses of wetlands, prairie and forests of the Missouri Valley were destroyed. Only the drawings and notes kept in the personal journals of explorers and naturalists reveal the history of the river.

—Jim Ducey, past chair, Nebraska Audubon Council, and executive director, Preserve Our Water Resources

Blue and snow geese at DeSoto National Wildlife Refuge along the Missouri River near Blair, Washington County

R. O. Koerner, Soil Conservation Service

Sand Hills Lake District Lured Wildlife and Scientists

Water brought the wildlife. Flocks of birds would nest in the summer or seek a resting spot during migration. Birds and other game brought people to the hills and valleys. Pawnee or Sioux Indians traveled the Sand Hills and knew where game could be readily taken.

The Sand Hills were also discovered by those seeking scientific gains. Researchers from the eastern states came on expeditions to explore the western "wilderness" of the United States. The first government exploration through the Sand Hills was the Warren Expedition. Dr. Ferdinand V. Hayden, the naturalist on that trip, prepared the bird list. Notes from a 1857 trek along the Middle Loup and then northward said breeding sandhill cranes were abundant. The trumpeter swan was another prominent species on the lakes and marshes of the region.

Two Yale University expeditions visited the Sand Hills in the 1870s. George "Bird" Grinnell, a bird enthusiast, along with other students led by Professor O.C. Marsh, a prominent paleontologist, came to study the Sand Hills. Their wagons left Fort McPherson and went north to the Dismal River. They looked for fossils in the river valley and then went back south. A second trip in 1873 went past the lake district to search for bones in the Niobrara Valley. Dr. Thomas Maghee, the trip physician, commented in his journal on the many lakes in the area.

The General Land Survey plotted the lake property claimed by settlers. Prominent land features shown on the survey maps showed the character of the land—its mostly rolling sand hills or ridges. Lakes and wetlands in the area are shown along with section lines. The survey team spent hours measuring and figuring that Big Alkali Lake covered about 460 acres. The mapping for Hackberry Lake showed this valley was mostly a marsh. Many lakes had names; other lakes, wetlands and marshes on the map went unnamed.

About a decade later, the Elias Stilwell family settled on Watt's Lake in 1880 and then nearby Hackberry Lake in 1890. The family shared their homestead with visitors who came to hunt the abundant waterfowl. Several families had land in the region where they were trying to establish a new home. Settlers marking the open land with their claims marked property but did not change the hills and lakes.

The flora of the hills of sand brought other students in July of 1902. Roscoe Pound and J.G. Smith were members of the Seminarium Botanicum at the University of Nebraska. Conducting a botanical survey, they walked through the lake district near Valentine. A cart pulled by their horse, Moses, carried food and supplies. Pound was walking the hills in search of plant specimens when he saw Dad's Lake from a hilltop. It was "a valley full of water—a magnificent sheet of water," he wrote. There were "24 good-sized lakes" visible from one hilltop.

University of Nebraska professors eventually came to the Valentine lake district in the summers of 1903 and 1911. Dr. Robert Wolcott, a professor in the entomology department, and Frank Shoemaker, an older student pursuing a degree in natural history, spent days wading through the water looking for bird nests. More than 50 different nests belonging to mallards, blackbirds, black terns and other waterbirds could be found hidden at and near the lakes and marshes. Shoemaker wrote about finding Forster's tern nests. The disturbed birds never uttered a syllable "except downright profanity. They are lovely and graceful birds—but such language," he noted, as they dived upon the human intruders.

Rivers Stilwell hosted both natural history expeditions at his ranch. The Stilwell's second house was used by hunters in the fall and fishermen in the summer. The first homeplace had two sod houses that were lost to the rising waters of Hackberry Lake. A two-story frame house was built north of the lake.

A Shoemaker photo depicts the sod houses and wagons and horses among the open land. Lake vistas show valleys of lakes and marshes. The Cherry County botany group shots show a lively party on a plant trip to the hills in 1911.

Harry C. Oberholser of the U.S. Biological Survey visited in 1920 to evaluate the value of the lakes for waterfowl. The biologist returned in October to make additional notes and participated in a fall waterfowl hunt. Ducks and other birds were abundant. There were at least eight hunting clubs noted at the larger Marsh lakes, Hackberry Lake and Dewey Lake. The Red Deer Gun Club is a prominent example of the hunting culture of the Sand Hills.

The Biological Survey report described most of the lakes and convinced the government of the importance of the lake district to waterfowl. Ranching was the primary land use in the lake district. Herds of cattle were born and taken to market. Hay cut from wetland meadows provided winter feed. Wildlife also continued to thrive, with many birds raising young each summer. Waterfowl hunting continued each fall when flocks of ducks rested on the water.

In 1933 most of the lakes and wetlands of this area were protected to create the Valentine National Wildlife Refuge. The importance of the wetlands for wildlife was realized, and most of that lake district is now managed for the benefit of birds.

—Jim Ducey, past chair, Nebraska Audubon Council, director, Preserve Our Water Resources

Ducks on a wetland in the Valentine National Wildlife Refuge near Valentine, Cherry County

Charles Flowerday, CSD

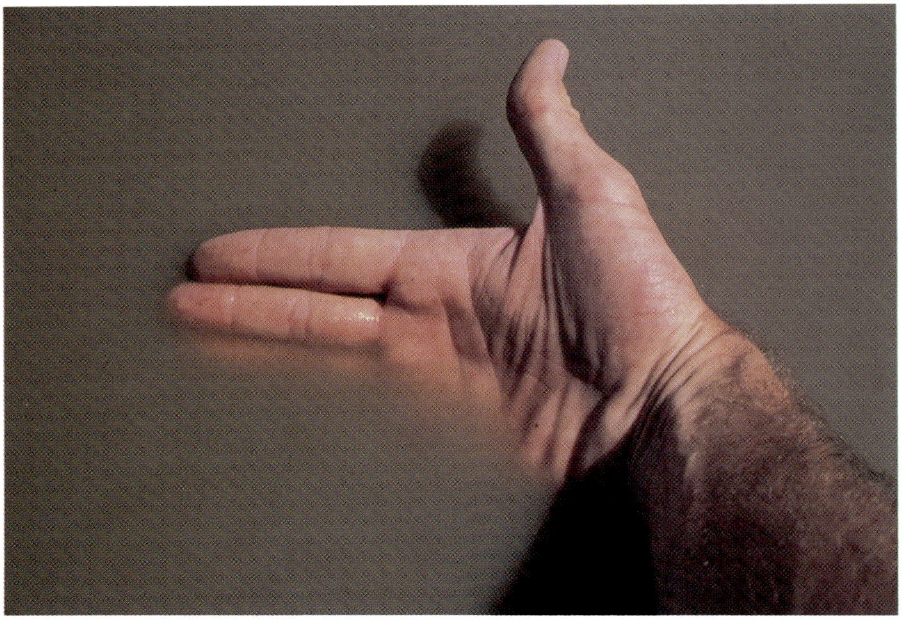

Nebraska Game and Parks Commission

Siltation from deposition of sediment is the number one nonpoint-source surface-water pollutant in the state and the nation. Nationally, agriculture contributes 64 percent of sediment pollution of streams, more than any other source.

(Continued from p. 169)
valley has declined by 65 percent since the river was controlled. This has happened through the loss of wetlands, where the accumulation of organic matter is high because decomposition is slow, and the cropping of the former floodplain, from which most of the above-ground plant products are removed annually.

Besides degraded habitat for aquatic wildlife and a general impoverishment of organic matter in the former floodplain, Pick-Sloan has had one more significant effect: the inundation of large tracts of Sioux peoples' most productive agricultural and timber lands, as well as "marginal" lands that were the source of traditional plant and game foods. C. Patrick Morris, in "Hydroelectric Development and Indigenous People," from *The Struggle for the Land*, says, "Three decades later, there can be little doubt that the Pick-Sloan plan is a major contributor to the 70-90 percent unemployment rates on the region's Indian reservations today."

Regarding another river, Christine Delich, in "River Portraits: The Nemahas," from *Nebraska Rivers* (1983), writes, "The story of the Nemaha rivers is one of trying to make complicated things simple; turning crooked lines into straight ones. Anyone who lived in the Nemaha Valley at the turn of the century lived with a Big and Little Nemaha and a Muddy Creek that wound back and forth through rich soils in what many described as, 'tortuous meanders' . . . loops in the river which cut . . . into the bank farther upstream or downstream, or deposited sediment to compensate for the energy gains and losses when flowing water changes course. They did this for centuries, and when it rained too much, they overflowed their banks.

"People grew tired of the floods . . ."

And as a result, ditching began in 1904. Altogether, 164 miles of stream were destroyed. Areas of the lower stretch of the Big Nemaha flooded anyway in 1939, 1941-49 and 1951, and the Army Corps of Engineers still estimates flood damage at over $3 million annually in the Nemaha basin. On the whole, flood risk in the basin is reduced, but the rivers are turbid and severely eroded, supporting few game species except catfish. And river-bottom plants like spiderwort, mouse-tail, brushed meadow garlic, rose mallow and wild iris haven't been sighted since the late 1950s.

The "need" to manage flooding seems so self-evident to most European-Americans, whose presence in Nebraska has meant intensive building on and cropping of the floodplains, that any sense of flooding as natural or desirable is hard to come by.

One biologist's account of a colleague's experience shows us another way of looking at a flood. Frog populations are generally thought to be in decline worldwide (although Nebraska populations seem at this date to be holding steady), and many biologists interpret this as an alarming bio-indicator. Amphibians are uniquely sensitive because they inhabit both aquatic and terrestrial environments, so are vulnerable to air, water and soil pollutants.

In April of 1991, a biologist surveying lands adjacent to the Platte was having no luck at all finding amphibians—they just weren't there. Then, heavy rains in combination with upstream impoundment releases resulted in uncharacteristic backflooding. "The amphibians went nuts!" according to the source. Many aquatic species, freshwater snails, spadefoot toads, chorus frogs and others, appeared as if from nowhere, in large numbers. In addition, fish fry of many varieties were abundant, some of whom fed other spawning populations, some of whom certainly made it to the river with at least a shot at adulthood.

The scene was a biologist's delight and raised fascinating questions. The relationship between groundwater and the former dominance of wet meadows along the Platte is not well understood. In former times, prior to irrigation pumping, would these high waters have been a seasonal feature? Is this flooding, atypical in modern times, a reassertion of an age-old natural process?

The high waters drained in about two weeks and left the fields much as before, with some possible improvement in subsoil saturation and maybe even a little groundwater recharge. As the water dried up and became more concentrated, the flooded area was a hotspot for bird and mammal feeding. And even in two weeks, many amphibians completed metamorphosis, ensuring one more generation anyway.

Lastly, regarding aquatic habitat, we need to examine the part of the Blue River basin known as the Rainwater Basin, or Rainbasin. It covers an area of about 4,200 square miles in 17 counties of south-central Nebraska where wind-blown silty soils (loess) line sink-like depressions. This process has produced clay-bottomed wetlands filled with runoff that are part of the historic richness of marshes and lakes in Nebraska.

Drainage and filling of these wetlands began in the early 1900s but accelerated significantly after World War II with improvements in earth-moving machinery. Road building and railroad building also contributed to the draining and filling of some wetlands. Others filled up with Dust Bowl silt. In 1936, when the U.S. Secretary of Agriculture determined that drainage was "an eligible conservation practice," until the mid-1960s, federal assistance for drainage was available through cost-sharing by the the Agricultural Stabilization and Conservation Service, with technical help from the Soil

Conservation Service. At present, although estimates vary, only about 20 to 30 percent of the original acreage of wetlands survives, of which less than half are protected today by state or federal wildlife agencies.

Nebraska's salt marshes, located in northern Lancaster and southern Saunders counties, have met a similar fate, chiefly through the growth of Lincoln and modifications of Salt Creek; channelizing, draining, filling, diking, diversion and in some cases pumping in of freshwater have in most cases destroyed these delicate areas, home to rare salt-tolerant plants, as well as habitat for waterfowl. Saline wetlands were never extensive, probably totalling less than 20,000 acres, of which only 1,200 acres remain, and these are for the most part fragments, none unaltered. Even through the 1980s, construction rubble was still being dumped in remnant salt marshes west of Lincoln.

Soil Conservation Service

This photo was taken after revegetation resulted in stabilization of the eroding bank in the critical area depicted below on Nine Mile Creek, Scotts Bluff County.

Loss of Biodiversity

From each of these stories of change and loss, a common theme emerges: a failure to see and foresee. Only in the long run does soil erosion by water become perceptible; the same is true for loss of organic matter on the prairie or along the floodplains of the Missouri. Wildlife was once so plentiful in Nebraska that it must have seemed impossible that the drainage of an acre here, a section there, could ever exhaust such bounty. The value of wetlands—especially regarding public benefits with economic consequences such as groundwater recharge, natural flood control, conversion of excess nitrogen into nutrients—is not obvious.

Certain changes, of course, are apparent to the naked eye, and probably the greatest of these is the replacement in the 20th Century of the broad, mostly open, braided prairie river with a broad riparian forest that follows the line of the Platte. With historic flows reduced by up to 70 percent at peak flows, by a little less than 50 percent for average flows, (through impoundments and diversion), the annual scouring of the sandbars and shores of the Platte no longer occurs. Nor are prairie fires, which formerly restricted the spread of many plant species, a factor any more. Many people see this new forest as a good thing, habitat for game species like whitetail deer and wild turkey, never more plentiful in our state. In fact, when efforts to preserve migratory bird habitat involve bulldozing cottonwoods along the channel, there is even some public outcry: "In a greenhouse world, how can there be too many trees?"

Streambank stabilization is one way to curb erosion that causes sediment pollution. This photo of Nine Mile Creek in Scotts Bluff County was taken before such efforts.

Soil Conservation Service

One consequence is that this new forest is functioning as an immigration corridor along which eastern and western species encounter and hybridize, contributing to the loss of biodiversity. As Steller's jays meet eastern jays, for instance, or eastern orioles meet western, or red meet yellow flickers, the populations inbreed, producing ultimately a single gene pool where there once were two. This has created some jays that are sterile, limiting reproduction.

Biodiversity, and its decline, is a very hot topic these days. One biologist called it the new buzzword and spoke of the history of such terms, words that have made it into our common tongue and words that did not. *Wildlife* and *environment* have both found wide acceptance. Ecology has not, perhaps because it sounds too technical and therefore threatening. Nobody uses the word *nature* very much anymore. Whether or not biodiversity gains general acceptance remains to be seen, but like each of the

Jon Farrar, Nebraska Game and Parks Commission

Prairie fires have been important in keeping shrubs and trees from encroaching on native grassland. Here, intentional burning of woody vegetation is part of managing the Audubon Society's Rowe Sanctuary for birds and other wildlife.

others, its meaning is simply *everything*, that system of systems, that interconnectedness that might be explained by paraphrasing a Biblical injunction usually thought of in social terms, "As you do to the least of these, so you do to me."

By design, this chapter has not said a great deal about cranes, either the whooping or sandhill varieties, nor about other endangered or threatened birds such as the piping plover, least tern, peregrine falcon and bald eagle, chiefly because the struggle to preserve their crucial migratory habitat gets a good deal of attention elsewhere. But as a parable on the complexity of the adaptations and interactions that are the very fabric of nature, let's examine one of the cranes' significant foods: mollusks.

Mollusks include snails and freshwater clams, also known as mussels. And again, modern researchers are hampered by a lack of historical data. There is no survey of pre-development mollusk populations on the Platte. After extensive sampling, biologists believe that, except for the Big Bend reach, there are no mussels in the main Platte channel. Whether they were ever there, they do not know, but they are in the side channels, backwaters and wet meadows. They attribute the lack of mussels east of Grand Island to a "lack of diversity of habitat along the river and to the increase in urban and cultural activities," according to the summary of the final report of the Platte River molluscan survey by Patricia Freeman of the University of Nebraska State Museum and Keith Perkins of Sioux Falls College. These wetlands are also important as a source of snails for the cranes. (Actually, there are few, if any, unaltered wet meadows along the whole of the Platte in our state.) Snails provide essential protein and calcium, and the decrease in wet areas and in snail populations could affect the crane populations using the river.

The mesh of the ecological web also includes mussels' relying on fish to disperse their larvae. Microscopic emissions that attach to the gills of fish are transported and eventually drop and with luck survive to colonize again. But many fish on the Platte are in serious trouble. Out of a historic total of about 70 species, 21 are rare or not present anymore. Thirty-five percent are gone or going. At the same time, biologists report an alarming increase in two species: the mosquitofish—a species introduced to control, naturally enough, mosquitoes—and red shiners—a native species. Both are very successful generalists that through a variety of strategies, including the eating of other species and their larvae, simply out-compete most others.

The catch is that many mollusks are host-specific, which means their larvae can only attach to the gills of certain fish. So, siltation, diversion and pollution displace more than fish or aquatic insects. In effect, when you lose a species of fish, you can lose one or more species of clams. Ultimately, the loss of one species could knock out a whole complex of species.

In fact, one way to fathom high-extinction (or high-extirpation) events is the way Wes Jackson has suggested, in terms of biological information. He maintains that, while we have created an explosion of human-made information, in a biological sense, we are in the middle of an information implosion. As we deplete biodiversity through the accelerated extinction of species and individual populations, we are in fact experiencing an information loss. That is, the lost genetic information is the loss

A red-winged blackbird near a wetland along Nebraska Highway 2 between Anselmo and Dunning, Custer County

Charles Flowerday, CSD

Soil Conservation Service

Some loss of biodiversity can be reversed through human management. River otters had been extirpated from (driven out of) Nebraska before the state Game and Parks Commission re-introduced them.

D. Cunningham, Nebr. G & P Comm.

Another example of a species successfully re-introduced to a region is the state Game and Parks Commission's efforts with beavers in southeastern Nebraska.

of an irreplaceable inheritance of nature's accumulated knowledge, more information than we could ever generate, comprehend or imagine.

What it seems we need to comprehend is that if the ecosystem is the fundamental unit of study, survival and value in natural science, and if we all are dependent on its resources, when do we avocate for our own rights (more narrowly understood) and when do we advocate for the ecosystem? In other words, who speaks for the land, the water and the non-human species?

It has been suggested by some that the state's water law, based on the first-in-use, first-in-right concept, wasn't designed to be fair but to prevent violence over disputes. Back in 1895, when the benefits of water management were being divided up, the non-human species didn't have a lobbyist, though some of them were certainly first in use, by 10 million years at least in the case of the cranes.

A Hamilton County wetland in the spring shows the essence of the dilemma over wetlands, depicting the conflict between agricultural use and "natural" benefits such as wildlife habitat, flood and sediment control and filtration of pollutants.

Soil Conservation Service

Sources: *Agriculture and the Environment—The 1991 Yearbook of Agriculture*, 1991, U.S. Department of Agriculture, U.S. Government Printing Office, Washington, D.C.; *Agricultural Sustainability in a Changing World Order*, Gordon K. Douglass, ed., 1984, Westview Press, Boulder, Colorado; *Biodiversity*, E. O. Wilson, ed., 1988, National Academy (of Sciences) Press, Washington, D.C.; "Conditions and Trends of Nebraska's Rural Land," Soil Conservation Service, 1986, Nebraska State Office; "Conserving the Ogallala: What Next?" David E. Kromm and Stephen E. White, 1985, Department of Geography, Kansas State University; "Conserving Soil" (curriculum enrichment materials for grades 6 through 9), Soil Conservation Service, Revised 1990; "Conserving Water in the High Plains," David E. Kromm and Stephen E. White, 1990, Department of Geography, Kansas State University; "Current Water Issues in Nebraska," 1982 Nebraska Water Seminar Series, Nebraska Water Resources Center, Institute of Agriculture and Natural Resources, University of Nebraska-Lincoln; "Diversity of Species as Crucial as Greenhouse Effect, Ozone Hole," guest editorial by Donella H. Meadows (adjunct professor of environmental and policy studies at Dartmouth College), Lincoln Star, April 23, 1990; *Entropy—Into the Greenhouse World*, Jeremy Rifkin with Ted Howard, 1989, Bantam Books, New York and Toronto; "Fifty Years of Soil Conservation—Keeping the Soil on the Farm," Jeff Obrecht and Pat McGrane (public affairs specialists), 1985, Soil Conservation Service; "Game and Parks Commission," *Nebraska Blue Book, 1989-1990*, State of Nebraska; "Going Wild with Soil and Water Conservation" (on crop and range management for conservation and wildlife habitat), Soil Conservation Service, 1985, Program Aid No. 1363; "The Groundwater Atlas of Nebraska," Robert Kuzelka and Darryll Pederson, project leaders, Charles Flowerday, ed., 1986, Conservation and Survey Division, Institute of Agriculture and Natural Resources, University of Nebraska-Lincoln; "Groundwater Levels in Nebraska, 1990," Gregory V. Steele, U.S. Geological Survey, and Perry B. Wigley, Conservation and Survey Division, 1991, Conservation and Survey Division, Institute of Agriculture and Natural Resources, University of Nebraska-Lincoln, and Nebraska District, Water Resources Division, U.S. Geological Survey; "Groundwater Quality Atlas of Nebraska," R.A. Engberg, U.S. Geological Survey, and Roy F. Spalding, Conservation and Survey Division, 1978, Resource Atlas No. 3, Conservation and Survey Division, Institute of Agriculture and Natural Resources, University of Nebraska-Lincoln; "Groundwater Quality of the Central Platte Region, 1974," Mary E. Exner and Roy F. Spalding, 1976, Resource Atlas No. 2, Conservation and Survey Division, Institute of Agriculture and Natural Resources, University of Nebraska-Lincoln; *A History of Nebraska's Fisheries*, David J. Jones, 1963, Nebraska Game, Forestation and Parks Commission, Lincoln; "History Relived on the Mighty Mo," J.P. Jackson, N.P. Stuckey, Missouri Department of Conservation, 1986, Norman G. Benson, ed., in *The Missouri River—The Resources, Their Uses and Values*, 1988, North-Central Division, American Fisheries Society, Special Publication No. 8 (from 48th Midwest Fish and Wildlife Conference, Omaha, Nebraska); "Low Water in the American High Plains," David Kromm, Natural Science, February 2, 1992; *Mammals of the Northern Great Plains*, J. Knox Jones, Jr., David M. Armstrong, Robert S. Hoffman and Clyde Jones, 1983, University of Nebraska Press, Lincoln and London; "A Manual on the Preparation of Special Groundwater Protection Area Action Plans," Robert Kuzelka, project leader, 1990, a contract report by the Conservation and Survey Division prepared for and published by the Nebraska Department of Environmental Control (Quality) and the Nebraska Association of Resources Districts; *A Microcosm of the Platte—A Guide to Bader Memorial Park Natural Area*, William S. Whitney and Jan Whitney, 1987, Prairie-Plains Resource Institute, Aurora, Nebraska; "North American Prairie," J.E. Weaver, The American Scholar, 1944; "Occurrence of Pesticides and Nitrate in Nebraska's Ground Water, 1990," Mary E. Exner and Roy F. Spalding, Water Center, Institute of Agriculture and Natural Resources, University of Nebraska-Lincoln; *One Earth, One Future—Our Changing Global Environment*, Cheryl Simon Silver with Ruth DeFries, eds., for the National Academy of Sciences, National Academy of Engineering, The Institute of Medicine, 1990, National Academy Press, Washington, D.C.; "On the Ethics of Agronomic Consulting," C.A. Flowerday and A.D. Flowerday, draft in review with Agriculture and Human Values; "Platte River Suitability Criteria-Habitat Utilization, Preference and Suitability Index Criteria for Fish and Aquatic Invertebrates in the Lower Platte River," Edward J. Peters, Richard Holland, Michael A. Callam and Donald L. Bunnell, Department of Forestry, Fisheries and Wildlife, IANR, UNL, 1989, Nebraska Technical Series No. 17, Nebraska Game and Parks Commission; "Some Aspects of Energy Flow in the Missouri River Ecosystem and a Rationale for Recovery," L.W. Hesse, Carl W. Wolfe, Nebraska Games and Parks Commission, Nancy K. Cole, Idaho State University, 1986, in Norman G. Benson, ed., *The Missouri River—The Resources, Their Uses and Values*, 1988, North-Central Division, American Fisheries Society, Special Publication No. 8. (from 48th Midwest Fish and Wildlife Conference, Omaha, Nebraska). "Soil Erosion by Water," Revised 1990, Soil Conservation Service, Agriculture Information Bulletin No. 513; "Soil Erosion by Wind," 1989, Soil Conservation Service, Agriculture Information Bulletin No. 555; *The Struggle for the Land*, Paul A. Olson, ed., 1990, University of Nebraska Press, Lincoln and London; "Sandhill Cranes . . . Wings Over the Platte," Nebraska Game and Parks Commission reprint from Nebraskaland, "Teaching Soil and Water Conservation," Revised 1986, Soil Conservation Service, Classroom and Field Guide Program Aid No. 341; "Toward a Unifying Concept for Agriculture," Wes Jackson, *in Agricultural Sustainability in a Changing World Order,* Gordon K. Douglass, ed., 1984, Westview Press, Boulder, Colorado; "Water—The Abundance, the Menace, the Challenge," College of Journalism Depth Report No. 21, 1990, University of Nebraska-Lincoln; *Water in Nebraska*, James Aucoin, 1984, University of Nebraska Press, Lincoln and London; *Water Quality Indicators Guide—Surface Waters*, Soil Conservation Service, 1991, U.S. Government Printing Office, Washington, D.C.;

Interviews with: Dave Bowman, U.S. Fish and Wildlife Service, Grand Island; A.D. Flowerday, Pioneer Hi-Bred International, Great Plains Division, agronomy services manager (head of education and research); Peter Jensen, retired, formerly Soil Conservation Service state range conservationist; Robert Kaul, professor of biological sciences (prairie plants), University of Nebraska-Lincoln; Clayton Lee, Soil Conservation Service assistant state soil scientist; Pat McGrane, Soil Conservation Service state public affairs specialist; Ed Peters, Forestry, Fisheries and Wildlife, University of Nebraska-Lincoln; Ken Pruess, entomologist, University of Nebraska-Lincoln; Julie Savidge, Forestry, Fisheries and Wildlife, University of Nebraska-Lincoln; John Sidle, U.S. Fish and Wildlife Service, head of small biodiversity projects throughout the state; Curt Twedt, biologist, Nebraska Game and Parks Commission; Kay Young, biologist, Chet Ager Nature Center, Lincoln, Nebraska.

Trout fishing in Long Pine Creek, August 1919, Brown County

Nebraska State Historical Society

Conservation Management and Environmental Protection

by Charles A. Flowerday
and Sally J. Herrin

A Profile (or Two) as Prologue

By mid-October, Carl Gangwish is a man with one eye on the ground and another on the sky. In the fall of 1992, harvesting replanted corn and soybeans—the second planting brought on by a late freeze—he wants to leave the corn in the field long enough to dry it out to near the moisture levels set by the local elevator. He also knows that a storm could delay harvest enough—waiting for the fields and crop to dry—that snow or hail could lay some of his crop down, and then the combine might not get as much as it would otherwise. Beans especially are vulnerable to dropping off or falling over and are harder to glean completely anytime. He weighs dockage at the elevator against the risk of severe weather doing more irreversible damage. So far the beans look good, darn good considering what could have been: 45-50 bushels per acre, and that on the rebound from nature's fickleness.

Gangwish, of rural Shelton, has close-cropped silver hair and a lined face in which his eyes still glint. They shine with the intensity of a man who still loves what he does, enough to make sure that it's done right—with care for the land, yet a careful business sense. In a seed cap, zippered, hooded sweat shirt pulled over a quilted vest and jeans, he rides from field to field in his pickup, talking by radio to his hired men. The corn looks fair at this point, no better. But it could have been much worse considering how the growing season bcgan. His fellow board member on the Central Platte Natural Resources District (NRD), Dick Mercer of rural Kearney, says, "The year '92, we ought to get rid of it, quick! It started at Halloween in '91 with one of the early bad turns." An early snow and ice storm threw harvest into chaos that year. And a late freeze in spring of 1992 made hollow promises of most everyone's cropping plans.

Gangwish is on the radio off and on through the morning, tapping on the calculator, getting the numbers down: rows planted and harvested, acres combined so far, how full the truck is, what the field hand thinks the moisture content might be, what the dockage could amount to.

Mercer too demonstrates an intensity borne of knowing that, if they're careful, if they hurry, another year's planning will come to fruition, that nature may give but can also take in a moment's notice. He is stocky, bespectacled, with a white, thinning crew cut. He wears insulated coveralls, the ever-present cap and a hooded sweat shirt while sitting at the desk in his office answering the telephone, the radio, the questions of hired men who come in wanting to know what to do with the just-broken piece of machinery.

It takes science and art, educated guesswork and as much hard data as you can lay your hands on. It takes a feel for the resources, what they can take and what they can't. And it all comes down to profit and loss like any other business.

But this one is so much more unpredictable. And often the producer feels caught in a bind between weather and prices, government programs and markets, an increasingly urban population that knows less and less about farming and a diminishing skilled labor force in rural areas, between sons and daughters who may or may not want to take over the operation and uncertain economic prospects even if they do.

In the midst of this has come a rising tide of environmental concerns that was seen by some farmers as an unwelcome, even unconstitutional, meddling and an even greater threat to profits than already existed. Many in this increasingly urban nation fail to realize that most producers average about a 3-5 percent return on each dollar invested (if that), a small profit margin compared to the 15-20 percent return set as a goal by many businesses. So one of the most recent binds has been the pressure to curb contamination of the water supply by agricultural chemicals while facing economic conditions that, especially through the 1980s, have turned increasingly dicey. This squeeze has been particularly tough on those growing irrigated row crops on what they call "lighter," meaning sandier, soils. Nitrogen is easily leached through these soils, washed out of the root zone into the water table. It was once assumed that if you cut the amount of fertilizer and pesticide applied, you'd probably hurt yields. Yet in some parts of the central Platte area, nitrates in groundwater were more than 20 parts per million (ppm) over the federal maximum contaminant level (MCL) of 10, making changes in farm management, especially in applied nitrogen, virtually inevitable.

Fortunately, a combination of factors—pressure from environmental groups, imaginative approaches to agronomic research and, interestingly, ongoing economic pressure on the farming

Harvesting center-pivot irrigated corn during the fall of 1981 on the Circle S-L Acres near Wallace, Lincoln County

Les Sheffield, UNL

Chuck Francis, IANR

Best management practices include a variety of techniques aimed at conserving soil and water and limiting chemical inputs. One alternative method of fixing nitrogen has been through crop rotations of legumes such as the sweet clover pictured here.

sector— has produced a way to slip the noose of economic hardship strung up on the tree of environmental regulations. Through deep soil and water testing for existing nitrogen, the progressive producer has discovered that, by cutting back on fertilizer, he or she can protect the environment, save money and still grow high-yielding crops.

The consensus among the agricultural community that emerged regarding the best way to tackle this threat to water qualilty, the very water they and their neighbors drink, was through an integrated approach to farm management known as "best management practices" (BMPs). These include shrewder crop-nutrient management (setting realistic yield goals, testing for and using soil and water nitrogen "credits" and other nutrient sources, especially on-farm inputs such as legumes and manures); using crop rotations to fix nitrogen and resist pests; integrated pest management (choosing pest-resistant varieties, timing planting to give the crop an edge over the pest, scouting the crop for the "threshold" of infestation after which a pesticide is economically justified and choosing the pesticide with the lowest leaching level); careful and limited irrigation, (including irrigation scheduling based on soil-moisture testing, ensuring as uniform an application of water as possible, leaving "room" for precipitation instead of saturating the root zone and preventing runoff through re-use pits and reduced tillage); and limited restrictions on chemicals applied, such as eliminating fall-applied nitrogen, especially on sandy soils.

While many of these were known and used before the 1970s and 1980s, this integration into one package with a single label signaled the beginning of a far-reaching change in agronomic practices, particularly on irrigated land. BMPs have been crucial in offering a reasonable alternative to previous practices. The increasing threats to water quality have brought on either the self-imposed (through an NRD) regulation and education of a groundwater management area (GMA) for quality, as with this NRD, or state-sponsored programs with similar goals such as the special protections areas (SPAs) for water quality in place in some areas. Both rely heavily on BMPs, coupled with increasing education about water quality, careful documentation of cropping practices and promoting or requiring water-use efficiency.

The Central Platte's GMA, a national model in local management of water quality, has included three phases of restrictions, each gradually more stringent. The most significant of these restrictions, and probably the regulation that has caused the greatest change in farming practices, Mercer says, is the restriction banning fall application of nitrogen on sandy soils (this affects those in a "phase I or II" area, the first two of three phases of increasingly restrictive limits on nitrogen); in phase III, fall-applied nitrogen is banned on all soils. This adds some pressure to farmers in the spring when planting requires long hours and cooperation from the weather, but it prevents leaching of nitrogen by getting it in the soil nearer the time the crop needs it. This program has slowed the increase of nitrates in most areas and has even reduced the level in a few spots, Mercer says.

This is the way Gangwish explains the new awareness: "We as producers found we had an inventory of fertilizer in our water, and so it was in our pocketbook; it was in our best interests to do it (adopt best management practices)." Regarding the establishment of a GMA in the Central Platte NRD, Mercer says, "It's too bad it didn't happen years ago. We cut down on the use of water and fertilizer, and it saved us money."

Pointing to the 98 percent participation in the GMA program, Mercer says that he feels very enthusiastic about the fact that the regulations in place today were developed by a task force of NRD directors and staff, as well as local producers and ag chemical dealers. (Gangwish estimated participation at 99 percent, with only 11 "cease and desist" orders being issued since the GMA was set up in 1988 and only one being fought in court.) The program is applied and policed by local producers and is based on the integrity of the producers' reports on their practices, Mercer adds.

In addition, water quality isn't the only concern of the NRD relating to water use. While the area hasn't experienced water-level declines as severe as some, nothing compared to the Upper Big Blue for instance, it promotes water-use efficiency as a way of limiting nitrates in groundwater. With water generally about 15-20 feet deep in the southern part of the NRD, Mercer says, "I'm ashamed to say this, but in this area, we didn't feel you could get enough water on the field. We'd get an inch of rain and keep the wells running." Such practices were possible because the shallow water table meant pumping costs were cheap.

Now Mercer keeps new gaskets on his gated pipe to prevent leaks and replaces as many gates as possible to reduce water loss. Just this year, he got surge valves put in several wells. *(See "Technology," p. 101.)* Gangwish has seven surge valves and says that, next to center-pivots, they are the best means of ensuring efficient water use. In addition, six years ago, he installed electric motors on his center pivots and medium-pressure spray nozzles that create a fine "umbrella" pattern to further conserve water. Gangwish has about 150 acres under one center pivot and 1,200 acres with surface irrigation from pumps. Mercer has 37 wells, most pumping groundwater through gated pipe, 6 of which are center pivots with low-pressure nozzles and 300 acres irrigated with diversions from the Kearney Canal. Gangwish owns a small herd of cattle, and Mercer has a feedlot. These are another way to use resources efficiently, as Gangwish grazes his on stalks and Mercer feeds some of his grain to his herd.

In addition to other water-efficiency measures, the GMA means the district promotes conservation tillage even more aggressively than it might otherwise. Both men use some form of reduced tillage. Gangwish uses ridge-till (planting a crop in ridges developed during the growing season and leaving residue undisturbed after harvest) on all his fields as of this year, and Mercer prefers to chisel plow, which leaves some residue on the surface to slow runoff and increase infiltration.

IANR

Deep soil testing, such as that depicted here by a University of Nebraska-Lincoln pickup equipped with a soil probe, has helped farmers save money and protect groundwater by determining and using nutrient "credits" already in the soil.

Mercer also notes that on his operation they used to broadcast atrazine (a common herbicide) every year on all the fields. That has changed. "All you have to do today is ride around the Central Platte (NRD) to see that a lot of the producers have simplified or reduced their use of pesticides," he says. Now, it's not uncommon to see a little sunflower, some grass or some shattercane in the fields.

"You can't imagine how proud I used to feel about the field because it was completely clean. The good thing is you know you're not the only one that's got a few weeds. It shouldn't interfere with the economics, but cutting back on herbicide is good way to save money," Mercer says.

Both men also sit on the NRD's Water Resources Committee. In fact, Mercer is the current chair and Gangwish is a past chair. The committee, Gangwish explains, tackles anything that has to do with water locally. Recently, it was involved in discussions with the groups sparring over the requirements for the Federal Energy Regulatory Commission's (FERC) relicensing of Kingsley Dam. The committee participated in formulating the proposal offered by Governor Ben Nelson for a block of water devoted to environmental purposes. This plan was endorsed by the Nebraska Public Power District and Central Nebraska Public Power and Irrigation District in October of 1992 as providing a suitable compromise with environmental groups concerning flows from Lake McConaughy and received the blessings of the NRD. Because of the importance of water to Central Platte producers, Gangwish says that the committee always invites the whole NRD board to attend committee meetings.

Both men say the rugged individualism (the "frontiersman" attitude, Gangwish calls it) that produced the perspective that "the water under my land is mine and I can do what I want with it" has mostly yielded to a sense that all citizens, urban and rural, have to work together to solve environmental problems stemming from water development. He says that the new environmental awareness has made it into farming practices because it been "farmer-acceptable, farmer friendly." "It's got to be cost-effective," he adds, or it won't be received as positively.

"We're very lucky to have the water we have and to be able to use it if that water goes for a beneficial use," Mercer says. "The bottom line is that the groundwater quality management area has been beneficial environmentally and economically. If agriculture is going to compete (with other uses) for the water, then we're going to have to prove that we're wasting the least amount possible," he adds.

Water Quantity and Quality

Gangwish and Mercer are only two producers, of course, and the Central Platte GMA is only one part of the state. But as a way making the issues come alive, we offer their experiences to emphasize the personal and particular way in which concerns about water quality (and quantity) have changed farm management, how the history of conservation has expanded to tackle these high-profile issues.

Of all the environmental issues related to water, water quantity and quality have dominated public consciousness because of the vast amounts needed for farming a semi-arid region with undependable rainfall and the population's need for clean drinking water. Agriculture accounts for nearly three-quarters of the total water use statewide, the largest of any use by far. And because point-source contamination (pollution at an isolated point) has been largely curtailed, nonpoint-source pollution is the greater water-quality concern. (Nonpoint pollution is dispersed contamination, agricultural or urban runoff or dispersed chemicals percolating to groundwater.) Groundwater quality concerns have also been in the spotlight because cleaning up contaminated groundwater is very costly and, sometimes, next to impossible.

Ironically, these may not be quite as pressing, not yet anyway, as the next water-related environmental problem to discuss, soil erosion by water. Such a statement should be qualified by the appreciation that droughts (perhaps an increasing threat with greater evidence for global warming) and better crop prices could well put pressure on groundwater supplies beyond what has been seen to date. In addition, since agricultural chemicals or their residues may show up decades after their application, we have no sure way of knowing the extent of future water-quality problems.

Even so, while the "history" of water-quantity and -quality protection, at least regarding most regulation, involves fairly recent developments, this is an area where environmentally conscious Nebraskans can find relatively good news. Due to the political foresight shown in the early 1970s in establishing the NRDs in this state, local control of water-quantity and, more recently, water-quality problems has promoted a relatively effective combination of educational and regulatory efforts aimed at solving these problems. This is not to suggest that all such problems are a thing of the past, only that, in most places, they are being addressed.

Groundwater Control Areas

In 1975, three years after the NRDs were established, the state legislature passed the Groundwater Management Act. Its primary intent was to slow groundwater declines in critical areas by authorizing an NRD to request the state Department of Water Resources (DWR) to designate all or part of it as a groundwater control area.

As made possible by the act, the Upper Republican Natural Resources District applied to DWR for designation as a groundwater control area in 1977. On August 1, 1977, the NRD above the Republican River—almost all of it—was declared a groundwater control district. The area includes the counties of Perkins, Chase and most of Dundy.

A review of the history of the district shows that by 1975 groundwater declines of 5 to 20 feet from predevelopment (about 1935) were common. Concern about conflicts among users and potential economic hardship due to continuing declines prompted the NRD to seek a control area. The first controls on irrigators included the following (cities, towns and domestic users are subject to restrictions also, but since agricultural use is the largest, only those on irrigators are listed here): flow meters on all wells by 1980; documentation of irrigated acres so as to allocate water; designating critical townships where the annual water-level decline exceeded 1 percent of the average saturated thickness in that township; and well-spacing requirements prohibiting new irrigation wells within 3,300 feet of any other well in critical townships.

In 1984 DWR approved regulations proposed by the Upper Republican NRD to protect groundwater *quality* in a quantity control area. The NRD required annual permits for each irrigation well that applied ag chemicals (chemigation) and stipulated a properly working check valve on all such wells. The regulations did not apply, however, to nonpoint pollution from ag chemicals.

The Upper Republican NRD has gone through a few allocation periods, the most recent of which began in 1988 and runs through 1992. It grants 75 acre-inches for the period, an average of 15 acre-inches a year, and irrigators can carry over no more than 45 acre-inches a year from a previous period. Designation of critical townships was revised to an an-

One of the many water-quality research and demonstration projects in Nebraska is the Management Systems Evaluation Area (MSEA) project near Shelton pictured here. This is one of five MSEA sites in the nation begun as part of the Bush administration's Presidential Initiative on Water Quality.

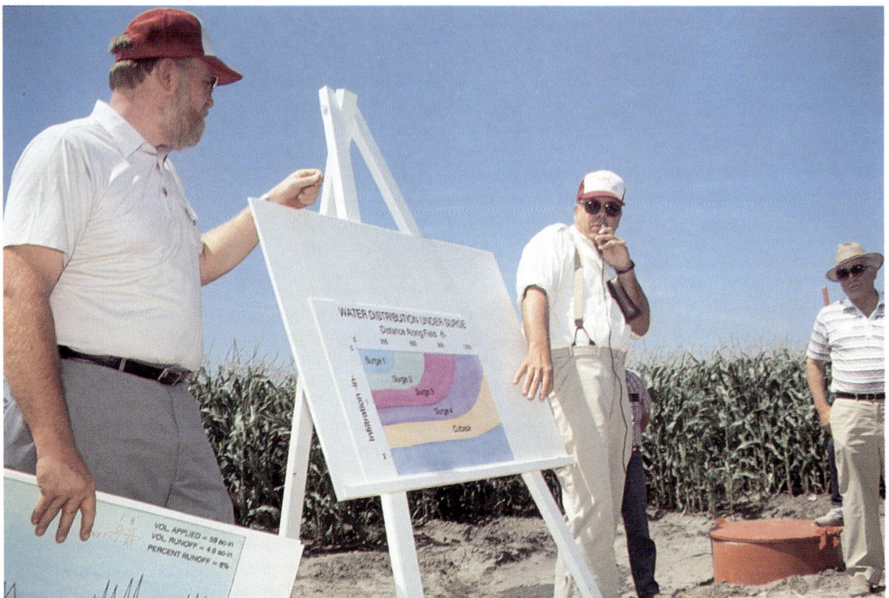

Nebraska Farmer

Pat McGrane, Soil Conservation Service

A water meter on an irrigation well for a center-pivot sprinkler system in Cuming County. Three natural resources districts require metering of irrigation wells. Two of these have had to allocate water.

nual decline exceeding *one-half of one percent* of the aquifer's saturated thickness, and they remain so for five years.

Shortly after the designation in the Upper Republican NRD, on Dec. 9, 1977, a quantity control area was designated in another region with major declines, the Upper Big Blue NRD. Counties affected included all or parts of Adams, Butler, Clay, Fillmore, Hamilton, Polk, Seward and York. The initial goal of the control area was to hold the decline rate to the average rate for the 1976-1978 period. Through conservation and due to above-average precipitation, that goal was met. The new goal became to hold the water level above the 1978 level.

Conservation measures promoted by the district (and similar to those encouraged by other control areas) are the following: 1) use of irrigation reuse systems; 2) the most economical use of groundwater; 3) crop rotations; 4) controlling runoff from center-pivot irrigation systems; 5) water storage and land treatment where needed; 6) irrigation scheduling; 7) using surface water where available; 8) maintaining pumping plant efficiency; 9) minimum tillage; 10) gathering all available information on water-use and efficiency; 11) installation of flow meters. (Encouraging use of moisture blocks are among the measures stipulated for the Little Blue control area, which will be discussed below.) The allocation in the Upper Big Blue is 48 acre-inches for a three-year period with a carryover of no more than 16 inches into the next allocation period.

Not long after the first two designations, in January of 1979, the northern part of the Little Blue NRD became a groundwater control area. This included parts of Adams, Clay, Fillmore, Nuckolls and Thayer counties. The NRD's goal became maintaining the same average water level as that of spring 1981, and allocation did not occur immediately. If the average district water level dropped lower than 18 inches below that baseline level, mandatory allocation would go into effect. Even before allocation limits, the NRD certified irrigated acres and flow meters and gathered yearly pumping data from groundwater irrigators.

In July of 1992, the Little Blue NRD declared that it would begin restricting the amount that groundwater users could pump. The allocation was to begin on Jan. 1, 1993.

Groundwater Management and Special Protection Areas

In addition to a groundwater control area, another way to regulate stress on the groundwater supply is through a groundwater management area. The Groundwater Management Act was revised in 1981 to become the Groundwater Management and Protection Act (GWMPA) to give the NRDs sole authority to request a groundwater-quality control area unrelated to de-watering of an aquifer.

In 1984, as an outgrowth of the Kerrey administration's Water Independence Congress, LB 1106 required each NRD to prepare a groundwater-management plan in which it would, among other things, stipulate a life goal for its aquifer. These plans were to be reviewed and approved by DWR. The GWMPA was revised again in 1986. One of the revisions said an NRD could propose a groundwater management area to *protect* water quality, rather than solely to stop deterioration. It also authorized the Department of Environmental Control (now "Quality") to designate a special groundwater-quality protection area.

Four NRDs currently administer groundwater management areas. The Central Platte NRD (parts or all of Buffalo, Custer, Dawson, Hall, Hamilton, Merrick, Nance, Platte and Polk counties) has had one for quantity and one for quality since 1988. The Tri-Basin NRD (Gosper, Kearney and Phelps counties) became a groundwater-quality management area in 1989, a small part of the South Platte NRD near Sidney (in Cheyenne County) became a GMA in 1990. In February of 1992, the board of the Upper Elkhorn NRD (parts or all of Antelope, Holt, Rock and Wheeler counties) voted unanimously to implement a groundwater-quality management area.

Restrictions in GMAs generally involve some combination of the following: requiring BMPs; attendance at educational programs about water quality; allocation of groundwater use; rotation of groundwater use; well-spacing requirements; and flow meters. As mentioned in the profiles above these regulations are administered by the NRDs, and they have the power to enforce them. As this book goes to press, the Lower Loup NRD is considering designating a GMA for a small part of its region.

The other option on nonpoint groundwater-quality problems is the special protection area or "SPA." This process begins at the local level. If a political entity, a city, county or state agency, believes that nonpoint groundwater contamination needs to be regulated, or if an NRD thinks it lacks the funding or expertise to adequately deal with nonpoint contamination, it can request the state, through the Department of Environmental Quality, to study the contamination in the area. Evidence of the contamination must accompany the request.

If an SPA is designated, the NRD has 6 months to draw up an action plan to address the problem, which must include educational programs for water users. In addition, the plan must include selected BMPs and/or any other "reasonable" measures needed to make the plan work.

Burlington-Northern 'Chemigation' Study: No Greater Risk to Groundwater When Proper Methods Are Used

Chemigation, a method of applying pesticide and fertilizer through sprinkler irrigation systems, added new problems to the study of groundwater contamination in the early 1980s.

Questions raised for water researchers at the University of Nebraska-Lincoln (UNL) concerned its effectiveness relative to conventional methods, its potential to contaminate groundwater, its effect on non-target species of plants and insects and its impact on nitrogen and irrigation-management systems.

Since about 40 percent of Nebraska's 27,000 center-pivot irrigators used chemigation technology as of 1986, researchers felt the questions were important, and officials of the UNL Institute of Agriculture and Natural Resources (IANR) began the search for chemigation-research money in 1984. Due largely to the efforts of D.B. "Woody" Varner, chairman emeritus of the board of directors of the University of Nebraska Foundation then working as a consultant, the Burlington-Northern Foundation committed $1 million to the 5-year study.

IANR proposed an interdisciplinary research and demonstration project to study chemigation technology and practices. The findings were to complement already established best management practices or other practices in irrigation, nitrogen and integrated pest management.

Work began at the university's South-Central Research and Extension Center near Clay Center in the winter of 1984 and spring of 1985. The site was chosen because it is representative of much of the state's irrigated agriculture.

The major findings, as reported in the final report of the Burlington-Northern Foundation Water Quality Project issued by the IANR in August 1990, state that chemigation is an effective method of applying chemicals when proper procedures are followed. The risk of groundwater contamination appears no greater than other conventional methods of pesticide application, the report states.

Most of the insecticide applied through chemigation stayed in the top 4 inches of soil and did not leach to the groundwater; and it was effective for targeted pest species but did not greatly affect non-target species of beetles, according to the report.

Nitrate fertilizer applied accumulated rapidly in soil where amount applied exceeded crop demand. Where excessive amounts were applied, leaching occurred. To control leaching, recommended rates of application and application procedures must be followed, according to the report.

One important problem with chemigation is the possibility of backflow of chemicals into the irrigation water source. The Burlington-Northern study showed that some devices for backflow prevention were very effective while others were not.

Overall, the study determined that chemigation is effective and does not increase groundwater vulnerability to contamination when used properly. Following recommended procedures and using proper equipment when applying fertilizer and pesticide are essential in preventing groundwater contamination.

—Brad Rundquist, weekend and copy editor,* Ft. Dodge (Iowa) Messenger, *and formerly CSD editorial assistant

One remarkable aspect of the SPA program is that, while some states have regulations protecting groundwater from pesticides, very few have such programs regarding fertilizer. Even while it remains the only state in the nation to refuse responsibility for the federal law that regulates pesticides—the Federal Insecticide, Fungicide and Rodenticide Act—Nebraska is on the forefront of nonpoint-pollution control regarding fertilizers.

Criticisms of the SPA program have focused on the limited funding for DEQ's studies and for its management of the program should the NRD refuse to; on the 1-year period within which DEQ must evaluate the causes of pollution and the need for its control; and on the fact that SPAs were designed also to prevent contamination where it is likely but are being enacted more sparingly, only where drinking-water standards are obviously being exceeded.

In addition to groundwater quantity and quality issues, surface-water quantity and quality have also assumed the spotlight in recent years. These issues represent important management concerns but will be discussed in elsewhere in this book. Surface-water quality problems are examined briefly in the section on aquatic habitat in this chapter *(see p. 191)*. Surface-water quantity problems are addressed in the chapter on "Politics and Policy" *(see p. 222 and 229)*.

Much other "good news" regarding conservation and protection of Nebraska's water resources probably could be discussed here, but in the interests of saving space, we will only note the following:

—Besides those organizations already mentioned, there is much regulatory, planning and data-gathering machinery at work in the state. Alphabetically, some of the agencies are: the Conservation and Survey Division, Department of Environmental Quality, Nebraska Association of Resources Districts, Nebraska Natural Resources Commission, Soil Conservation Service, U.S. Geological Survey and the Water Center of UNL, (to mention only some of the main public agencies). A number of non-profit citizen-run environmental and water-users groups also offer their voice to the management debate. *(These are profiled at the end of the "Politics and Policy" chapter.)*

—There is hope that dialogue is alive. Not too far back, the Crane Trust and a group of irrigators worked out a deal (in a version of the Prairie Bend proposal) to impound flow near Overton for release when the cranes and other birds needed it in late winter, after which it was slated for diversion into the Blue Basin. This deal did not go through, but the surprising bedfellows suggest that possibly a statewide tradition of basic civility is holding up and could create promising, resourceful alliances. The recent discussions surrounding the relicensing of Kingsley Dam, continuing a precedent set during the Platte River Forum For the Future, have also provided evidence that developmental and environmental groups can find common ground.

—The Nebraska Groundwater Foundation stands out for its unparalleled success at getting the word out on all sides of water-related issues. They are all booked up for the Children's Groundwater Festival, an event that has tremendous teacher support and makes being informed and getting involved seem exciting to students. A nationwide Groundwater Protection Network is being set up through the foundation to take advantage of heightened interest in groundwater issues nationally.

—The Burlington-Northern chemigation study showed that with properly working backflow check-valves, testing for nitrogen in soil and water and proper timing, using irrigation systems to apply fertilizer and pesticides is not more harmful to groundwater than traditional application. It may, in fact, augur a more efficient method of applying water and chemicals.

—Tourism may offer alternative economic development that promises relatively benign and non-consumptive use, depending on numbers of users and their management, of course. This industry continues to grow. A recent *Omaha World-Herald* article put tourism at $1.7 billion annually, one of the state's larger industries. It should provide multiplier-effect dollars in rural areas. In 1988, according to a state Game and Parks Commission estimate, the total economic effect, including multiplier effect (money spent and re-spent), coming from bird-watchers along the central Platte in the spring of 1988 was $1.26 million. Former Crane Trust executive director John VanDerwalker estimated that 10,000-12,000 people came crane watching in 1988 (and spent $70 each, which is then multiplied by 1.8, a figure suggested by the visitors' bureau), and that within five years the number could total many tens of thousands annually.

Soil Loss by Water

Surprising as it may be to some, while water is essential to crop production, it is also one of the biggest threats to the long-term productivity of the land. A single raindrop strikes the soil like a mini-bomb. The 1991 USDA Yearbook of Agriculture, *Agriculture and the Environment*, says it this way: "Every year, throughout most of the United States, more than a quadrillion (1,000,000,000,000,000) drops strike each square mile of land with the impact energy of thousands of tons of TNT." It then compares the annual impact energy of rain falling on the state of Mississippi every year to the explosive power of 1 billion tons of TNT!

Even so, whenever the "few" try to get the "many" concerned about soil erosion, those same "many" may well wonder, *What is all this fuss about saving the "dirt"*? Common as it may be, it is actually alive and sustains all of us, and as such, it deserves our respect. To appreciate what soil really is we need to see it as "a protective living geomembrane through which energy, nutrients and water pass as they nourish land-based life," again according to *Agriculture and the Environment*.

It goes on to say, "Climate, parent material [soil-forming rock], topography, living organisms, and time are the factors of soil formation. . . . Soils are complex physical, chemical, and biological entities. The biological activities are influenced by climate, and together they interact with earthy materials in a landscape over time. . . . Fertility is the most important inherent characteristic of soils."

It also warns, "The cultivation of soils for cropping alters many surface features of soils—plowing and disking disturb the soil structure and heavy machinery packs the particles together and reduces the pore space and the [water] infiltration rates."

The fact remains easy to ignore, but it is persistent as the storms themselves: rain falling on unprotected soil detaches soil particles and moves them downhill by runoff. This erodes the productivity of the soil, and the water is lost for crop production as well. Organic matter on the surface, growing or dead (such as stalks or stubble), is one of the most cost-effective ways of fighting soil loss by water. In addition, besides supplying essential nutrients, organic matter incorporated into and decomposing in the soil adds to its structure and cohesiveness, allowing water to better infiltrate and supplying gums that hold soil particles together.

Basically, all soil conservation efforts related to water erosion involve keeping vegetation or crop residue on the field to stabilize soil and slow runoff or involve checking runoff through structural practices such as terraces, grassed waterways, small check dams and runoff ponds. If these practices cut erosion to a rate that is considered tolerable without harming production capacity (different for each parcel of land but often generalized at 5 tons per acre per year), then that land is considered adequately treated. This tolerance for loss is called a "T" value, and the Nebraska Natural Resources Commission reports that 3.2 million acres of the state's cropland are eroding at a rate between 1 and 2T; an additional 3.6 million acres are eroding at greater than 2T. This acreage represents 34 percent of the state's cropland and accounts for 80 percent of the total soil loss in the state.

History of Soil Erosion Management

Emerging awareness of the problems of water and wind erosion caused the federal government to set up a Division of Soil Erosion in 1908 in the U.S. Department of Agriculture (USDA) Bureau of Chemistry and Soils. Congress, however, didn't appropriate much money for studying the problem until 1929, when it allocated $160,000 for 10 erosion research centers. The next year, Hugh Hammond Bennett, director of the chemistry and soils bureau, would publish a circular called "Soil Erosion, A National Menace" to rally support for his proposal to put young men to work for the federal government in soil-conservation programs.

More far-reaching action came in late March of 1935. As Bennett was testifying before Congress about soil erosion, a huge dust cloud—about which he had been informed and for which he had

Sheet erosion from farming up and down the hill near Columbus, Platte County. Farming on the contour is a little less convenient but does a much better job of saving soil and water, especially when combined with other conservation practices.

Soil Conservation Service

timed his testimony—blotted out the sun above Washington. Congress quickly passed the bill establishing the U.S. Soil Conservation Service (SCS), and on April 27, 1935, President Franklin Roosevelt signed it.

Already established in 1933, the U.S. Soil Erosion Service was then transferred from Interior to Agriculture and was renamed SCS. At the same time, the Civilian Conservation Corps, a U.S. Forest Service program, was also moved to SCS. And in 1936, the Agricultural Adjustment Administration established a program to share the costs with farmers of shifting from surplus row crops to grasses and legumes. This was the predecessor to the agency that currently provides financial assistance to farmers for conservation, the Agricultural Stabilization and Conservation Service (ASCS).

Then, as now, decision-makers and federal technicians preferred an emphasis on working cooperatively with local landowners through education and moral persuasion to large-scale regulation. This was encouraged as a neighbor-to-neighbor approach and prompted an interest in forming local soil conservation districts to help this approach work.

Nebraska State Historical Society

Civilian Conservation Corps workers planting shrubs and trees on slopes and gully banks in Nance County during the 1930s. These were to serve as permanent control measures after the temporary structures seen in the gullys had been removed.

Support for this idea came through federal legislation on May 13, 1936, after the USDA proposed a "Standard State Soil Conservation Districts Act." Designed as a model for the states to help them create local conservation districts, which would then manage state and federal erosion-control efforts, it allowed them to make agreements with local landowners, to offer technical and financial help, to propose land-use regulations and to pursue erosion research and demonstration projects.

On the state level, soil-conservation programs in Nebraska date back to the 1880s, when the agricultural experiment stations began doing soils studies. The first soil survey for the state was published in 1903 on parts of Stanton, Madison, Pierce and Wayne counties, and as early as 1908 the *Nebraska Farmer* magazine urged that farmers pursue soil conservation practices. The extension service also began a statewide educational campaign to fight erosion in 1920 that included published studies and promotion of conservation measures.

In 1935, the state Soil Conservation Advisory Committee was formed to coordinate soil conservation efforts and facilitate the establishment of conservation districts. Two years later, the legislature enacted LB 533, creating the state Soil Conservation Committee to eventually take over the duties of the Soil Conservation Advisory Committee. The first committee included W.H. Brokaw, director of the Nebraska Agricultural Extension Service, W.W. Burr, director of the Agricultural Research Experiment Stations, and George E. Condra, director of the Conservation and Survey Division. The bill was vigorously debated, and many farmers and farm groups, believing that soil erosion should be fought by individuals not organizations, testified against its passage, even implying it smacked of socialism. And while the final bill retained much-contested provisions for setting up land-use regulations, the authority to enforce those regulations was eliminated.

Even so, the movement for local conservation districts bore fruit in Nebraska in 1938, when the first successful referendum to create a soil conservation district passed in northeastern Nebraska in February of 1938. And by Dec. 16, 1949, when the Perkins County district was established, Nebraska had become the first state west of the Mississippi to organize all its land into soil conservation districts.

The affiliated cause of water conservation—which then meant controlling floods and stemming runoff—was fostered by the state legislature's passage of PL 366. This bill facilitated the formation of watershed conservancy districts and established the Nebraska Soil and Water Conservation Committee, which allowed those organizations to sponsor watershed projects. This was in keeping with federal legislation in 1936 that authorized USDA to implement runoff retention programs in upstream watersheds and, in 1954, through the National Watershed Protection and Flood Protection Act, empowered local watershed projects by providing cost-sharing money and allowing local districts to acquire land and maintain and operate small watershed projects.

In 1961, the state Soil and Water Conservation Committee became the Nebraska Soil and Water Conservation Commission, which became by statute the comprehensive soil and water planning agency for the state in 1964. It was renamed the Natural Resources Commission (NRC) in 1972. It has had a major role in water planning since the 1980s, and in 1983 it began coordinating with SCS the drafting of the state "Soil and Water Conservation Strategy," a comprehensive conservation plan, a copy of which is available from the NRC.

The commission's duties include:

> providing assistance to the Water Management Board [now defunct] and governing subdivisions with water resources responsibilities [particularly NRDs], facilitating coordination and cooperation between local, state and federal agencies, and planning, developing, and promoting the implementation of a

comprehensive program of resource development, conservation, and utilization for Nebraska soil and water resources. The commission's major planning program is the State Water Planning and Review Process. The commission is also directed by statute to develop and maintain a Natural Resources Data Bank, which stores thousands of individual pieces of natural resources information,

This is according to the *Base Document of the Nebraska Soil and Water Conservation Strategy*, published by the NRC in 1991.

As mentioned above, in 1969, the legislature directed that the commission's Natural Resources Data Bank supply the state and its natural-resources agencies with current information about land receiving adequate treatment against soil erosion. This database tells us that, as of 1986, 62 percent of all the state's land was receiving adequate treatment. Nonirrigated cropland needed treatment on 41 percent of its acreage, and irrigated cropland needed treatment on 47 percent.

To further assist the cause of soil and water conservation, the legislature established in 1977 the Nebraska Soil and Water Conservation Fund, which by 1992 had provided more than $25 million in state funds for conservation. Provided for by the state Soil and Water Conservation Fund, the Nebraska Soil and Water Conservation Program has been managed by the commission since its inception in 1977. This program was established to help share the costs of installing conservation practices to stem runoff and conserve water.

The 1985 and 1990 Farm Bills

In addition to local, state and federal programs, much of which are oriented toward the more expensive structural land treatment, the most influential governmental action to date affecting soil loss by water, and soil conservation generally, has been the passage of the landmark 1985 farm bill and its continued conservation emphasis in the 1990 farm bill.

The 1985 farm bill, formally called the Food Security Act (FSA), along with amendments to the Clean Water Act in 1987 and the Conservation Program Improvements of the 1990 farm bill, were enacted to bring USDA farm commodity programs in line with emerging environmental goals. The concern was that federal income tax incentives, commodity programs and speculative land-value hikes were driving increases in cropped acres more than market incentives. One provision of the FSA involved crop production begun on highly erodible land before the bill's passage. To continue to be eligible for farm-program benefits, producers had to submit plans for conservation practices on this type of land and have them approved. Another provision, called "sodbuster," regulated conversion of grassland to cropland on highly erodible acres without an ap-

A terracing demonstration, probably a joint effort of the University of Nebraska and the Soil Conservation Service in eastern Nebraska during the late 1940s

IANR

Civilian Conservation Corps workers building terraces, Butler County

Nebraska State Historical Society

Dr. Tom M. McCalla, bacteriologist with the University of Nebraska, discusses the role of bacteria in soil and water conservation at the 4-H Leader Training Conference at the College of Agriculture in Lincoln, Jan. 28-29, 1948.

B. C. McLean, Soil Conservation Service

R. W. Hufnagle, Soil Conservation Service

Soil Conservation Service district conservationist Ed Saylor of Hebron (holding paper) discusses conservation's benefits with (from left) Einar Anderson, a farmer from Edgar; H. E. Swedburg, a banker from Clay Center; M. E. Kitzinger, a farmer from Harvard; E. N. Hertel, a farmer from Clay Center; O. J. McDougal, the county agent from Clay Center; and H. W. Kitzinger, a business man from Harvard, probably during the late 1940s or early 1950s.

A. C. Fox, Soil Conservation Service

Teachers attending the second Conservation Education Workshop at Doane College, May 28 through June 9, 1951, received instruction on soil conservation, grassland management, wildlife conservation and other related topics. Here O. J. McDougal (left), extension soil conservationist, and William F. Rapp, director of the conservation workshop, Doane College, demonstrate how water moves sediment.

proved conservation plan. Plans for cropping highly erodible land, whether begun before or after 1985, had to be in by 1990 and be implemented by 1995. Yet another called "swampbuster" regulated draining or filling wetlands for the purpose of producing farm-program crops with the same restrictions as sodbuster.

The point of these restrictions was to curb conversion of wetlands and fragile grasslands to produce crops already in surplus. Failure to comply eliminated producers from the benefits of the voluntary federal farm program. Government benefits, such as price supports, storage facility loans, federal crop insurance, disaster payments, loans authorized by the Farmers Home Administration and storage payments from the Commodity Credit Corporation were prohibited.

In addition, the Tax Reform Act of 1986 eliminated or reduced tax credits for cropping of marginal land or wetlands. And some of the provisions of the FSA were "fine-tuned" in the 1990 Food, Agriculture, Conservation and Trade Act, particularly a perceived loophole in swampbuster that eliminated a farmer's eligibility for benefits only if the wetland was converted to produce a farm-program crop. Substituted was a gradient of restrictions for any conversion of a wetland, regardless of purpose.

The remaining part of the federal push to limit cropping of marginal lands was the Conservation Reserve Program (CRP), which provides annual rental payments for 10 years and helps pay for keeping a permanent cover on the land. Among the various federal programs to idle a total of 78 million acres at their peak in 1988, the CRP enrolled 25 million of those acres and, unlike other acreage-retirement programs, such as Soil Bank, stipulated explicitly that marginal lands, not just any acres, be idled for environmental reasons. As of the 12th enrollment period in 1991, 1.3 million acres in Nebraska were in the CRP. As this book goes to press, Congress had decided not to authorize enrollments for 1993.

While the 1985 and 1990 farm bills have gone farther than any previous farm legislation to advance the cause of conservation, they have had their critics. Some of these have focused on the CRP. One criticism is regarding the short-term (10-year) nature of enrollment. Some have suggested that allowing economic uses of CRP land, such as haying, grazing and tree cropping, could ensure protection beyond the 10-year period. Another has been that the low bids relative to land values in the Corn Belt have resulted in many more acres of well-protected, but potentially erosive, wheat-growing land being enrolled than Corn Belt cropland with real erosion problems—land particularly vulnerable to water erosion. (The land rental is based on a low bid for the whole county.) In light of this, a one-time "corn bonus" was offered to increase enrollment of Corn Belt land.

Yet another concern has involved the potentially negative effect on the local rural economy as sales of farm equipment and inputs drop with acreage retired. This resulted in a limit of 25 percent of any county's cropland being enrolled. (Waivers allow up to 35 percent in some counties.) Other concerns have been regarding absentee owners "farming the CRP" to guarantee a 10-percent return and with CRP rentals "outbidding" tenant farmers, reducing opportunities for young farmers. In December of 1987, a provision was added to limit the bid to the local prevailing rental.

Another problem has been with the entire bill, not just CRP. The cyclical nature of agriculture means that the conservation provisions are only as strong as the incentives in the voluntary farm program. If farm prices rise and farmers decide not to use the program, these measures become ineffective. Some have urged that permanent easements (rent in return for restricted use) be offered on key tracts of highly erodible land.

Perhaps the most serious criticism has involved the loosening of the soil-loss tolerance levels to which conservation plans were tied. Originally, the law said that, except where compliance caused extreme economic hardship, the conservation plan was to bring soil loss to T levels (discussed above). The exceptions were to be cut to 2T. However, final regulations in late 1987 and early 1988 inserted language stipulating that on land already in production (not sodbuster), "alternative conservation systems" could be used to meet compliance goals for the economic hardship cases. This meant primarily the use of reduced tillage, but not necessarily in combination with other conservation measures, which makes it most effective. On May 3, 1988, this change was applied to all producers, and the T-level stipulation was switched to a merely qualitative reference to "a substantial reduction in erosion" without quantifying the acceptable loss. In Nebraska, the state Erosion and Sediment Control Act says that erosion must be reduced to T levels, and this has caused some NRDs to not approve the adoption of so-called "alternative conservation" plans and to post disclaimers in NRD offices explaining the difference between state and federal compliance.

Another fundamental criticism, partially addressed in the 1990 farm bill, is that farm-program benefits are based on historical cropping patterns, called "base acreage," and thereby, penalize crop rotation. Critics have pointed out that these keep economic and ecological costs uncoupled and that the bills make farming one of the most regulated businesses in

Farmers use many different forms of reduced tillage depending on their cropping needs, but whatever the method, leaving at least 60 percent residue cover cuts water erosion by 80 percent.

IANR

James Lafleur, Soil Conservation Service

A no-till drill plants in wheat stubble in Lincoln County. The nation's first no-till planter, the Buffalo Till Planter, was invented in the 1950s in Columbus.

the country. These critics say studies have shown that without farm-program incentives, sustainable or alternative cropping systems (such as those detailed below in the section on sustainable farming) are both economically and ecologically superior. This is according to studies chronicled in *Paying the Farm Bill,* a critique of the farm program put out by the World Resources Institute. The 1990 bill included a provision that allowed for 15 percent of crop acreage to be flexible and still count as base acreage.

In addition, in the Integrated Farm Management Program Option of the 1990 bill, a provision allows for adoption of "resource conserving crop rotations." Farmers planting these crops—legumes, legume-grass and legume-small grain mixtures or mixtures of all three—do not forfeit this acreage from their base. Other changes were that set-aside requirements can be waived if such reductions are spread over the course of an entire rotation and that cross-compliance regulations can be waived so that program crops can be included in the rotations. This program option was extended with a signup limit of 3-5 million acres, but the World Resources Institute has suggested offering this for all acres.

Regardless of modifications made in the 1990 bill, the World Resources Institute, among others, has advocated the following policy alternatives, prioritized by impact and scope: 1) that farm-income support be uncoupled from production levels so as to eliminate the reliance on chemical-intensive, monocultural farming and encourage more sustainable practices; 2) that ag chemicals be taxed to incorporate natural-resource costs into the true costs of farming practices; and/or that greater flexibility in crop production be allowed through the farm program. These changes, they believe, would also eventually raise farm income and help cut the federal deficit.

Reduced Tillage

Beyond specific programs and regulations, among the most promising developments in recent years has been the increasing adoption of conservation, or minimum, tillage. While conservation tillage was being researched as early as the 1930s and was practiced by a few in the 1950s, it did not gain widespread popularity until the late 1960s and early 1970s, particularly after the oil embargo of 1973.

Because of the fuel, and money, saved in foregoing extra trips across the field, farmers became more and more receptive to reduced tillage. Besides saving fuel, time and money, minimum tillage conserves soil by leaving a cover of crop residue on the surface. USDA researchers estimate that leaving 30 percent or more residue cover on the soil can cut soil loss by water enough to meet the minimum definition of conservation tillage, in other words, enough to do some good. With corn or sorghum, 60 percent residue cover cuts water erosion by 80 percent on most moderately sloping land.

The three most popular modes of reduced tillage are called *no-till*, *ridge-till*, and *mulch-till*. "'No-till' means planting a crop in the undisturbed residue of an old crop; 'ridge-till' means planting a crop in ridges that were developed during the growing season by cultivation and left undisturbed since harvest; and 'mulch-till' is planting a crop where the total surface has been disturbed but at least the minimum required residue remains after planting," according to *Agriculture and the Environment*.

This transformation of the clean-tilled fields that were the pride of the traditional farmer using a moldboard plow has not come easily, but it has been dramatic enough over a few decades that the June 29, 1992, issue of Time magazine called it a "revolution on the farm." Bill Richards, chief of SCS, says that estimates put the acreage currently under some form of reduced tillage at one quarter of the 281 million acres of the nation's cropland. But he predicts that by the end of the century, 80 percent will be under what he calls "residue management."

In Nebraska, conservation tillage is used on about 45 percent of the state's cultivated land. Since reduced tillage generally costs far less than structural practices, new planting or cultivating equipment notwithstanding, it has been widely promoted by SCS as a means of helping farmers meet conservation-compliance requirements.

A major environmental concern with minimum tillage has been the notion, probably mistaken, that this kind of residue management requires more pesticides, especially herbicides, to kill weeds and insects. Although some additional pesticide use may occur during the first few years of reduced tillage, in general, studies have shown that the producer can change the timing of application, rather than increase the amount, to achieve the equivalent pest control. In fact, it is not uncommon after a few years for farmers to report that they are using less, not more, pesticides with residue management.

Another more traditional technique that has received less attention than reduced tillage is contour strip-cropping. This practice interrupts the erodible span with a close-sown crop, usually wheat, rye, oats or barley. It can be combined with terraces and grassed waterways to move runoff around the hill toward a stable outlet. Strip cropping helps conserve moisture as well, as the close-sown barrier aids infiltration of water.

Sustainable Farming Systems

With the popularity of conservation tillage and the recent public concern about groundwater quality, a problem has sur-

R. W. Hufnagle, Soil Conservation Service

Contour strip-cropping of corn and small grains, probably wheat, on the Dusek farm west of Ravenna, Buffalo County, July 19, 1938

faced regarding the increased infiltration provided by minimum tillage. While limiting nonpoint-source pollution of surface water by keeping water, ag chemicals and sediment on the field, greater residue cover can contribute to more nonpoint pollution of groundwater as fertilizer (particularly nitrate-nitrogen) and pesticides or their residues go into solution and find their way to the water table.

One way out of this seeming double bind, as well as a means of conserving soil, water and money spent on off-farm inputs, is the alternative presented by low-input, sustainable agriculture. Also known as "alternative agriculture," "integrated crop management," or just "sustainable agriculture," it aroused some controversy at first, but on-farm case studies and a growing body of credible research support the "good-news" claims of its proponents. Some have preferred to drop the "low-input" tag and refer to it simply as sustainable agriculture, perhaps to avoid the impression that low input means low output. What low input refers to is the lower levels of purchased, off-farm inputs, particularly synthetic fertilizers and pesticides.

Initially seen by some as a return to the days of low yields before commercial fertilizers and fence-row-to-fence-row production, sustainable agriculture emphasizes animal and green manures, and crop rotations, particularly of legumes, as alternative means of adding nutrients and controlling pests. It also promotes biological control, intercropping (seeding wheat between soybeans, for example) and limited mechanical control of weeds and pests, thorough testing of nutrient "credits" in soil and water, limited use of pesticides and use of those that break down quickly. It stems in part from the organic farming movement, but tempers this with sophisticated research and management options that focus on the highest possible net income, rather than gross yield. Substituted for purchased-input-intensive (capital-intensive) farming is a management-intensive philosophy that emphasizes integrated crop and livestock management and a systems approach to meeting society's food-safety and environmental needs, as well as a farmer's economic needs.

Sustainable farming practices vary from farm to farm but commonly include:

> —Crop rotations that mitigate weeds, disease, insects, and other pest problems; provide alternative sources of soil nitrogen; reduce soil erosion; and reduce risk of water contamination by agricultural chemicals.
>
> —Pest-control strategies that are not harmful to natural systems, farmers, their neighbors, or consumers. This includes integrated pest management techniques that reduce the need for pesticides by practices such as scouting, use of resistant cultivars (varieties), timing of planting, and biological pest controls. . .
>
> —Increased mechanical and biological weed control; more soil and water conservation practices; and strategic use of animal and green manures (legumes used to add nutrients and organic matter to the soil).
>
> —Selection of synthetic chemicals [that pose] no significant hazard to humans, animals or the environment,

This, again, is according to *Agriculture and the Environment.*

As of 1992, state and federal research dollars for sustainable agriculture and its components totaled more than $100 million in the USDA Agricultural Research Service alone. Through the Cooperative State Research Service, an arm of USDA that works closely with state experiment stations, another $76 million is being spent on sustainable systems research.

A key to good management of sustainable farming systems—in keeping with "thought-intensive," (rather than "capital-intensive," farming) as it is sometimes depicted, is access to detailed information on soil moisture, soil and water nutrients, precipitation, cropping history, pest levels and energy costs, to name a few. To meet this need, a number of research and educational programs, as well as a number of computer databases, are helping farmers make the transition to more sustainable agriculture.

Some of these are: centers for sustainable agriculture at major research universities in the Midwest, such as the program in agroecology at the University of Illinois, the Leopold Center in Iowa and the Center for Sustainable Agricultural Systems at UNL. Other private research centers are studying ways to maximize crop residues, develop more drought-tolerant varieties, maximize ground cover and on-farm inputs through rotation of legumes or through perennials, as well as looking at low-energy and renewable-resource technologies.

In addition, other informational programs are: the extension service's Sustainable Agriculture Research and Education program, which gathers information on integrated pest management, soil testing and fertility, pesticide management, soil and water conservation and water quality; the National Agricultural Library, featuring an Alternative Farming Systems Information Center, a clearinghouse authorized by the 1985 farm bill, which offers a bibliography on sustainable farming, a directory of education programs and online computer searches for research and educational grants; and the office of Appropriate Technology Transfer for Rural Areas in Fayetteville, Arkansas, which gets federal funding to answer any questions related to groundwater contamination, food safety, cutting chemical use and raising profits in agriculture. Much computer software is also available to help farmers manage inputs and profits more wisely.

Rangeland and Sustainability

Grazing can be an appropriate production system anywhere the land is unsuited to cropping, too hilly for instance. Livestock, properly managed, can allow producers to make a living and conserve soil and water. Controlled grazing—which can include short-term, high-intensity rotational and deferred grazing, all methods for minimizing stress on grassland—takes a number of forms, but like reduced tillage, it offers a means of keeping adequate cover on the land while managing it to produce.

While sustainable agriculture is just emerging on cropland, it has been largely practiced in another form in one of Nebraska's unique rangelend ecosystems, the Sand Hills. Other areas where proper grazing methods are practiced represent relatively sustainable agriculture, but the Sand Hills present particular problems if the land is poorly grazed or cropped (such as wind erosion and nonpoint-source pollution), so they offer a lesson in good land use and controlled grazing.

As a result of this management, the Sand Hills native prairie is about 90 percent intact. And though buffalo and domestic cattle graze on different kinds of grasses, modern grazing management systems, including fencing and the dependable water provided through a system of wet meadows, windmills and irrigated hay, have produced a type of sustainable agriculture in that region. Most cattle producers now understand that their real crop is grass, which they market through the medium of cattle.

Because the soils of the region are so fragile, serious mistakes like over-grazing or most attempts at row-cropping are immediately apparent in the form of blowouts, forcing a change in the producer's behavior. While many now concede that the introduction of brome grass (a native of Siberia imported in the 1880s) was probably a mistake—brome tends to inhibit native grass species but has shallow roots and so is a poor stabilizer of soils—the plant cover in the Sand Hills is in places superior to that at the turn of the century. In fact, blowouts are so well-managed that the blowout penstemon, a beautiful blue wildflower, is now endangered.

In addition, tourism in the Sand Hills and other range and pasturage areas, including recreational harvest of animals (hunting and fishing)—should help to improve the understanding and image of beef cattle, which, when fed on perennial grasses, are an environmentally viable way to capture solar energy for human consumption. In fact, while not extremely popular yet, lean, grass-fat beef could help meet the needs of beef producers and the demands of the emerging health consciousness in the general public.

Aquatic Habitat

Sediment Pollution

After an informed minority has harped on soil erosion for 60 years, while not yet a household word, the problem does get political attention and may even be somewhat better appreciated by the public. What is probably not as well understood is the effect of poor land-use practices on aquatic habitat.

The two main sources of damage to aquatic habitat are sediment from erosion and runoff polluted by fertilizers or pesticides. Of these, degradation caused by erosion (at least where running water is concerned) is far more severe and widespread. *(These pollution problems for aquatic plants and animals are examined in "Environmental Quality and Changes in Ecosystems," p. 168.)*

Short of protected status for the stream and surrounding land, the most effective management strategies are the menu of conservation strategies outlined in the section above on water erosion, particularly BMPs. And, fortunately, it would seem that more and more farmers are willing to adopt such tactics. This kind of management helps minimize both pollution by sediment and by agricultural chemicals.

Another, less publicized problem for stream water quality is streambank erosion. This is usually the result of livestock overgrazing and trampling the banks or of water deflected from its normal flowpath. Fencing livestock off the banks, except for a 75-100 foot gap, allows the bank to revegetate. Replanting cottonwoods, Russian olive, cedar and willows and reseeding with reed canarygrass, sedge or other plants will restabilize the bank. Water deflected from its course usually is caused by logs, brush piles or other debris. If not diverting water, these provide important habitat for insects, fish and other species. They should be removed, however, if they are causing streambank erosion or other flow patterns that enhance deposition of sediment.

The classic demonstration project in Nebraska regarding many aspects of water pollution—and sediment pollution in particular—is the Long Pine Creek watershed near Ainsworth. This watershed was having serious surface water and groundwater problems in the late 1970s. Water in town and rural domestic wells exceeded federal maximum contaminant levels for nitrates. Groundwater and surface water were becoming contaminated with pesticides, and sediment-laden runoff was fouling streams and degrading a world-class trout habitat. Bacteria in stream water, probably from livestock waste, was a hazard to humans and wildlife as well.

In 1981, the watershed was selected as one of 21 in the United States to participate in the experimental Rural Clean Water Project (RCWP). ASCS, SCS, the Game and Parks Commission, the Middle Niobrara NRD and the Cooperative Extension Service (CES) all cooperated in applying for RCWP designation and helped administer the program. Through the program, farmers and ranchers were eligible for $1.3 million in cost-sharing money to help them implement BMPs to control nonpoint pollution.

The extension service picked up the information and education duties and found that, even after farmers were encouraged to enroll in the program and 75-percent cost-sharing was provided, they were reluctant to participate during the depressed agricultural economy of the early 1980s. Eventually, however, contracts were written for 71 percent of the critical area. CES also assumed responsibility for nitrate and pesticide management, teaching and demonstrating ways to cut chemicals without sacrificing yields. Farmers throughout the watershed began adopting these practices. In addition, an integrated pest management association was established that still operates.

SCS provided technical assistance and wrote water-quality plans for each farm that incorporated BMPs. The North-Central Nebraska Resource Conservation District contributed by installing critical area treatments along roadsides to stop erosion. The Middle Niobrara NRD installed grade-stabilization structures at the headwaters to curb headcutting. SCS, the NRD and CES helped DEQ collect water-quality samples, and the Game and Parks Commission paid 25 percent of the producer's costs for practices that provided benefits to wildlife, most of which involved streambank stabilization. The Ainsworth Irrigation District provided machinery and labor for the installation of a secondary storage reservoir. While the program encountered various kinds of difficulties, in general it stands out as a model of interagency and landowner cooperation on a land and water management problem.

Obviously, aquatic habitat comprises any body of water that supports life, but for reasons of brevity, we have emphasized mostly streams and wetlands. Lakes, reservoirs and farm ponds also are subject to quantity and quality prob-

Nebraska State Historical Society

These ducks on a Sand Hills wetland were photographed by G. E. Condra, the first director of the Conservation and Survey Division.

lems and require specific management practices, but these are less extensive in the state and many of the management strategies for them are the same as for streams and wetlands. Surface-water *quantity* problems threaten habitat but generally in ways that are more obvious (dewatering often means a loss of habitat), and this has received much publicity regarding Platte and Republican river flows and corresponding reservoir levels *(see "Politics and Policy," p. 222/229).* Preservation of wetlands is another important surface-water quantity issue, and, as such, its history deserves some attention.

Wetlands Protection

In the past 20 to 30 years, scientists and environmentalists have begun to analyze, focus on and promote the various ecological and economic benefits to the public provided by the nation's wetlands. Countering the notion that wetlands were wastelands was not easy, but gradually a consensus has emerged regarding the ecological services they supply. Some of these include: waterfowl habitat for breeding, nesting, resting and feeding; habitat for other birds, including threatened and endangered species; habitat for furbearing mammals and other species; marine fish and shellfish production; freshwater fish production; timber production; flood control; pollution filtration; shoreline stabilization; and recreational and educational benefits.

In light of the recognition of such benefits, our society at large has begun to demand a halt to conversion of wetlands for farming or urban development. But for the private landowner pinched by environmental regulations and no better than mediocre crop prices, such a policy seems to single out agriculture unfairly to bear the the cost of a public good. A review of the history of wetlands policy in this country and state makes this conflict a little more understandable.

History of Wetlands Policy

In 1849, 1850 and 1860, the passage of the Federal Swamp Lands Act allowed 15 western states to "reclaim" about 65 million acres of wetlands. As it turned out, not all of these acres were wetlands, and some of the wetlands were not reclaimed. Nevertheless, this federal policy reflected the attitudes of European immigrants who believed that to clear forests and drain swamps would help create a strapping nation of yeoman farmers. Enough land for all to farm was seen as essential to the proper workings and economic well-being of a free, democratic society. To speak of the need to preserve wetlands would have seemed ridiculous in light of the vastness of the new country and its resources. Besides, often wetlands were seen as a mosquito-breeding, disease-spreading nuisance.

One hundred years later, the disappearance of Nebraska's wetlands was enough of a concern to the state Game and Parks Commission that in 1948 it began acquiring them in the Rainbasin region of south-central Nebraska. What had happened in the meantime? Among other things, Dust Bowl winds had blown sediment into many wetlands, filling some and diminishing others. Road and railroad building eliminated some. Many more were drained or filled for farming. And some were concerned enough to take action. In 1934, the federal Migratory Bird Hunting Stamp Act passed due to a decline in waterfowl. The profits from the stamp sales were used to buy land for refuges. And in 1937, Ducks Unlimited was founded, dedicated to the preservation of waterfowl habitat. But the main force at work in wetlands policy was the federal government. About the same time that one federal law was trying to secure more habitat, in 1936, the federal Soil Conservation and Domestic Allotment Act allowed producers to receive half the cost of the drainage work and free technical advice from SCS.

So we see the apparent contradictions in federal policy and social attitudes about wetlands that was to come to a head in section 404 of the Clean Water Act of 1974 and in the "swampbuster" provision of the 1985 farm bill. While this conflict merely represents a society with diverse needs and interests, one real irony here is that the increased gross domestic product made possible (at least in part) by the extra acreage farmed no doubt contributed (at least indirectly) to the extra leisure time that meant more recreational hunters were bagging more ducks amid declining numbers of waterfowl due largely to overhunting and diminishing habitat.

This conflict accelerated during the post-war period as large earth-moving equipment allowed for easier drainage and filling of wetlands. Accordingly, the first legislation restricting use of USDA money to drain wetlands was passed in 1962 in Minnesota and South Dakota. Several years later, other states, including Nebraska, were also included. By 1967, under the River and Harbors Act of 1859, the federal government empowered the U.S. Army Corps of Engineers to protect wetlands. An outgrowth of the U.S. Fish and Wildlife Coordination Act, this required any federal agency to consult with the U.S. Fish and Wildlife Service (FWS) before altering a waterway. FWS estimates that between the mid-1950s and the mid-1970s, about 458,000 acres of wetlands a year were lost. About 80 percent of the loss of inland wetlands was due to drainage and filling for farming.

Part of this loss resulted from the building of five large mainstem dams along the Missouri River during the 1950s and 1960s. Approximately 388,000 acres of riparian wetlands and oxbow lakes were destroyed by flooding. Many more of these began to dry up due to the channelization of the river that was also part of the plan that built the dams *(see "Politics and Policy," p. 212)*.

By 1972, through the Clean Water Act, the Corps and EPA were given joint authority to regulate wetlands; this authority was made more specific in the section 404 provision of the federal Water Pollution Control Act (known as "the Clean Water Act") of 1974, which mandated, as part of regulating the fouling of the nation's waters, that all landowners receive a permit from the Corps before altering a wetland. Legal recognition was given to the benefits wetlands provide in controlling runoff and flooding.

Phase II of section 404 was implemented in 1977, after the jurisdiction of the Corps over "the waters of the United States" was interpreted to mean all waters. Originally, the Corps did not interpret this phrase to include any water beyond navigable waters and their adjacent wetlands. The Natural Resources Defense Council filed suit, and the court held that the Clean Water Act provided for this expanded coverage. Opponents lobbied for an amendment that passed exempting certain farming, forestry and ranching practices. (The conflict lived on!)

A color-infrared aerial photograph taken over the Fairfield area, Clay County, on Aug. 28, 1984. Wetlands appear as irregular, oval shapes in blue-green and black. Cropped fields and other healthy vegetation appear bright red. Fallow fields are gray-green.

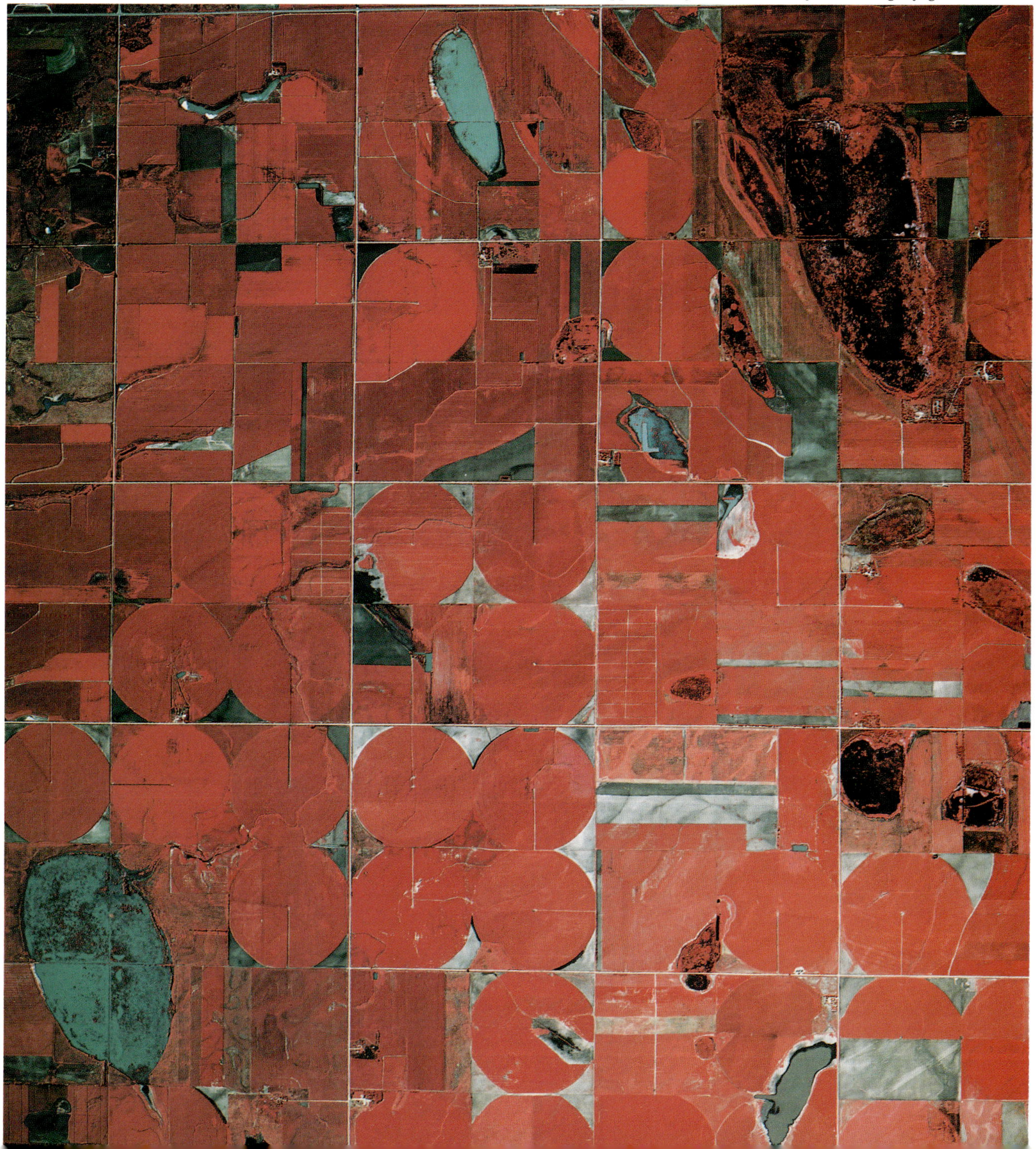

CALMIT, CSD

In December of 1980, EPA set down guidelines for section 404 that created a strong presumption against alteration of wetlands. In 1982, the National Wildlife Federation and other environmental groups sued the Corps over the regulation of small isolated wetlands. The Corps had sought a nationwide permit on such isolated wetlands. A settlement in 1984 allowed for permits on those more than 10 acres, a nationwide permit for those less than one acre and an intermediate procedure for those in between.

In the 1980s, the two most significant mileposts regarding preservation of aquatic habitat in Nebraska have probably been the passage of the 1985 farm bill, with its swampbuster provision, and the North American Waterfowl Management Plan. The 1985 farm bill, discussed above in the section on water erosion, also authorized a Wetland Reserve, similar to the CRP. Signups for this reserve program have taken place in Iowa, Minnesota, South Dakota and Missouri but to date have not occurred in Nebraska.

A coordinated effort between Canada and the United States, the $1.5 billion North American Waterfowl Management Plan seeks to restore waterfowl populations to the 1970-1979 levels (about 100 million fall-flight birds) by enhancing and protecting habitat in both countries. The Nebraska Rainbasin was singled out for a specific management plan—the Rainwater Basin Joint Venture waterfowl habitat plan—because of its importance as migration habitat along the Central Flyway. The Nebraska plan involves the state Game and Parks Commission, FWS, the Nebraska Association of Resources Districts and Ducks Unlimited and is expected to take 10 to 15 years to implement. It has three main goals: to protect, restore and create an additional 25,000 wetland acres; to secure enough reliable water for wetlands that a minimum of one-third of protected wetlands provide sufficient water quantity, quality and distribution to meet the needs of migratory waterfowl; and to develop wetland-enhancement strategies to improve existing habitat and optimize use of wetlands by migratory birds and threatened and endangered species.

In addition to the Rainbasin, Nebraska is blessed with two other major and a number of minor complexes of lakes and wetlands. These major areas are the riparian wetlands and river habitat of the Big Bend Reach (Lexington to Grand Island) of the Platte River and the Sand Hills wetlands and lakes in north-central Nebraska. The other complexes include the Missouri River habitat, the salt marshes of the Salt Valley, the alkali wetlands and lakes in the Panhandle, the Niobrara River habitat, particularly in the west, and the wetlands of the southwestern High Plains.

Another wetlands mitigation effort, the Nebraska Wetlands Priority Plan, is an outgrowth of the 1986 federal Emergency Wetland Resources Act, enacted to intensify cooperative efforts to conserve and acquire wetlands, and the National Wetlands Priority Conservation Plan of the Department of the Interior, which has similar goals and seeks to integrate these into state outdoor recreation plans. Developing uniform, scientifically based criteria is another goal of the plan. The priorities established have to do with three criteria: wetlands loss in critical ecoregions; wetlands threats—of loss or degradation; and wetlands functions and values—relating to diverse and important functions, unique functions and the broadest range of functions.

In seeking to mitigate loss of the Missouri River habitat, the Papio-Missouri Rivers, Lower Platte South and Nemaha NRDs are cooperating with the U.S. Fish and Wildlife Service, the Corps of Engineers, the Soil Conservation Service, the National Park Service, the state Game and Parks Commission, as well as with city, county and tribal governments, on a project to restore and/or create riparian wetlands, mostly through restoration of ox-bow lakes and management of wing dikes along the Missouri. In addition, as was mandated in the original Pick-Sloan Plan, the Corps independently has received authorization to mitigate loss of aquatic habitat caused by the plan, but has received no appropriations to date. There is some urgency here because of the threat of most of these drying up and becoming much too expensive to restore.

Another piece of farsighted management strategies regarding aquatic ecosystems that has not been well explored, but one that may well deserve more attention, is the need to find alternative ways to harvest wetlands. Wetlands are one of the most productive ecosystems on the planet. But our society has not done a superior job of learning how to harness that productivity without harming other wetland values. Alternative income-producing uses for wetlands have been, in various parts of the nation, wild rice, crustaceans (crawfish mostly), rough fish as feed, game fish, timber, cranberries, blueberries or trapping of fur-bearing mammals and leasing of duck blinds. More research could be devoted to helping producers find ways to harvest alternative crops from wetlands, and more emphasis could be put on persuading them that this could be profitable.

The last significant management strategy regarding preservation of wetlands to be mentioned here, but hardly the final in an exhaustive list, is the emergence of citizen monitoring groups. These are groups of educated volunteers who monitor mostly water quality and, to a lesser extent, quantity, by learning how to take samples, measure water levels and gage streams. While these groups have sprung up mostly around lake and river monitoring, some groups in a few states have organized to monitor wetlands. Wisconsin and Minnesota have such groups. None to our knowledge exists in Nebraska, but the interest and energy might be available.

Aquisition of wetland acreage by the Game and Parks Commission, FWS, the Nature Conservancy, the Audubon Society, the Platte River Trust and Prairie Institute also represent important efforts to stem the loss of riparian and upland wetlands. *(These are listed in the "Profiles" section at the end of the two chapters on environmental issues.)*

Biological Diversity

What many of these management issues point to is an appreciation for a broader picture: humans using water and other natural resources not in isolation but in a specific context, one that involves a myriad of other species, plant and animal, and all other earth systems.

Regardless of what we thought we were doing, we may have been acting as if our productive base were really narrower than it is. Not that living on this planet is as simple as uncritically embracing all of nature as our "friend," but as our understanding of the complexity of natural systems broadens, we confront ethical, economic and environmental issues that require renewed interest, if they are not actually new in themselves. One of these, emerging with quickening intensity, especially with regard to species-intensive regions such as rainforests, but in general as well, is biological diversity (usually termed "biodiversity").

From a strictly economic standpoint, two forces in our value systems come into conflict that have a direct impact on biodiversity and its loss. The first is that specialization of production generally lends a competitive advantage. A global exhange economy puts pressure on the production system to specialize—to produce what the region or company can do well, exchanging surplus production for other goods. In neoclassical economic terms, everybody gains. Most economists would also readily acknowledge that this system does fairly well in assigning value to private goods (or benefits) but runs into problems regarding public benefits.

Nevertheless, the purely economic problem—which captures much of but not all of the larger ethical quandary—is that time fosters the fruition of certain actions by offering more information than we currently have, primarily information about the consequences of our actions. Therefore, actions that preserve the maximum flexibility possible allow for a greater chance of exploiting that information. They are to be prized. Any decision that closes off future possibilities, such as an extinction (or extirpation, for that matter, or any degradation of habitat or watersheds that eliminates, for economic or environmental reasons, other future uses), abandons that flexibility. In addition, because technologies and political and economic pressures change, as well as tastes and preferences, this flexibility assumes a prominent place in market decisions.

Another complication with economic, ecological and ethical implications is that the way production efficiencies and the gains from trade are calculated—such as the potential losses to future generations through the use or depletion of resources in exchange for present gains—may be biased by an inaccurate discount rate. The discount rate is essentially the interest rate that puts a premium on a benefit now over a benefit deferred. Since many resources, even those we think of as renewable, soil, trees or a wetland with good quality water, may be barely renewable in human time, this conflict has an important bearing on how we pursue economic and environmental policies.

Furthermore, and perhaps this is the biggest wild card, the biggest joker in the deck, is that the potential flexibility inherent in biological resources due to genetic variability, particularly with the rise of biotechnology, is probably impossible to even roughly calculate economically. The other especially ticklish dilemma in examining loss of biodiversity, economics and ethics is that, for the most part, while economists can become experts at explaining what value people *do* place on an ecosystem (or species), they are not experts in explaining what value people *should* place on an ecosystem or species. This is usually reserved for areas of expertise that border on or delve into ethics: the humanities, philosophy or religion.

We have then, a relatively new approach to understanding all these issues that requires an integration of the natural and physical sciences, the social sciences and the humanities, including philosophy and religion. Probably the best news, in Nebraska and elsewhere, is that the need for this interdisciplinary work is coming into focus and is being heralded as the *wave of the future*, and perhaps, the primary *way to the future*.

Regardless of how much of a thicket the issues surrounding loss of biodiversity seem, when entertaining such questions, one concept emerges supreme. It is so much ballyhooed in recent times that it has almost become trite. But worn out or not, "the ecosystem" is not merely a cliche. It has become the fundamental unit of our understanding of the natural world. An understanding of biodiversity means understanding how to preserve and protect habitat for a wide variety of species. These are ecosystemic issues.

Ecosystems vary in size and complexity. Composed primarily of all the living things (the biological community) and their surroundings (the physical environment) and the relationships among them in a given area, they are the fundamental units of food and energy movement in the natural world. The consumption and transformation of energy, beginning with photosynthesis making sunlight into chemical energy stored in plants, is basic to what ecologists study. Animals who eat plants, and animals who eat the plant-eaters, move this energy through various levels of the food chain, or web, one that is always begin woven. Microorganisms help break down dead material and waste, returning it to nutrients in the soil, making the energy available to a lower level in the web. Generally, the more complex the food web is, the more stable ecologists consider it to be. Any destabilizing of these relationships, such as elimination of habitat leading to extinction, can have a destabilizing influence on the rest of the system. Just how destabilizing is a matter of much debate.

Regarding the debate over whether biodiversity lends stability to ecosystems, G. Carlton Ray of the University of Virginia, in an essay on "Ecological Diversity in Coastal Zones and Oceans," in the book *Biodiversity*, says, "More to the point is whether characteristic diversity confers some predictability to ecosystems. Behind this important question lies our definition of a system. Ecosystems are far from chance physical-biotic associations or mere hueristic creations [created for purposes of discovery], they are functional units in every sense of that term. But defining them presents challenges . . ."

Management of Non-protected Areas

Preserving natural communities is a more daunting task in Nebraska than in much of the West (where large tracts are federally owned), because most of the land is in private hands, so there are many hundreds of thousands of property-owners who are decision-makers, urban and rural, instead of a handful of governmental entities. But generally, there are two ways to make people behave responsibly in relation to their local ecosystems: education and law.

The first attempt to legislate regarding threatened and endangered species *and their habitat* was the passage of the Federal Endangered Species Act (FESA) in 1973. Previous legislation had provided for protection of species if they were being hunted or fished into extinction on federal lands, but with FESA, the intent was to protect their critical habitat. The crucial section 7 stipulation of FESA says that any federal agency planning any action is required to consult with the FWS to ensure the action is not likely to harm habitat critical to a threatened or endangered species. FESA's big test came in 1977 during a legal dispute over the Tellico Dam in Tennessee, the construction of which was stopped when it was determined that building the dam would jeopardize habitat critical to the snail darter, a three-inch fish unknown elsewhere at the time. Ironically, when the snail darter was later discovered at another spot, construction on Tellico was completed.

Following this confrontation, opponents to the legislation got an amendment passed in 1978 that inserted an exemption clause into FESA. Exemptions are studied by a three-person review board, which makes a recommendation to the Endangered Species Committee, a seven-person body made up of the Secretary of Agriculture, the Secretary of the Army, the Chairman of the Council of Economic Advisors, the administrator of EPA, the Secretary of the Interior, a representative from the National Aeronautics and Space Administration and a representative from the state in question. The committee must grant an exemption if the there are no reasonable and prudent alternatives; the benefits clearly outweigh the benefits of

Jon Farrar, Nebraska Game and Parks Commission

Native, warm-season grasses and woody vegetation on a ridge top on the Nature Conservancy's Rulo Bluffs preserve in Richardson County

possible alternatives and the action is in the public interest; the action is of regional or national significance; and the committee sets up mitigation requirements to minimize harmful effects on critical habitat.

The Nebraska Endangered Species Act was passed in 1975 with language similar to the federal act but with one important difference. Its language is what legal scholars call "absolute." Similar to the federal law, no state agency can pursue an action that, after consultation with the state Game and Parks Commission, is determined to jeopardize critical habitat. But the state law says the consultation is to ensure that this action *does not jeopardize* such habitat, rather than ensuring that the action *is not likely to jeopardize* critical habitat. The full implications of this difference have yet to be determined, but it is clear that the state provision is stronger than the federal. It is also clear that both have played important roles in Nebraska's water-development history, stopping projects outright, leading indirectly to their demise or forcing establishment of mitigating habitat or habitat organizations, such as the Platte River Trust. These policy decisions are explored in "Politics and Policy"*(see p. 223)*. In addition, the federal legislation is slated to run out in 1993, and no one can guarantee it will survive in its present form.

Beyond endangered species legislation, when contemplating ways to stop loss of biodiversity, preserves, refuges and other protected areas are usually the first efforts that come to mind. These are often valuable as models for researchers to study and, more importantly, as outdoor museums, zoos and encounter centers. (*The major protected areas in the state are listed at the end of the two environmental chapters.*) But in general, if anything can be done to stop widespread decline in biodiversity, it must go beyond a few isolated tracts of land.

One of the first orders of business must be to develop an adequate inventory of the plains biological resources, and, most important, an adequate understanding of the interdependence of species and earth systems, a functional, not a merely taxonomic, understanding. One starts with taxonomy, however, just as when learning a new subject, one starts with the vocabulary.

In its literature explaining some pilot projects in biodiversity along the central Platte River, the U.S. Fish and Wildlife Service likens this lack of an adequate inventory to trying to run a business without a good grasp of how much or what raw materials you're working with, no documentation of the manufacturing process and no management information system. Since this is no way to manage a business, or an ecosystem, FWS has been cooperating with other state and regional organizations to start such an inventory along the Platte.

Working with the University of Nebraska State Museum, the Nebraska Game and Parks Commission, Iowa State University, the Audubon Society, the Nature Conservancy and the Platte River Trust to establish coalitions and share people and effort, FWS has "invested" $150,000 in seed money and reaped hundreds of thousands of dollars in benefits toward habitat restoration, inventories and educational efforts. FWS explains that this effort is crucial because the river is "a haven of biodiversity," a place where species from the north, south, east and west come together.

Interestingly, this is true of much of Nebraska's grassland ecosytems. Whereas relatively few species of plants or animals are endemic to the Central Plains grasslands, many species from surrounding ecosystems coalesce, combine and occasionally provide unique remnant populations. This is one reason the Niobrara River Valley was promoted for and received designation as a Wild and Scenic River. Through the efforts of The Nature Conservancy and the American Rivers society, in cooperation with U.S. Senator J.J. Exon and the FWS, as well as other groups and some local landowners, a 76-mile stretch will be preserved. Of special concern was a 20-mile stretch that is a unique biological mix and a remnant of the moist, cool Pleistocene world, home to some glacial relicts, particularly plants.

In light of this lower endemism (being native to an area), there are fewer than 15 true grassland species listed or proposed as threatened or endangered. However, the ecotypic variation—genetic variation within species as they adapt to subtle variations in ecosystems—has probably decreased due to the loss of native grasslands to cropping, resulting in a loss of genetic diversity.

In view of the need to expand management options for biodiversity beyond the protected areas, management of intensely cropped and inhabited areas needs a look. Fortunately, for much of the Midwest and the eastern Great Plains, many management strategies, some new and some old, have evolved for enhancing habitat for wildlife and the species on which they depend. This generally depends on leaving some cover on the land. Even stalks and stubble, as with minimum tillage, provide greater food and cover for wildlife. So the use of a package of soil and water conservation measures developed for agriculture, including but not limited to reduced tillage, can have benefits for biodiversity. Others include buffer strips of grasses and annuals along cropped fields and wooded areas and in fence rows, which provide important "edge" habitat. Selective use and banding of herbicides and insecticides allow for greater survival of plant and insect populations, upon which most wildlife feed. Many of the practices of sustainable agriculture have great benefits for species diversity.

Regarding the "good news" about pesticide use, perhaps the most dramatic is that organochlorines, such as DDT, aldrin, dieldrin and heptachlor, have gradually declined in the food chain since many were banned or severely restricted once it was found their residues became concentrated in the food chain and were especially damaging to the reproduction of raptors and other birds, thinning their egg shells. Though other less long-lasting pesticides remain legal, in general, the benefits of sustainable agriculture discussed in the section on water erosion will contribute mightily to greater health and numbers of plants and animals, especially where they encompass control of chemical use and sedimentation. Smaller fields, crop rotations (which increase cover) and greater use of pasture, windbreaks and strip cropping all aid species diversity.

Wetland preservation efforts, as discussed above, also have a large positive impact on preserving aquatic plant, insect and molluscan species and the birds and mammals that feed on them. (*These benefits and the efforts to protect wetlands from conversion are discussed in the section above on aquatic habitat.*) For more information on how farm and city folk can contribute to greater numbers and diversity of grassland and woodland species, the Nebraska Game and Parks Commission has published the *Wildlife Habitat Improvement Guide* that details extensively how to manage range, pasture, cropland, woodland and urban areas for habitat benefits. These also generally benefit "less glamorous" species, such as insects, rodents and amphibians and reptiles, by emphasizing strategies that benefit game and non-game wildlife.

In addition, one of the mostly theoretical, but potentially far-reaching innovations with implications for biodiversity is the research going on at the Land Institute in Salina, Kansas. Researchers at this private, nonprofit research farm are looking into the cropping of a diverse group of perennial grasses and legumes as a way of pioneering the future of agroecology (a new term for agriculture, emphasizing cooperation with, rather than control over, nature). Director Wes Jackson and co-workers have modeled their test plots on the native prairie in an attempt to create a more nearly sustainable cropping method *(see "Environmental Quality and Changes in Ecosystems," p. 168)*. They call attention to the fact that native perennials are the plants naturally adapted to this ecosystem; they are what the land "wants" to grow. What we need to do is to learn, they say, how to harvest them efficiently and then reap the greater benefits in dramatically reduced soil loss and chemical use (as many desirable species crowd out undesirables—weeds), increased water conservation (as most are adapted to semi-arid climates) and more suitable habitat for other species.

A Paradigm Shift

Finally, probably the most significant, or at least overarching, development in conservation management involves a development in intellectual history known as a shift in paradigms. A paradigm can be understood as any unifying and organizing principle guiding a culture or people, a complex of related ideas that motivates and informs behavior. Much of what has informed the industrial revolution has been a model of being-in-the-world adapted from 19th-Century physics that was set aside by physicists early in the 20th Century. However, it continued to supply a major metaphor (or model) for the organization of Western society well into this century. Westerners have been especially good at breaking down natural, physical and chemical phenomena into smaller constituent parts so as to manipulate these phenomena, which are in reality cybernetic systems.

This way of looking at and manipulating reality, known by philosophers as "reductionism," is a mechanistic model that often assumes the whole is the sum of its parts. It is mostly deterministic (oriented toward predictable results through experiments that can be replicated) and has allowed for a great deal of short-term predictability in science and technology. It has been less effective in its capacity to predict broader, longer term consequences. Its biggest flaw ecologically involves its inability to deal with side effects, or as they are known in economics, "externalities," because they are usually not covered in anyone's cost accounting or pricing.

In physics, the discovery of molecular interaction and atomic structure fostered the notion that certain key mechanical principles governed all physical phenomena (sometimes referred to as "billiard-ball physics"). The corollary in chemistry was the discovery that all of the universe was composed of a few dozen elements making up the periodic table. In psychology its analogs were the "black-box," stimulus-response mechanisms of behavioral psychology and Freud's division of the psyche into the voracious appetites of the id and the hypercontrolling nature of the superego, requiring the mediation of the ego and the various "drive" mechanisms of hunger, sex and power to move this "ghost-in-the-machine" model. In sociology and economics, the corollary was the bureaucratic hierarchy modeled on the military, economies of scale and the specialization of the workplace, which produced the assembly-line approach to production, each worker responsible for a specialized, replicative task. In simplification lay mastery. The results of this mentality in agriculture were the monocultural methods of so-called "factory farming," with its emphasis on specialization of cropping, large, expensive equipment, economies of scale and extensive use of purchased, (specialized)

Charles Flowerday, CSD

Robert Manley, Nebraska historian, shows children at the 1992 Children's Groundwater Festival where the nearby Platte River is. Manley uses storytelling, song and period dress to create an an interdisciplinary experience of history and environmental education. Interdisciplinary approaches to environmental education, formal and informal, are likely to play a prominent role in the future of conservation management.

chemical inputs sometimes applied with the linear logic that ''if some is good, then more must be better.''

The theory of relativity and the world of subatomic physics opened the door to a much less deterministic, much more dynamic view of the physical world (''Is light is a wave or a particle—or both?'' ''string theory,'' and subatomic activity that was a dance of dozens of particles and antiparticles, for example). The notion of cultural relativity in anthropology represented an incursion into the perceived superiority of Western values and ethics. In psychology, Jean Piaget, the father of cognitive psychology originally trained as a biologist, who termed himself a ''genetic epistemologist,'' discarded stimulus-response theory and Freud's drive mechanisms and put activity at the center of behavior, mapping its development through various stages that integrated more and more sophisticated, interactive schemes of knowledge. Among his more important contributions were observations that logical patterns of thought emerge in increments over time, logic being the equilibrium toward which mental structures tend, and that the interaction between organism and environment extends into the realm of knowledge.

More recently, social scientists and organizational psychologists have called attention to the importance of subcultural systems in society at large, and, regarding the workplace, to non-hierarchical, more responsive ''clusters'' and working groups that combine and recombine as conditions and tasks change. Even evolutionary biology, which initially proposed that each species evolved to fit an individual niche, has been transformed by the notion that species coevolve, each dependent on many others. In addition, sociobiology and psychobiology have made inroads into integrating our knowledge of ourselves as social animals into our understanding of the natural world.

The implications for agricultural and natural resources is that sustainable agriculture and its close relatives represent an expression of this evolution in farming. This is not to say we are at the pinnacle of intellectual development in this or any other area, but rather that we are more willing to admit that all reality is composed of dynamic systems, ''clusters'' like ecosystems with fuzzy boundaries and cybernetic effects, that integrating agroecology into the natural environment represents a more ''objective'' understanding of the realities of the total environment than merely trying to poison pests to extinction (after which, other pesticide-resistant strains can develop), unearth soil (which expends ecological capital faster than it can be accumulated) and solely breed plants producing more and more dry matter (which no doubt increases yields but which may have reached its limits and which produces surpluses that decrease prices, another side effect).

What alternative agriculture would offer here is a focus on multiple strategies of pest control, especially biological ones (particularly rotation), maximum cover possible on the land to reinvigorate it and management for net-income instead of strategies focused on gross yield. Obviously, all social change occurs slowly, and none of this should imply that agricultural chemicals or intensive, monocultural farming should or will disappear soon, only to suggest that change is the only constant, but one that humans can help manage.

What we are highlighting here is an evolution in the history of Western thought as it has been exemplified in a limited mechanical model borrowed from 19th-Century physics to an organismic model based on the synergistic and interactive responsiveness brought to light by biological and ecological insights. It is not surprising that this should also find expression in the management of agriculture and natural resources. Instead of the linear, the cumulative, the replicative, we are more willing to embrace a world where the idiosyncratic, the site-specific, the information-intensive and narrative approaches to knowledge predominate. We are more willing to discard our need to assume an independent and omniscient perspective and are coming to grips with our objective status as subjects, as creatures who are dependent on and cooperators with the earth and its natural systems. In other words, it is only recently that industrial-age humans have begun to ask, ''*What does the land itself, the ecosystem, require?*''

Sources: *Agriculture and the Environment—The 1991 Yearbook of Agriculture*, U.S. Department of Agriculture, 1991, U.S. Government Printing Office, Washington, D.C.; *Base Document of the Nebraska Soil and Water Conservation Strategy*, Nebraska Natural Resources Commission, 1991; *Biodiversity*, E. O. Wilson, ed., 1988, National Academy (of Sciences) Press, Washington, D.C.; ''Conditions and Trends of Nebraska's Rural Land,'' Soil Conservation Service, 1986, Nebraska State Office; ''Going Wild With Soil and Water Conservation,'' Soil Conservation Service, 1985, Program Aid No. 1363, U.S. Government Printing Office, Washington, D.C.; *Nebraska Long Pine Creek Rural Clean Water Program—Ten Year Report, 1981-1991*, Agricultural Stabilization and Conservation Service in Brown County, 1991, Ainsworth, Nebraska; ''Groundwater Levels in Nebraska, 1990,'' Gregory V. Steele and Perry B. Wigley, 1991, Conservation and Survey Division, Institute of Agriculture and Natural Resources, University of Nebraska-Lincoln, and Nebraska District, Water Resources Division, U.S. Geological Survey; ''The Impact of Crane Watching,'' Jon Farrar, in Nebraskaland, v. 69, no. 2, March 1991, p. 8-23.; ''Law of Wetland Regulation,'' William L. Want, 1990, Clark Boardman Ltd., New York; ''A Manual on the Preparation of Special Groundwater Protection Area Action Plans,'' Robert Kuzelka, project leader, 1990, a contract report by the Conseration and Survey Division prepared for and published by the Nebraska Department of Environmental Control (Quality) and the Nebraska Association of Resources Districts; ''Missouri River Corridor Project'' (brochure), Lower Platte South, Nemaha and Papio-Missouri Rivers natural resources districts, undated; ''Nebraska's Policy Response to Groundwater Contamination'' (adapted from *Nebraska Policy Choices, 1987*, Center for Applied Urban Research, University of Nebraska at Omaha), Mary E. Exner and Roy F. Spalding, in Resource Notes, v. II, no. 2, Fall-Winter 1987-88, p. 19-21; ''Nebraska Wetlands Priority Plan (draft),'' adapted and expanded for inclusion in the 1991-1995 Nebraska State Comprehensive Outdoor Recreation Plan, Nebraska Game and Parks Commission; ''New Directions in Nebraska Water Policy,'' J. David Aiken, in Nebraska Law Review, v. 66, no. 1, 1987, p. 8-75.; ''NRD Faces Hard Choices After Brief Reprieve,'' Pat Larsen, in Resource Notes, v. 1, no. 1, Spring 1986, p. 2-3; ''Occurrence of Pesticides and Nitrate in Nebraska's Ground Water, 1990,'' Mary E. Exner and Roy F. Spalding, Water Center, Institute of Agriculture and Natural Resources, University of Nebraska-Lincoln; *One Earth, One Future—Our Changing Global Environment*, Cheryl Simon Silver with Ruth DeFries, eds., for the National Academy of Sciences, National Academy of Engineering, The Institute of Medicine, 1990, National Academy Press, Washington, D.C.; *Paying the Farm Bill: U.S. Agricultural Policy and the Transition to Sustainable Agriculture*, Paul Faeth, Robert Repetto, Kim Kroll, Qi Dai, Glenn Helmers, 1991, World Resources Institute, Washington, D.C. ''Rules and Regulations for the: Upper Republican Natural Resources District Ground Water Control Area (1991); Little Blue Natural Resources District Ground Water Control Area (1986); Upper Big Blue Natural Resources District Ground Water Control Area (1990),'' Nebraska Department of Water Resources; ''Rules and Regulations for the Implemenation of the Groundwater Management and Protection Act in the: Central Platte Natural Resources District (1988); Tri-Basin Natural Resources District (1989); South Platte Natural Resources District (Sidney Groundwater Managment Plan; 1990),'' Nebraska Department of Water Resources; ''Rainwater Basin Joint Venture Implementation Plan (draft), 1992, Rainwater Basin Joint Venture Management Board; ''Special Protection Areas: A New Nonpoint-source Management Option in Nebraska,'' Richard L. Ehrman, Martha L. Link and Jeffrey J. Gottula, in Journal of Soil and Water Conservation, v. 44, no. 2, March-April 1990; ''Toward a Unifying Concept for Agriculture,'' Wes Jackson, 1984, *in Agricultural Sustainability in a Changing World Order*, Gordon K. Douglass, ed., Westview Press, Boulder, Colorado, ''Twenty-two Years of Progress in Soil and Water Conservation in Nebraska,'' 1959, The Nebraska Soil and Water Conservation Committee and The Nebraska Association of Soil and Water Conservation Districts; ''Sustainable Agriculture—A Brighter Outlook for Fish and Wildlife,'' Ann Y. Robinson, 1990, Izaak Walton League of America, Arlington, Virginia; Unsettled Ground: An Assessment of the 1985 Soil Conservation Provisions and Future Policy Recommendations,'' Devorah Lanner, with Patricia Funk, Marty Strange, Kristie Thorp, 1989, Center for Rural Affairs, Walthill, Nebraska; ''Water Resources Development in Nebraska, 1991,'' U.S. Army Corps of Engineers, Missouri River Division; ''Wildlife Habitat Improvement Guide,'' v. 69, no. 1, January-February 1991, Nebraskaland special issue.

Interviews with: Carl Gangwish, irrigated crop and livestock producer, rural Shelton, Nebraska; Dick Mercer, irrigated crop and livestock producer, rural Kearney, Nebraska.

Profiles of Protected and Less Developed Areas (Alphabetically):

Audubon Society, National:

Lillian Annette Rowe Sanctuary: Gibbon, Nebraska. The sanctuary encompasses 2,219 acres of tallgrass prairie, riparian woodlands, wet meadows and Platte River channel. It serves as a main staging area for 500,000 lesser sandhill cranes and occasional whooping cranes on their annual spring and fall migrations. It seeks to preserve and maintain habitat for these birds and other wildlife. Some of their functions include: the control of woody plants on river sandbars vital for the cranes; restoration of native prairie on upland sites; and conducting various crane viewing programs during the spring. Access is limited for the rest of the year.

Niobrara River Sanctuary: Rock County, Nebraska. This sanctuary covers 218 acres, including a half-mile stretch of riverfront along the south bank of the Niobrara River. The sanctuary includes native prairie, 100-foot cliffs dropping off from wooded draws, and a creek through a 200-foot forested canyon. Cranes, prairie chickens, eagles, peregrine falcons, long-billed curlew, upland sandpipers and other wildlife find this habitat ideal for their needs. The Niobrara River Sanctuary seeks to preserve plant and animal communities, especially endangered species. Access is only by permission.

Bader Memorial Park Natural Area: Located 3 miles south of Chapman in Merrick County, Nebraska, it is part of Bader Memorial Park. Eighty acres are designated as the natural area within the 200-acre park and includes a stretch of the Platte River, riparian woodlands, sand pits and tallgrass prairie. From the natural area, visitors can observe the different stages of island formation and the invasion of woody vegetation—from recently exposed sandbars to forested islands and land along the riverbank.

The natural area is located at the east end of an approximately 100-mile stretch of the Platte known as the Big Bend and lies about 15 miles downstream from the heaviest concentrations of sandhill cranes (Grand Island to Kearney) during their annual spring migration and stopover.

Fish and Wildlife Service, United States:

Crescent Lake National Wildlife Refuge: Central Garden County, Nebraska. This refuge offers visitors a wide array of enjoyable recreational activities. Wildlife observation, hiking and photography are encouraged. A number of Sand Hills lakes occur within the refuge and a wide variety of wildlife, particularly birds, fish and amphibians and reptiles are generally present. These include antelope, bass, bluegills, Canada geese, chorus frogs, cormorants, coyotes, deer, eared grebes, garter snakes, grouse, horned larks and meadowlarks, painted turtles, pelicans, pheasant, sandhill cranes, sandpipers, snow geese, various toads, teal and Wilson's phalarope.

Fort Niobrara National Wildlife Refuge: Four miles from Valentine; The 19,122-acre Fort Niobrara National Wildlife Refuge provides habitat for many types of wildlife, but the refuge is managed mainly for the 300 buffalo, 40 elk and 275 Texas longhorns.

In 1879 Fort Niobrara was completed to help keep peace between the settlers and the Sioux Indians. But during its 27-year existence, the fort was essentially quiet. It was abandoned in 1906 and used as a remount station for cavalry. By 1912, all operations ceased, and the fort was dismantled. All that remains is a red barn, some old foundations and some earth works. In 1912, Mr. J.W. Gilbert of Friend, Nebraska, offered a few buffalo, elk and deer to the federal government if land could be set aside. The land of Fort Niobrara was so designated and became the Fort Niobrara National Wildlife Refuge.

The rolling Sand Hills and breaks along the Niobrara River provide excellent habitat for a great variety of wildlife and plant life.

When most people think of sanctuaries and refuges as a means of preserving biodiversity, they probably imagine natural processes taking their course without human intervention. However, many require careful, even difficult, management choices to mitigate losses. Here a worker burns woody plants to provide habitat for whooping cranes, sandhill cranes and other migratory birds at the Rowe Sanctuary near Gibbon, Buffalo County.

Jon Farrar, Nebraska Game and Parks Commission

In winter, bald and golden eagles hunt. Whitetailed deer and turkeys can be seen along the river. Mule deer forage among the open hills.

Wind shapes the Sand Hills where hardy grasses grow. The low precipitation, sandy soil and high wind prevent the establishment of trees. Numerous species of beautiful and hardy wildflowers grow here. A few pronghorn antelope can occasionally be seen. A prairie dog town greets tourists near the Visitor Center. The prairie dog town also serves as home for burrowing owls and snakes.

Rainwater Basin Wetland Management District: South-central Nebraska; The Rainwater Basin of south-central Nebraska covers approximately 4,200 square miles within the counties of Gosper, Phelps, Kearney, Harlan, Franklin, Adams, Hamilton, Clay, Polk, Butler, York, Seward, Fillmore, Saline, Nuckolls and Thayer. The land varies from flat to gently rolling loess plains. Other habitats, such as grasslands, freshwater wetlands and riparian habitat along drainages and creek bottoms are also within the basin.

The Fish and Wildlife Service manages 45 Waterfowl Protection Areas, encompassing 18,000 acres, within the basin. These areas are especially important for the spring migrations. Waterfowl numbers usually peak in late March to early April, and normally include 90 percent of the total mid-continent population of white-fronted geese. The Rainwater Basin management district also seeks to preserve pheasants, quail, rabbit, squirrel and deer.

Valentine National Wildlife Refuge: Near Valentine, Nebraska; During the 1930s conservationists recognized the need to protect a portion of the Sand Hills and their wildlife. In 1935, this 71,516-acre refuge was set aside. Many lakes, marshes and hilly grasslands provide necessary habitat for numerous types of wildlife. Blue-winged teal, mallards, pintails, gadwalls, redheads, ruddy ducks and shovelers nest here in large numbers. During the spring and fall migrations, 150,000 ducks of many species can be found here.

More than 260 species of birds use the refuge. Herons, terns, shorebirds, pelicans, and numerous songbirds use the refuge. Sandhill cranes fly over during their spring and fall migrations. Winter weather brings the bald eagle and golden eagle to hunt the prairie. Whitetail deer are found in the marshes and small woodlots, and the mule deer frequent the open hills. Muskrats and beaver build homes in the marsh and lake edges. Coyotes, mink, raccoons, skunks forage in the Refuge.

Fontenelle Forest and Neale Woods Nature Centers: The Fontenelle Forest Association, a private concern founded in 1913, established the Fontenelle Forest and Neale Woods Nature Centers. Located on the edge of Bellevue along the Missouri River, the Fontenelle Forest Nature Center is a 1,300-acre nature preserve with woodlands and marsh habitats. The forest has more than 17 miles of hiking trails and is home to at least 200 species of native birds and mammals. The land is also the site of the first trading post in Nebraska, established in 1822, and numerous ancient Native American dwellings. Situated north of Omaha, and straddling Washington and Douglas counties, the Neale Woods Nature Center is a 300-acre sanctuary comprised of woodland and prairie habitats on hills overlooking the Missouri River Valley.

Forest Service, United States:

Nebraska National Forests and Grasslands: Northwestern and north-central Nebraska; Nebraska has more than 351,000 acres of national forests, located in the central and western sections of the state. The Nebraska National Forest and Samuel R. McKelvie National Forest are two of 156 national forests administered by the U.S. Forest Service. The Nebraska National Forest at Halsey is the largest human-planted forest in the country, encompassing 22,000 acres of numerous conifer and hardwood varieties dating from 1903. The Bessey Nursery, the oldest active federal tree nursery, produces 3 million seedlings each year. The McKelvie National Forest, a human-planted conifer forest, covers 2,200 acres. The Oglala National Grassland, located in extreme northwest Nebraska, is one of only 19 national grasslands administered by the U.S. Forest Service.

The Multiple Use-Sustained Yield Act requires the forest service to maintain a balance between competing uses: production, (timber, cattle grazing, mineral and energy exploration, etc.); recreational uses (camping, hiking, wildlife viewing, etc.); and still preserve the essential character of these Forest Service lands.

Pine Ridge National Recreation Area: A spectacular 6,600 acres of undeveloped scenery, these rugged sandstone buttes emerge from the surrounding ponderosa pine forest dissected by small, clear streams in hardwood bottoms.

Soldier Creek Wilderness: This wilderness area encompasses 7,794 acres, and has a well developed trail system for both hikers and horsebackers. In July 1989, the 48,000-acre Fort Robinson fire burned extensively within this area. Since the area is being allowed to recover naturally, visitors can observe the remnants of the fire for many years to come.

Toadstool Campground: This small campground within the Oglala National Grassland features unusual geologic formations and valuable fossils. Also, a reconstructed sod house provides an interesting look back to when homesteaders used the only building material available.

Lincoln Parks and Recreation Department:

Chet Ager Nature Center/Pioneers Prairie Interpretive Center: Dedicated in 1963 to Chet Ager, superintendant of Lincoln's Parks from 1931 to 1940, the center was a 40-acre bird and wildlife sanctuary in the southwestern corner of Pioneers Park officially called the Chet Ager Bird and Wildlife Nature Study Center. All totaled, the center now has 56 acres with 2.5 miles of nature trails. The nature lodge and activities building have natural history exhibits and live animal displays. Plans are underway to develop the Pioneers Prairie Interpretive Center. Located at the west end of Pioneers Park, this area will incorporate much of the Chet Ager center and add new areas of native grasses, wildflowers and indigenous wildlife.

Wilderness Park: While not a true wilderness area as the term has come to be used, this park is still a place where little development has taken place (some hiking, biking and horseback-riding trails and a fitness trail, as well as a cross-country ski trail) and where much of the native ecosystem has been allowed to develop with less disruption from humans than in most parks. Purchased under the Open Spaces Act with federal funds, the 7-mile stretch along the Salt Creek bottoms southwest of Lincoln has been managed under an agreement between the City of Lincoln and Lancaster County as a minimally developed area.

Nebraska Natural Resources Districts:

Most natural resources districts in the state operate wildlife habitat management areas in conjunction with the Nebraska Game and Parks Commission. These are too numerous to list, but the state Game and Parks Commission or the Nebraska Association of Resources Districts, both headquartered in Lincoln, can provide information on these.

Nebraska Nature Conservancy:

The Nature Conservancy is a nonprofit, tax-exempt international membership organization dedicated to the preservation of natural diversity. Its global mission is to find, protect and maintain the earth's rare species and natural communities by preserving critical habitat. While some conservancy lands are transferred for management to other conservation groups, both public and private, the conservancy owns and manages some 1,400 preserves—the largest private system of nature sanctuaries in the world. The Nebraska Chapter of the Nature Conservancy, chartered in 1989, is a statewide organization with 3,100 members. The conservancy has assisted in the protection of more than 65,000 acres in the state. Of that number the conservancy retains ownership of more than 55,000 acres. Below is a list of the ten conservancy preserves and projects in Nebraska.

Graves Ranch: An 840-acre preserve in Garden County protects the world's largest known population of federally endangered Hayden's (blowout) penstemon, also known as blowout bluebells and native only to Nebraska.

Arapaho Prairie consists of 1,298 acres in Arthur County, about 25 miles from the town of Arthur. It is a natural prairie preserve of typical southwestern Sand Hills prairie—mixed and short grasses. Used for Sand Hills prairie research and education in conjunction with nearby Cedar Point Biological Station, it is managed by the UNL School of Biological Sciences from the Cedar Point Biological Station for teaching and research purposes.

Niobrara Valley Preserve: 54,000 acres in Brown, Keya Paha and Cherry counties and open to the public. The property extends for more than 20 miles along the Niobrara River. Here eastern deciduous and Rocky Mountain forests intermingle, and the site is also home to three distinct grassland communities: mixed grass, Sand Hills prairies and tall grass. The preserve also protects unique springbranch canyons on the southern canyon wall.

Platte River: 1,168 acres at several roost sites along the Big Bend area of the river, retained by the conservancy; an additional 6,000 acres managed jointly with the Platte River Whooping Crane Trust. It provides habitat for the spring staging of the sandhill cranes. Least terns and piping plovers also nest on the site's exposed sandbars. The acreage provides habitat for ducks, geese and a variety of wading birds.

Prairie Marsh: 160 acres in the Rainwater Basin, managed in partnership with the Nebraska Game and Parks Commission. The wetlands provide migratory bird habitat. The conservancy purchased the land and later it was acquired by the commission and turned into a state wildlife management area.

Willa Cather Memorial Park: 640 acres near Red Cloud in Webster County, open to the public. The acreage was purchased to protect southern mixed-grass prairie.

Pawnee Prairie: 320 acres in Pawnee County, acquired by the conservancy and transferred to the Nebraska Game and Parks Commission for use as a state wildlife management area. The tract of tallgrass prairie is home to upland game species and other nongame wildlife.

Rulo Bluffs: 284 acres in Richardson County; it protects southeastern upland forest. The property features some unique plants: yellow lady's slipper, rock elm and one-flowered cancer root. Wild goose plum and white oak also have been found at the site.

Cuming City Cemetery: 11 acres in Washington County, first acquired by the conservancy and later sold to Dana College. It protects tallgrass prairie and serves as a study site for Dana teachers and students.

Scotts Bluff National Monument: 40 acres in Scotts Bluff County, acquired as a monument addition by the conservancy for the National Park Service.

Nine Mile Prairie: Nine Mile Prairie is located inside the Lincoln city limits on West Fletcher Avenue next to the Municipal Airport.

Consisting of about 180 acres of virgin tallgrass prairie, it is valuable because it has not been plowed or otherwise disturbed. J. E. Weaver, a world-reknowned University of Nebraska agronomist, used the site for his life-long study of prairies. Nine Mile Prairie is managed by a committee appointed by the chancellor of the University of Nebraska-Lincoln. The land is owned by the University Foundation. Its primary use is research, but it is open to school groups and the general public by appointment.

Park Service, United States:

Agate Fossil Beds National Monument: Central Sioux County, Nebraska; Except for grazing livestock on the gentle hills, Agate Fossil Beds National Monument remains relatively pristine. The area is mostly grassland, with species such as prairie sandreed, blue grama, little bluestem and needle-and-thread. Various prairie flowers, such as lupine, spiderwort, sunflower and penstemon offer a wonderful multi-colored display. Cottonwoods and willows along the river provide habitat for birds and various animals such as mule deer, pronghorn, coyote, cottontail and the prairie rattlesnake.

At Agate are concentrated fossils of animals in beds of sedimentary rock formed approximately 20 million years ago through compression of mud, clay and erosional material deposited by water and wind. The Agate Beds, named from their proximity to rock formations containing agates, are under the grass-covered Carnegie and University hills. From the tops of these two hills, you can look down and see the Niobrara River some 200 feet below. Scientists believe that three-quarters of the fossil-bearing parts of the hills are unquarried. The Miocene fossil mammal bones are very abundant, covering numerous species, are very well preserved and contain numerous complete skeletons. The monument was established June 5, 1965, and comprises 1,970 acres. The nearby Stenomylus Quarry of 60 acres is included.

Homestead National Monument Native Prairie: Near Beatrice, Nebraska; The tallgrass prairie, containing numerous plant and animal species, is a very complex ecosystem. Once this prairie covered millions of acres; now it blankets only a few isolated pockets. A program to restore the prairie, one of the first in the country, began here in 1939. Today, many techniques, such as a controlled use of fire, are used to maintain tallgrass prairie. The monument is about 5 miles northwest of Beatrice.

Scotts Bluff National Monument: Near Scottsbluff, Nebraska; This monument contains 3,000 acres of prairie habitat and lies on the North Platte River 3 miles southwest of Scottsbluff. Scotts Bluff is a cross-section of the High Plains, rising 500 feet above the present Great Plains. From this 60 million-year-old monument, geologists have determined the origin of various sedimentary materials, deposited on the ancient plains by wind and water, and the approximate age of each layer.

Geologists have also looked at the erosion of the High Plains some 4 or 5 million years ago. The land began to erode at a faster pace than the formation of new strata. Certain concretions, in various patches near the surface, were more durable than the surrounding material. Known as the cap rock, this cap has saved Scotts Bluff from the same erosion as the surrounding area.

Scotts Bluff Monument contains mostly short and mid-length grasses, with a colorful variety of wildflowers. On northern slopes are Rocky Mountain juniper and ponderosa pine. Nesting in these trees are swifts and cliff swallows in summer, and magpies and kestrels all year. Rabbits, mice, pocket gophers, prairie dogs and squirrels live in the sod, trying to avoid foxes, badgers, coyotes and a variety of snakes. Deer still inhabit the area, but the once abundant buffalo, bighorn sheep and pronghorn are gone because of human encroachment.

Platte River Whooping Crane Habitat Maintenance Trust: Grand Island, Nebraska: Formed in December 1978, the trust is a private non-profit organization dedicated to the preservation of migratory bird habitat along the Big Bend reach of the Platte River (between Overton and Chapman). Specifically, the trust's mission is to protect and maintain "the physical, hydrological, and biological integrity of the Big Bend area so that it may continue to function as a life-support system for the Whooping Crane and other migratory species which utilize it." The trust also conducts research, manages land and water resources, purchases land and water rights, and conducts litigation in order to protect resources.

With financial assistance from the Nature Conservancy, the trust has purchased more than 8,000 acres, including 16-1/2 miles of riverfront. Its personnel have restored 9 miles of crane roosting habitat and 200 acres of grassland.

The Platte River is a major stopover for migratory birds using the Central Flyway. The broad Platte River, wet meadows and abundant food supply bring millions of birds annually. Six endangered species use the Platte: whooping crane, least tern, piping plover, bald eagle, eskimo curlew and peregrine falcon. In all, more than 300 species use the Platte, of which, 130 nest in the Platte River Valley. Seven to nine million ducks and geese, and one-half million sandhill cranes use this area during their spring migration. Noted naturalist Roger Tory Peterson includes the Platte River as one of the twelve best birdwatching areas in the nation.

Wild and Scenic River Designation on the Niobrara: In May 1991, 70 miles of the Niobrara River were designated by the federal government as wild and scenic under the Niobrara Scenic River bill.

This designation protects the river's natural values by limiting future development along this stretch but allows existing uses to continue.

Salt Marshes in Salt Valley: Nearly 1,000 acres of land containing remnant saline wetlands are protected by the Nebraska Game and Parks Commission on the Jack Sinn Memorial Wildlife Management Area on Rock Creek south of Ceresco. In addition, two other saline wetland areas near Ceresco are protected under perpetual conservation easements by the Lower Platte South Natural Resources District. Arbor Lake, a saline wetland just north of Lincoln, is managed for the city by the Nebraska Game and Parks Commission.

—Duane Mohlman, CSD editorial assistant; and Devorah Lanner, CSD editorial staff assistant, and editor, SCS Midwest National Technical Center

N. H. Darton (?)

The North Platte River during a high-flow period near Scotts Bluff National Monument from north of the river. The photo probably was taken by Nelson H. Darton, early geologic mapper of Nebraska, during the 1890s, before the building of Pathfinder Dam in Wyoming helped control such flows.

R. W. Hufnagle, Soil Conservation Service

A classic conservation practice, the contour plowing on this field holds rainwater in furrows, curbing erosion. Eight inches of rain fell on the J. B. Claasen farm near Beatrice on June 8, 1941, after more than 5 inches had fallen the day before, and, according to the Soil Conservation Service, little soil was lost.

Politics and Policy

by Steve Gaul

Public Power: Tri-County and Related Projects

Construction of the spillway for Kingsley Dam at Lake McConaughy

Nebraska State Historical Society

Stretching more than 3 miles across the North Platte Valley, Kingsley Dam impounds water reaching more than 20 miles up the North Platte River, the long expanse known as Lake McConaughy. If central Nebraska were a living organism, the Platte River would be its main artery and McConaughy its pacemaker. The reservoir stores and releases water for surface-water irrigation over a vast downstream area. An intricate system of canals, laterals and regulating reservoirs aids that delivery. The water-use system spawned by McConaughy ultimately provides vital groundwater recharge, a wealth of recreational benefits and reasonably priced hydroelectricity. It has changed the nature of both the river and the human enterprises downstream, providing the lifeblood for the agriculturally based economy of much of central Nebraska. But if McConaughy constitutes a pacemaker, the operation to install it was a long and difficult one.

Construction of an unidentified large-scale water development project somewhere in Nebraska; probably construction of the conduit moving water out of Kingsley Dam; March 21, 1938

Nebraska State Historical Society

To most Nebraskans the name McConaughy conjures the image of a lake, but in 1914 Charles McConaughy was one of a handful of people determined to turn the vision of a supplemental-water project into a reality. McConaughy, a grain merchant and mayor of Holdrege, had helped organize the "Tri-County Supplemental Water Association" to build support for a project to provide water in Phelps, Kearney and Gosper counties.

Interestingly, his first idea was to build subsoil moisture in the area by applying water from the Platte during its flood stages and other times in spring and fall. Then, as now, cost to farmers was a concern with dams, and this approach could avoid some of the costs of dams and other schemes.

The group had marshaled the political force necessary to obtain a U.S. Reclamation Service (now U.S. Bureau of Reclamation) study of the potential project, but then Charles Pease, chief engineer of the Tri-County Survey for the federal Reclamation Service, issued a damaging and highly questionable report that found water requirements of the project would interfere with previous appropriations. Although the report weakened public support for the project, McConaughy and George P. Kingsley, a Minden banker and project proponent, had just begun to fight.

They enlisted the support of Nebraska State Engineer George Johnson in 1915 and in 1922 organized the "Central Nebraska Supplemental Water Association." The group quickly pressed for a new survey of the Tri-County Project and in 1922 and 1923 made its first filings for water rights. Funding was obtained for a U.S. Bureau of Reclamation

Construction of the pipeline to move North Platte River water south of the South Platte River for irrigation; Oct. 23, 1937

Nebraska State Historical Society

survey, and in 1924 that survey showed that a project was feasible. Instead of subsoil storage of moisture, the report recommended irrigation in the growing season from water stored in two proposed Plum Creek reservoirs. It also included plans for hydroelectric-power production. The report was a landmark in the project's history. Project proponents now had a federal analysis to support their aspirations.

In 1925 Nebraska's Senator George Norris, a Republican from the McCook area, submitted a bill to the U.S. Senate for construction of Tri-County. The bill passed the Senate but failed to pass in the House of Representatives. Although it was 8 years before supporters took their next major step, when the state enabling act for public power and irrigation passed, interest in the project never died. Indeed, the history of the Tri-County project from 1914 until it began operating in 1941 is one of repeated setbacks followed by new political action and surveys. At no period was there inactivity, and this brief account can only sketch some of the complex and tortuous political roads that Tri-County followed.

Tri-County promoters were certainly never operating in a political vacuum. After 1925 Tri-County had three major obstacles to overcome. These were opposition from private power interests, opposition from promoters of other projects and competition for federal financing. However, the project was ultimately to have powerful allies—drought, the Depression and the New Deal.

Sadly, in October of 1929, long-time project proponent George Kingsley died. However, his son, Dr. Donald W. Kingsley, and Ralph O. Canaday were to help carry the torch into the 1930s. The decade witnessed increased opposition as the proposed Sutherland Project gained the allegiance of many irrigation proponents. Today, the Sutherland-project facilities are operated by Nebraska Public Power District (NPPD) and are closely coordinated with the operations of Tri-County, which is now formally known as the Central Nebraska Public Power and Irrigation District (CNPPID). However, in 1930 the opportunities to coordinate the two proposed projects were not as apparent as was the difficulty of

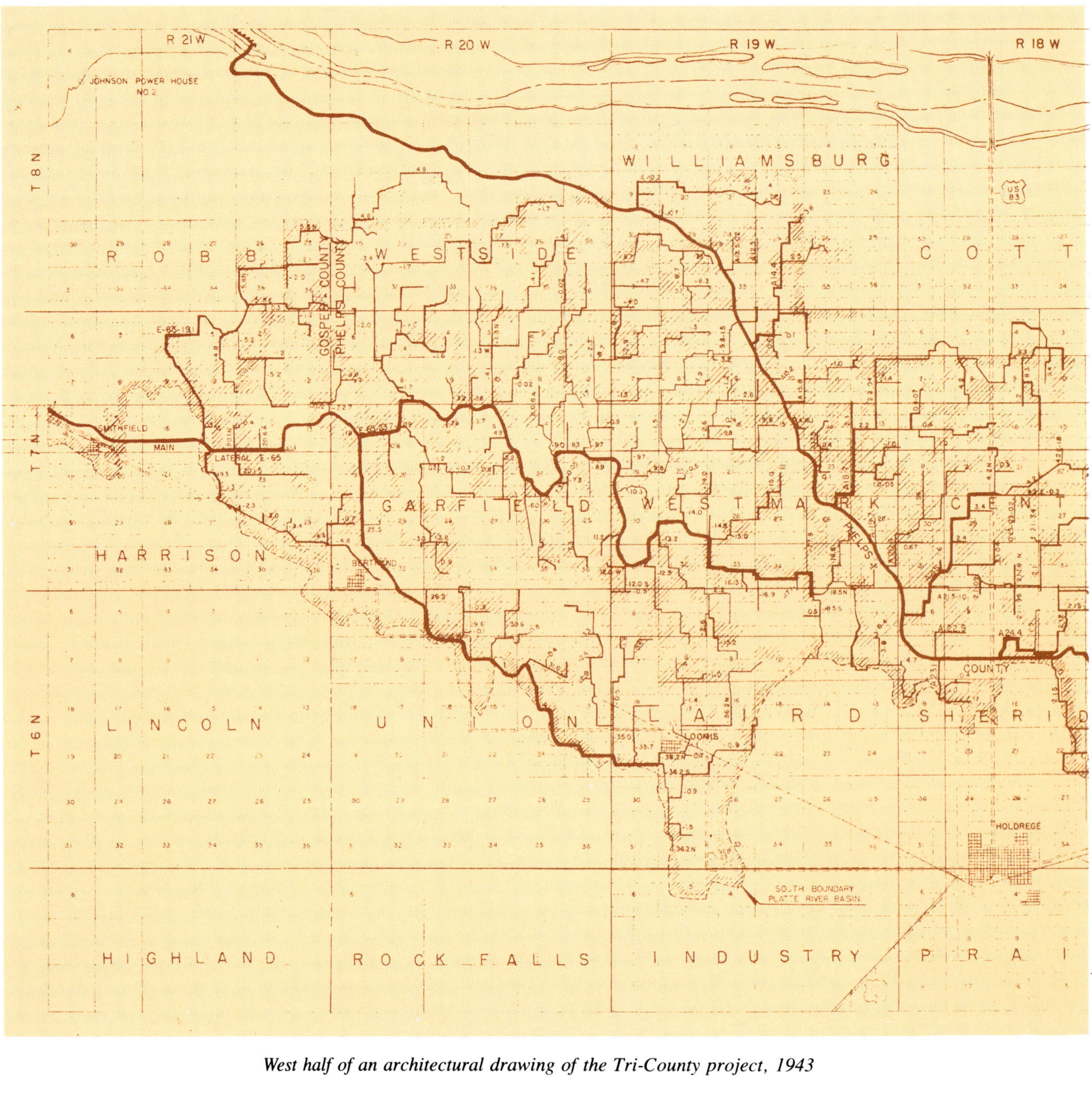

Central Nebraska Public Power and Irrigation District

West half of an architectural drawing of the Tri-County project, 1943

obtaining state water rights and federal financing for both.

The next breakthrough for Tri-County came in 1933. President Franklin D. Roosevelt was inaugurated, the federal Public Works Administration (PWA) was formed and the Nebraska Public Power Enabling Act was signed into law. On July 24, 1933, Central Nebraska Public Power District, or Central, as it is often called, was granted approval to organize. Through the first part of 1933, Sutherland and Tri-County had tried to negotiate their differences. Now, Tri-County and Sutherland backers broke into open competition.

Sutherland had the lead since that project had been partially processed by the predecessor of the PWA. On July 17, Tri-County filed a protest with the PWA asking that the Sutherland application be held up and the two projects studied together. Eleven days later Sutherland filed a protest against Tri-County's state water rights. The PWA denied Tri-County's protests and on November 3 approved the Sutherland Project. However, on November 17, the Nebraska Advisory Board to the PWA endorsed the Tri-County project and days later sent its application to Washington. Still, Sutherland proponents were to put up stiff opposition to Tri-County's passage through the federal bureaucracy.

Despite the immense progress of 1933, spring of 1934 found Tri-County at the brink of defeat. The PWA had decided to reject the project. But in early April, they sent two engineers to examine all Nebraska projects. One of those engineers, Major C. R. Oldberg, came to believe the project's dam site should be moved from the proposed Upper and Lower Plum Creek tributary sites in Gosper County to a Platte River storage site near Keystone. PWA officials postponed their decision while the application was revised.

Even the revised application was to be rejected by the PWA Financing and Engineering divisions. But in the best bureaucratic traditions, PWA's Electric Power Board of Review and Power Division were also to play significant roles in project approval.

Senator Norris intervened personally with President Roosevelt in June of 1934.

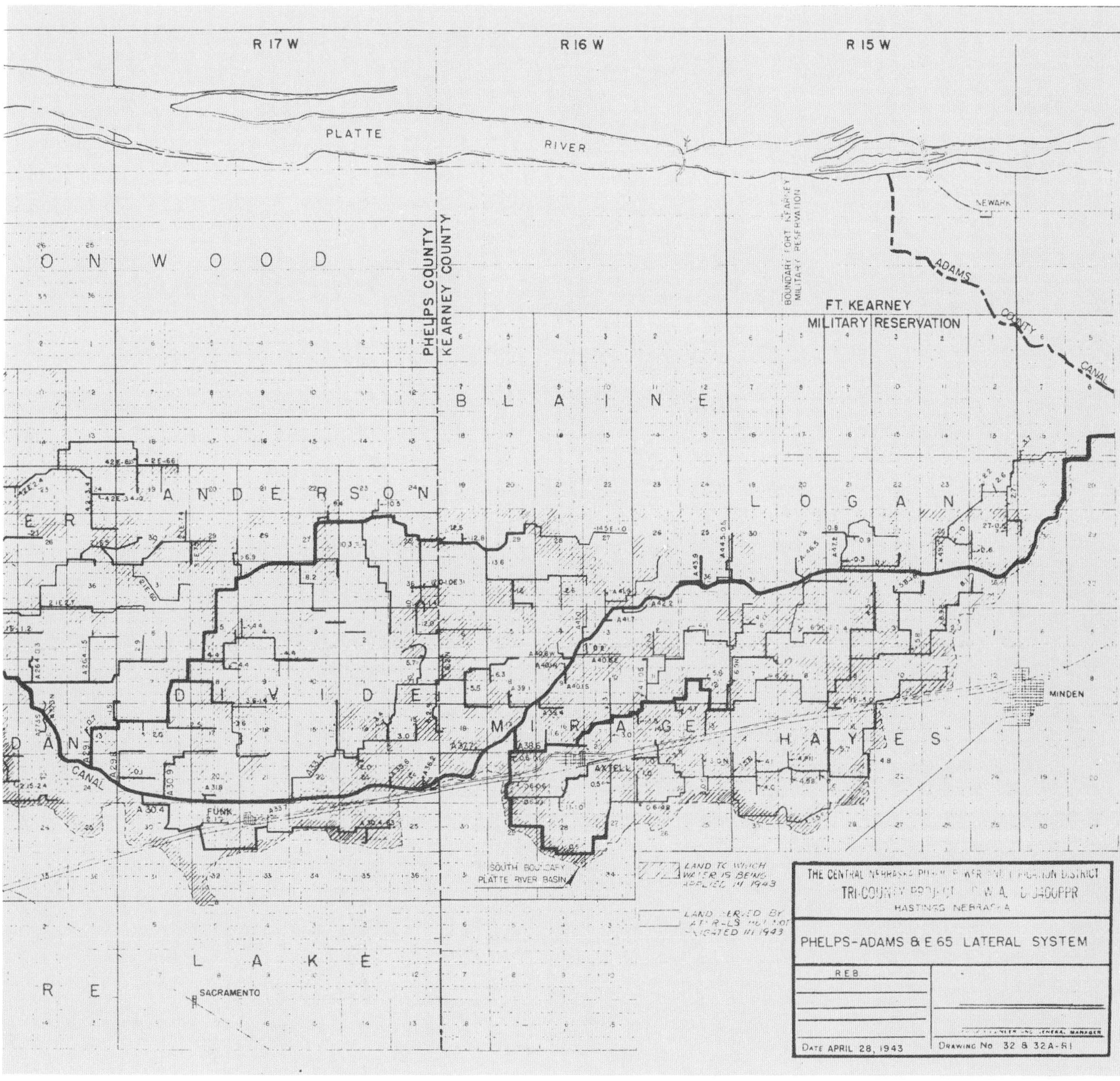

Central Nebraska Public Power and Irrigation District

East half of an architectural drawing of the Tri-County project, 1943

Although a Republican, Norris had considerable influence with Roosevelt. In 1932 he had publicly supported Roosevelt against his own party's incumbent, Herbert Hoover. Nevertheless, Roosevelt noted that "Nebraska's allotment of Public Works monies has been for some time over its quota" and said that he hoped the next Congress would consider the project. In January of 1935 another amended application was submitted to the PWA, and on August 1, still another Tri-County application was submitted, this time to PWA's newly organized Power Division.

Norris intervened forcefully with Secretary of Interior Harold Ickes. Ralph Canaday wrote, "Norris got his ire up and told Ickes that he was about done playing with him and from that time we had action." On August 21, Ickes asked that one of his assistant administrator's best men be assigned to speed the action on Tri-County.

George Norris's victory creating the Tennessee Valley Authority was in doubt until the U.S. Supreme Court upheld its constitutionality. The court said that, although it competed with private utilities, its sale of electricity grew naturally out of the government's right to build dams. This cartoon by John P. Robertson was published in the Washington Evening Star *Feb. 18, 1936 (adapted from* George W. Norris: The Triumph of a Progressive, *by Richard Lowitt, 1978, and reprinted courtesy of the University of Illinois Press, Urbana, Chicago, London).*

Journal-Star Printing Company

Charles W. McConaughy

George Johnson and J. D. Ross worked late into the night of August 23 and submitted the final application to the office of Secretary Ickes on the morning of August 24. The application contained the Plum Creek Reservoir but not the Keystone site. Before the day was over, they had seen Roosevelt at the White House and the application had received preliminary approval. The announcement was delayed two days, but then it was time for celebration.

Despite the rivalry with Sutherland, newspaper accounts were generally positive. An account in the *Hastings Tribune* trumpeted: "Well, it looks as though the Tri-County boys have scored a touchdown, doesn't it? Good Ol' George Norris carried the ball right through the middle of the line. Kingsley and Canaday sure ran some fine interference.

"Did you ever see two players get up off the ground so many times and go on to bowl over the opposition? They bounced off stonewalls so often I believe they must be made of rubber.

"Now all we need is the extra point. I hope Kingsley and Canaday handle the ball and let George do the kicking."

However, the "extra point" was contested. Arthur Mullen, a tireless worker for Sutherland and Nebraska's other U.S. Senator, Edward Burke, met with Roosevelt and presented strong objections to development of Tri-County. Burke was a Democrat and Mullen, Democratic national committeeman from Nebraska, had been Roosevelt's floor leader at the Chicago convention of 1932. But the President had deep respect for Norris, a progressive with a stubbornly independent mind. On September 18, Roosevelt gave final approval to the grant and loan for the project, the announcement coming on September 26. The extra point was good! Tri-County had won.

Only two obstacles remained to placing the victory in the books. One was the need to acquire state water rights. These were granted on November 4. But with Tri-County's long and tortured his-

Journal-Star Printing Company

George E. Johnson

Journal-Star Printing Company

George P. Kingsley

tory, someone was bound to contest the right and try to stop the project. Both occurred, but Tri-County prevailed. The five Nebraska private power companies that brought suit to stop construction may have sensed their own demise.

They achieved only one partial victory. A Nebraska Supreme Court opinion (Osterman vs. Central Nebraska Public Power and Irrigation District) handed down in 1936 refused to cancel Tri-County's water right as a whole but established that water could not be diverted out of the basin. This eliminated more than half its planned irrigation territory, including nearly all of Adams County, which by that time had been added. A part of the alternative was to build the Keystone reservoir and use power revenues to make up for some of the lost irrigation. December was also the month in which the loan and grant agreement between the district and the federal government was signed. At that time, the on-again, off-again Keystone reservoir site was reinserted as the "pacemaker" of the project.

The power companies were to continue to fight the project, but construction on the Phelps County Canal began on March 13, 1936. The project was three-quarters completed by the end of 1939. On Jan. 5, 1941, Norris flipped on the switch on the Jeffery Canyon Power Plant. He announced, "Here for the first time we are going to put the Platte River's waste water to use for the benefit of us all, to moisten our soil and produce cheap electricity." Kingsley Dam was dedicated July 22, 1941, and Tri-County was officially completed in 1943. Tri-County did run into early financial problems and required federal assistance in refinancing. But the economy of central Nebraska was never to be the same; the economic benefits to the area have been undeniable.

Through incredible persistence and determination a group of Nebraskans had wrought a stunning achievement. The politics they endured at home and in Washington can only be described as Byzantine. But their success is splashed across the face of central Nebraska. Names like Canaday, Johnson, Norris, McConaughy and Kingsley are known today mostly for public facilities named after them. But they were Nebraska giants in their time.

And what became of C.W. McConaughy, the visionary who worked so hard to bring his dream to life? Unhappy that the Keystone site had been chosen over the Plum Creek reservoirs, McConaughy resigned from the Tri-County board of directors on Dec. 7, 1935. "I take this action with a broken heart," he said. McConaughy died in April of 1941 while the reservoir bearing his name, and over which he resigned, was filling. His grave is in Lafayette, Indiana, but his monument covers 20 miles of the North Platte Valley.

* * *

With the completion of the Tri-County and Sutherland projects, Nebraskans had accomplished a mammoth task. That task should not be seen in isolation. Project backers had been working to develop major water storage on the Platte since before the turn of the century. The first major step was Congressional passage of the Reclamation Act in 1902. The act provided for construction of dams and irrigation systems in the western states.

In 1903 a proposal to provide irrigation in the North Platte Valley became one of the first six projects authorized under the act. Lake Alice and Lake Minatare were completed near the North Platte River in 1913 and 1915, respectively. Pathfinder, the first major Wyoming dam on the North Platte was completed in 1909. By the time McConaughy and Kingsley began campaigning for a supplemental-water project for the central Platte area, the benefits of both irrigation and water storage were very evident in the North Platte Valley.

Nebraskans were also refining the appropriation system they formulated in 1895. Several court cases in the early 1900s had clarified the rights of riparians

(Continued on p. 209)

THE WHOLE STORY

Not Irrigated Irrigated

WHY WE WANT A RESERVOIR

The "Kearney Ditch" has the oldest appropriation on the North Platte River, but what good is an appripropriation without water?

Nebraska State Historical Society

This undated depiction of the need for surface-water development projects shows what water storage meant to many Platte Valley farmers during irrigation's early days.

The Water War of 1935 Pitted the State Against Irrigators

In the summer of 1935, not enough water was available in the North Platte River to provide for all irrigation projects and, in accordance with Nebraska state law, those who held water rights with priority dates later than Oct. 17, 1889, were ordered closed. The priority of the 18,000-acre Mitchell Irrigation District was dated June 20, 1890.

Robert H. Willis, head of the state Bureau of Irrigation, Water Power and Drainage—now called the Department of Water Resources, sent a notice to all affected districts on July 10, 1935. Mitchell district officials ignored the order, contending the state of Nebraska could not regulate their district, as they had secured a water right under Wyoming law and diverted North Platte River water west of the state line in Wyoming. Mitchell officials had been making that claim for some 25 years, despite a 1931 Nebraska court ruling that said the district was located within the state and therefore could be regulated by it. The Wyoming government supported the district's decision to remain open and continued to deliver water.

Another notice was sent by Willis on August 7, this time informing district officials that a dam would be built if they refused to comply. The head gates remained open.

Keeping his promise, Willis ordered dam construction the very next day, and work began. But it was not easy. Men hired to build the dam waded into the canal, putting sandbags in place one at a time. Just as they began to make headway, district officials would open wide the headgates, and the onrush of water washed out the work of many hours.

Workers and district officials battled this way for six days. Finally, on August 14, the dam was complete.

On August 24, nearly 100 of the district's water users came out of the valley and cut a new channel by-passing the dam. A motion was filed in district court the next day charging district officials with contempt of a court order.

Mitchell Irrigation District directors and state officials met on August 26 to discuss the conflict. The district was given 48 hours to close their headgates, and water users were asked by district officials to sign a pledge stating they would not interfere with the construction of the dam and would accept Nebraska water administration laws. Irrigators refused to sign, and the headgates remained open.

Willis appealed to Governor Robert L. Cochran. Cochran reluctantly sent 180 guardsmen from Scottsbluff and Gering armed with riot guns and tear gas to protect the dam and help shut off the new channel.

"The scene near the dam was strongly reminiscent of war times. Armed guardsman watched the workers, whose bodies cast long shadows in the fitful light caused by the oil flares," said the *Scottsbluff Star-Herald* on Aug. 30, 1935.

Cochran also put the district under martial law; a military tribunal was established, and newspapers were ordered not to print anything against the guard, the state or the country. No one was permitted within a mile of the dam. Later, the act was said to be the climax of 40 years of disagreement about water-rights laws. The *Scottsbluff Star-Herald* called the day one of "sorrow" that would be a "blot" on the history of the county.

The guardsmen remained for five days, but no incident was reported. They were sent home on September 5, and local sheriffs were given the duty of protecting the dam.

District officials were resigned to having their say in court, but they fared no better there. Their claim to be outside Nebraska jurisdiction was denied by the U.S. Supreme Court on March 31, 1936. On May 31 of that year, the state ordered the Mitchell District closed until they applied for and were granted a Nebraska water right, which they finally received in November 1940, with a priority date of June 20, 1890.

—Brad Rundquist, weekend and copy editor,* Ft. Dodge (Iowa) Messenger, *and formerly CSD editorial assistant

Future site of Pathfinder Dam, Wyoming

Nebraska State Historical Society

Nebraska's Greatest Water Project . . . In Tennessee?

"Through all the history of water runs the history of George Norris"
—Mari Sandoz

George Norris was not an exceptional orator. He was unobtrusive and unpretentious. In a time when political parties were of increasing importance, he did not follow a party line. But he was effective. Norris was an outstanding parliamentarian, a bastion of integrity and a towering figure on the national political scene. In 1933, in his 21st year as a U.S. Senator from Nebraska, as the state and nation faltered, Norris was hitting his stride.

As a member of the progressive wing of the Republican Party, George Norris of McCook had long been a champion of the common people. Now Depression, unemployment and plunging crop prices were decimating living standards and paving the way for legislation he had long supported.

In addition to his progressive leanings and his less-than-flashy style, another irony was that a senator from Nebraska, a state that now ranks second in the nation in irrigated acreage, would be the moving force behind a major water project in a humid region. Seven southeastern states are now served by the Tennessee Valley Authority (TVA). Norris introduced the Senate bill for the TVA and was largely responsible for shepherding it through Congress. But Norris' fight for public power went back long before 1933.

In 1913 Norris supported a bill to require direct public distribution of power from a reservoir in the Hetch Hetchy Valley near Yosemite National Park in California. In 1921 a dispute of over whether the Muscle Shoals holdings in Alabama, including partially constructed Wilson Dam, should be leased to private interests, was turned over to the Senate Agriculture Committee, of which Norris was chairman. In May of 1922 Norris introduced a bill providing for government ownership of all Muscle Shoals property. Years of political wrangling ensued, pitting Norris against the vetoes of presidents Coolidge and Hoover. But Norris never gave up. A new president won the day for his cause. In an April 10, 1933, address to Congress, President Franklin D. Roosevelt not only supported public ownership of Muscle Shoals but proposed legislation expanding the concept and applying it to the entire Tennessee Valley. The next day Norris and Representative John McSwain of South Carolina simultaneously introduced bills in the Senate and House proposing creation of the TVA. Roosevelt signed the Muscle Shoals Bill on May 18, 1933, and gave the pen to Norris.

Today the TVA is still probably the federal government's most ambitious attempt to develop the resources of a region. Its detractors have seen it as an example of socialism and questioned whether it was worth the cost. But its benefits have been visible. The harnessed rivers have provided vast sources of hydroelectric power and low electric rates to consumers. A 9-foot-deep channel 650 miles in length has provided navigation. A multitude of lakes have created tremendous recreational opportunities. The project has helped reforest more than 1 million acres. In short, the TVA is one of the greatest water developments in the world. It is also a monument to the man many historians consider one of the best senators our nation has ever known, Nebraskan George Norris.

—Steve Gaul, water resources planner, Nebraska Natural Resources Commission

(Continued from p. 207)

versus appropriators. (Riparian use basically means the landowner adjacent to the stream can use the streamflow as long as it does not interfere with any downstream use. This was inherited from England and worked relatively well there and in the eastern United States, especially with regard to water power. Appropriative rights, a system that evolved in the western United States, means that first in time is first in right.) And in 1923 the South Platte Compact divided waters of the South Platte between Nebraska and Colorado.

The political climate of the era was still tinged with the ideals of a populist

(Continued on p. 211)

Pathfinder Dam, Wyoming

Nebraska State Historical Society

North Platte Decree Began in Argument Arising Out of 1934 Drought

When the tremendous drought of 1934 was at its peak, Wyoming water users held back all the water they could, leaving little for Nebraska irrigators downstream.

Although there had been shortages before and argument arose between the two states with every dry spell, these would die down when the rains came. In 1934, though, the rains didn't come and Nebraska demanded a fair share of the water.

The demand was made at a conference held in Wyoming that summer. Paul F. Good, the Nebraska attorney general; R.H. Willis, head of the Bureau of Irrigation; and Robert L. Cochran, state engineer and future governor, asked Wyoming officials to recognize senior rights to water in Nebraska. It was the contention of the Nebraska delegation that projects with senior rights should get water before subsequent rights in Wyoming. Water should be appropriated on the "first-in-time, first-in-right" basis regardless of state lines, they said. In fact, the U.S. Supreme Court had ruled that states using the same river for irrigation must share the water according to distribution rights. That ruling was made, ironically, when Wyoming demanded more water from Colorado.

But Wyoming officials said that with no compact or decree, it would not close down subsequent appropriations in favor of prior Nebraska appropriations. Nebraska's request for water was "refused in such terms that it became very, very evident that no relief could be obtained otherwise than by suit," Good later said.

In October 1934, Good received permission from the U.S. Supreme Court to file suit, with the purpose of forcing Wyoming to recognize Nebraska priorities.

In December that year, Good, speaking in North Platte at a meeting of the Nebraska State Irrigation Association, said, "If we were in Europe, or almost any other part of the world right now, we would be engaged in a war with Wyoming, probably. I can imagine our troops having first occupied Torrington, advancing up the river to Casper and then having captured the Pathfinder Dam and assuming control, all for the purpose of protecting our rights over a dispute that has arisen."

Nebraska also asked the court to make Wyoming change the priority date of a project being constructed. A permit for the construction of the Casper-Alcova irrigation project, now called the Kendrick project, was filed in September 1934. Even so, the Wyoming state engineer entered a priority date of Dec. 6, 1904 for it. His reasoning was that in 1904 an application was made for an irrigation project in the area and the new application was but an amendment to the older application. However, the new application changed the land description, point of diversion and point of storage and switched the area to be irrigated from south and east of the river to north and west. Nebraska officials wanted the date changed since there were 60,000 to 100,000 acres of irrigated lands in Nebraska with priorities after 1904 but prior to 1934.

Wyoming failed to answer the suit but instead filed a motion to dismiss it on the grounds that Nebraska had not brought Colorado into it. Colorado also uses water from the North Platte for irrigation.

Before Nebraska filed an answer to the motion, Wyoming agreed to change the priority date of the Casper-Alcova project, although it made no difference to them since they refused to recognize Nebraska priorities anyway.

Nebraska filed an answer to the motion to dismiss in March 1935 and the U.S. Supreme Court overruled the Wyoming motion in April. Wyoming finally filed an answer to the original suit.

On Oct. 14, 1935, the Court appointed a Minnesota attorney as special master. A special master is appointed by the court in original actions between states to receive evidence and hear testimony and arguments. At the end of all the hearings, the special master issues a report or recommendations to the full Supreme Court. Parties involved may file exceptions to the report and briefs are submitted. The court then hears oral argument and makes a decision.

Soon after the appointment of the special master, Wyoming's lawyers decided to enter an amended answer. They sought to further delay the case by asking that Colorado be brought in. They also claimed in the new answer that Nebraska had no right to more water because it already let too much go to waste with no major dams or reservoirs on the North Platte River. They also said they intended to prove that Nebraska lands east of Oshkosh had no need for irrigation.

Wyoming officials agreed that priority should rule within a state but beneficial use should govern interstate rights. They said they could make better use of the water and should, therefore, be able to keep more of it.

The first testimony was finally heard in July 1936, as it had taken time for Colorado lawyers to prepare. Colorado officials were mainly concerned with retaining as much water as possible for the North Park region and future development.

The U.S. government entered the case in 1938 to assert its ownership of all unappropriated water.

When all was said and done in December 1941, the testimony filled nearly 30,000 pages and made use of about 1,200 exhibits. The Minnesota attorney filed his report in September 1944.

The Supreme Court handed down its decision on Oct. 8, 1945, expanding on its previous decision in the Wyoming vs. Colorado case. In Nebraska vs. Wyoming, the court fully defined the concept of equitable appropriation.

The Court placed restrictions on the amount of North Platte River water Colorado irrigators could divert, store or export and limited the amount of land that could be irrigated in the North Park region. The Court placed similar restrictions on Wyoming. In effect, the Court required Wyoming to allow a sufficient flow of water to pass in the North Platte River during the

(continued on next page)

The North Platte Decree represents the first legal compact between Nebraska and Wyoming regarding North Platte River flows. Wyoming irrigators held back all the water they could during the 1934 drought, forcing a legal battle. Here, C.M. Corr, 12 miles north of Riverton, Wyoming, uses electric conduit pipe as siphon tubes in the summer of 1941. This was one of many materials farmers tried for siphon tubes.

Nebraska State Historical Society

irrigation season. Under the decision, Nebraska was entitled to 75 percent of that flow.

In 1951, Colorado officials notified those in Nebraska and Wyoming that it would seek to have restrictions imposed by the 1945 decree lifted or modified. They said irrigators in the North Park region needed to put additional acres under irrigation to maintain the economy. A stipulation was made in October 1952 authorizing the construction of the federal Glendo project. It also allowed Colorado to divert water sufficient to develop an additional 10,000 acres.

Nebraska filed suit against Wyoming in 1986, alleging the 1945 decree would be violated if Wyoming was allowed to construct its planned dam and reservoir on Deer Creek, a North Platte River tributary. Wyoming claimed the project was exempt from the decree because water stored would be used for municipal rather than irrigation purposes.

Additional issues surfaced shortly thereafter. Wyoming contended that streamflows in the Laramie River, another Platte tributary, were not regulated by the 1945 decree. Wyoming officials said they could, therefore, use as much of it for irrigation as they chose. Nebraska claimed that in accordance with the decree, it was entitled to 75 percent of the Laramie flow.

Wyoming also questioned Nebraska's right to divert North Platte River water into Lake Alice, Lake Minatare and Winters Creek Lake, located in northeastern Scotts Bluff County, and argued that Nebraska farmers living near the Nebraska-Wyoming border were not being regulated in accordance with limits placed on them under the 1945 decree. In effect, Wyoming officials were trying to gain the use of more water by requiring Nebraska irrigators to use less. Nebraska asked the U.S. Supreme Court to declare a Dec. 9, 1904, priority for the lakes to store 46,000 acre-feet of flow during October, November and April.

The U.S. Supreme Court appointed a special master to preside over the negotiations in June 1987. Oral arguments were heard in June 1991.

In preliminary rulings issued in February 1992, the special master ruled that Nebraska does have a right to divert water into Lake Alice, Lake Minatare and Winters Creek Lake and that there are no restrictions under the 1945 decree as to how much water Nebraska farmers along the Nebraska-Wyoming border can divert. The special master also ruled that Nebraska is entitled to historical flows in the Laramie River during both the irrigation and non-irrigation seasons in accordance with the 75 percent for Nebraska and 25 percent for Wyoming split established in 1945. The special master did not rule on construction of the Deer Creek Dam.

Arguments to the rulings were heard on March 9, 1992. The special master is now preparing a final ruling and will send his recommendations to the U.S. Supreme Court. The Court's ruling will likely be made sometime in 1993.

Also, in March 1991, Nebraska filed a motion with the special master requesting that the 1945 decree be extended to regulate the appropriation of non-irrigation-season streamflows. The master denied the motion, and Nebraska filed a similar motion directly with the U.S. Supreme Court later that year. The motion is currently pending before the Court.

—Brad Rundquist, weekend and copy editor,* Ft. Dodge (Iowa) Messenger, *and formerly CSD editorial assistant

(Continued from p. 209)

movement that had faded to a more benign progressivism. The support for the common farmer and the distrust of large businesses and monopolies that had characterized populism were still evident. However, progressives had replaced angry rhetoric with more practical ways of getting government to help the common worker. Water projects represented such a way for the average farmer. This was certainly one explanation for the politics of George Norris. But even more important was Norris' abiding interest in public power.

Norris holds a special place in any history of Nebraska and certainly in the history of Nebraska water and power development. He began his 40-year legislative career in 1903, initially serving in the U.S. House of Representatives. By 1942, when his bid for a sixth term in the U.S. Senate was defeated, Norris was one of the most prominent and respected men in American politics, and for good reason. His record included sponsorship of legislation resulting in: the Tennessee Valley Authority (TVA) *(see vignette on TVA, p. 209);* putting the Rural Electrification Administration (REA) on a permanent footing; the 20th Amendment to the U.S. Constitution, which abolished the lame-duck session of Congress; and labor reform (the Norris-La Guardia Anti-Injunction Act). Norris also played a major part in Nebraska's creation of the nation's only unicameral, non-partisan state legislature. His long-term interest in public-power issues was reflected in the interest shown in public power by members of the Nebraska legislature.

While it is difficult to find Norris' fingerprints on all of the major water legislation affecting Nebraska in his lifetime, his work provides the backdrop for much of what happened. Although Norris played no direct role in passage of the Nebraska Public Power Enabling Act, much of his legislative career was dedicated to public power. His advocacy of legislation creating a Mississippi River Valley Authority was not successful, but the cumulative effect of public-power systems in Nebraska bears some resemblance to a little TVA. When the Tri-County Project needed help at the federal level, Norris provided it. Although Tennessee receives 98 percent of its power from public sources, Nebraska is the only state in the nation with an all-public power system. It is no coincidence that Nebraska is also a state that spawned and was shaped by the career of George Norris.

While the backing of Norris and the dogged persistence of supporters such as McConaughy and Kingsley must be seen as reasons for Tri-County's success, other major reasons stem from the national politics of the time. The Depression had ushered in the greatest program of public-works spending in U.S. history. The federal administration was very sympathetic to water projects, both for their direct benefits and as a means of public-sector employment. Nebraskans had been hit with both the Dust Bowl drought years and a massive flood in the Republican Basin in 1935, events that likely were not lost on the federal administration. Nor had Nebraska hurt itself with the administration in its own political manueverings. Passage of the state Public Power Enabling Act in 1933 set the stage, election of Democratic governors for the terms between 1931 and 1941 helped and legislative championing by Senator Norris secured Nebraska's influence in Washington.

To obtain funding for construction, Tri-County supporters overcame powerful opposition. Initial opposition was based on a belief that there was insufficient flow. Subsequent reports did not support that view. Private-power interests opposed Tri-County, but within a few years they no longer existed. Some supporters of the Sutherland project opposed Tri-County both because it could compete for funds and might compete for water. There were water and funds enough for both, it turned out. Some politicians were skeptical of the project. Many were later convinced. Even conservative farmers were doubtful, but many were later to become project supporters.

The completed Tri-County project had

U.S. Army Corps of Engineers

Glenn Sloan and Lewis Pick

a way of making believers out of people. It employed people in the Depression and made electric power available at reasonable rates. It raised farm income and provided lakes for recreation. Most of all it provided green fields in an area that had suffered the ravages and horrors of the Dust Bowl.

The benefits of Tri-County, the persistence of its supporters and the importance of federal financial backing were not lost on those who were to advocate water-development projects in the coming decades.

Progress and Complacency, 1944-1966

A shotgun wedding was taking place on a fateful October day in Omaha in 1944. The betrothed were understandably reluctant about the ceremony. It was no love match. But on October 16, recently promoted Brigadier General Lewis Pick of the U.S. Army Corps of Engineers and William Glenn Sloan, assistant commissioner of the U.S. Bureau of Reclamation, were about to have their names and their water-development plans wed into a single entity.

The Pick-Sloan Missouri Basin Plan would set the water-development agenda for the Missouri River basin, including Nebraska, for years to come. The previous year Omaha had been devastated by a major flood, so it made an especially appropriate site for the merger. Interestingly, Pick was off in Burma building the Ledo Road (sometimes referred to then as "Pick's Pike") and could not attend this marriage ceremony, but he certainly understood the necessity.

As division engineer for the Corp's Missouri River Division, then-Colonel Pick had proposed and promoted a Missouri Basin water-development plan that placed a relatively high priority on navigation and flood control and would leave the Corps in control of many projects and project functions. Sloan and the Bureau were preparing a plan that placed relatively greater emphasis on irrigation and upstream power, as well as other "beneficial uses" such as silt control, fish and wildlife preservation and recreation.

The shotgun to marry these two proposals was "held" by President Roosevelt, who had supported the idea of a Missouri Valley Authority. Since neither agency found that idea acceptable, they had finally agreed to negotiate in Omaha. Their compromise marriage was to be ratified by Congress and signed into law barely two months later as part of the Flood Control Act of 1944.

The Pick-Sloan compromise was a marriage of convenience as well. As "dowry" both agencies demanded and received almost all of the projects proposed in their plans. The areas of compromise involved split allocation of duties on some projects and providing sufficient water for both navigation and irrigation. The marriage generally gave the Bureau jurisdiction for reservoirs and other facilities on tributaries and preserved Corps responsibility for components on the main stem of the Missouri. In total the plan called for 98 reservoirs storing 85 million acre-feet of water plus miles of levees and floodwalls throughout the basin. By 1988, 55 of those lakes were either completed or under construction.

The Pick-Sloan Plan was to have profound effects throughout the Missouri Basin. The main-stem dams on the Missouri were to flood hundreds of square miles in the upper basin and provide benefits from hydropower, flood control, recreation, navigation, and, in some places, irrigation. Although the cost of construction was high, reservoirs throughout the basin have provided numerous benefits.

In Nebraska the offspring of the Pick-Sloan marriage helped turn the era between 1944 and 1969 into what some

U.S. Army Corps of Engineers

The terrible Missouri River flood of 1943 prompted fast action on a flood-control and irrigation plan for the Missouri River basin called the Pick-Sloan Plan. Here, the flood has inundated a refinery and farmland near Omaha.

would consider a golden age of surface-water development. Pick-Sloan reservoirs or flood-control structures completed during this period included the Omaha Levee Project (completed in 1949), Harry Strunk Lake (1949), Enders Reservoir (1951), Harlan County Lake (1952), Swanson Reservoir (1953), Lewis and Clark Lake (1955-1964) and Sherman Reservoir (1963). Other major federal reservoirs or flood-control structures completed or underway in the same period included Box Butte Reservoir (1949), Hugh Butler Lake (1962), Merritt Reservoir (1964), the Missouri River Stabilization and Navigation project (continuing), the rural parts of the Missouri River Levee System and the Salt Valley lakes in the Lincoln area (1962-1967). The U.S. Department of Agriculture (USDA) Soil Conservation Service (SCS) also completed a large number of small watershed structures provided for in the Small Watershed Protection and Flood Control Act of 1954. By 1992, almost 450 structures and 92 miles of channel had been built in Nebraska through that program.

Although the Pick-Sloan Plan was phenomenally successful in building projects, it was and remains to some extent a wish list. As of 1982, less than 10 percent of the acres planned for irrigation had actually been developed. Many of the originally planned reservoirs have not been developed. In recent years, some of the projects developed or proposed have been criticized regarding environmental impacts or project economics. However, the sheer volume of projects developed proves that some wishes can come true.

Even so, the acclaim hasn't been universal; critics have characterized Pick-Sloan as more a nightmare than a dream come true. They cite destruction of waterfowl habitat, barge traffic levels that are far lower than anticipated, innundation of Native American and other fertile bottomlands and mishandling of tribal rights, project costs and the potential effects of trapped sediment on long-term use of the reservoirs. In short, detractors see the plan as a drain on the federal treasury that had negative consequences for the environment and for certain minorities. But few people, supporters or detractors, would deny that Pick-Sloan has had a major impact on the Missouri Basin.

In some respects, other water-resources activity of the 1944-to-1966 era pales beside the stark political and engineering accomplishment of the federal projects. Indeed, the lasting political

Journal-Star Printing Company

Harry D. Strunk

achievement of the time may well have been the effective effort by Nebraska's project promoters and political figures to see that the projects authorized by Pick-Sloan received federal appropriations. Yet, since 1964, the Calamus Reservoir is the only major dam completed in Nebraska under the Pick-Sloan program. But before that time, project construction was the rule, especially in the Republican Valley.

* * *

Harry Strunk stood 6 feet, 3 inches tall and weighed only 145 pounds in 1949. But that lanky frame was making a world of difference in the Republican Valley. As editor of the *McCook Gazette*, Strunk had a natural audience. As a resident of the valley devastated by the 1935 flood, he had a cause. Strunk was not one to let his work interfere with his public obligations. When he built a newspaper building in 1926, he had the phrase "Service is the rent we pay for the space we occupy in this world" carved on its stone face.

In 1940 the Republican Valley Conservation Association was formed to work for flood control and use of the Republican River and its tributaries for irrigation and recreational purposes. Strunk became its president. From that time forward Strunk lobbied for Republican Valley projects constantly and effectively. "If you want a man to haunt a house, call on Harry D. Strunk," quipped Michael D. Smith, commissioner of the Bureau of Reclamation. The Republican Valley Conservation Association was well organized, with some 30 communities banding together and paying dues. The group even managed to place a lobbyist in Washington, D.C.

By 1947 construction was underway on both Harlan County and Enders reservoirs. Then, in 1947 another major flood took 13 lives and gave Strunk the opportunity to push for the other reservoirs. He printed a special edition with pictures of the disaster and the headline, "To Hell with The People!" Copies were sent to every member of Congress and every federal agency. Harry Strunk Lake, Bonny Reservoir (in Colorado), Swanson Reservoir, Hugh Butler Lake, Norton Reservoir (in Kansas) and Lovewell Reservoir (in Kansas) followed. If the Republican Valley has more reservoirs than other areas of Nebraska, Harry Strunk is one of the reasons. The support of his organization and Nebraska political figures were certainly others. Credit must also be given to the Pick-Sloan plan.

* * *

If the conditions of the 1930s ignited a continuing interest in water-project construction, they also made it clear to many Nebraskans that soil and water conservation was vital. In 1937 the Unicameral adopted the state version of a model national act that provided for creation of soil conservation districts. As members of the first state Soil Conservation Committee, George Condra, W. W. Burr and W. H. Brokaw helped the state forge ahead with creating the districts. In December of 1949, Nebraska became the first state west of the Mississippi to have all of its agricultural lands organized into conservation districts. The Soil Conservation Committee evolved into the Soil and Water Conservation Committee in 1951, then, in 1969, into the Soil and Water Conservation Commission and finally, in 1972, into the Nebraska Natural Resources Commission, the state agency now responsible for soil and general water-resources planning.

In 1945 the North Platte Decree allocated waters in the North Platte basin between Wyoming and Nebraska *(see vignette on North Platte Decree, p. 210)*. But the era was certainly not without action by the Nebraska Unicameral. In 1947 a state reclamation law was passed. However, 1957 was the pivotal legislative year.

Don Thompson was to become president of the Republican Valley Conservation Association in 1961, holding that position for 23 years. But in 1957 he was a very active state senator. Thompson was co-sponsoring legislation to divide the Department of Roads and Irrigation and create a new state agency, the Department of Water Resources (DWR), the agency that still administers and manages water rights in Nebraska.

Thompson was also co-sponsoring two other pieces of legislation that showed a strong recognition of the growing importance of groundwater. By this time, Nebraska irrigated most of its acreage by groundwater, and Thompson wanted the state to know the location of the irrigation wells. It was a first step towards potential regulation of groundwater use, and many landowners recognized it as such. At one point, people concerned about the bill filled the rotunda of the capitol. The other groundwater bill Thompson introduced required a minimum 600-foot spacing between irrigation wells. All three bills passed. A bill assigning relative preference in groundwater use passed in the same session. Clearly the legislature had become concerned about groundwater use.

In 1963 that concern was expressed again when Thompson co-sponsored and the Unicameral passed bills defining groundwater and allowing pumping of water from pits located within 50 feet of a natural stream. More important was approval of a bill that allowed cities and villages to transport groundwater. Lincoln and Omaha now move much of their

Interstate Water Compacts Help Settle Surface-water Disputes

Interstate compacts are, in simple terms, agreements or contracts between two or more states, in this discussion, concerning water in rivers. Unlike agreements between individuals, however, the United States Constitution requires that all compacts receive the consent of Congress. Unless Congress gives its consent to a compact, the compact is not valid. While compacts are actually part of each states's laws, congressional approval causes the compact to be part of the body of federal law as well.

Compacts have been used by states to resolve many different types of interstate problems. It wasn't until the enactment of the Colorado River Compact in 1922, however, that a compact was used to divide the waters of an interstate stream. Because litigation was viewed as a cumbersome and expensive way to divide or apportion the waters of an interstate stream, compacts like the Colorado River Compact were widely accepted. Nebraska jumped quickly on the compact bandwagon when it adopted the South Platte River Compact in 1923.

The South Platte River Compact was entered into between Nebraska and the state of Colorado. According to the compact, certain Colorado irrigators are to be regulated for the benefit of Nebraska when the flow of the South Platte River near Julesburg, Colorado, drops below 120 cubic feet per second. Unfortunately for Nebraska, the compact fails to regulate large upstream water users in Colorado. As a result, the South Platte generally has flows into Nebraska that are quite low.

After entering into the South Platte River Compact, Nebraska did not enter into another interstate water compact until 1942. In that year, Nebraska signed the Republican River Compact with the states of Colorado and Kansas. This compact divided up the waters of the Republican River based upon a complex accounting method. Unfortunately, the accounting method was not clearly addressed in the compact itself. As a result, some controversy has recently arisen as to the accounting method and the true intent of the compact. Efforts are presently underway to resolve the controversy.

Perhaps Nebraska's most overlooked compact is the Wyoming-Nebraska Upper Niobrara Compact. This compact was signed by Nebraska in 1962. Its intent was to divide the waters of the upper Niobrara River and engage in a study of the surrounding groundwater. Virtually no activity concerning this compact has occurred since its creation (a good reason for its being overlooked).

The last water compact to be entered into by Nebraska was the Blue River Basin Compact. This compact was entered into between Nebraska and the state of Kansas in 1971. While the main focus of the compact is to divide the waters of the Blue rivers between the two states, the compact is unusual for several reasons. First, it requires Nebraska, under certain circumstances, to regulate a few groundwater irrigators near the two rivers for the benefit of Kansas. Under existing Nebraska statutes, no such regulation is permitted. But because the compact is also a federal law, Nebraska must yield to the federal law and regulate groundwater users under terms of the compact. To date groundwater irrigators have not been regulated.

The second unusual aspect of the Blue River Compact is that it addresses water quality concerns. Specifically, it requires the states to make efforts to investigate and address interstate sources of water pollution. At the present time, Nebraska is examining possible water quality issues raised by Kansas pursuant to the compact.

Since 1971, Nebraska has not entered into any other interstate water compacts. Given the increasing demands on the state's water resources, existing water compacts may need to be renegotiated with the other states in order to meet current or future needs. Compacts remain, however, an important tool of interstate water dispute resolution.

—Don Blankenau, attorney, Nebraska Department of Water Resources

Journal-Star Printing Company

Warren Fairchild

municipal water supply in accordance with this bill.

By the mid-1960s the luster of surface-water development was being clouded by the shadow of other issues. Post-war Nebraska water policy has an almost eerie parallel to the mood of the nation itself. The can-do philosophy exemplified by the war effort and manifested in Pick-Sloan carried into the 1960s. The two decades following the war saw tremendous progress in the development of Nebraska's water resources. But progress can carry a price tag of at least some complacency. In addition, groundwater development was changing the nature of water use in the state. Nebraskans were also beginning to think about how their actions would affect the environment. Before the 1960s were over, the national mood would be far more skeptical and less unanimous. Accordingly, Nebraska water policy would find itself beset by new issues and controversies.

Reorganization, 1966-1972

More than two decades after passage of the resolution that fostered them, Warren Fairchild would be praised for his political courage while receiving an award based partially on his work to create Nebraska's natural resources districts (NRDs). But in 1968 it probably didn't feel like courage. It may have felt more like the insanity required to brave hell on a warm day. Fairchild was executive secretary of the Nebraska Soil and Water Conservation Commission, and the districts that helped select his commission members comprised a majority of those to be reorganized. A total of 154 special-purpose districts were eventually to be merged into 24 natural resources districts. Not surprisingly, many of the board members of those 154 districts were unhappy.

On paper the idea had support. In 1966 the delegates to the 26th annual conference of the Nebraska Association of Soil and Water Conservation Districts had passed a resolution calling for consideration of legislation reorganizing districts along hydrologic boundaries. The resolution also urged making them of sufficient size to fund and aid operations, granting them powers to carry out comprehensive land- and water-development programs and governing them by locally elected officials.

The trouble was that the support was thin. Most board members in the districts being consolidated probably felt that they had been doing a good job.

In an address to the 1967 conference, Governor Norbert Tiemann, Republican from Wausa, called for "a functional realignment of local water agencies in Nebraska." A resolution passed at that conference said a committee should be appointed to report at the next meeting so action could be taken during the 1969 legislative session.

But in 1968 the moment of truth had come, and a battle clearly loomed. Fairchild was a popular figure. He now used that popularity to lobby for reorganization. The business session where the resolution to reorganize was introduced has been described as "fiery." But Warren Fairchild was a man who was used to the heat. The resolution passed: 57 for, 42 against.

The spotlight turned to the legislature. On April Fools Day of 1969, four state senators, including Jules Burbach of Crofton and Maurice Kremer of Aurora, introduced LB 1357 to consolidate the 154 special-purpose districts into NRDs. This was no joke. Gone would be soil

Jules Burbach speaking to the Nebraska State Legislature

Journal-Star Printing Company

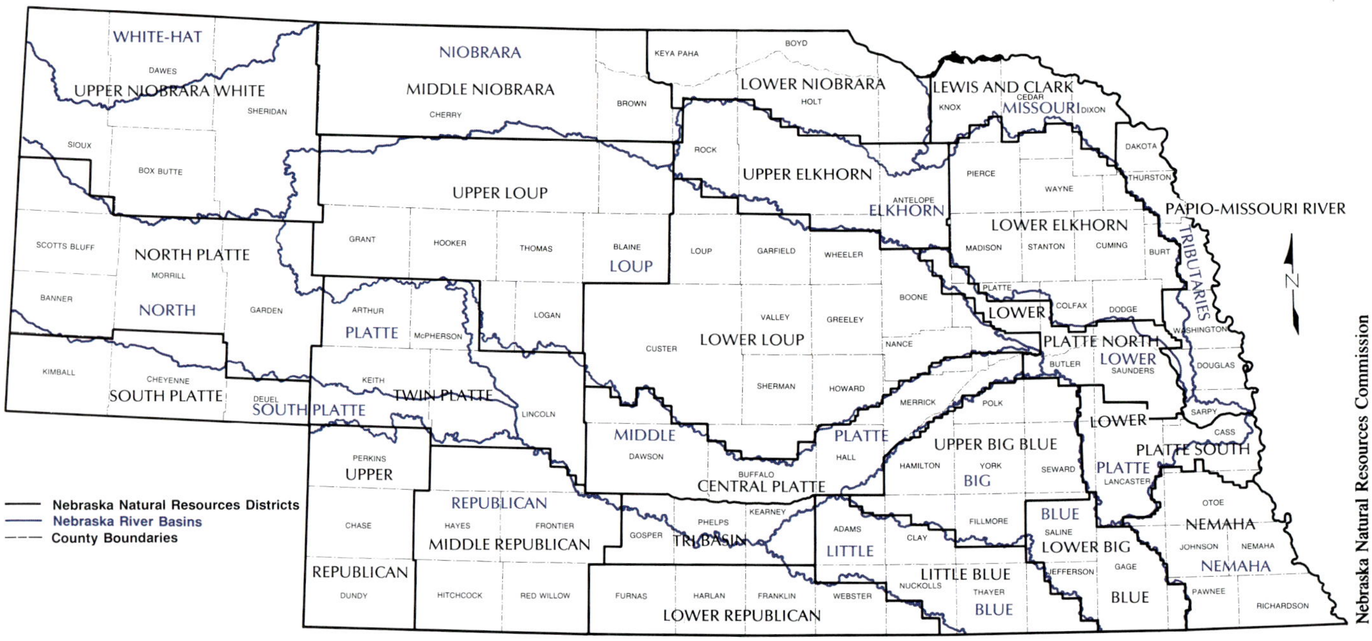

Nebraska natural resources districts and river basins

and water conservation districts, watershed conservancy districts, watershed-planning boards and watershed districts. Members of some soil and water conservation districts organized to oppose the legislation and hired a lobbyist.

But it was too late. On Sept. 16, 1969, a resolution to oppose the reorganization outlined in LB 1357 was defeated at the annual meeting of the state Soil and Water Conservation Commission. Two days later, the Nebraska Unicameral passed LB 1357 by a vote of 29 for, 9 against and 11 not voting.

The members of some soil and water conservation district boards were very unhappy. The opposition ironically centered in the same southeastern part of the state that 30 years earlier had opposed the formation of those very soil and water conservation districts. Over the next few years an opposition group, partially funded by some of the districts being abolished, lobbied against the reorganization and ultimately filed a lawsuit challenging the constitutionality of the NRD law. The expenditure of district funds on opposition activity received an adverse state attorney general's opinion, but that controversy was never fully resolved. In April of 1974, the Nebraska Supreme Court upheld the constitutionality of the NRD system.

Look around the nation, and you will not find another local NRD system like Nebraska's. Anywhere! In Nebraska locally elected boards, receiving half their funds from local property taxes, can address a wide spectrum of natural resource needs. The list of responsibilities is almost without precedent in a local special-purpose agency. It now includes duties related to soil conservation, flood control, water supply, water conservation, drainage, pollution control, fish and wildlife habitat, recreation facilities and forestry and range management.

Through time, the legislature has increased the duties of the districts and, thereby, allowed them to maintain local control over many natural resource problems. NRDs can now regulate how much water a landowner uses and the amount and timing of fertilizer application. Districts can require conservation measures be applied to land that is eroding beyond limits they set. Districts administer distribution of soil and water conservation funds to farmers. They can and do sponsor a wide variety of water-development projects. They can monitor both quantity and quality of groundwater. Through wildlife-habitat and tree-planting programs, they are helping to remake the face of Nebraska. Through educational programs, they are helping to teach Nebraskans young and old about the benefits of conservation.

Are the districts better than the old system? From this observer's point of view, the answer is a resounding yes. The present 23 districts have the ability to address a wide variety of natural resource needs as they arise. The legislature has granted districts authorities for groundwater management areas, special groundwater-quality protection areas, chemigation (applying agricultural chemicals through irrigation water), and the Sediment and Erosion Act for a reason. The districts can and usually do work.

Districts have a local property tax base large enough to enable them to accomplish more than the separate districts. Combining functions and area also allows economies of scale while maintaining a local touch. They can address most natural resource problems from a broader point of view than a single-purpose agency. Although the districts are sometimes criticized for their number of board members, there are still substantially fewer board members than there would have been with 154 special-purpose districts.

To be sure, the districts have problems. The inequalities of property taxation leave some districts "rich" and others "poor," even though they may have nearly equal-sized areas needing management. This helped lead to the merger of two districts in 1989 and other mergers, while not imminent, appear possible in the long term. Another issue that has surfaced is that only eight districts currently elect board members on a subdistrict basis. Other whole districts elect candidates from subdistricts based partially upon area. The result is a process that some feel favors candidates from rural areas.

Another problem is confusion at the ballot box. District boards currently range from seven to 21 members. The voter confronted with up to 11 races in separate NRD subdivisions may feel some understandable frustration. In addition, there is some citizen apathy about the district concept. A 1979 poll found that 73 percent of adult Nebraskans could not remember having ever voted in a district election. Perhaps worse, a 1980 poll showed 41 percent were unaware of their existence. Some citizens who would prefer more vigorous regulation of environmental problems may feel the districts have not done enough. Districts can regulate the amount of water a landowner uses and the amount and timing of fertilizer application, but these powers are only used to a limited degree in a minority of districts. Others, however, may feel districts have already gone too far down the regulatory road.

Finally, the NRD concept probably sacrificed some of the close local contact the 154 special districts had with their constituents. On some local boards, it was possible for a landowner to know most or all board members. They were neighbors. NRD boards, while still local in nature, cover a significantly wider area.

Even so, as Warren Fairchild sweltered in the political heat of the late 1960s, he would have found it comforting to know the fate of the seed he helped plant. Successive Nebraska legislatures have carefully tended the NRD concept, and today that work is bearing fruit.

* * *

Norbert Tiemann was a lame duck. But sometimes lame ducks support ideas that fly. In June of 1970, Governor Tiemann appointed a nine-member committee from outside state government to evaluate Nebraska's environmental-protection laws. By November, Tiemann knew he would be leaving the statehouse in January. But he had the report from the committee. The 59-page report concluded that a new state agency with expanded authority and financing must be created if Nebraska was to have a successful program to address the state's total environment.

Nebraska's first pollution-control efforts had been with the Department of Health. But since then, those efforts had become scattered among various agencies dealing with air, water, solid waste and radiation. Tiemann latched onto the chance for reorganization. He called the report a "landmark" and said it would be translated into legislation. "Landmark" is a word that may be overused regarding politics and policy, but in this instance Tiemann was right.

The report itself pointed to a major push for Nebraska's changing approach to the environment. Congress had created the U.S. Environmental Protection Agency earlier that year based on President Richard Nixon's recommendation. The report cited it as a prototype. The politicians sensed the public demand. Voters had become environmentally conscious.

State senators Loran Schmit of Bellwood and Willard Waldo of Dewitt introduced LB 939, the Nebraska Environmental Protection Act, together in February 1971. At about the same time, senators Richard Proud of Omaha and Wally Barnett of Lincoln introduced a competing bill, LB 879. LB 879 was drafted by the committee that Tiemann had appointed. It also had the support of the new administration. It gave more power to the governor rather than an appointed council. It also gave broader powers to the new department. However, LB 939 was amended to address some additional concerns such as powers over solid waste. It had strong legislative support and passed the Unicameral on May 25th by a vote of 42 to 7. Its appropriation bill had passed earlier.

The surprising thing about the bill was that little fanfare accompanied its passage. Following the 1971 session of the Unicameral, the *Lincoln Star* published a list of seven bills "which may emerge as the landmark achievements" of the session. The Nebraska Environmental Protection Act was not among them, nor did its passage receive prominent newspaper coverage.

One reason may have been that the controversy surrounding the bill was not so much about whether to have a Department of Environmental Control as over how strong to make it and who should be in charge. The *Omaha World-Herald* noted, "The measure has been criticized by environmentalists because of the makeup of the council and a provision which prohibits the state from setting pollution standards stronger than the federal government."

Governor J. James Exon, a Democrat from Lincoln, signed LB 939 on May 26. He was not without some misgivings. A *Lincoln Star* editorial noted, "A displeased Governor Exon Wednesday for practical reasons signed into law LB 939, a 'compromise' environmental protection bill creating a new Department of Environmental Control. Should LB 939 be struck down, the 'present limited efforts in the field of environmental control' undertaken by the Water Pollution Control Council and the Air Pollution Control Council could go unfunded, the governor said, because their funding has been transferred to the new department through companion legislation." The editorial was critical of the composition of the Environmental Control Council, noting that "eleven of its sixteen members will come from interests which are most responsible for pollution." It concluded that "The measure changes the framework of environmental protection efforts, but not much of substance is in it."

Twenty-one years of experience have shown that whatever the substance of the bill, the framework was essential. The following year Congress would pass the Water Pollution Control Act Amendments of 1972 (the first Clean Water Act), and Nebraska would be ready to respond.

The renamed Nebraska Department of Environmental Quality (changed from "Environmental Control" in 1992), has provided a single point of reference for regulation of most environmental concerns, whether regarding land, air or water quality. It has also served as the conduit for federal environmental-protection funds. Those funds have provided direct aid for physical improvements such as wastewater-treatment facilities and have helped fund the state's planning and enforcement efforts. The department's work to regulate point sources of surface-water pollution has been notably successful. Through legislation authorizing special groundwater-quality protection areas, it is now engaged in the more difficult battle against nonpoint sources of contamination from agricultural chemicals. The department is likely to continue to touch off controversy surrounding its actions because of the issues it addresses. But few would question the need for the agency.

* * *

If the reorganization of Nebraska water politics between 1966 and 1972 was a three-legged stool, the first leg was legislative reapportionment. There had been no realignment since 1937, and at the dawn of the 1960s, rural forces dominated the Unicameral. When the legislature did submit a compromise redistricting amendment that voters approved, it was later struck down by a federal court. The court held that a provision basing legislative districts in part upon physical area was invalid.

It was to be 1966 before the judges approved a new arrangement that more closely conformed to a one-person, one-vote standard. That arrangement quickly resulted in relatively more seats for urban areas. Nebraska's rural-to-urban population shift has since resulted in even more urban legislative seats, a trend that seems likely to persist. In combination with the creation of NRDs and the Department of Environmental Control, redistricting created the new power structure and organizational framework that would address the natural resources challenges of the coming decades.

The Environment, Economics and Emotion: Confusion and Complexity, 1972-1993

Maurice Kremer didn't blaze the Oregon Trail in a covered wagon or settle a sparsely populated prairie. But he was a Nebraska pioneer. In 1940 Nebraska's frontier was groundwater, and Kremer was about to brave that frontier. During a trip to Fort Morgan, Colorado, Kremer witnessed the success of irrigation and briefly considered moving to Colorado.

But he was a Nebraskan, and soon he was drilling a well on his home turf. "It took them two weeks to drill the well, and people came from all around to watch. I can still hear them saying, 'This guy's going to go broke.' What they didn't know is that I was already broke," he once remarked. Kremer put down three of the earliest wells in now heavily irrigated Hamilton County. He was highly successful. By 1981 his original 80-acre farm had grown to a 1,700-acre operation with 3,000 head of cattle.

In 1940 groundwater irrigation was beginning to take hold in Nebraska. By year's end, 1,500 new wells were installed, making a total of 3,050 wells irrigating 183,000 acres. Some Nebraskans never looked back as groundwater became the fuel that ran much of Nebraska's agricultural engine.

But Kremer did. Kremer recognized that irrigators were creating "a new Nebraska because of the availability of water—a standard of living we otherwise wouldn't have known." But he also identified with the thought, "maybe we've gone too far."

Kremer was to become a Nebraska state senator in 1963, and in the crucial decade of 1973 through 1982 chaired the legislature's Public Works Committee. Under his leadership, Nebraska faced complex and sometimes soul-searching questions as the state addressed its environmental limitations. He was uniquely suited to the job.

A religious man, Kremer was deeply concerned with doing what was ethical. Speaker of the Legislature Richard Marvel was to say, "The first thing you have to say about Maurice Kremer is that he is a man of integrity. No one ever has or will question his integrity or honesty." Kremer had also seen both sides of natural resource use. He knew of the unprecedented prosperity groundwater development had brought to his area. But he also remembered the Dust Bowl. He recounted riding with his father toward Aurora on a country road. Their headlights were on even though it was daytime. A dust storm was piling topsoil up along the fence rows like snow. To the day he died, Maurice Kremer considered himself a conservationist.

In weighing the benefits of development and the dangers of going too far, Kremer valued information to unravel the complexities of Nebraska's water resources. He was instrumental in giving direction to the state's water-planning efforts. But above all, Kremer understood the need for compromise. In uncertain waters, it is sometimes best to steer a middle course. In any case, that is often all that works. Kremer explained, "Not that I want to be a compromiser to the extent that you don't do anything, but most legislation is a system of compromise. You take what you get, and be thankful you get that. I think you'll probably get more done that way." Kremer did. In 20 years in the Unicameral, Kremer sponsored the bill creating NRDs, helped found the Nebraska Resources Development Fund, led the fight to create Ground Water Control Areas, introduced the 1967 bill that resulted in the framework study for a state water plan and chaired the committee introducing a 1978 resolution that directed state agencies to study water-policy issues. While Kremer believed strongly in the benefits of surface-water projects, his legislative record also shows

Journal-Star Printing Company

Maurice Kremer

Biological Systems Engineering, UNL

Opposition to large surface-water development is usually associated with urban citizens concerned about environmental values and subsidized water for surplus crops. But in an increasingly complex political landscape for water development, dissent has also come from rural areas, as seen in this sign, site and date unknown.

an attempt to understand resource limitations. Many called him "Mr. Water."

The Ground Water Control Area legislation of 1975 was Nebraska's first major excursion into the realm of groundwater regulation. As usual, Kremer accomplished it through compromise. When another committee introduced a bill less restrictive to groundwater users than that introduced by the Public Works Committee, the practical Kremer was willing to accept a compromise version that eventually became law.

Kremer retired from the Unicameral following the 1982 session and died in 1990. But if Nebraskans are willing to appreciate it, his legacy lives on. At first glance that legacy seems to be his pioneering work with NRD legislation and his championing of groundwater legislation. But his real legacy was in his approach. Kremer could fully appreciate the benefits of water development and yet carefully consider its liabilities. He recognized the complexity of water issues, and he fully respected views from all parties. He was supremely aware of his responsibilities to future generations, and he was willing to compromise. Kremer was a water statesman. And Nebraska was fortunate to have had such a statesman help it enter an era of immense change.

* * *

From the dedication of Kingsley Dam in 1941 to the creation of the Department of Environmental Control in 1971, surface-water development had occupied front and center stage in Nebraska water politics. But by 1971 its star was being eclipsed. Fundamental changes had occurred in the state's economy, in resource knowledge and use and in its electorate. An increasing proportion of Nebraska's citizens lived in urban areas, and the number of people engaged in agriculture continued to fall. Nebraskans were more mobile and had more leisure time. The environmental movement had not bypassed Nebraska; citizens were increasingly concerned with environmental issues. Moreover, the technical ability to detect environmental problems was growing.

The first glimmer of a national move toward environmental regulation had also appeared. The U.S. Environmental Protection Agency was born in 1970, and the Clean Water Act appeared in 1972. There were soon to be federal incentives and mandates for states to regulate activities affecting the environment.

However, the most basic change was in the nature of irrigation in Nebraska. In 1939, groundwater accounted for only a small percentage of the irrigated acres in Nebraska. By the early to mid-1950s, the state had probably passed a major milestone. By then, more acres were irrigated by groundwater than by surface water. In 1970, about 70 percent of Nebraska's irrigated acres were irrigated by groundwater, and the state was on the doorstep of a center-pivot irrigation boom. The water politics and policy of the state were focused on surface-water development, but the nature of the state's irrigated economy had changed dramatically. By 1990 Nebraska would have the second highest number of irrigated acres in the nation (California is number one), and nearly 40 percent of the state's cropland would be irrigated.

Agriculture remains the preeminent Nebraska water user. In 1985 it accounted for nearly three-quarters of total water use and 97 percent of consumptive water use. However, water has been important to other users as well. In recent decades the needs of domestic users, industrial users, municipalities, livestock producers, navigation, power interests, recreators and fish and wildlife have all assumed a higher profile. By the early 1970s it was becoming apparent that policy would need to be adjusted to reflect the changes. Soon one more change was to occur. Citizens would begin to question more vigorously both the costs and benefits of government-financed water projects and how those benefits were calculated. Who should receive the benefits and who should bear the costs were to become major political issues.

During the last 20-some years, Nebraska's governors and legislature have responded to these changed conditions with a virtual avalanche of notable legislation. From World War II almost until the bill creating NRDs in 1969, the federal government was an overwhelming factor in Nebraska water policy. Securing federal funding for surface-water projects had been the major job for state and local project promoters and political figures. Now the state, sometimes with federal prodding, was to assume a far more substantial role.

One of the earliest forays into the new policy realm actually came in 1967. One step ahead of the federal government, the Unicameral adopted the Nebraska Flood Plain Regulations Act. The law provided for outlining of floodplains and regulation of land use in floodways. Final authority for regulation was to rest with the state Soil and Water Conservation Commission if local political subdivisions failed to act. This allowed the landowner to receive federal flood insurance when the National Flood Insurance Act passed in 1968.

In many ways this act set the pattern for water policy that was to follow. The state would provide information or education (in this case through determining floodplains), and pressure would be applied to local governments to act. Federal money would ultimately be available as a carrot, and if local governments failed to act, state government could carry the stick.

Within the state bureaucracy, the calls for change were also being heard. In response to a 1967 legislative resolution, the Soil and Water Conservation Commission issued its "Report on the Framework Study of Nebraska's State Water Plan" in 1971. The idea of a state water plan was certainly not one spawned in isolation. The preparation of the report was supported in part by federal funds and agencies and nurtured by the continued interest of Maurice Kremer, Jules Burbach and other state legislators.

The result was an informational document that curiously had one foot in the future and one in the past. The foot in the past was certainly the plan's continued emphasis on surface-water development. Basin-specific parts of the plan are in some respects a state-level refinement of the "wish-list" approach provided in the Pick-Sloan Missouri Basin Plan. But the framework study also provided a view of some of the water issues that were to engulf the state in the coming decades. It identified stream reaches with potential for designation as wild, scenic or recreational rivers. It analyzed water demands and made recommendations regarding flood control, environmental values, fish, wildlife and recreational resources, data and planning needs and, finally, water-development needs.

The major contribution of the framework study was as a data source. It provided a snapshot of the state's water resources in a readable format that continues to be useful more than 20 years later. But even as that snapshot was being taken, Nebraska's water policy was changing. On the very day the framework study was delivered to the legislature, Governor Exon signed the Nebraska Environmental Protection Act and created the Department of Environmental Control.

In 1979 the water-planning effort of the state was changed from a plan to a process, thereby recognizing changing conditions. State legislators also directed that state planners address the water-policy issues the legislators had been facing. One of the issues that had taken much of their time was groundwater.

Groundwater Development Policy

If the 1971 framework study proved less than prophetic on the future of some issues, the likelihood of groundwater legislation was surely one of them. Timing was part of the reason. In 1970 Nebraska sat poised on the brink of an eight-year frenzy of irrigation-well installation ignited in part because of new areas made irrigable by center-pivot technology and attractive by high crop prices. This was to accelerate legislative concerns about declining groundwater tables and water quality.

In 1975 those concerns helped lead the Public Works Committee to introduce a bill allowing NRDs to establish groundwater control areas. The bill had been slowly crafted with hearings and committee work over an extended period. Passage of the legislation established the pattern of local control over groundwater usage that persists to this day.

The new law, The Groundwater Management Act, allowed the NRDs to form control areas to conserve groundwater supply and to control or prevent runoff. To achieve those ends, groundwater use could be restricted, more restrictive well spacing could be required or a system of rotation in the use of groundwater could be employed. In extreme cases, a moratorium on new wells could be adopted. The law allowed local districts to begin proceedings to start control areas, but the actual designation was to be made by the director of DWR. Therefore, the state retained oversight on local action. Over the ensuing years the director declared three control areas and turned down three others.

The 1975 bill was the first of a series of legislative actions that led to what today is known as the Groundwater Management and Protection Act. Groundwater was already subject to regulation by the state. However, this was the first statute to put teeth in that policy.

In 1982 the law was expanded to allow for creation of Groundwater Management Areas; these could be declared by the NRDs after completion of a groundwater management plan. The plan was to include a goal for the life of the groundwater reservoir. In addition to the regulatory powers granted to control areas, the Groundwater Management Areas could employ mandatory well metering, but they were not allowed to put moratoriums on drilling of wells. Furthermore, as it has been amended, the law allows measures to protect water quality, specifically educational programs designed to protect water quality and stipulations about best management practices (BMPs), practices aimed at soil and water testing, timing of agricultural-chemical application and other soil and water conservation measures.

In 1984 legislation passed requiring all districts to complete a groundwater management plan. By 1991 three groundwater management areas had been declared. All three have rules that regulate regarding water quality, and one can also regulate to protect water quantity.

In examining their groundwater-quantity problems, the NRD directors have faced a labyrinth of questions: How long should an area's groundwater supply last? At what level, if any, should groundwater levels be stabilized? How cost effective is education versus regulation? How much improvement can be achieved by better water-use efficiencies alone? What is the effect on the economy and individuals of denying water use? Will groundwater be worth more later than it is now? How long does natural recharge take, and what is the potential of artificial recharge?

The answers to these questions are not simple and neither is the job of the districts. Have Nebraska's groundwater control areas and groundwater management areas been effective in dealing with groundwater depletion? The jury is still out on that question. Mandatory well metering and education are approaches that are already being tried. By 1992, only one NRD, the Upper Republican, had actually limited the amount of water irrigators apply. However, a second NRD, the Little Blue, took steps to limit use in 1993. Still other districts have established average groundwater levels that will "trigger" restrictions. A few dry years could conceivably pull those triggers and result in more extensive regulation.

* * *

Federal agriculture and taxation policies also had a way of generating political controversy in the post-1970 era. These were not, strictly speaking, water-development policies. However, by affecting the profitability of various types of agricultural investment, they affected the level of irrigation development. Crop price-support programs, capital-gains taxation and accelerated depreciation all likely played a role in irrigation-investment decisions. They are perhaps the unsung benefactors of groundwater development in Nebraska.

Federal and state conservation policies have also played a role in some groundwater-development decisions. The well-installation boom of the 1970s resulted in some irrigation development on the fragile soils at the fringes of the Sand Hills. Provisions of the Food Security Act (1985 farm bill) required that to remain eligible for federal programs, farmers must have a conservation plan to reduce soil losses to required levels. The 1986 state Erosion and Sediment Control Act also inhibits use of improper conservation measures on fragile soils. Since the Sand Hills is a vast area of fragile soil containing about half the state's groundwater, soil conservation policies could play a key role in limiting the amount of development the area might receive in times of high crop prices.

Municipal Water Needs

Municipal water supplies probably account for (depending on what figures one chooses) only about 1.5 to 5 percent of total water use in Nebraska. But the value of that water is far out of proportion to its amount. In some other western states, municipalities are buying up agricultural water rights. They simply can afford to pay a great deal more than agricultural users.

Groundwater supplies more than three-quarters of the people served by Nebraska's public water-supply systems. The 1963 City, Village and Municipal Groundwater Permit Act was a major step forward for municipalities because it allows cities and villages to transfer groundwater off of overlying land. But passage of the Federal Safe Drinking Water Act in 1974 and a similar state act in 1976 constitute perhaps the most important legislation related to public water supplies. The Nebraska Safe Drinking Water Act requires the director of the Department of Health to adopt drinking-water standards and requires permits to operate public supplies. The department trains operators of water-treatment plants and provides laboratory service at a minimum cost. Municipal users can also receive protection from other state programs, including the wellhead-protection program, special protection areas (SPAs) for groundwater quality and other provisions of the Groundwater Management and Protection Act.

Recently, introduction of legislation providing municipal access to surface water rights for well fields dependent on the river for recharge has provoked opposition from other users, particularly irrigation interests, and has so far been defeated. This would benefit municipalities with wells on riverbanks and islands. Expanding municipal rights through the legal system of water-use preferences has also been proposed but not enacted. But since an ever larger proportion of Nebraska's population lives in cities, that population is wielding increasing political clout.

Water Quality

Think of raw sewage pouring from a pipe and into a stream, or of hazardous chemicals flowing from an industrial site, the conventional images of contaminated water. If that is your picture of water pollution in Nebraska, think again. To put Nebraska's immense water-quality challenges in perspective, you should first consider a major success of the last 20 years.

Surface-water pollution from an identifiable point source, such as discharge out of a pipe, has been comparatively easy to control. Nebraska has effectively controlled most point-source pollution of surface water. The National Pollutant Discharge Elimination System, industrial pretreatment and sewage-treatment construction grants have sharply limited this type of pollution. In contrast, pollution that comes from no specific point (nonpoint-source), such as agricultural chemicals or urban runoff, has proven more difficult to address.

What types of water-quality problems do Nebraskans face? The Department of Environmental Quality has noted that "major sources of groundwater contamination in Nebraska have included: agricultural activites, industrial facilities, leaking underground storage tanks, oil or hazardous substance spills, solid-waste landfills, wastewater lagoons, brine disposal pits, septic systems and other sources." In other words, groundwater contamination can come from a wide variety of activities. It also can affect almost the entire spectrum of human use of water. Surface-water quality problems in Nebraska stem largely from nonpoint-sources, mostly agricultural runoff, with domestic and industrial point sources accounting for less than 10 percent of total impairment to stream use.

Since 1971 water quality has skyrocketed to the top of Nebraska's water-policy agenda. One of the reasons for that rising level of concern is the increased opportunity for pollution from agricultural chemicals due to the major expansion of groundwater development. Other reasons include increasing knowledge of the health effects of pollutants, improved ability to detect water-quality problems and increased public awareness of water-quality problems. In the immediate post-war era, Nebraska also witnessed a major increase in the amount of fertilizer and pesticides applied per acre. In combination with increased irrigation and improved seed varieties, this helped stimulate Nebraska's agricultural production boom. But it also increased the importance of wise chemical management by the agricultural producer.

Since the 1970s, the nation as whole has become much more concerned about groundwater quality. An expanding web of federal laws and incentives has helped spur state governments to further efforts. But Nebraska has also taken some very significant action on its own.

Most of these actions came in 1986 when the Unicameral provided for regulation of chemigation, passed legislation allowing creation of SPAs and added groundwater-quality regulations as an option for Groundwater Management Areas. Although the state maintained some authority on SPAs, such actions were in keeping with other legislative moves that delegate authority to NRDs.

The Chemigation Act requires farmers applying chemicals through irrigation systems to have a proper check valve and other safety devices and to have a permit from the NRD. Applicators are required to pass a training course with a written exam. NRDs conduct inspections.

The SPA program assigns to the Department of Environmental Quality responsibility for studying areas where any political subdivision asks it to investigate if nonpoint-source groundwater contamination is occurring or is likely to occur. If problems are identified, the department can designate an SPA. Once designated, the NRD is charged with developing an action plan that usually includes education and regulations to improve management of groundwater quality. The department must approve the plan and oversee its preparation and implementation.

The Groundwater Management Area statutes give NRDs even more authority. As amended in 1991 the law now requires them to address water-quality problems and their solutions in the groundwater management plans by July 1, 1993. If a management area is created primarily for water-quality purposes, the district is required to specify actions it will take to address groundwater contamination so as to meet its groundwater-reservoir life goal and has to implement those actions. Required actions can include analysis of water and deep soils for fertilizer and chemical content, BMPs and educational programs.

The authorities given by Nebraska's environmental-protection statutes go much further than those delegated to NRDs. A partial list of duties related directly to water quality that are the responsibility of the Department of Environmental Quality includes wellhead-protection areas, the National Pollutant Discharge Elimination System, the Underground Injection Control program, the Underground Storage Tank program, the Superfund program for contaminated sites and requirements for the Resource Conservation and Recovery Act. (The Superfund program was created within the U.S. Environmental Protection Agency [EPA] to provide funding to clean up contamination caused by the improper storage or disposal of hazardous waste and to address emergencies involving such waste. The Resource Conservation and Recovery Act [RCRA] was designed to give states authority to properly manage hazardous waste from creation to disposal.)

Although point sources of pollution are less widespread than nonpoint-source problems, there are concerns, especially about past activities. Nebraska currently has 10 sites on EPA's Superfund national priorities list of most seriously contaminated sites. Former grain-storage facilities and former military ordnance facilities have been among the sites of hazardous-waste problems. Underground storage tanks are located throughout the state and are also potential sources of contamination. Spills can be guarded against, but in an imperfect world, they most likely will occur.

Solid-waste programs only recently have received much attention in Nebraska. One concern about land disposal of waste is that potential contaminants be prevented from reaching the water supply. Only lately have changes removed the exemption of small communities from landfill-licensing requirements. In terms of news-making potential, one waste-disposal issue has been a very potent source of political controversy in Nebraska: the proposed low-level nuclear waste storage facility in Boyd County.

Some programs, while not regulatory in nature, are designed to help control water pollution. These include the Nebraska Groundwater Quality Protection Strategy, the Nebraska Nonpoint Source Management Program, stream inventory and classification programs, clean lakes activities and various water-quality assessment programs.

Other programs are strictly federal. In recent years there has been increasing debate over whether Nebraska should assume responsibility for the Federal Insecticide, Fungicide and Rodenticide Act. At present, the supervision of pesticide application through this act is under federal jurisdiction in Nebraska; it is the only state in the country not to assume this responsibility.

Many of the existing state programs are in fact responses to provisions in the federal Clean Water Act. Passage of the Clean Water Act, its amendments and the related flurry of Nebraska legislation are a major story in themselves and beyond the scope of this more general water history. The question of whether those actions have gone "too far" or "not far enough" can best be debated elsewhere. However, a general discussion of the dilemmas at least one water-quality question has posed for Nebraska is useful in explaining Nebraska's past policy conflicts and its future options. That question is: "How do we address nonpoint-source contamination from agricultural activity?"

* * *

In the last 20 years, an increasing number of samples from Nebraska wells have shown evidence of contamination by nitrogen fertilizers and, in some cases, pesticides. These problems tend to be greater in irrigated areas where over-irrigation can flush contaminants to the water table more quickly. State efforts such as SPAs and state-authorized local efforts such as Groundwater Management Areas have attempted to regulate the amount and timing of chemical applications, but in some instances, the core of a problem remains.

In a few areas with shallow water tables and sandy soils, even better fertilizer-application practices will not stop chemical leaching. Furthermore, the government's dilemma is which areas to regulate. Should areas where nitrate levels in the groundwater already exceed federal standards be regulated first, or is that like "closing the barn door after the horse has left"? Since those areas already exceed standards, should "triage" be applied and contaminated areas receive less attention while others are regulated before they rise above federal standards? Even if regulations are applied, contaminants already in the ground may cause contaminant levels in some areas to continue to rise even after chemical application is slowed or stopped.

In the face of these difficulties, what are the problems with allowing contamination? Water with levels of nitrate higher than federal standards is unsafe for babies and some livestock. There are also suspicions about carcinogenic effects of nitrates, but little hard evidence. Evidence of health problems with pesticides varies but is mostly related to cancer-causing potential and is generally of greater concern in the long run and in large enough concentrations. Nitrates will generally persist in the water once introduced, although proper conditions can lead to lower levels of nitrate in some areas. Pesticides usually break down into other chemicals after a time, but questions remain about the health and ecological effects of the chemicals that result.

The state response to all this uncertainty has been large-scale research and demonstration efforts and smaller regulatory efforts, primarily at the local level. Groundwater Management Areas have been declared in the Central Platte NRD, the Tri-Basin NRD in central Nebraska, near Sidney, and the Upper Elkhorn NRD board has voted for one. Other areas are being studied. These programs mostly target education first, then regulate areas with the greatest problems. Other educational efforts have revolved around numerous demonstration and research programs in different areas of the state. The hope is to research the most effective techniques for preventing nonpoint pollution and disseminate those techniques to local farmers.

Application of farm chemicals is only one of a myriad of water-quality questions Nebraska has addressed in recent decades. The tradeoff is between growing areas of worsening contamination and regulations that can be expensive and difficult for the state or NRDs to enforce. Regulation also can place varying degrees of financial burdens on farmers.

The Clean Water Act and associated state actions are likely to have a very high profile in future years. The relationship between clean water needs and the practicalities of farming and irrigating will require careful thought as laws are amended and enforced.

Surface-water Development Policy

If 1975 heralded an age of state regulation of groundwater, it also foreshadowed a cooling of Nebraska's love affair with surface-water development. On Nov. 4, 1975, voters rejected the proposed Mid-State Project, which would have provided 140,000 acres of irrigation water, as well as groundwater stabilization, flood control and recreation in Buffalo, Hall and Merrick counties. Mid-State was a major irrigation project proposed by the Bureau of Reclamation, and environmentalist objections played at least some role in its defeat.

Interestingly, only the year before, proponents of small-scale water-development projects had gained a major victory. In 1974 the Unicameral passed a measure authorizing a Nebraska Resources Development Fund to provide state funding for water projects. Although use of the fund has since been limited to projects costing less than $10 million, it was and remains the primary state funding for surface-water development. NRDs have sponsored the vast majority of projects using the fund. State funding began at about $1 million per year, rose to about $2.9 million in the early 1980s and has been about $2 million in recent years. In total the fund has contributed to 47 projects and spent nearly $25 million. Project purposes have been recreation, flood control, sediment retention and recharge. Although this was not the only attempt to amass state money for surface-water projects, the Nebraska Resources Development Fund remains the only state program that has actually provided such funding.

Despite the success of this small-scale state fund, a host of environmental and political concerns were to lay obstacles in the path of any larger proposed projects. The exhaustive list is almost exhausting in length: instream-flow needs, endangered-species needs, fish and wildlife habitat, wetlands regulations, in-state water rights rivalries with other projects, interstate rivalries, natural area preservation, recreation needs, water availability, land-treatment needs, water-quality problems, low crop prices, lack of government funds, lack of appropriate sites, debates over project evaluation standards, debates over the equity of governmental funding and subsidies and the rise of organizations, mostly environmentally oriented, that are generally opposed to large water-project construction.

Is the sun setting on large-scale surface-water development projects in Nebraska? The bottom photo of a Central Nebraska Public Power and Irrigation generating station near Grand Island evokes such a thought. If so, the sun continues to "rise" over groundwater development, which has become the premier source of irrigation water since 1950, now supplying about 85 percent of the state's irrigated acreage. This is evoked in the top photo of a center-pivot arm.

Soil Conservation Service

Nebraska Farmer

The environmental movement had come to Nebraska. If 1975 was the year that Mid-State died and the state addressed groundwater regulation, it was also the year the Unicameral passed the Non-Game and Endangered Species Conservation Act. A similar federal act had passed in 1973. Nowhere were habitat needs for endangered species to become a more complex issue than on the Platte River.

Proposed Platte River Projects

Water projects now had to pass a more stringent environmental muster, and that meant competing projects and interests had a tool they could use against each other. The water-short, environmentally sensitive Platte River was now a Gordian Knot of competing water interests. The amounts of water needed for endangered species and wildlife were open to debate. The amount of water available for individual projects was open for debate. The North Platte Decree and the amount of water allocated to Wyoming and Nebraska were open to interpretation. The relative economic benefits of each of the various projects proposed for the Platte were arguable. The environmental benefits or damages of each proposed project were difficult to determine. Within a proposed project, the balancing of irrigation, power, recreation, flood-control and environmental interests was becoming increasingly difficult. Furthermore, the legal issue of water rights did not hinge on the relative merit of a project.

The addition of environmental factors had made the already complex situation on the Platte River positively labyrinthine. At one time or another during the 1970s, sponsors applied for Nebraska water rights for six major projects that would use Platte River water (Prairie Bend, Twin Valley, Landmark, Catherland, Perkins County and Enders). Potential upstream projects such as Two Forks and the Narrows Unit in Colorado, as well as Grayrocks and Corn Creek in Wyoming were also being considered. Studies on the projects were in various stages and so the available information varied dramatically. Plainly, not enough water was available for all proposed projects.

Given this web of complexity, the thought of some observers may have been to cut the knot. It would certainly have been possible for the legislature to decree that no state money would be available until a state agency had completed a plan that would determine the "best" use for Platte River water. However, a number of factors worked against this approach. The complexity of choices meant an acceptable state effort would likely take a great deal of time and money. It would also involve policy and technical judgments that would certainly be subject to controversy. Even if the state did come up with a plan, there was little certainty the legislature would provide funds for it. In the meantime local sponsors would want to go ahead with their projects. Would the state actually want to oppose their requests for planning or construction funds?

Nebraska has kept its local-sponsor approach to projects. This has led to tensions between sponsors of proposed Platte River projects similar to the tensions between sponsors of the Sutherland and Tri-County projects in the 1930s. There are also continuing tensions between developers and environmental interests. Environmental organizations are especially concerned with preserving wildlife habitat for endangered species such as the whooping crane, bald eagle, least tern and piping plover, as well as with overall biodiversity—the value of many species complementing one another.

The Grayrocks Settlement

The first milepost of Nebraska water-development politics in the environmental era was the Grayrocks Settlement of 1978. The REA granted a loan guarantee and the Corps of Engineers issued a construction permit for a power dam on Wyoming's Laramie River in the late 1970s. Nebraska then filed suit alleging that the REA and the Corps had failed to comply with the requirements of the federal endangered species act and the National Environmental Policy Act of 1969. Furthermore, water interests dependent upon the North Platte River, as well as Lake McConaughy, would be affected. The state of Wyoming was not a party to the proceedings.

Ultimately, a settlement was reached in which the project sponsors agreed to limit their water use, provide water releases at specified times, replace up to 11,250 acre-feet of water withdrawn by an irrigation district and provide $7.5 million for a trust fund for whooping crane habitat. The agreement was a landmark in efforts to preserve endangered species on the Platte.

But lurking just below the surface were even wider issues. How much, if any, Platte River tributary water was Wyoming entitled to under the North Platte Decree? As of 1992 Nebraska and Wyoming are in court on the issue. How should environmental mitigation efforts stemming from the project be financed? In this case, power companies would need to pass costs to consumers, including many in Nebraska.

The Grayrocks settlement allowed a project using Platte River water to gain an exemption from the requirements of the federal endangered species act, but at a price. Other projects would not be so fortunate. In 1991, Colorado's Two Forks Project fell victim to an Environmental Protection Agency veto based upon endangered species requirements. It was ruled that Nebraska's Catherland project would jeopardize endangered species and thus not meet requirements of the state endangered species act. This, among other concerns, meant that it ultimately failed to receive a water right. Even so, project features can be added to mitigate environmental damages and protect endangered species. The Prairie Bend, Twin Valley and Landmark projects all adopted this approach. However, mitigation measures often add to project costs, and federal economic feasibility standards are becoming more stringent.

Relicensing of Platte River Facilities

While some sponsors were struggling to build new projects on the Platte, the governing bodies of existing projects began the process of relicensing their facilities. In 1984 CNPPID and NPPD began working on relicensing before the Federal Energy Regulatory Commission (FERC). The original licenses had been authorized for 50 years. The Depression-era sponsors of the Tri-County and Sutherland projects would probably have been interested in and might have been troubled by how times had changed. More than fighting each other, the project backers are now battling environmental groups and FERC.

The conflict centers on how to best allocate water releases to provide for irrigation, fish and wildlife habitat, hydroelectric power generation and recreation uses. In this case, requirements of the federal endangered species act have resulted in a preliminary federal decision that a Nebraska project be operated in a different manner than was required in the 1940s. The draft of the FERC alternative chosen in January of 1992 said it would "reduce drought protection and power generation a modest degree, while moderately improving reservoir-based recreation . . . and substantially improving the health of the Platte River aquatic ecosystem and the wildlife species that depend upon it." The reduction in power generation would ultimately be absorbed through higher electric rates.

Toughened Economic Requirements for Federal Project Funding

Tougher environmental requirements are only the first of a series of hurdles the federal government has placed before surface-water development projects since 1970. In the 1980s the Reagan administration substantially increased the discount rate used in determining the feasibility of federal projects. (The discount rate determines the value of future benefits. In other words, we would all rather have benefits now instead of later. The numbers used have varied through time. *See "Economics and Finance," p. 123*). As a result, meeting federal economic-feasibility requirements became much more difficult. Even if those requirements are met, federal projects must now clear a hurdle requiring far more state and local financing than was needed prior to 1970.

In the early 1990s, Congress was also considering legislation that would ban landowners from receiving both subsidized water and crop subsidies. A major area of contention was whether the legislation should apply only to new projects or to all federal projects. In the 1950s it probably often seemed there was truth in the maxim that "water flows towards money." By the early 1990s there was precious little money to which water could flow.

The O'Neill Saga

In this chilly atmosphere for developers, two major federal surface-water projects marched to very different fates in the late 1980s. The North Loup Project, with dams on the Calamus River and Davis Creek, is now a reality. It provides irrigation water for 53,000 acres and excellent recreation facilities. The O'Neill Unit and Norden Dam have faded, to be replaced by a National Wild and Scenic River designation for a stretch of the Niobrara River.

The saga of the O'Neill project presents a summary of the obstacles modern Nebraska water projects can face. Norden Dam and its 19-mile long reservoir would have impounded 6,300 surface acres of water on the Niobrara midway between Valentine and Springview. Canals were to deliver water to irrigate 77,000 acres near Springview, Atkinson and O'Neill. Other benefits were for recreation, fish and wildlife and flood control. The project was authorized as part of the Missouri River Basin Project in 1954 and reauthorized in 1972. Its feasibility report was approved in 1965, and in federal fiscal year 1976, the project even received federal money to start construction.

Then all hell broke loose for project proponents. In March 1977 a court injunction by the Save the Niobrara River Association halted construction until an environmental impact statement had been adequately supplemented. No one knew it at the time, but Norden Dam was down for the count.

Project supporters had pointed especially to a favorable cost-benefit ratio, recharge benefits, new recreational opportunities and general economic stimulation as reasons for building the project. Project opponents questioned the methods used to derive the cost-benefit ratio, felt the benefits went to too few landowners, questioned whether the investment was sensible given other agricultural subsidies, disputed geologic findings and dam safety, suggested that a biologically unique area would be damaged and objected that a part of one of Nebraska's best canoeing rivers and most scenic areas would be flooded.

The delay to answer environmental questions allowed project opponents to build support. It simultaneously allowed construction costs to escalate. Opposition to the project grew. Soon national environmental organizations lent their support to project opponents. "Save the Niobrara" was a slogan that could be found on tee shirts throughout Nebraska. Project opponents were convincing many Nebraskans that the area was a scenic resource that should not be flooded.

A supplement to the environmental impact statement was filed in 1978, but the court found that two issues had not been adequately addressed and so did not halt the injunction. In 1979 the U.S. Fish and Wildlife Service issued an opinion that the project was likely to jeopardize the existence of the whooping crane. This led to more studies. A supplement to the environmental impact statement was issued in July 1981. It included a letter from the Fish and Wildlife Service lifting the jeopardy opinion. But by now it was too late.

Opposition led to a 120-day Bureau of Reclamation study of alternatives to the O'Neill Unit. That study was completed in October of 1981. The Nebraska congressional delegation then obtained funds for a state-led study of a recharge alternative, completed in January of 1985. The state proposed that detailed investigation of the well field/groundwater-recharge proposal continue under the O'Neill Unit Authorization.

On March 11, 1985, acting Commissioner of Reclamation Robert Olson wrote Senator Exon that the bureau's review indicated the state's recharge plan had substantial merit and more feasibility studies and a design study should begin. And he added, "As a result of environmental controversy and escalating construction costs, the original O'Neill Unit does not have sufficient support to warrant continued investigations by the Bureau of Reclamation. Therefore, I have deferred all ongoing investigations on the Norden Dam Plan for the O'Neill Unit." Norden Dam was dead.

However, proponents' hopes for the O'Neill recharge alternative and the small Springview portion of the O'Neill Unit remain alive to this day.

Platte River Forum for the Future Got Feuding Parties Talking

An adage in the West is, "Whiskey is for drinkin,' but water is for fightin'."This describes quite well the waters of the Platte River since the early settlers of Nebraska began diverting water for irrigation before the turn of the 20th Century to the current day.

However, a unique approach was pursued in 1982 under the leadership of Governor Charles Thone's Natural Resources Coordinator and the Nebraska Natural Resources Commission to resolve conflicts on the future uses of unallocated water in the Platte through what was called "The Platte River Forum for the Future." The technical leadership for this effort was provided by a team skilled in adaptive environmental assessment processes from the U.S. Department of Interior Fish and Wildlife Service office in Fort Collins, Colorado.

The forum provided an opportunity for involvement by many organizations and interests—some which were competing but all of which were concerned with Platte River waters. An innovative approach, it was to involve a series of alternating workshops and research. Following preliminary work, a week-long workshop was held in August 1982 at the Law Enforcement Training Center in Grand Island, Nebraska. Forty-three individuals representing a wide range of interests and technical specialists participated in an intensive, live-in activity.

A multi-objective, mathematical model was built with input from these people consisting of sub-models for surface water and groundwater, as well as for agriculture, municipalities, industry and wildlife and recreational uses. It also included an economics sub-model.

For perhaps the first time, widely varied interests were brought together in a dormitory/lab environment where they became personally acquainted and searched for compromise face-to-face. This initial effort enhanced the overall understanding of all participants of the river as a system and also called attention to the need for additional data on a variety of topics.

The Colorado team served only as technical model builders, facilitating the process without interjecting personal opinions, and trained a Nebraska team to take over the operation of the model. The model was further perfected, updated data was incorporated, and in early 1983 scenarios of various Platte River water uses were run for state water-policy decision-makers.

For a variety of reasons, including a change in governors in early 1983, "The Platte River Forum for the Future" never achieved its initial promise. However, the talks between irrigation interests, environmental groups and other key participants that began in the early 1990s probably owe some of their inspiration and instigation to this earlier attempt to get the crucial interests talking. This latter set of talks has been prompted by the desire to create "a Nebraska solution," rather than letting the federal government, in this case, the Federal Energy Regulatory Commission, dictate the terms of an agreement. Perhaps in the hope that the unfinished business of the Platte River Forum might eventually be more than a mere footnote to the history of the river, the final paragraph of the Nebraska Natural Resources Commission report on the process in January of 1985 summed up its progress this way:

"Our efforts in the Forum may never be recognized as a benchmark accomplishment but to the extent they may contribute to a more rational, informed approach to understanding the river and conflict resolution on proposed changes, we take some satisfaction."

—Verlon K. "Tony" Vrana, formerly executive vice president, Soil and Water Conservation Society; also formerly head of the planning division of the Nebraska Natural Resources Commission

The Battle over Scenic River Designation for the Niobrara

In September of 1985, Senator Exon introduced a bill to designate a 76-mile stretch of the Niobrara, including the area that would have been flooded by Norden Dam, as a national scenic river under the National Wild and Scenic Rivers Act, which would limit further development along that stretch. Coming on the heels of the defeat of Norden, the bill generated immediate controversy. It was introduced at the urging of some local landowners, but there were still questions about the level of local support. A subsequent landowner-opinion poll was inconclusive. County governments and NRDs in the project region generally opposed the designation. A statewide poll by the *Omaha World-Herald* and a differently worded one commissioned by the Nebraska Water Resources Association, a pro-development organization, produced conflicting results on statewide preferences.

Soil Conservation Service

American coots on Funk's Lagoon in Phelps County, a Nebraska Rainbasin wetland

A vigorous fight ensued, much of it over what a designation would actually mean for the area. Supporters saw the bill as a way to prevent undesirable development along the river while allowing existing uses to continue. They emphasized that no more than 5 percent of the land in the river reach was subject to condemnation for easements. Opponents viewed the scenic river bill as precluding water-development options and killing all hope of ever reviving Norden Dam. But primarily they saw it as outside interference in landowners' rights in an area where landowners believed they had done an excellent job of stewardship.

By March of 1990, Senator Exon and each of Nebraska's three representatives had bills before Congress. Congresswoman Virginia Smith, representing most of the scenic river area, fought long and hard to preserve local options. Her bill called for a study of the scenic river area. However, in January of 1991 she retired from Congress.

The following May, President George Bush signed the Niobrara Scenic River bill into law. It designated 70 miles of the originally proposed 76-mile segment as a federal Wild and Scenic River. Six intervening miles were to be designated if Congress did not authorize and appropriate funds within five years to construct a water project on that segment of the river. That left the door open at least temporarily for the Springview part of the O'Neill Unit. The bill also called for a study of the suitability of establishing a national park near the scenic river; designated a 39-mile segment of the Missouri River and the lower portions of the Niobrara and Verdigre Creek as Recreation Rivers; called for study of a national recreation area in Boyd and Knox counties; and placed limitations on land acquisitions.

The Norden-Dam-to-scenic-river transformation was symbolic of the larger changes in Nebraska and national water politics. Environmental benefits had proven more politically important than the benefits provided by project construction. In 14 years the government had come full circle from financing the start of dam construction to regulations nearly precluding most development.

The Water Independence Congress

Governor Robert Kerrey's 1983 Water Independence Congress was a notable attempt to reach compromise between environmental and development interests. The governor, a Democrat from Lincoln, had appointed 40 Nebraskans of diverse backgrounds to develop recommendations and a consensus on water policy. Their report served as the major push for passage of LB 1106 in 1984. That bill created a Water Management Board to, "identify, propose, support, advocate, resolve conflicts regarding and expedite water development projects . . ." To fulfill this tall order, a water management fund was created to finance projects of more than $10 million. The environmental side of the equation was that the bill contained provisions allowing the state Game and Parks Commission and NRDs to apply for instream-flow rights. The bill also contained provisions making it mandatory for NRDs to complete groundwater management plans.

At the time it seemed to be major legislation indeed. However, the Water Management Fund never received any significant funding, and in 1991, the legislature abolished the board. It had received only one funding request during its five-plus years of existence.

The Game and Parks Commission has succeeded in gaining an instream-flow appropriation to protect fishery values in a reach of Long Pine Creek. And in 1992, the Central Platte NRD was granted an instream-flow appropriation for the Platte River. The process for proving instream needs and obtaining a water right is lengthy. However, the effect of this part of the law may continue to grow. The Game and Parks Commission is planning to seek instream water rights on other streams. Overall, LB 1106, which at first glance seemed a major step forward for surface-water development, may, in the long run, turn out to be an environmental-protection law.

Federal Wetlands Regulation

Meanwhile, environmental legislation was proving even more potent at the federal level. By the early 1990s federal wetlands regulation had become a major national land- and water-use issue. In general wetlands are areas saturated by water often enough to support a prevalence of vegetation adapted to saturated soil conditions. They are usually defined in terms of characteristic soils, vegetation and hydrology. But providing a definition or method of determining wetlands risks stepping into a political firestorm, at least in part because wetlands are usually ponding up or draining and drying out and therefore constitute a "boundary condition," the state of a resource mostly in flux.

Wetlands provide not only fish and wildlife habitat but recreation, improved water quality, flood-damage protection, water supply, groundwater recharge and the harvest of wetland products. They are often considered important as the base of an entire ecological system rather than as a single land-use value. Questions in Nebraska and elsewhere have revolved around what constitutes a wetland, how much regulation of private land is proper and how much money should be spent on governmental purchase of wetlands. According to Fish and Wildlife Service estimates, by the 1980s Nebraska had lost about 35 percent of the wetland acreage it had in the 1780s.

In recent years an array of federal wetlands programs have been established. The major federal programs for regulating wetland activities are Section 404 of the Clean Water Act and the "swampbuster" provisions of the farm bills. The 404 program regulates discharge of fill material into waters for a variety of uses, including conversion (draining or filling) of wetlands. It is administered by the U.S. Army Corps of Engineers and the Environmental Protection Agency. The USDA has administered the wetland-preservation provisions of the 1985 and 1990 farm bills. The 1985 farm bill contained "swampbuster" provisions that denied federal farm-program benefits to landowners who converted wetlands to land producing farm-program crops; the 1990 bill denied many benefits for any conversion of wetlands but introduced a graduated set of penalities depending on the severity of the offense and allowed for some mitigation through creating or restoring wetlands.

If this alphabet soup of agencies sounds confusing, that's because it is. At times landowners have become frustrated by the combination of limitations placed on the use of their land and having to deal with many federal agencies to find out what they can do. Some feel the regulations constitute an illegal "taking" of the land.

Even deciding on a single definition of a wetland for use by all agencies has proven difficult. President Bush announced a policy of "no net loss" of wetlands in 1988, but as of 1992 federal agencies had not yet agreed to a final manual for defining and classifying wetlands. Along with regulatory programs, the federal government also maintained a program of wetland acquisition through the Fish and Wildlife Service.

Interstate Issues

Interstate water disputes are newsworthy by their very nature. States have conflicts with each other on many issues, but water seems to create confrontations with abysmal regularity. Nebraska is no stranger to such controversy. Since 1970 interstate disputes have raged over Wyoming's Grayrocks Dam, the North Platte Decree, the operation of the Missouri River main-stem reservoirs, the operation of Nebraska's Harlan County Lake, a South Dakota-sponsored proposal to use groundwater for a coal-slurry pipeline, and Colorado's proposed Narrows Unit and Two Forks Reservoir. The proposed reservoir on Deer Creek in Wyoming has produced a controversy over whether it is governed by the North Platte Decree. Nebraska says, "yes," Wyoming says, "no," and the federal arbitrator known as a special master says, "maybe."

In addition, water-transfer proposals and the Sporhase Supreme Court case have raised questions about how water should be allocated between states. A related topic over which lawsuits could develop is water rights of the Indian tribes along the Missouri. Even water quality may become an issue. Kansas has protested pesticide levels sometimes found in the Big Blue River. In the past interstate conflicts have been addressed primarily through rulings in the U.S. Supreme Court (such as the North Platte Decree), voluntary agreements (such as river-basin compacts) or Congressional legislation.

To properly allocate surface water, Nebraska has, since 1923, entered into four separate river-basin compacts. These allocate waters of the South Platte River basin (including Lodgepole Creek), the Republican River basin, the Blue Rivers basins and the Upper Niobrara River basin.

Interstate disagreements are natural regarding a commodity that flows across boundaries. Water is the lifeblood of the economy and environment in western states. Therefore, it is hardly surprising that a partisan atmosphere has developed around the water rights of each state. Some Nebraskans may still bemoan that the state gave away too much water to Colorado in the South Platte Compact of 1923, but it was a voluntary agreement. Fights over the allocation of North Platte water led to the U.S. Supreme Court's North Platte Decree in 1945. Considering the value of the water, it is not entirely surprising that the states are now litigating over how that decree should be interpreted. Congressional passage of environmental legislation such as the Clean Water Act, the National Environmental Policy Act and the Endangered Species Act have given additional grounds on which to oppose projects and may be the major battleground of many interstate water battles to come.

Water Transfers

Water transfer is an issue that has long bedeviled Nebraska. In 1936, the Nebraska Supreme Court's decision in Osterman vs. Central Nebraska Public Power District prohibited interbasin transfers of water. That decision was not overruled until 1980. There are now different standards for out-of-basin transfers of surface water that provide some protection for the basin of origin. Groundwater generally cannot be transferred off the overlying land. But there are exceptions for municipal, large-scale industrial and some other uses.

The great fear many Nebraskans have about transfers is that some of Nebraska's areas of abundant water supply will be bled dry by surrounding areas with lesser endowments of water. Others see an opportunity to use water more efficiently and thereby benefit the state's economy.

The 1987 Unicameral directed the Water Management Board to study the issue. The board's final report contained draft bills that would have permitted more water transfers if conditions designed to protect the area losing the water could be met. However, those bills failed to pass.

Through the years there have been proposals for large-scale surface-water diversions of rivers such as the Missouri. There has been talk of large-scale diversions of part of the groundwater underlying the Sand Hills for everything from Denver municipal use to agricultural use in other states. Given current circumstances these types of large scale long distance schemes don't appear even close to being economically viable. However, the controversy will likely continue because economic conditions change and poor economic returns have not always prohibited water-project construction.

Surface-water and Groundwater Relationships

If transfers of water hold some fascination for the public, the subject of conjunctive use of groundwater and surface water is almost numbing in its appeal and daunting in its complexity. The physical relationships that exist between groundwater and surface water can be complicated and variable. But under Nebraska law they generally have little legal relationship. This means that groundwater pumping may affect surface flow and the surface-water user will have no recourse. Similarly, groundwater users in some areas may be dependent on surface flows to recharge their source of water but have no protection from additional surface-water use.

Major municipal water systems inducing river recharge to well fields for at least part of their supply include Omaha, Lincoln, Fremont, Grand Island, Kearney and Sidney. The Unicameral has discussed legislation to help these cities obtain surface-water rights to protect their source of supply. In 1983 the Unicameral passed legislation recognizing artificial recharge of groundwater as a beneficial use and allowing new surface-water projects to assess charges on those who benefit from the recharge attributed to those projects. The 1992 Unicameral gave serious consideration to bills that would have allowed municipalities to obtain surface-water rights for their induced recharge into aquifers near rivers. This would place some groundwater users in the appropriative system. It seems likely the issue will receive further discussion in upcoming sessions.

Water Conservation and Water-use Efficiency

Water conservation has a double meaning in the West. To some it means "store more water." To others it means "use less water." Both viewpoints have validity. In a water-short area, keeping water from flowing away does indeed preserve some of it for future use. Generally that means building dams or inducing groundwater recharge. However, avoiding overapplication and/or inefficient storage, transportation or use of water also can save water.

The effect of water-efficiency measures depends on where they are established. If an area is not likely to experience a water shortage, the water "saved" will not necessarily be put to use. However, lowered energy costs, less chemical and fertilizer leaching and reduced soil erosion can mean a substantial benefit even where water savings themselves are not important. State regulation of the efficiency of surface-water use has proven a difficult subject to address. In part, this is because one appropriator may use what another "wastes." Nonetheless, various proposals have been made to encourage more efficient water use while allowing those conserving to keep part of the "savings."

In the last 20 years, state and local governments have intensified efforts to see that Nebraskans use water more efficiently. Because agriculture uses the lion's share of Nebraska's water, efforts have been concentrated in that sector. Nebraska statutes place some limitations on excess use of irrigation water. However, the most meaningful work in recent years has been education and demonstration efforts. From an irrigator's standpoint, these are essential because under-irrigation can result in an immediate decline in yields and reduced profits. In addition, Ground Water Control Area and Groundwater Management Area regulations can require water me-

The relationship between surface water and groundwater has become an increasingly important hydrological and political issue, particularly as many cities and towns draw drinking water from wells in aquifers near rivers. This is a Lincoln Water System pumphouse beside the Platte River near the city's Ashland well field.

Reprinted from Resource Notes, CSD, v. III

ters that measure the water used or even regulate the total amount of water used to grow a crop.

The Republican Basin Revisited

By the 1990s the system of reservoirs that Harry Strunk fought so hard to see built was experiencing major problems. The reservoirs have provided a wealth of recreational opportunities, plus flood-control and irrigation benefits, to the Republican River valley. But inflows to the reservoirs began dropping, and by the 1970s, there were serious concerns. Irrigators were not receiving expected supplies and recreation-based businesses became worried about reservoir drawdown. In the case of Harlan County Reservoir, the Corps of Engineers has been faced with the problem of allocating a scarce supply to competing irrigation and recreation interests. The decreasing inflows are probably due to a number of factors. Fish and wildlife also depend upon the water of the Republican Valley and have their defenders.

Increasing use of conservation measures has helped hold water on the land, saving water, soil and money, and prevents runoff to streams. Groundwater-irrigation development has probably lowered water tables and resulted in less groundwater contributing to streamflow. Cyclical precipitation patterns also figure in the equation. There is debate as to how much of the problem is due to any one cause, but all agree there is a problem.

Recent Policy Trends

So how has Nebraska water policy evolved in the last 20 years? Five separate trends can be identified. These involve:

1) The rise of greater responsibilities for NRDs resulting from a preference for local control;
2) An increasingly regulatory emphasis;
3) A shift in policy emphasis from water quantity to water quality;
4) Reduction in political and financial support for water-development projects; and
5) A growing role for fish, wildlife, recreational and general environmental concerns such as biodiversity and scenic values.

Most of these trends arise from the need to protect the environment. While the battles have often been portrayed as "environmentalist" versus "developer" confrontations, that image is only partially correct. Barring technological advancements, sustaining economic development will mean sustaining the natural resource base that development uses. Not all Nebraskans place the same value on all aspects of the environment. However, the idea of being able to sustain economic use of the state's water has strong appeal to many Nebraskans. Indeed, almost every NRD adopted an infinite life goal for its groundwater reservoir in its groundwater management plan.

Of all the environmental legislation passed in the last two decades, the state and federal endangered species acts seem likely to have the most effect on future surface-water projects. Water is a basis of life and therefore has a major effect on the prospects of the human species and all other species. Water projects will often affect species protected by the endangered species acts. The acts may be the most effective legal tool for those opposing projects, no matter what their motivation for opposition.

While the trends of the last 20 years revolve largely around the role of environmental concerns versus development, there is one undercurrent that involves project economics and financing. When the era of large water projects began in the 1930s, government subsidies were seen as a way to provide employment, rescue agriculture and "jump-start" the economy. The success of the Tri-County project may have been that it was perceived as doing all of these.

The political mood of the 1980s and 1990s has been very different, and not just for water-project construction. Government subsidies are now more often seen as a drag on the economy. Agricultural surpluses and subsidies make construction for more agricultural production unacceptable to some. There are questions about the degree to which those receiving subsidies are in need. In an era of federal spending restraints, funding for water projects has suffered. Furthermore, environmental laws have provided a tool for opponents who may oppose a project primarily on economic grounds. Finally, the entire method by which the economic benefits of water projects are calculated has been called into question, making the process of building political support even more difficult.

Whatever the outcome of the fights over the environmental and economic effects of water projects, it is plain that many important political fights are likely to revolve around use of Nebraska's groundwater. Prior to 1957, groundwater was a seldom-mentioned topic in Nebraska water politics and policy. Now it is an everyday policy issue. Groundwater now supplies about 85 percent of the irrigated acreage in the state and supplies drinking water for more than 80 percent of Nebraska's population (although partially from induced recharge from surface water). Its importance has grown steadily since the 1940s, and the trend seems likely to continue.

The last 20 years have been an era of confusion and emotion in Nebraska water policy. The reasons for that lie not so much in Nebraska's institutions or policy makers as in the type of issues that have surfaced. How do you value a scenic vista? How do you value a whooping crane? In assessing the effects of groundwater contamination, how do you value risk to a human life? How do you make a wise judgment on a proposal if health or environmental data are inconclusive?

The values of human life and the environment don't fit neatly into an economic framework. In any case, there is argument over how to calculate economic benefits. In the midst of this confusion, the proper roles of government, private property rights and the rights of

Children tasting the difference between saltwater and freshwater at a demonstration on water quality put on by the U.S. Army Corps of Engineers at the Children's Groundwater Festival in Grand Island, 1992.

Charles Flowerday, CSD

future generations must be weighed. No wonder there have been conflicts. No wonder there has been confusion.

Conclusion

It has been said that "Whiskey is for drinkin' and water is for fightin'." Water is a resource that moves across boundaries, changes form and readily takes on contaminants. It can have great value. Its movements are complex. It is custom-made for "fightin'." In a state as dependent upon its water as Nebraska, the surprise isn't that there are disagreements on water issues. The surprise may be that there are not more disagreements.

Nebraskans are faced with difficult questions as they enter the mid-1990s and look toward the 21st Century. How do we evaluate the economic versus environmental consequences of our water-policy decisions? Who should bear the costs and who should receive the benefits of alternative water policies—what is fair? How do we balance long-term versus short-term concerns? Who should make our water-resources decisions?

Reasonable people can answer these questions very differently. This indicates the next 100 years of Nebraska water politics may be as interesting as the first 100 were.

If promoters, politicians or preservationists sometimes react to water in a mercurial manner, perhaps it reflects the nature of the invaluable substance that set off all the controversy. In describing this protean substance, novelist Tom Robbins may have best captured the changeable "character" of the commodity that has changed much of Nebraska politics:

> "Water—the ace of elements. Water dives from the clouds without parachute, wings or safety net. Water runs over the steepest precipice and blinks not a lash. Water is buried and rises again; water walks on fire and fire gets the blisters. Stylishly composed in any situation—solid, gas or liquid—speaking in penetrating dialects understood by all things—animal, vegetable or mineral—water travels intrepidly through four dimensions, sustaining (Kick a lettuce in the field and it will yell "Water!"), destroying (The Dutch boy's finger remembered the view from Ararat) and creating (It has even been said that human beings were invented by water as a device for transporting itself from one place to another, but that's another story). Always in motion, ever-flowing (whether at steam rate or glacier speed), rhythmic, dynamic, ubiquitous, changing and working its changes, a mathematics turned wrong side out, a philosophy in reverse, the ongoing odyssey of water is virtually irresistible."

Sources: "Central District's Hydro Division" (brochure), Central Nebraska Public Power and Irrigation District, undated; "Central District's Irrigation Division" (brochure), Central Nebraska Public Power and Irrigation District, undated; *Cadillac Desert—The American West and Its Disappearing Water,* Marc Reisner, 1986, Viking Press, New York; "Compromise Enacted," Lincoln Star editorial, May 27, 1971; *Even Cowgirls Get the Blues,* Tom Robbins, 1976, Houghton Mifflin, Boston; *George Norris—Gentle Knight of American Democracy,* Norman L. Zucker, 1966, University of Illinois Press, Urbana and London; *George W. Norris—The Making of a Progressive, 1861-1912,* Richard Lowitt, 1963 Syracuse University Press, Syracuse, New York; *George W. Norris—The Persistence of a Progressive, 1913-1933,* Richard Lowitt, 1971, University of Illinois Press, Urbana, Chicago and London; *George W. Norris—The Triumph of a Progressive, 1933-1944,* Richard Lowitt, 1978, University of Illinois Press, Urbana, Chicago and London; "Ground Water From Windmills to Comprehensive Public Management," Ralph Fischer, Richard Harnsberger and Jarret Oeltjen, *in* Nebraska Law Review, v. 52, no. 2, 1973; "Harlan County Lake" (brochure), U.S. Army Corps of Engineers, 1987; *History of Nebraska,* James C. Olson, 1966, University of Nebraska Press, Lincoln and London; "A History of Nebraska's Natural Resources Districts," Hazel M. Jenkins, 1975, Nebraska Natural Resources Commission; *Irrigation Pioneers—A History of Tri-County to 1935,* Gene E. Hamaker, (1958 as dissertation) published in 1964, Warp Publishing Company, Minden, Nebraska, for Central Nebraska Public Power and Irrigation District; Laws of Nebraska, various years; Letter to Nebraska Senator J. James Exon, Robert Olsen (Acting Commissioner of Reclamation), March 11, 1985; Lincoln Star, various issues with stories on Maurice Kremer; "Mirage Flats, Nebraska" (brochure), U.S. Bureau of Reclamation, Department of Interior, 1984; "Nebraska Ground Water Resources," U.S. Geological Survey, 1984, U.S. Geological Survey Water Supply Paper No. 2275, *in National Water Summary;* "Nebraska Surface Water Resources," U.S. Geological Survey, 1985, U.S. Geological Survey Water Supply Paper No. 2300, *in National Water Summary;* "Nebraska Water Supply and Use," U.S. Geological Survey, 1987, U.S. Geological Survey Water Supply Paper No. 2350, *in National Water Summary;* "Nebraska 1990 Water Quality Report, Nebraska Department of Environmental Control (Quality); *Nebraska, No Place Quite Like it!—20 Years of Memories,* Harold Hamil, 1985, The Lowell Press, Kansas City, Missouri; "Nebraska's State Water Plan—Report on the Framework Study," 1971, Nebraska Soil and Water Conservation Commission; Nebraska Revised Statutes, various chapters; "The Niobrara—A Proposal for Scenic River Designation," Nebraska Natural Resources Commission, August 1986; "North Loup, Nebraska" (brochure), U.S. Bureau of Reclamation, Department of Interior, 1984; "North Platte River Projects, Wyoming-Nebraska" (brochure), U.S. Bureau of Reclamation, Department of Interior, undated; "The Pick-Sloan Missouri Basin Plan—Conference Proceedings," Missouri Basin States Association, August 10-11, 1983; "The Pick-Sloan Missouri Basin Program and Individually Authorized Projects—Seminar Proceedings," Missouri River Basin Commission and the Western States Water Council, October 28, 1980; "Powerful Pleasure and Plentiful Power" (brochure), Central Nebraska Public Power and Irrigation District, undated; "A Profile of Harry Strunk," Allen D. Strunk, unpublished paper; *Public Power in Nebraska—A Report on State Ownership,* Robert E. Firth, 1962, The University of Nebraska Press, Lincoln; "Republican Valley Development," Robert Kutz, unpublished paper, Nebraska-Kansas Projects Office of the U.S. Bureau of Reclamation; "Rights to Nebraska Streamflows—An Historical Overview with Recommendations," Ralph Fischer, Richard Harnsberger and Jarret Oeltjen, Nebraska Law Review, v. 52, no. 3, 1973; "Sargent-Farwell, Nebraska (brochure), U.S. Bureau of Reclamation, Department of Interior, 1982; "Twenty-two Years of Progress in Soil and Water Conservation in Nebraska" (booklet), Nebraska Soil and Water Conservation Committee and the Nebraska Association of Soil and Water Conservation Districts, 1959; *Water in Nebraska—Use, Politics, Policies,* Aucoin, James, 1984 University of Nebraska Press, Lincoln and London; "Wetlands—Their Potential for Profit: Alternative Usages of Wetlands Other Than Conventional Farming in Iowa, Kansas, Missouri and Nebraska" (brochure), U.S. Environmental Protection Agency, November 29, 1991.

Interview with: Zack Hill on municipal water use.

Through most of the post-settlement history of water in Nebraska, the scenic and aesthetic properties of water were largely taken for granted or ignored. Now, due to citizens' demands, these values too must be figured into the political equation determining water benefits. This photo shows Snake River Falls in Cherry County.

Charles Flowerday, CSD

Water Policy Fought Out in the Law: The Evolution of Nebraska Water Law

by J. David Aiken
Extension Agricultural and Water Law Specialist, UNL

The evolution of Nebraska water law reflects changing public attitudes and values regarding water and the environment over the last 100 years. During the first period of its evolution—from the 1890s to the 1970s—development of surface water for irrigation was the primary policy objective. Beginning in the 1970s, environmental concerns led to new federal laws dealing with environmental impacts of federal projects and protection of endangered wildlife species and plants. These new laws conflicted with earlier irrigation-development policies and resulted in court battles stopping Nebraska irrigation projects.

Nebraska groundwater management policies dealing with depletion and, later, groundwater quality, began in 1975. Rights for instream flow for fish, wildlife and recreation were authorized in 1984, reflecting the new environmental consciousness as well as an attempt to accommodate competing irrigation and environmental values.

Future policy challenges for the next 100 years include state groundwater regulations, state pesticide regulations, instream-flow rights, surface water-groundwater conflicts and water marketing.

Surface-water Development

Prior Appropriation

In 1895 the Nebraska Legislature adopted prior appropriation ("first in time is first in right") as the state's surface-water law. The irrigation code made the state responsible for establishing and administering surface water rights, which was essential to successful irrigation development. That year the legislature also authorized the creation of irrigation districts to finance irrigation projects. The 1902 Reclamation Act provided critical federal funding for western irrigation projects, and Nebraska irrigators in the Scottsbluff valley received water from the first federal reclamation project.

Trans-basin Diversion

In the 1930s the Central Nebraska Public Power and Irrigation District proposed to build Kingsley Dam, creating Lake McConaughy to irrigate land in the Platte and Republican River valleys. In 1936 the Nebraska Supreme Court ruled that Platte River water could not be used to irrigate land in the Republican River basin. This prohibition against trans-basin diversion continued until 1980, when the Nebraska Supreme Court reversed its 1936 decision. The controversial trans-basin diversion prohibition slowed water-project development on the Platte, and consequently provided indirect protection of Platte River instream-water uses.

The Environmental Revolution

By about 1970 the United States was undergoing a profound change in its public policies towards natural resources and the environment. The new philosophy focused on protecting the environment from pollution and upon resource protection and preservation rather than economic development. This was sometimes expressed in terms of the need to consider quality, as well as quantity, of life. In light of this change, two new laws had special significance for Nebraska irrigation projects: the National Environmental Policy Act, which required preparation of environmental impact statements for federal projects, and the federal endangered species legislation, which prohibited federal agencies from taking any action harming endangered wildlife or their habitat. The new environmental laws led to court battles in Nebraska stopping or delaying irrigation projects.

Norden Dam

The Norden Dam on the Niobrara River was proposed to supply water to irrigators depleting local groundwater supplies. A federal court ruled in 1977 that the project's environmental impact statement was deficient in failing to consider a number of issues, including groundwater quality. While a new impact statement was prepared, project opponents convinced Congress to stop funding the project. Eventually, in 1991, the Niobrara River (including the dam site) was added to the federal Wild and Scenic River system.

Grayrocks Dam

The Grayrocks Dam on the Laramie River (tributary to the North Platte River in Wyoming) was proposed to provide cooling water for an electrical plant on the Laramie. Grayrocks would also have reduced North Platte River flows into Nebraska to the detriment of Nebraska irrigators and hydropower producers.

In 1978 a federal court ruled that the project's environmental impact statement should have considered the downstream impacts of reduced streamflows from Grayrocks in Nebraska (not just in Wyoming), especially the potential impact on critical whooping crane habitat in the central Platte River basin. The power plant developers consequently agreed to release water from Grayrocks for Nebraska irrigators and established the $7.5 million Platte River Whooping Crane Habitat Maintenance Trust. The trust has since become a major player in Platte River basin water issues.

Catherland Project

The Little Blue Natural Resources District (NRD) proposed the Catherland project to supply Platte River water to Blue River irrigators depleting local groundwater supplies. Catherland involved a trans-basin diversion, but in 1980 the Nebraska Supreme Court ruled that the diversion was legal, *overruling the 1936 Kingsley Dam decision.*

In 1982 the Nebraska Supreme Court further ruled that the impact of the Catherland project on Platte River endangered wildlife species and their habitat must be considered before the state could issue the project water rights. After continued controversy regarding endangered species and instream flows, the Nebraska Supreme Court in 1988 finally dismissed the Catherland water rights on a legal technicality.

Instream Appropriations

The court battles that derailed the Norden and Catherland projects were disputes regarding whether the irrigation projects would harm instream flows for fish, wildlife and recreation. After the Catherland-endangered species decision, Governor Bob Kerrey convened the Nebraska Water Independence Congress in 1983 to develop policies for resolving the irrigation-environmental impasse. The congress recommended establishing instream-appropriation statutes, which were legislatively adopted in 1984. Now water rights can be acquired in Nebraska in order to leave water in the stream (rather than divert or store the water) for fish, wildlife and recreation. In 1991 new legislation was adopted authorizing the use of water stored in a reservoir for instream flows.

Groundwater Management

Groundwater Depletion

Intensive well irrigation has led to groundwater supplies being depleted in several areas of Nebraska more rapidly than they are recharged from precipitation. Many irrigators prefer to deal with depletion through a dam for irrigation, but such projects are difficult to implement due to financial and environmental constraints.

Another option for dealing with depletion is regulating groundwater withdrawals to lengthen aquifer life. Groundwater management statutes were adopted in 1975 which authorize NRDs to regulate groundwater withdrawals and well drilling in state-designated groundwater control areas. Three control areas have been designated, and groundwater withdrawals are metered and restricted in the Upper Republican NRD. In January of 1993, restrictions on withdrawals are to go into effect in the Little Blue groundwater control area in the northern part of that NRD. Similar regulations may be established in NRD groundwater management areas. The Upper Republican program is a model for other states and NRDs dealing with depletion.

Groundwater Quality

The 1975 groundwater management legislation was later expanded to include groundwater-quality protection. NRDs may regulate irrigation practices and agricultural chemical use to protect groundwater quality in state-designated special groundwater-quality protection areas. Two protection areas have been designated, and irrigation and fertilizer restrictions have been established in southern Nuckolls County. Similar regulations have been established in NRD groundwater-quality management areas. NRD nitrogen management programs are models for other states dealing with groundwater contamination from fertilizer use.

Future Policy Issues and Recommendations

Several unresolved Nebraska water policy challenges remain, including state groundwater regulations, state pesticide regulations, Platte River instream flows, surface-groundwater conflicts and water marketing.

Groundwater Depletion

Some areas of Nebraska are experiencing groundwater depletion, but the local NRDs have regulated well withdrawals to extend aquifer life. New legislation should allow state groundwater regulations to control depletion where local NRDs have not addressed the problem.

Pesticide Regulation

Nebraska is the only state not administering federal pesticide regulations. This makes Nebraska ineligible to develop a state pesticide-management plan protecting groundwater quality. Pesticide plan preparation and implementation is required by federal law in order to continue using pesticides most likely to contaminate groundwater (such as atrazine). New legislation should authorize state administration of federal pesticide regulations, as well as require preparation of the state pesticide-management plan to protect groundwater quality from pesticide use.

Instream Flows

When instream-appropriation legislation was enacted in 1984, there was some hope that this would reduce litigation regarding instream flows and new irrigation projects. This hope has not been realized. One limitation of the instream-appropriation law limiting its usefulness in resolving Platte River instream-flow controversies is that unappropriated water must be available to meet the total instream flow needs—partial instream appropriations are not possible. This creates difficulties on the Platte, where instream flow needs are great, but unappropriated streamflow is not available to meet those needs without additional water from other sources (for example, dams). Instream-appropriation statutes should be modified to authorize partial instream appropriations and the use of water stored in dams or other sources to meet total instream flow needs.

Surface water-groundwater Conflicts

Several Nebraska municipalities (including Omaha, Lincoln, Grand Island, Kearney and Fremont) depend on Platte streamflow to recharge aquifers near the river that supplies municipal wells. These cities, however, have no surface water rights to protect those wells. While Nebraska statutes do recognize that irrigation projects may result in groundwater recharge, our water law does not define how to resolve disputes between competing surface water and groundwater users.

All other western states apply surface water law to groundwater hydrologically connected to a stream. However, Nebraska groundwater irrigators historically have opposed adopting such legislation. While there is general agreement that Nebraska surface water and groundwater law should be integrated, water users disagree regarding whether the governing law should be surface water law ("first in time is first in right") or groundwater law ("sharing during times of shortage").

Water Marketing

Buying and selling water rights is called water marketing. Surface-water rights may be sold in Nebraska but only with tight restrictions; the purchased water rights may be used only for the same purpose (for example, irrigation) and then only within the same river basin. The modern trend for dealing with municipal water shortages in the West is for cities to buy the needed water from irrigators. In some western states, cities may also pay for lining irrigation canals and obtain a part of the water saved. Both methods are much less expensive than building a new dam to supply municipal water needs.

Making surface-water rights marketable would also provide much needed flexibility in resolving Nebraska water disputes. One alternative to court battles over instream flows could be to purchase irrigation water rights for instream uses.

Implementing these policy changes will help move Nebraska water law and policy into the next century.

Profiles:

C. Petrus Peterson: From the Nebraska House to the National Reclamation Association

C. Petrus Peterson devoted his entire adult life to promoting water resources development and conservation.

Born in 1880, he obtained his law degree from the University of Nebraska in 1909 and practiced in Wahoo and Lincoln. He was elected to the Nebraska House of Representatives in 1915 and served through 1918. He served in the state Senate from 1919 to 1941, the year the Unicameral legislature was formed. He was a senator in the Unicameral from 1941 through 1957.

A few minutes before adjournment of a January 1944 meeting of the Nebraska legislature's Water Resources Committee, Harry D. Strunk, a newpaper editor from McCook and the state representative of the National Reclamation Association, made a plea for the establishment of a state reclamation association. Peterson, chair of the committee, put the question to motion and Strunk seconded it; the motion carried.

From 1945 to 1947 and again from 1949 to 1965, Peterson served on the board of directors of the Nebraska Reclamation Association, now known as the Nebraska Water Resources Association. In 1946 and 1947 he was association president.

He said in 1946, "Our supreme task as an association is to challenge the attention of our people to the imperative demand of nature that these basic resources be carefully preserved and adequately developed."

In 1947, and again from 1959 until his death in 1965, he was a member of the board of directors of the National Reclamation Association, holding the distinction of being the only Nebraskan ever to be president of the national association, a post he held from 1952 to 1955. He is a member of the National Reclamation Association Hall of Fame.

—Brad Rundquist, weekend and copy editor,* Ft. Dodge (Iowa) Messenger, *and formerly CSD editorial assistant

Harry D. Strunk: Newspaper Man Claimed the Cause of Reclamation

Born on a farm near Pawnee City on Dec. 2, 1892, Harry D. Strunk's first ambition was to become an artist or musician. However, he was forced to quit school in the eighth grade to help support the family. That year, at age 14, Strunk's father took him to the local newspaper office, and he was hired as a printer's devil at $2 a week.

After working a year for the *Pawnee City Republican*, Strunk moved to the Fairbury paper for $7 a week. He returned to Pawnee City shortly and, at age 16, took over publication of the paper, as the editor had taken sick.

After three months, Strunk moved on to Powhattan, Kansas, to publish for an absentee owner. Shortly, at age 17, he took over publication of the *Norton Daily Telegram* in Norton, Kansas.

Later, he returned to Nebraska and worked at several small weeklies before he got a job at the *McCook Tribune*. After nine months of work there, Strunk and a colleague opened a printing shop. The shop drained their money and, out of desperation, they launched a third paper in the town of 4,000, the *Red Willow County Gazette*. The year was 1911; Strunk was 19. In 1914, the paper's name was changed to the *McCook Daily Gazette*.

Strunk became interested in reclamation of the state's natural resources in 1928, devoting many editorials to the subject. That year, he worked with Republican Valley community and business leaders to organize the Twin Valley Conservation Association for the purpose of controlling water and preserving soil in southwestern Nebraska.

Strunk organized a Flood Congress in 1929. The congress, which met in McCook, was attended by governors and representatives from 14 states. The framework for what was later called the Pick-Sloan Plan was established at that meeting.

After the tremendous Republican River flood of 1935 *(see vignette on Republican River flood, p. 58)*, Strunk's efforts intensified and almost all of his time was devoted to the cause of reclamation. He expanded the Twin Valley association into the Republican Valley Conservation Association and was made its president. The association sent lobbyists to Washington to demand action. Strunk continually needled officials, congressmen and even presidents Harry Truman and Dwight Eisenhower.

He was made state representative of the National Reclamation Association in 1940 and, at an association meeting in Denver, recommended a coordinated plan of development for the Missouri Basin as drafted at the Flood Congress in 1929. Plans for such development were scrapped when World War II started in 1941, but Strunk was determined to again take up the fight at war's end. Pick-Sloan legislation was passed in 1944.

That year, Strunk helped organize the Nebraska Reclamation Association—now the Nebraska Water Resources Association.

The Republican Valley was the first to receive Pick-Sloan money, and dirt began to fly two years after the war ended to construct five dams and reservoirs in Nebraska. Those reservoirs are now commonly referred to as Nebraska's "Great Lakes." The first reservoir, completed on Medicine Creek, is named Harry Strunk Lake.

Strunk died Aug. 5, 1960, one month after the last of the Republican Valley projects was begun.

—Brad Rundquist, weekend and copy editor,* Ft. Dodge (Iowa) Messenger, *and formerly CSD editorial assistant

Stan Matzke: Assistant General Manager of Central and Legislator

If variety is truly the spice of life, Stanley A. Matzke lived a potpourri.

Along the way, Matzke worked as a hardware salesman, farm implement dealer, bookkeeper, assistant bank cashier, lawyer, county attorney, building and loan association official, farm editor of a daily newspaper and farm reporter for a radio station. He was also director of the Nebraska Department of Insurance, assistant general manager of the Central Nebraska Public Power and Irrigation District (CNPPID) and served four terms in the Nebraska State Legislature. In the legislature and at CNPPID, Matzke worked hard to promote the development of the state's water resources.

Born on a Seward County farm on July 27, 1898, Matzke enlisted in the Army immediately after graduating from Goehner High School in 1917. He never made it to the battlefields of World War I, however, as he contracted a serious case of influenza and didn't recover until war's end. After military discharge, he worked as a tradesman, banker and bookkeeper in Milford and Goehner. On a dare, Matzke enrolled in the University of Nebraska Law College and received his degree in 1924. He had a practice in Milford for a time and soon moved to Seward to serve as county attorney for Milford County. He practiced law in Seward until 1945.

He later said, "I never really intended to become a lawyer and I didn't enjoy it when I was one; too much detail."

Matzke was elected to the state legislature in 1940 and re-elected in 1942. During his first term he drafted legislation that established the Nebraska Trade School at Milford.

Governor Dwight Griswold appointed him director of the Nebraska Department of Insurance in 1945. He left that job after a year to become farm editor at the Nebraska *State Journal*. He wrote for the paper until 1951, helping it win a Pulitzer Prize for public service in 1948. As secretary of the Salt-Wahoo Watershed Association from 1950 to 1953, Matzke promoted the new idea of the watershed approach to water management. Matzke became one of the leading promoters of the Salt Valley Flood Control project, which eventually created several lakes around Lincoln.

George E. Johnson, general manager of CNPPID, hired Matzke as his assistant in 1953. Matzke served as the district's public relations specialist, organizing tours for state officials and publishing information about the district.

He once said, "It has been my pleasure to have had intimate contact with the program which has made Nebraska one of the leading irrigation states in the nation."

He retired from CNPPID in 1963 to pursue a third term in the legislature. He served in 1964 and again in 1966.

While at CNPPID, Matzke became involved in the Nebraska State Irrigation Association, serving as secretary from 1957 through 1963 and treasurer from 1957 through 1967. With the Nebraska Water Resources Association, he served as director from 1962 to 1966, vice president in 1964, president in 1965 and executive director from 1966 until his death in 1969. He received the 1968 Headgate Award from the Four States Irrigation Council, which includes Nebraska, Colorado, Wyoming and Kansas, for his outstanding contributions to water control, conservation, development and use.

—Brad Rundquist, weekend and copy editor, Ft. Dodge (Iowa) Messenger; formerly CSD editorial assistant

Maurice Kremer: 'Mr. Water' Sponsored 24 Bills on H_2O

During his 20-year tenure in the Nebraska State Legislature, Maurice Kremer came to be known as "Mr. Water."

The Aurora senator sponsored at least 24 bills related to the management of Nebraska's water resources from 1962 to 1982. One of the most important created the state's 24, now 23, natural resources districts (NRDs).

The NRDs, broadly based local conservation agencies, were developed in 1972 to monitor, manage, protect and regulate the use of water and land in areas with common soil and water characteristics. They were created by combining many local resource organizations and districts, particularly soil and water conservation districts. Nebraska is the only state in the union with such a political subdivision.

Kremer, who died Feb. 10, 1990, at age 82, credited his water management ideas to the fact that he lived through the drought of the 1930s and saw the necessity for irrigation in central and western Nebraska.

Kremer was born on a farm near Milford, and his family moved to Hamilton County, near Aurora, when he was six. He attended the University of Nebraska College of Agriculture in 1930 and 1931 but never obtained a full college degree, as the terrible drought and ensuing depression cut short his education. He did, however, receive a 2-year degree.

Kremer became one of the first irrigators in Hamilton County when he installed a well in 1940. He later helped organize the Hamilton County Irrigators Association and became chairman of the group. The association succeeded in its goal to bring fertilizer to the county, as trainloads of ammonium nitrate were delivered between 1944 and 1946. The association began to measure groundwater levels in the mid-1940s, and therefore has the longest continuous water-level records of any organization in Nebraska.

The Hamilton County Groundwater Conservation District was later organized with Kremer's support and replaced the irrigators' association in the late 1940s.

Kremer believed strongly in farm conservation practices and constructed terraces, reuse pits and shelter belts on his land. By the mid-1980s, he owned eight wells.

For years, Kremer's associates urged him to run for state office, but he was reluctant. He finally did so only after the Aurora Chamber of Commerce petitioned for his candidacy.

In the legislature, Kremer's involvement with water management grew. He chaired the Agricultural Committee for four years and the Public Works Committee for 10. He was a key player in the passage of the original Groundwater Management Act of 1975, which allowed for the creation of local groundwater control areas, and in establishing the state's water planning and review process. He also helped develop water-management guidelines for the NRDs and processes for groundwater control and protection. As a member of the Education Committee, he worked to create the UNL Institute of Agriculture and Natural Resources and the state's community college system.

Kremer was awarded the Kiwanis Award for Distinguished Service in 1986 and the Nebraska Groundwater Foundation's Maurice Kremer Groundwater Achievement Award in 1987. A lecture series sponsored by the University of Nebraska-Lincoln Water Center is named in his honor.

—Brad Rundquist, weekend and copy editor, **Ft. Dodge (Iowa) Messenger,** *and formerly CSD editorial assistant*

Profiles of Organizations (Alphabetically):

Center for Rural Affairs: Advocacy for Small Farms and the Environment

For the past two decades from its offices in a renovated grocery in Walthill, Nebraska, the Center for Rural Affairs has provoked public thought about issues affecting rural people on the plains with an eye toward the fortunes of small farms and the environment.

So when center-pivot irrigation boomed in Nebraska during the 1970s, it was no surprise that the young organization of rural social activists declined to join the chorus of enthusiasm for the spreading technology. This was especially true for the development in the nearby eastern Sand Hills.

Born in 1973 in the midst of the state's irrigation bonanza, the private, non-profit, rural research and advocacy group took a sobering view of the center-pivot phenomenon. Formed by rural Nebraskans alarmed about the decline of family farms and rural communities, the center raised questions about the long-term social, economic and environmental costs of irrigation development.

Among the center's early projects was *Wheels of Fortune*, a 1976 study of the impact of large-scale irrigation on land ownership in Nebraska. It found that more than groundwater flowed through the center-pivot systems: irrigation also delivered tax breaks and promised increased land values to those investing in the technology. The study noted that absentee investors and corporations were cashing in on these benefits. Land-ownership data compiled by the center for the region covering Antelope, Brown, Pierce, Holt, Madison and Rock counties showed that investor-owned farms accounted for 30 percent of all pivot systems.

Center staff members and its urban-rural board warned that many investors with few ties to the communities near land they owned had irrigated fragile soils especially prone to wind erosion. They argued that the trend toward absentee ownership in Nebraska's irrigation belt undermined the social stability of rural areas and led to the abuse of natural resources for short-term economic gain.

By the early 1980s, the irrigation boom turned to bust when the farm economy worsened. Nebraska's Sand Hills were in the media spotlight; the nation's major newspapers ran photos of idled pivots and blowing sand. Publications by environmental groups did the same. At the same time, the center was pursuing efforts to eliminate tax shelters attractive to non-farm investors and place restrictions on corporate ownership of Nebraska farmland.

In 1982, along with other farm, labor, religious and environmental organizations, the center tapped into a deep vein of statewide public sentiment and won at the ballet box. Voters passed Initiative 300, an amendment to the state constitution banning farm corporations and limited partnerships from owning farmland.

Although the victory was not as clear-cut at the federal level, an 8-year campaign paid off by ending incentives for investors to farm the tax code. Under the Tax Reform Act of 1986, Congress repealed or scaled back tax measures that encouraged the conversion of native rangeland and pasture to irrigated crops.

At present, with a staff of 19, the center has shifted a major part of its efforts toward federal farm-program reform to protect both family farmers and the environment. Through the Midwest Sustainable Agriculture Working Group, a grassroots coalition of farm, religious, social justice and environmental groups formed in 1988, the center lobbied Congress to include provisions in the 1990 farm bill to curb groundwater contamination, conserve topsoil and promote sustainable farming practices.

Modest progress was made on resource stewardship in the 1990 farm legislation, according to Kris Thorp, who leads the center's Conservation and Environment Project. For example, the center is now promoting new USDA programs that provide incentives to rotate crops and protect water quality. In cooperation with the working group, the center opened a Sustainable Options Hotline to counsel farmers regarding the new Integrated Farm Management and Water Quality Incentives programs.

But in addition to these new voluntary initiatives, groundwater protection strategies for Nebraska must include more fundamental changes in federal farm policy to have a widespread impact on the rural environment, the center says. For example, government farm commodity programs push farmers to grow a few cash crops rather than use a diverse crop rotation. Farm program rules result in heavy use of commercial fertilizer and pesticides, which threaten groundwater quality and the health of rural people, the center believes.

The center also helps develop local leadership that will find solutions for rural environmental problems. Through a series of center-sponsored workshops across Nebraska, farmers and rural residents have discussed farm chemicals and groundwater quality. Participants recommended specific actions to address the issue on their own land and at local, state and federal levels. Long-range workshop goals include getting farmers, ranchers and other rural residents to take individual responsibility for problems in day-to-day operations and helping them get involved in setting public policy on issues that affect their lives.

—Devorah Lanner, paralegal/communications specialist, Nebraska Center for Legal Services; formerly CSD editorial staff specialist; formerly public affairs specialist and policy analyst, Center for Rural Affairs

National Audubon Society: Making the Platte a National Priority

The National Audubon Society manages two wildlife santuaries in Nebraska and sponsors an annual conference on the Platte River. In addition, it has tried to influence policy on Platte River flows to provide for instream flows for wildlife.

One santuary, probably best known, is the Lillian Annette Rowe Sanctuary in Gibbon, Nebraska. It covers 2,219 acres of tallgrass prairie, riparian woodlands, wet meadows and Platte River channel, serving as a main staging area for 500,000 lesser sandhill cranes and some whooping cranes on their annual spring and fall migrations. The other is the Niobrara River Sanctuary in Rock County, Nebraska, which covers 218 acres, including a half-mile stretch of riverfront along the south bank of the Niobrara River. The sanctuary includes native prairie, cliffs dropping off from wooded draws and a creek through a forested canyon. Cranes, prairie chickens, eagles, peregrine falcons, long-billed curlew, upland sandpipers and other wildlife use this habitat.

But in addition to the santuaries, during the late 1980s, the society declared the Platte one of its top national priorities, which included attempts to influence decisions regarding the relicensing of Kingsley Dam and apportionment of Platte River flows by filing statements with the Federal Energy Regulatory Commission. It also holds an annual conference on the river, the Spring River Conference, held in Kearney, which has attracted international speakers and guests. The 1991 conference was called, "People, Water and Wildlife: Human Population Impacts on Cranes." The Nebraska Audubon Council represents the six chapters in the state, which are in Grand Island, Kearney, Omaha, Scottsbluff and South Sioux City.

—Charles Flowerday, CSD editor

Nebraska Sustainable Agriculture Society: Demonstrating Sustainability

The Nebraska Sustainable Agriculture Society (NSAS) promotes the adoption of economically viable agricultural production practices that ensure the sustainability of natural resources.

Based in Hartington, the society seeks to teach farmers and landowners about practices that reduce or eliminate the need to apply pesticides and herbicides; make use of farm-based resources; replace purchased inputs with management strategies and skills; conserve and protect natural resources; and ensure an adequate and healthy food supply for the future.

Such practices include biological control of weeds and insects, the planting of windbreaks for crop protection and wildlife habitat enhancement, crop rotation to maintain soil fertility and to reduce erosion, improvement of soil structure to create better water-holding capacity, the use of alternative energy sources such as wind and sun power, and conservation tillage, contour plowing and contour strip cropping to reduce soil erosion.

The society, formally established in 1987, teaches producers mainly through workshops and farm tours. Forty-five workshops and 13 tours were held in 1991. Farm tours, according to the society, are the best way to spread the word about sustainable agriculture because participants can see real farmers using the techniques.

In addition, some local NSAS chapters make available various publications and video tapes relating to sustainable agriculture.

Each year, NSAS awards an Agricultural Stewardship Award to a farmer using sustainable practices. The society, which has about 300 members, also publishes a quarterly newsletter and directory of sustainable agriculture producers and experts.

—Brad Rundquist, weekend and copy editor,* Ft. Dodge (Iowa) Messenger, *and formerly CSD editorial assistant

Nebraska Water Resources Association: Water Users Settled Disputes, Banded Together to Ensure Supplies

The Nebraska Reclamation Association, now known as the Nebraska Water Resources Association, rose out of a January 1944 meeting in Omaha of the Nebraska legislature's Water Resources Committee, which had been appointed to examine a dispute between the Central Nebraska Public Power and Irrigation District and a group of concerned water users.

At that meeting, Senator C. Petrus Peterson *(see profile on Peterson, p. 232)*, chair of the committee, moved to establish the state reclamation association based on the recommendation of Harry D. Strunk *(see profile on Strunk, p. 233)*, a McCook newspaper editor and director of the state chapter of the National Reclamation Association. Strunk seconded the motion and it carried.

During the early years of the association's existence, much argument occurred among and between public and private water interests. The Nebraska Reclamation Association allowed those diverse interests to come together to air their differences. The association was, in effect, a group of mediators serving to foster cooperation. They were successful, to a large extent, as the number of disputes dwindled in the first few years of association efforts.

The association's name was changed in 1972 to reflect the expanded water interests of the organization. The Nebraska Water Resources Association (NWRA) is now affiliated with the National Water Resources Association.

NWRA's goals are to protect state water rights and existing irrigation and other water uses; to promote development of supplement water supplies to ensure the future availability of water for all needs; to conserve water by management, including irrigation scheduling, streamflow regulation, flood abatement, surface-water storage and supplementing surface and groundwater supplies; to strengthen the agricultural base of Nebraska through irrigation development; and to expand the economy of Nebraska through agricultural development and food production, as well as to represent fish and wildlife and recreation interests.

Association membership is composed of individual irrigators, irrigation districts, municipal and rural water systems, power districts and utilities, natural resources and water conservancy districts, agribusiness, financial institutions, construction and service industries, manufacturing and supply firms, professional and engineering groups, retail businesses, individual hunters and anglers and their interest groups and outdoor recreation interests and businesses. NWRA, a non-profit organization, is funded solely through membership dues. The association now has about 200 members.

—Brad Rundquist, weekend and copy editor,* Ft. Dodge (Iowa) Messenger, *and formerly CSD editorial assistant

Nebraska Well Drillers Association: Leadership for the Water Well Industry

Timing was right in the late 1920s for a focus on groundwater in Nebraska. A new science—groundwater hydrology—was slowly becoming established at the state and federal levels and desperately needed data. Well drilling as an industry was emerging out of the dark ages of "water witching" and "well digging." The close relationship between certain diseases—notably typhoid fever, cholera and diphtheria—and poorly constructed wells had been established. And early investigations suggested the potential of a vast groundwater resource in the state.

Dr. George E. Condra, dean and director of the Conservation and Survey Division (CSD) of the University of Nebraska, grasped the opportunity to provide service to the industry and to the people of the state. Condra, a visionary and a dedicated scientist, began two complementary activities. In 1929 he established a cooperative test-drilling and water-level monitoring program with the U.S. Geological Survey, and, with the help of a number of interested drillers, organized a statewide association of well drillers.

In January 1929, Condra invited 12 water-well drillers to the university campus to organize the association. Among those present at the first meeting were Harry Brown of Blair, Frank Cole of Loomis, Andrew Olson of Oakland, Fred D. Salmon of Wakefield, Jay Webb of Hastings, and Frank Williams of Belvidere. Brown was named interim president of the group, thereafter called the Nebraska Well Drillers Association (NWDA). Held in old Nebraska Hall on February 28, 1929, Condra set forth in the first meeting the goals of the association and advised the members that Nebraska was only the sixth state to form an association of water-well drillers. He helped establish the format of an annual convention, including a trade show and "Dutch Lunch" sponsored by the manufacturers and suppliers of drilling equipment. Seventeen manufacturing and supplier representatives attended that first meeting. The manufacturers and suppliers have been an integral part of NWDA events through the years.

The key element of that two-day convention was a "short course." The short course envisioned by Condra was designed to educate members about the geology and groundwater characteristics of the state and techniques of well drilling and construction needed to obtain a safe and adequate water supply. From the beginning a spirit of cooperation and sharing of ideas and techniques among the drillers was encouraged. Minutes of early meetings suggest that some members viewed idea sharing as a revelation of trade secrets. That barrier has gradually broken down, and over time cooperation and idea sharing among members became the strength of the organization.

CSD continued to sponsor the NWDA until 1982, when the association hired a part-time executive director and began to chart its own destiny. However, the division continues to work closely with NWDA, co-sponsoring its continuing education efforts across the state.

NWDA provided leadership for the water-well industry in developing industry standards. With the help of the University of Nebraska and several state offices, the drillers developed and adopted minimum standards for irrigation wells in 1957 and, in 1965, minimum standards for sanitary domestic wells. The water-well standards and contractors' licensing act of 1986 requires that well drillers and pump installers earn credits for educational activities attended and must pass an examination to receive a license. These requirements have spurred NWDA to increase the number and scope of these sessions.

The annual convention and trade show, held each February in Lincoln, is the highlight of the NWDA year. Hazardous weather has had little effect on the hardy well driller, accustomed to working under the most disastrous conditions, so the conventions are well attended. World War II travel restrictions resulted in the cancellation of the conventions of 1943, 1944 and 1945. But except for those three years, members and supporters of the drilling industry gather each February to exchange information, to see the latest products and to greet old friends. Through the years, NWDA has become one of the largest and most successful of state water well drillers associations in the country.

—Marj Seidel, formerly CSD administrative assistant, and Vince Dreeszen, CSD hydrogeologist emeritus, formerly CSD director

Nebraska Wildlife Federation: Protecting Wildlife and More

The Nebraska Wildlife Federation is an affiliate of the National Wildlife Federation. Both work to educate individuals and organizations to conserve natural resources, protect the natural environment and build a globally sustainable future.

The groups state that the quality of human life is inseparable from the existance of clean air, rich soil, pure water, mineral wealth and the abundance of plant and animal life.

The Nebraska group conducts workshops for educators, distributes educational materials to private and public school children, grants a yearly scholarship for $500 at the annual Greater Nebraska Science Fair, and sponsors a statewide essay contest for 4th through 6th graders.

In addition to its educational function, the Nebraksa federation assumes a legislative advocacy role. The group helped establish the Pine Ridge Wilderness Designation in Dawes County and the Soldier Creek Wilderness Area in Sioux County, promoted the Niobrara Wild and Scenic River Designation, is working with other groups toward protection of the Platte River for wildlife habitat, supports the national Endangered Species Act, defends the Clean Air and Water acts, and supports the sodbuster and swampbuster provisions in the national farm bills.

The National Wildlife Federation was established in 1936, the Nebraska affiliate in 1971. The Nebraska group has about 2,000 members. The national group has nearly 6 million members and an annual budget near $70 million.

—Brad Rundquist, weekend and copy editor,* Ft. Dodge (Iowa) Messenger, *and formerly CSD editorial assistant

Nebraskans First: Advocacy for Conservation and Groundwater Users

Nebraskans First is the state's newest water-development interest group. Founded in 1991, Nebraskans First represents the special interests of the state's groundwater irrigators.

By keeping an eye on groundwater-related bills in the state legislature, group organizers hope to keep their members informed, get them involved with groundwater issues and give them a voice in future regulatory decisions. They will also urge members and other groundwater irrigators to use more and better water-conservation practices. Nebraskans First was not formed to oppose groundwater regulation, its members say. Rather, it seeks to foster complete discussion before such restrictions are implemented.

Of immediate concern to the group is the reduction in groundwater recharge because of plans for conservation practices by local surface-water irrigation districts. By lining irrigation canals and laterals, irrigation districts have greatly reduced the amount of water lost to the soil and, eventually, to the groundwater reservoir. Such reduction has meant less water for pump irrigators.

Also of concern is legislative discussion about the relationship between groundwater and surface-water flows. The two systems are naturally related but that relationship can be complicated and varies from site to site. Nebraska law generally does not recognize the relationship except for projects that incidently or intentionally recharge groundwater. Legal recognition could place many new restrictions on groundwater pumping near streams to help ensure mandated in-stream flows.

—Brad Rundquist, weekend and copy editor, **Ft. Dodge (Iowa) Messenger,** *and formerly CSD editorial assistant*

Preserve Our Water Resources Association: Sand Hills Group Expands Scope

The Preserve Our Water Resources Association (POWR) strives to preserve any and all resources related to Nebraska water. Nebraska is in an enviable position of having ample quantities of good, clean groundwater and surface water. Very few other places are so fortunate.

The people of Nebraska, past and present, have made both wise uses and wastes of Nebraska's water resources. Certainly, conservation and preservation have not necessarily been the bywords of Nebraskans when pumping water from a seemingly endless supply or draining meadows that destroyed a wetland resource. Without more recent, enlightened scientific information, the various past practices were not only understandable, but were the best available practices at the time. Even now, considerable disagreement exists among all users of Nebraska water concerning the "highest" and best uses of our water.

During the late 1980s, there arose something of a movement among a very few Nebraskans to try to promote the sale of Nebraska water to out-of-state interests. The Sand Hills region was often singled out as the potential site for well fields that would provide marketable water. The Preserve Our Water Resources Association was initially formed in response to this "water sale" movement. The organization is mostly ranchers and residents of the Sand Hills. These folks know of the extremely fragile ecology of the Sand Hills and reacted uncharacteristically strongly when being made "sacrificial lambs" by those not directly involved in trying to make a living, and moreover a way of life, in the Sand Hills. These people know also that any substantial sale or cumulative exportation of Sand Hills water would likely result in serious consequences for not only the Sand Hills area, but also for more populous eastern Nebraska. The membership of the organization, showing the widespread concern for protecting water, has expanded to as far away as Virginia.

The POWR Association participates in the Sand Hills Task Force. POWR also works to promote active wetland conservation and protection of water quality. These efforts are to implement measures that will preserve the precious and vital water resources and associated ecology of the Nebraska Sand Hills.

—Bill Tumblin, president and founder, Preserve Our Water Resources Association

One main difference between surface-water development before and after the late 1930s was that social pressure, combined with advances in engineering and earth-moving technology, meant projects of much larger scale were possible, as seen in this brief photo essay of older and later projects. The hardships of the depression, the desire to irrigate more acres and a cry for flood control led to the construction of much larger dams than previously had been built. These needed large amounts of federal money and, therefore, caused a great deal more political commotion than the (mostly) privately funded projects of an earlier time, seen below and at the top of the next page.

Department of Water Resources

A concrete diversion dam on Sand Creek, Keith County, probably near the turn of the century. Small diversion dams for irrigation were the most common surface-water projects just before and just after 1900.

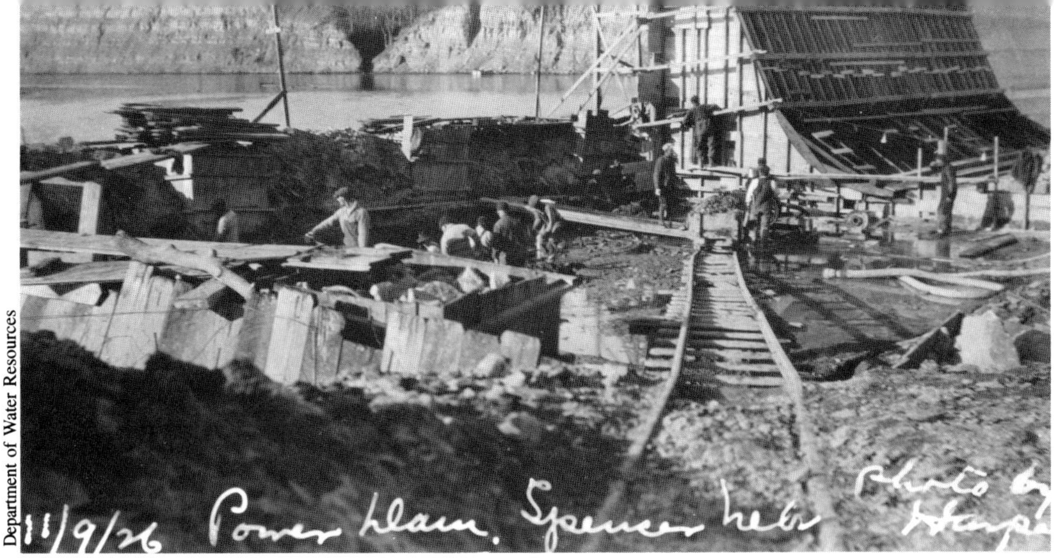

Department of Water Resources

Looking northwest upstream during construction of the Whiting Bridge Project and Spencer Power Dam by the Northern Nebraska Power Company, Nov. 9, 1926. Soon into the new century, somewhat larger projects, some developed for power generation as well as irrigation, were undertaken.

Department of Water Resources

Although it produced one of the smaller reservoirs developed just before and after World War II, these photos of the construction of Enders Dam, above and below, show the great difference in scale of the later surface-water development projects. The steel forms in place for the outlet works would eventually create an 84-inch conduit.

Department of Water Resources

A Farm Family

"Living with the Water"

By Don McCabe

Wiping the sweat from his brow, William Royal Vasey dropped his spade, stood back and eyed his work. The 10-acre timber claim was finally planted. The rows of seedling cottonwood, boxelder, ash and American elm stretched before him under the warm, midday sun.

It was the spring of 1880. Vasey had staked his claim and his hopes in the Platte River valley, 6 miles east of the small, struggling settlement of Cozad. Gazing across the Nebraska prairie, he recalled the back-breaking year since he first arrived. His journey west by covered wagon had begun from his home in central Iowa. William's spirits were higher this spring. His wife, Elizabeth, had finally arrived by train to join him on their new homestead.

And they were not alone. By 1880, homestead fever was spreading quickly across the 18-mile-wide valley around Cozad. More communities were rising from the grasslands, and not even the sporadic confrontations between cattlemen wanting open range and "sodbusters" wanting to fence their fields could slow the march westward.

"Elizabeth, come see the trees," he yelled toward the unfinished three-room, wood-frame house.

At last, she thought. Now he had time to finish the house. It would take at least one more trip by wagon to the North Loup River for the remainder of the wood for their home. Scrub brush grew along the river, but the vast, flat Platte valley offered little in the way of trees. The view was unobstructed along the 8-mile distance eastward to Plum Creek—present-day Lexington.

Mr. Vasey, an Englishman born in Canada, would next begin turning over a few acres of the rich prairie on the 160-acre homestead. He acquired 80 acres through the Homestead Act and another 80 through the Timber Culture Act of 1872. To obtain the latter, he was required to plant 10 acres of trees. Unlike some settlers, who never did plant the promised trees, William nurtured his grove, carrying water weekly from an open homestead well, hoping it would someday remind him of the trees on the family's rolling Iowa farmland.

When William planted his first crop of corn and wheat that year, he also sowed the seeds of a family-farm operation that continues successfully into the third and fourth generations in Dawson County more than a century later. In honor of this longevity, the Vasey homestead received a 100-year Pioneer Farm Family award from Ak-Sar-Ben in Omaha in 1979.

Royal William Vasey, the youngest of William's three sons, took over the 160-acre homestead in 1920. That same year, Royal married Hilda German, one of 13 children of a neighboring Danish family. They then moved to a different farmstead 2 miles west, where they bought an adjoining 80 acres, expanding their farm to 240 acres.

Bill Vasey, Royal's son, and his wife, Mary Jane, till the same soil today, plus another 1,200 acres of leased land. Bill bought the original homestead land in 1970 when Royal retired.

Today the farmstead where Bill was born and reared is Vasey Farms, Inc., a modern farmstead and hub of a 1,400-acre alfalfa and irrigated corn operation. Bill, an animal husbandry graduate of Colorado State University, sold off the family livestock in the 1970s to concentrate on producing corn and alfalfa. Chris Henry, Bill's son-in-law, and his wife, Nancy, are partners in that corporation and intend to continue the family farming legacy into the 21st Century.

It's a typical Platte River valley farmstead, neatly organized with two large

William Vasey family and others in front of the Vasey homestead near Cozad, date unknown: (from left) Royal, Edgar, Edna, an unidentified man, William, Kent, Elizabeth

Vasey family collection

machine sheds, a farm shop and a grain-handling center that consists of seven round bins along what the family calls the "old Lincoln Highway," an earlier term for U.S. Highway 30. A house stands at each end of the farmstead. Chris and Nancy and their son Matthew live nearest the highway in the house where Royal and Hilda raised their family. Bill and Mary Jane live on the north end of the farmstead in a house that includes two rooms from William Vasey's second home on the original homestead 2 miles east.

More than a century of farming has produced sweeping changes in the Vasey operation, the most dramatic of which is irrigation. Canal and well water have transformed the rich soils in this stretch of the Platte Valley into one of the most productive regions of Nebraska.

Vasey family collection

William Vasey family, 1898; in back: (from left) Edgar, Edna, Kent; in front: (from left) William, Royal, Elizabeth

Dawson County today ranks as one of Nebraska's top two cattle-feeding counties, with more than 450,000 head on feed. This cattle-feeding empire is anchored here by subirrigated alfalfa and irrigated corn. At one time, Dawson County promoted itself as the "alfalfa capital of the world," with its scores of alfalfa-drying plants. Only a handful of the plants remain, but the county still produces more alfalfa than any other in Nebraska.

"My grandfather no doubt realized the fertility of the flat prairie land he settled but he never realized the bountiful water supply beneath the land for pump irrigation," Bill says, "nor did he envision the canals that would deliver Platte River water to crops."

The Vasey family farm history closely parallels the evolution of irrigation through the decades in Dawson County. William, in the late 1890s, joined on as a member of the Farmers and Merchants Irrigation Co., a private ditch company today known as the Dawson County Canal System. It was one of the first canal projects to bring Platte River water to drought-prone farmland in Dawson County. In 1915, on the homestead farm, Royal dug one of the first Dawson County irrigation wells to supplement the unreliable irrigation-ditch supply. Bill later leveled and shaped the land and drilled more wells to irrigate more cropland. Half of the irrigated land he and Chris Henry farm receives water from the Cozad Ditch Company.

The story of irrigation and its benefits can be told in much of Nebraska but would be no more dramatic than in the semi-arid central Platte Valley and points west, where rainfall is too unpredictable for consistent crop production. The Vaseys settled just east of the 100th Meridian, which runs directly through Cozad. Rain didn't "follow the plow" then, as many 19th Century promoters claimed. Droughts wiped out crops and settlers in the process. Many, their dreams ruined, packed up and returned east after only one or two years.

William lived through those droughts and the grasshopper plagues that accompanied them and was discouraged enough to once threaten to leave. Royal, his son, told a nephew in an interview in 1979 that only through Elizabeth's encouragement did his father stick it out. Royal says of his mother, "She was a good pioneer and she had a mind of her own, and when times used to get pretty hard, some would leave. I always felt Dad was about ready to leave sometimes, but she never would. She'd never give up."

A few years of generous rainfall in the 1880s brought new hope, but due to overproduction, prices for corn, wheat and livestock plummeted. A depressed economy nationwide left many a Nebraska farmer on the verge of collapse.

Long-term prosperity, if it was to occur for Dawson County farmers between Cozad and Lexington, rested on the waters of the Platte River to the south. In 1890, a drought year, the first attempt to plan an irrigation canal southwest of Lexington, near the present system of the Cozad Ditch Company, was abandoned because of strong opposition from farmers unaccustomed to irrigation. It didn't help that 1891, "the year of mud and moisture" as some called it, brought precipitation well above average.

Dry years returned in 1893 and 1894, prompting Lexington businesses and farmers to form the Farmers and Merchants Irrigation Company in 1894. Working feverishly, crews completed the main canal by July of 1895. On June 27, 1895, the first headgate was opened on the system.

A hand-dug well and pump on the Vasey homestead, 1915

Vasey family collection

In October of 1894, nearly 100 horse teams and 125 men labored with slips to carve out the main ditch. (Slips were horse-drawn buckets resembling wheelbarrows without the wheel and with the front edge flattened and sharpened that were manually controlled like old walking plows). A rider on the rig had to have skill and strong arms to pull the levers to raise and lower the bucket. Three large graders, each with four men and 16 horses, also were used on the canal. Individual irrigators employed horse-drawn, single-row ditchers to dig laterals from the main canal. The operation required endless trips, back and forth, to form the lateral. The chore eased slightly with later ditchers that came with double wings to scrape out both sides of the lateral at once.

In those early years, the water saved corn crops that undoubtedly would have withered without it. But it brought no quick oasis to all farmers. The unruly Platte River, spreading out its floodwaters in the spring, often lapped at the Union Pacific tracks in Cozad. In August, its flow often slowed to barely a trickle, leaving canals and laterals empty in the process.

William joined on with the ditch company several years later. At that point, it was, as Bill remembers his grandfather and father saying, a matter of "living with the water" all day. A spade became almost a permanent extension of an irrigator's arm as he opened the lateral bank, allowing water into several rows, then closed the bank and moved on to the next set of rows. Inevitably, a hole just closed would break, partially draining the lateral. As water flowed to low spots in the row, they mudded up a ridge to block the water, then dug across rows in an often futile attempt to direct water to higher spots in the field. If you were blessed with water, hard work was the expected sacrifice.

"It was a plain fact of life then: everything was tough," Bill says. "But it was worth it. My dad (Royal) told us kids that irrigation was a life-saver, a gift from heaven for (him) and his father. He told me as a kid that he thanked his lucky stars that his dad picked this spot to settle."

As a lad, Royal worked alongside his dad and two brothers in the fields and carried water to crews digging parts of the Farmers and Merchants Irrigation system. Later, he and his brothers shelled corn and did other jobs for neighbors to supplement the family income.

In spite of the long hours "living with the water" and tending livestock, Royal found time to fish in the Platte and play baseball on teams of local farmboys. In late summer, he often netted catfish trapped in pools in the nearly dry Platte River bed. As more farmland was turned over by the plow, he began collecting buffalo skulls.

Life changed dramatically for Royal after the turn of the century when his dad died and one brother, Kent, was killed in a shooting accident. He quit school in the seventh grade to help his remaining brother, Edgar, farm. Several years later, in 1915, Royal decided the farm needed another irrigation source to back up the unreliable ditch supply. The problem then was that just when corn needed water the most, the canal and laterals were dry.

Immediately west of the site where William built his first house, Royal and a friend, a former oil well driller, dug a well by hand, using a sand bucket and a horse-powered derrick. Interviewed for a 1971 newspaper article, Royal says, "We hit blue clay at 55 feet, and in those days we didn't have the equipment to go through, so we stopped." They put down two holes and placed an 8-inch pipe in each hole. The two wells were connected to a centrifugal pump located in an 8-foot-deep pit. Originally, an International Mogul tractor powered the pump. Bill figures the static water level was about 19 feet when his dad dug the well. By the 1990s, after decades of seepage from canals and laterals, the water level is even closer to the surface.

For travelers along the Lincoln Highway, or U.S. 30, the well served as more than a source of irrigation water. Royal often watched as cars pulled to the side of the road and their occupants got out to draw a drink of the cool water.

That well continued to produce about 1,200 gallons per minute until the mid-1970s and was then considered the longest operating irrigation well in Nebraska. The old Viking pump pulling water from the well now has its place in local history, on display in the Dawson County Museum in Lexington.

Royal's marriage to Hilda German in 1920 was a big occasion, particularly for the Germans since Hilda was the oldest girl in a family of 13. But their married life began heavy in debt. Royal borrowed money to buy the 80 acres adjacent to their new farmstead and, just before a hefty payment was due, hog cholera wiped out all but three of a 400-head swine herd. They then had to rely on faith and perseverance.

Hilda provided much of that. Born in a musical family, she brought in extra income by offering piano lessons. Each year she canned quarts of fruits, vegetables and meat. Her goal was 1,000 quarts a year. She also sold eggs.

Through the trying times, the church served as the center of their social and spiritual life. Royal and Hilda attended the Church of Christ in Cozad, where family and church dinners were frequent events.

The Vaseys and relatives also enjoyed summer picnics on special family occasions. And, thanks to William Vasey, they had the tree grove on the original homestead for those picnics. In fact, the Vasey Grove, as it was known, was the site of many Dawson County July Fourth picnics. At one such picnic, more than 5,000 people gathered in the grove.

Hilda remembered those times more than the hardships. "People wouldn't have the faintest idea of what we had to do without (back) then, how hard we worked. Everything had to be done by hand, water carried in, water carried out, fuel in and fuel out and rendering lard,

(Continued on p. 243)

Arlene and Betty Vasey cool off in an irrigation ditch, 1926.

Vasey family collection

Milo German Made the Siphon Tube Practical, Revolutionized Irrigation

Milo German never disliked work exactly. It just wasn't a passion. He always figured there had to be an easier way to do any job, often taking an unconventional approach.

Take the time that Milo, while farming south of the Platte River near Cozad, drove to an alfalfa field in his new 1935 Ford. His younger brother was raking alfalfa, and Milo told him to unhook the rake from the tractor and attach it to the car. They jumped in the Ford and sped up and down the field turning hay while listening to the World Series on the car radio the rest of the afternoon.

That idea spawned a haying invention. He rigged a haysweep on the rear of a tractor that placed the weight of the hay on the rear tractor tires. Reverse gear provided more speed. Also, when the tractor pivoted on its rear tires, it had a shorter turning radius.

Milo's desire for finding an "easier way" led to many inventions, often unorthodox, some of which were successful and some failures.

But his place in irrigation history is assured through a simple plastic tube he developed and popularized that, in the 1940s and 1950s, revolutionized ditch watering for farmers in Nebraska and throughout the country. It was the siphon tube, a curved plastic tube that drew water over the bank of an irrigation "lateral" and into corn rows. Soldiers at Fort Sidney probably used siphon tubes to water gardens in the early 1870s, but German saw a way to make them marketable and popular, pioneering a trend that transformed irrigated agriculture nationally.

It was in 1942 that Milo, while a partner in a John Deere dealership in Cozad and working on other contraptions, began manufacturing siphon tubes in the dealership shop. Three years later, in 1945, he launched Nebraska Plastics, Inc. in Cozad to market the siphon tube, using COZAD as the brand name. Milo's son, Rex, currently is president of the company that now manufactures plastic irrigation pipe and plastic fencing materials.

Milo found himself in a tireless battle, first with prototypes and then, after building the first models, with farmers, convincing them the all-too-simple device would work.

A California inventor at about the same time claimed to have come up with the siphon tube, but that claim is disputed by the German family.

It's an understatement that irrigators deserved the siphon tube but surprising it took so long coming. With the siphon tube, a farmer could set his irrigation water and actually leave the field for several hours to take care of other chores. This simple tube allowed irrigators that luxury and also shortened the amount of time needed to get water through a set of rows, which in turn lessened the amount of soil washed down the rows. It also saved labor, allowing irrigators to junk those troublesome lathboxes set in the lateral bank to allow water into rows.

(continued on next page)

An unidentified farmer setting some of Milo German's first siphon tubes. While German probably was not the first to use siphon tubes for irrigation, he made the first plastic tubes and mass produced them, transforming irrigation.

Nebraska Farmer

"The siphon tube was the first significant innovation in irrigation systems in many decades, if not centuries, and it was the harbinger of a flood of (irrigation) innovations which were to occur after World War II," according to a family history written about Milo.

More efficient gated-pipe systems and center-pivot sprinklers gradually replaced most of the siphon tubes, although they remain in use on many fields irrigated from canal systems. But it's estimated that in the late 1940s and 1950s, siphon tubes were used on the largest share of acreage irrigated by flooding in the United States.

Milo's road to refining his breakthrough invention was sometimes a rocky one. Even after forming his company, he and his employees struggled for more than 10 years to find the right material for the siphon tube. One of those employees, Virg Kostman, who worked with Milo in early siphon tube experiments and later in refining the device, is still employed at Nebraska Plastics. It wasn't until several months before his death in a Sand Hills plane crash in 1957 that Milo finally made the necessary refinements for a successful, weather-resistant siphon tube.

Born in 1904 on a farm east of Cozad, Milo worked for his father on the farm and in a livestock-trading business, as well as for neighbors. Restless, he left home and school at age 15 to ride the freight trains in Wyoming, where he worked for a year in the oil fields. Chicago was his next stop, where he worked as a Sears and Roebuck shipping clerk.

Returning to Cozad, he married in 1924 and, with the help of his wife's grandfather, farmed until 1941, when he and a friend purchased a John Deere dealership in Cozad.

As a young farmer, he pondered the problems of ditch irrigation. Lath boxes didn't exist then. Milo and his neighbors spent the better part of the day in the fields, mucking through mud to open lateral banks with a spade and adjusting the flow in each row.

How Milo came up with the idea for the siphon tube isn't clear. One of his farm employees at the time figures Milo latched onto the idea while using a garden hose to siphon water from a tank to hog troughs. A boiler flue pipe, bent in the shape of a half circle, was used on his own farm in his first stab at siphoning water over lateral banks.

Milo then decided to use aluminum, but during World War II, rationing made that impossible. On one of his frequent trips east, Milo found samples of plastic tubing made by a Detroit company. He ordered some for experiments. Next, he discovered that a Tennessee company called Tennessee Eastman was making straight plastic tubing. Intrigued, Milo ordered a railcar of that tubing sent to Cozad.

The plastics industry was in its infancy then, and new materials, Milo found, were difficult to shape. To bend the plastic tubing sections, he filled them with sand and dipped them in hot water. It was a frustrating experience, but Milo was encouraged enough that he installed his own plastic extruder.

Hearing of his early experiments at the John Deere dealership shop, several area farmers came in to watch, and most of them laughed, Rex recalls.

But Milo was a natural and persistent salesman. "There are a whole lot of farmers around here who say Milo brought out early-model siphon tubes for their dads to try," Rex says. "Most (of the farmers) thought it was crazy. But when the tubes were properly set and farmers got the hang of using them, they were astonished at how easily the tubes worked."

Rex remembers that the Tennessee Eastman material was a clear plastic. "He sold many of them in this area in Nebraska and they were accepted," Rex says. "But he then marketed them in Arizona and the Imperial Valley in California. There, the sun made them brittle and they broke. That experience about broke us."

As the plastics industry developed in the next several years, Milo began manufacturing siphon tubes made from a different plastic, polyethylene. But these siphon tubes straightened out in the sun. "It required hours to heat these tubes so they wouldn't straighten out," Rex says. "Mountains of these straightened polyethylene tubes came back to us, under warranty," Rex says. "We were still struggling to a certain extent."

In the mid-1950s, Milo secured a loan from a friend and, with the financing, solved the problem. He died at age 53 in 1957, never fully receiving the credit he deserved.

Nebraska Plastics eventually sold its siphon-tube manufacturing equipment to concentrate on plastic irrigation pipe and on fencing materials. However, the impact of Milo's invention was incalculable for irrigators in the 1940s and 1950s and even later.

"He (Milo) was unconventional in about everything he did," Rex says. "When he got an idea, he plunged into it."

—Don McCabe, managing editor, Nebraska Farmer; selected information taken from* Milo V. German: Inventor and Developer of the Siphon Irrigation Tube, *by son Douglas K. German

(Continued from p. 241)

putting up meat. But we were happy to do it," she told a nephew in a 1979 interview.

Then there were the children, three girls and two boys. Bill, born in 1930, was an "efficient kind of boy," according to his mother. "He was the type that organized and planned."

Bill and younger brother Bob were FFA members, and Bob later served as state FFA president. But as boys, Bob wasn't always interested in helping Bill with chores. Hilda related a story of one of those occasions. Unwilling to finish a job ordered by Bill, Bob climbed to the top of the family windmill and refused to come down. "Bill was afraid of heights and couldn't climb up to get him," their mother recalled.

In their free time, the boys played ball with their dad, an ardent baseball fan. Their parents passed on to Bill, Bob and their three sisters—Arlene, Bette and Patricia—a strong faith, and the church remained a focal point of family activities.

But it was more work than play. Every summer morning, Royal, Bill and Bob left for the fields before breakfast, spade in hand, to tend the ditch water and their one well. "The cropland still wasn't entirely irrigated in those years," Bill says. "Fields weren't leveled yet and we had to get water on them from several directions."

There were cows to milk, cattle and hogs to feed, and the chickens. "In the early 1930s, Mom convinced Dad to build what was in those days an ultramodern chicken house that, at its peak, housed 1,000 hens," Bill says. "We sold eggs for years. We kids helped but the chickens were mostly Mom's job."

For Royal, his sons and other valley farmers, the ditch water suddenly became more dependable after the early 1940s. Kingsley Dam had been built, Bill remembers, and the Dawson County Canal and Cozad Ditch Canal benefited from more consistent flows due to Lake McConaughy's storage. They had water in late summer when they needed it most.

"We've never had to worry about the canal supply ever since that time." Bill says. Delivering water to fields also became easier about this time. Milo German, Hilda's brother and a Cozad inventor, had developed the siphon tube and began popularizing and selling it in 1942. It revolutionized ditch irrigation well beyond Nebraska. No more would irrigators have to toil with fragile lath boxes or the rubber railcar brake hoses they buried into the lateral banks to direct water to the corn rows. While some historians have documented the use of metal siphon tubes by Fort Sidney soldiers to water gardens in the early 1870s, German made and marketed the first plastic siphon tube, starting a trend that dramatically changed the rural landscape.

Bill Vasey and Chris Henry by a surge valve

Don McCabe

Many initially scoffed at German's simple siphon tube, doubting it would ever work. But it did, and irrigators quickly became proficient in the art of setting the tubes, to the point that siphon-tube-setting contests were once part of the Dawson County Fair and the annual Cozad Hay Days, Bill recalls.

In 1953, Bill earned a college degree and entered the Army as a commissioned officer. "I was lucky enough to serve in Germany rather than Korea," he says.

It was no accident Bill enrolled in college. It was a must, at least to Royal and Hilda. One of Royal's most vivid memories was the day he left grade school to farm full-time. Royal and Hilda vowed that their children would receive college degrees. All five attended college, and four earned degrees.

"They would not have allowed any of their five children not to go to college," Bill says. "It was sort of like brainwashing."

Discharged from the Army in 1955, Bill debated about pursuing a master's degree in order to teach. In the end, he stayed on the farm, gradually adding more leased land, more alfalfa and corn and more irrigation wells.

"Dad (Royal) now was able to begin easing out of the operation," Bill says. "Dad was never really interested in farming, at least the machinery and cropping end of it. My returning to the farm worked great. Dad essentially quit farming then and stayed with the livestock. I quit the livestock and did the cropping."

Two years after his military discharge, Bill married Mary Jane Bovee, a "city girl" from Cozad who says she knew nothing about farming. "I always heard farmers complaining if it rained or if the sun shined or if it was cloudy," she remembers. "Now I'm right in there complaining with them."

Some of the city girl remained. One of her fondest memories was when Royal decided to sell off the hog herd after he fully retired in 1970. "It was the greatest day in my married life," she says with a grin, "when we got out of the hog business."

As the Vasey operation expanded, so did the family. Bill and Mary Jane reared three children on the farm and still managed time for outside activities. Bill recalls the time brother Bob, while a veterinarian in Kansas, flew his own plane back home for a visit.

"I got the flying bug, took lessons and bought my own plane." A long stint in the Nebraska Flying Farmers followed, including one year as state president. Son Doug, who in 1992 was studying for his doctorate in economics at the University of Nebraska-Lincoln, and daughter Susan Moore, now of Boise, Idaho, each were involved in teen offices at the state and international levels of the Flying Farmers organization.

Many assume the continuing trend to fewer and larger farms will foster a more detached farm population, more interested in the bottom line and less in the community. But this is not so for the Vaseys. Bill and Mary Jane are fixtures at the Church of Christ in Cozad. Bill is an elder and sings in the choir, which is directed by Mary Jane.

Mary Jane never has trouble keeping current on community activities, either. As executive director of the Cozad Chamber of Commerce, she helps direct them. Bill keeps on top of federal farm-program rules and helps others do the same as a member of the Dawson County committee of the Agricultural Stabilization and Conservation Service.

As in most Nebraska towns, school athletics are a source of community pride. It was particularly so for the Vaseys when Nancy participated in basketball, volleyball and track for Cozad High School. In the latter sport, she earned honors as a high school All-American shot putter.

It's easier now for Bill and his son-in-law Chris, who gained experience working for the Vaseys during the summers while in high school and college. The scene early each summer morning is much the same as it was nearly a century ago. "Our summers, for the most part, are spent watering crops," Bill says. "You live with the water less than my dad or grandfather did, but you still live with it."

Before breakfast, Bill, Chris and a full-time employee, Ray Jacob, head to the corn fields in different directions in their pickups or all-terrain vehicles to set dams and siphon tubes and open gates on irrigation pipe. The same two-hour routine occurs in the evening. "It (irrigating) is the first thing in the morning and last thing in the evening."

There are 14 wells, plus the open ditches from the canal systems. Siphon tubes are still needed for the canal-fed ditches. Irrigation wells deliver water by gated pipe to most of the remaining acreage.

Alfalfa consumes most of their time

(Continued on p. 246)

Dorothy Garrison: North Loup Irrigator Recounts an Irrigation History

Dorothy Garrison loves the land and the water that nourishes it. And after nearly 52 years of tilling the soil and irrigating crops, she has strong feelings and vivid memories of both.

Born in the Nebraska Sand Hills, Dorothy has lived all her life in Garfield County. In 1928, when she was 13, she and her family moved from their ranch 18 miles north of Burwell into town so she could attend high school. In 1934, she married her high school sweetheart, Lemuel Jefferson Garrison, known as "L.J.," and moved to his family's farm just southeast of Burwell.

Nineteen thirty-four was memorable, and not just because that was when Dorothy and L.J. began married life. It was also the year a terrible drought settled in to ravage the countryside. The Garrisons knew that if they were to survive on their farm, irrigation was a necessity.

In 1890, when L.J.'s grandfather purchased the farm, there had been some makeshift irrigation by landowners out of the North Loup River at the Sioux Creek diversion dam. But the ditches and canals hadn't been maintained, and by 1934, the systems had long since been abandoned.

The drought put pressure on for regular and reliable systems of irrigation that only government projects could provide. The senior Garrisons were in the forefront in urging national politicians to approve such projects.

Being some of the few Democrats in the county, the Garrisons had worked on Roosevelt's election campaign. Dorothy recalls her mother-in-law using the back of some of the leftover FDR campaign posters to draft letters to Washington urging the approval of the irrigation project.

The North Loup Valley Public Power and Irrigation District finally was approved in 1934, and canal work was completed in 1940. So began the Garrison's long history of developing and using irrigation on their land. It was work that encompassed the time and effort of Dorothy, L.J., and their three children.

"For row-crop irrigation on this farm, we went from ditches cut open at each row, (which was) pure disaster, to lath boxes, to siphon tubes, to gated pipe. Each change had to be considered carefully because money was involved. But each change made the work easier and was better stewardship," she recalls.

In the ditches that first served the Garrison fields, water had to be raised to the right level to accommodate the slope of the land. At first, gunny sacks were partially filled with dirt and laid in the ditches, one on either side of the slope, with one across them. "No great solution," Dorothy said of this method, "because the sacks rotted away or mice gnawed holes in them. Also the holes dug to supply the filling made likely places to trip and fall."

Next came earthen dams in the ditches with apparatuses called "rabbit boxes," which were one-by-sixes nailed together to form square tubes with a one-by-two bent around it to help hold it in the dam.

Canvas dams and, later, plastic dams helped streamline the moving of water through the ditches, each having unique features that controlled the flow of the water with a little less labor from the irrigator.

Moving the water from the ditches to the crops also followed a trial and error progression of techniques and tools that started first with holes cut in the ditches followed by the use of lath boxes, then siphon tubes, and later, gated pipe.

To water alfalfa and pastures, the Garrisions bordered and flooded the areas. A homemade plank "V" was used to create ridges, and a land plane leveled between ridges. Initially, holes cut in the supply laterals allowed the water to run free. Later, large siphon tubes improved the process.

The end of summer and irrigating season meant gathering and cleaning the dams, along with collecting lath boxes and then, later, the siphon tubes. Dorothy said this process sometimes brought surprises as the boxes often became havens for mice and snakes. "When a snake decided to bail out, there was fancy footwork on the part of the carrier."

With the advent of help from the U.S. Department of Agriculture Soil Conservation Service in the late 1940s, more technological advances became available to the Garrisons. SCS personnel surveyed the land, made sure laterals were correctly placed and eventually developed plans for permanent check dams in the laterals. In 1958-1959, the Garrisons began installing these cement structures on their land. The Garrison's son, Le, became an "expert" at moving dirt, leveling, cutting and filling. Their son Len redesigned the plans for the aprons above and below the "checks" when they proved unworkable. L.J. and Dorothy were the "check and apron constructors." When completed, Dorothy recalls that "it was about as near heaven as we figured we'd ever be."

When the Garrisons retired from farming in 1976 and rented out their farm land, Dorothy said the renters "did not appreciate our handy methods," and so eventually they installed gated pipe, which Dorothy admits was "easier and better once the pain of the investment subsided."

But even today, Dorothy and L.J. continue to lend a hand and watchful eye to the task of irrigating the tenant's crops. Part of Dorothy's daily routine during the growing season is to help check the fields and gated pipe. To do this, she and L.J. ride a bench-seated, three-wheel all-terrain vehicle.

After the many years of irrigating, it appears Dorothy Garrison has not lost her sense of wonderment over the good things water provides. Each day she says she looks forward to "getting out and working with the water, getting everything ready to give new ground a drink."

As much as she appreciates the benefits irrigation projects have provided, she is often rankled by those she feels have exploited the intent behind it and the local natural resources. Such is the case as she recalls events and circumstances surrounding the

(continued on next page)

Dorothy Garrison and Lemuel Garrison on their all-terrain vehicle. Like many farm families with irrigated acreage, the Garrisons have passed through many different kinds of irrigation technology.

Kathy Sullivan

development of the Calamus Dam. She doesn't mince words in her anger over the outcome of the project.

In her eyes, the Calamus Project "ruined a beautiful valley that supported sub-irrigation and varying agriculture. It brought heartache and resulted in unkept promises." On top of that, she believes it added to an already unmanageable government deficit. The Garrisons were part of an outspoken but unsuccessful group of landowners organized as "The Farmers and Ranchers to Save the Loup and Calamus."

Today, Dorothy continues to live out her philosophy regarding the wise and frugal use of natural and human resources. In June of 1992, she finished a 12-year stint on the board of directors of the Center for Rural Affairs in Walthill. During that time the center published a book of her "sayings and times remembered from life in the Nebraska Sand Hills." The book, entitled "A Borrowed Horse Never Tires," is peppered with commentary, anecdotes and quotes of her wit and wisdom.

—Kathy Sullivan, free-lance writer and formerly extension communications specialist

(Continued from p. 244)

between irrigations. Throughout the summer, they windrow and put up nearly 3,000 tons of alfalfa a year, most of which is marketed within a few miles to the Hi-Gain Feed Lot just outside of Cozad. The subirrigated alfalfa benefits from a water table that has been recharged due to decades of seepage from the canal and lateral systems in the valley.

With its alfalfa and corn production, Dawson County could handle even more feedlot cattle, Bill believes. "The best thing for this valley would be another 200,000 head of cattle in feedlots from Gothenburg to Lexington," he adds. "We've got the grain and alfalfa to feed them. We ship thousands of tons of alfalfa pellets a year from this area. If we had 200,000 more cattle, we could feed the alfalfa right here."

This is a close family, living a comfortable but far from extravagant lifestyle. Nancy and Chris want to continue the success Bill and Mary Jane have had. But Bill and his son-in-law see some dark clouds on the horizon for the future of this family farm. Sitting at the kitchen table, they talk of the mounting demands for Platte River water and wonder aloud what impact wildlife and environmental regulations may someday have on Platte River water rights held by irrigators and irrigation districts.

"I consider myself interested in what's happening in and to the world," Bill says, "but sometimes we're dealing with people, environmentalists, who aren't concerned about agriculture and its future.

"It's something I get too riled up about. We have hopes of this family going on for a long time. But when we talk about regulations protecting a few hundred whooping cranes and two birds, the piping plover and least tern, that few people ever see, then I think it's a threat to our livelihood and the work of generations of people and what they've dreamed of and worked for and how they make a living."

In Bill's view, irrigation has created a "completely recycled water system" in this portion of the Platte River valley, a system in which groundwater pumping takes advantage of decades of groundwater recharge attributed to the canals and laterals, leaving a balanced aquifer supply. "That could be harmed if we should have some of our ditch system shut down," he adds.

With irrigation water available to them since they began farming, it's difficult for Bill and Chris to imagine their operation or the area's economy without it.

"The economy is tied to irrigation," he says. "And, with land values based on irrigation, the school systems here are top-notch."

Irrigation and farming practices will have to change, however, Bill concedes. In the long run, Bill fears a drinking-water problem here unless more caution is used in irrigation-water and nitrogen-fertilizer use. But to date, the Central Platte Natural Resources District, based in Grand Island, has not detected a serious groundwater nitrate problem in this portion of Dawson County, nothing like the nitrate contamination farther east in Buffalo, Hall and Merrick counties.

"I think we still have to be concerned; we know so little about what's actually taking place in the soil regarding nitrogen movement to water. It's going to take a lot of education."

Through research and demonstrations, Bill and Chris have made big strides in lowering commercial nitrogen use. "At one time, we had the philosophy that the more nitrogen you applied, the more corn you raised. Of course, that's not true."

A professional crop consultant tells Bill and Chris the amount of commercial nitrogen to apply on each field after soil samples determine the nitrogen available to corn from the soil. Research plots have made a believer out of Bill. "We've had a nitrogen test plot here for 3 years, conducted by the University of Nebraska and

Bill Vasey holding Matthew Henry and (from left) Mary Jane Vasey, Nancy Henry and Chris Henry

Don McCabe

the natural resources district. For the past two years, no nitrogen has been applied to the plot and we've harvested as much as 198 bushels an acre.''

The family also has altered its view on water use. ''One thing we have all changed our minds on is that we've been a lot more cautious about how we use (the water),'' Mary Jane says. ''We do have a lot of water here, but we want to make sure it lasts for future generations.''

Royal Vasey died in 1989 at the age of 98. Hilda died a year later, at age 91. In addition to the values they imparted to their children and grandchildren, they left behind a memento that links the third and fourth generations with William and Elizabeth, who settled the land in 1879. It's a painting, made from a photograph of the original homestead house taken during the 1880s. Each of the copies Royal gave his children is framed with pieces of cedar from the foundation of the first house before it was taken down.

Sources: *Battle of the Bridges,* Rex German and Russ Czaplewski, 1988, Dawson County Historical Society in cooperation with the Nebraska Council for the Humanities; The Dawson County Historical Museum, various materials; Nebraska Farmer, various clippings; clippings from various Nebraska newspapers; *Son of the Gambling Man,* Mari Sandoz, 1960, Bison Books reprint, University of Nebraska Press, Lincoln and London; Unpublished family history of Royal and Hilda Vasey, 1979; transcript of an interview with Royal and Hilda Vasey by nephew Douglas K. German.

Interviews with: William Vasey family, Cozad, Nebraska.

A 1915 Viking No. 6 pump once owned by Royal Vasey, now in the Dawson County Historical Museum (above) and a Vasey hand-dug well and pump (below)

Don McCabe

Vasey family collection

Mads, George and Vance Anderson: Three Generations of Irrigation

By Brad Rundquist

Weekend and Copy Editor, Ft. Dodge Messenger
Formerly Editorial Assistant, CSD

Standing between the Anderson farm and water from the North Loup River in 1895 was a 45-foot bank. That bank, indeed a formidable obstacle, formed a boundary that did more than mark the division between dryland and irrigated agriculture for the family; it stood in the way of an improved chance for survival and the promise of a better life. It was a boundary that had to be crossed.

Mads Anderson and his sons hadn't been on the farm long when they conquered the bank with their water wheel. Mads, who had first farmed near Dannebrog and later moved to St. Paul to open a grocery store and wholesale grocery business, moved his wife and sons to the farm on the confluence of the North and South Loup rivers in 1895, saying the city was no place to raise a family. In 1898, the boys and their father built and installed the 18-foot water wheel, and it worked. They were able to lift water up over the bank and onto the 25-acre field. The water from the Loup was kept in a small storage reservoir, which provided the boys a fishing and swimming hole. For two years the wheel stood, and the Andersons no longer needed to look to the sky for water to keep their crops alive.

But nature was not so easily tamed. When the ice broke on the Loup in the spring of 1900, the wheel was wrested from its place on the bank and carried down the river.

Mads, a native of Denmark, and his boys set out in search of their contraption, and found it not far downstream. Determined theirs would be an irrigated farm, the men took the wheel apart, carried it in pieces back to the farm, and reassembled it.

For two years more the wheel stood, until the ice again swept it away in 1902. This time, however, the wheel was carried some 8 miles downstream, where it lodged against a railroad bridge near Palmer. Ice quickly began to pile up around the wheel, exerting tremendous pressure on the bridge supports. Palmer officials told Mads he would have to pay for the bridge if it collapsed.

Though the bridge held and the wheel was removed, Mads never put it back up. At that time, Mads and family, facing the uncertainty of a dryland farming future, said that one day a machine would be invented to lift water from the river over the bank.

Thirty-eight hard years passed before the Anderson farm was again watered by the Loup. When it happened, a wheel wasn't needed, as a water-lifting machine—the centrifugal pump—had been invented.

While on the farm, Mads decided he needed a machine to help prepare a better seed bed. In 1908 he invented and patented the land roller, and in 1911 he moved to Hastings to start the Western Land Roller Company. The company grew and prospered with sales of his new invention. The company also made hay stackers and sweeps. In 1926 they began to produce 6-inch centrifugal pumps.

In the meantime, George Anderson, Mads' son, had bought the farm and in 1923 moved his family to Hastings. There he joined brothers Arthur and Charles to help run the family business. Mads had retired and turned his stock over to his children.

George retained ownership of the farm and, in 1926, put in a 6-inch pump below the 45-foot bank. Using a model 15-30 International Harvester Company tractor for power, water was pumped up over the bank and used to irrigate some 70 acres.

Vance Anderson, George's son, had spent some time the year before helping an uncle in Oregon irrigate. He was sent to George's farm in 1926 to help with the new pump and to teach Glen Paulson, the tenant, about irrigating. The system worked well. Two years later, some 3,000 feet of 16-inch pipe were laid in a hand-dug trench from the river up to the side of the valley. Two 6-inch pumps were installed with a 40-horsepower motor directly connected to each. The 2,200 gallons per minute were used to water parts of three farms.

Since the system required a great deal of maintenance, Vance was sent to care for the pumps. He pitched a tent by the river and looked after them during the irrigation season. High water and ice always played havoc with the suction pipes, which ran out into the river.

In 1934, Western Land Roller started to manufacture deep-well turbine pumps. The two centrifugal pumps were removed in 1936. A trench was dug in the river bank and a turbine pump was placed on either side of the T-shaped trench. Since no parts of the system were directly in the river, high water and ice problems were eliminated.

In 1944, the river flooded and the North Loup shifted its course three-quarters of a mile upstream, leaving the pumps dry. Vance tried to divert water from the river back to the pumps but ice in the spring always took out the diversion system.

Since he could no longer depend on surface water for irrigation, Vance began to prospect for groundwater. Several test wells were drilled, but none produced more than 250 gallons of water per minute. Finally, a suitable spot was found.

The 174-foot well produced 1,000 gallons per minute from a formation only 90 feet down. A pump with a 30-horsepower motor was installed. Since then, the pump and motor have never had to be repaired or pulled. The well remains the only 1,000 gallons per minute well for 20 miles up or down the river, and it is still used to irrigate the farm that Mads bought in 1895 to escape city life.

Drawing of the Loup River water wheel built by Mads Anderson in 1894. It cost $500 and irrigated 15 acres. The wheel had the unfortunate "habit" of getting dislodged by ice floes in the spring. It was abandoned eventually.

Nebraska State Historical Society

Future Control of Water Resources

by L. Kent Wolgamott

To understand the politics of water in the 1990s and beyond, some fundamental changes in the last two decades must be recognized. These changes permanently altered the way water is addressed as public policy and added a host of new players to the struggle over the use of the resource.

First, the federal government, for all intents and purposes, got out of the dam-building business in the 1980s, at least in Nebraska. By establishing high cost-sharing percentages for states to meet in order for federal dollars to flow into a water project, the federal government essentially put a major development outside the reach of states with small populations like Nebraska. Lacking the broad tax base of a more populous state or the money generated by a severance tax on a resource as in Wyoming, Nebraska probably has seen its chance for major water development disappear in the last decade.

Part of the federal change was undoubtedly based on the ever-rising deficit, but it was also fueled by a belief by many in Congress that large amounts of money should not be spent to build irrigation projects that would only increase crops already in surplus. This attitude is likely to continue, creating a near-permanent roadblock for the big dams that came to symbolize water development in the 20th Century. That attitude could change with a dramatic alteration in the world food supply or an extended period of drought, but those are factors outside the control of any government.

The death of the Norden Dam in the early 1980s and the demise of all the proposed projects for the Platte River—except the recharge-based Prairie Bend II—is evidence enough that the era of the big dams is likely over for good. Those struggles also established the second permanent change in the politics of water, the rise to prominence of environmental organizations.

In the 1970s and 1980s, environmental groups became equal players in the politics of water. Using federal legislation, primarily the Endangered Species Act, and grassroots lobbying efforts, the environmental groups succeeded in their long-term goal of preserving a stretch of the Niobrara, which received federal Wild and Scenic River designation in 1990. They also have successfully made preservation of flows for endangered species a central part of the strategy for managing the Platte River, earning themselves a place at the negotiating table in the process.

That position will only be enhanced in the environmentally conscious 1990s, especially as emphasis necessarily switches from construction of water projects to water preservation, both in quantity and quality. And the environmental positions will also likely be strengthened as Nebraska's urban populations grow, particularly Lincoln and Omaha, while the rural population continues to decline. That continuing demographic trend will inevitably lead to a stronger urban majority in the legislature and a subsequent tilt in battles over where the water should go.

As we enter the 1990s, the rural/urban split is primarily about taxation. But as the value of water as the state's most substantial natural resource becomes more and more apparent, a difference in philosophy over how that resource should be used could easily pit farmers against

Control of Nebraska's water is not solely in the hands of Nebraskans, as evoked by this photo of Seminoe Dam, Wyoming, 1978.

Department of Water Resources

Canoeing on the Niobrara River near Valentine

city folks to the detriment of both.

Water is likely to earn that value in the market within a few years as the third and least-discussed change of the 1970s and 1980s—the Sporhase case—has its impact on the politics of water. This case established Nebraska's abundant groundwater supply as an article of commerce, virtually ensuring that it be sold across state lines. In 1986, Senator Loran Schmit introduced a bill in the legislature permitting the sale, laying groundwork for the inevitable. But since that time, the issue has disappeared. It will return, perhaps with an intensity few can now imagine.

If ever-expanding Denver is unable to obtain water by building the Two Forks project, it could easily cast its eyes toward the Sand Hills, where hundreds of millions of acre-feet of water sit, available for purchase under Sporhase. Coal-rich Wyoming is now sitting on a bundle of money to build Deer Creek, its proposed dam on the Platte, and could very easily suggest that water it takes from the river be replaced from a supply it purchases in the Sand Hills. And Denver could take a similar tack with its efforts to get more municipal water.

While a massive project to pump water from the Sand Hills to Colorado, or into the Platte to replace water taken for a dam, is not economically feasible today, as water becomes scarcer across the West, such transfers could become more and more appealing. This is an area in which Nebraska must take action to guarantee the resource is used in the state's best interest or see those decisions made by federal regulatory agencies and the courts, the other major players in water politics.

One only needs to look at the years-long controversy over relicensing of the operations of Lake McConaughy by the Federal Energy Regulatory Commission to see the impact federal agencies have on water in the state. Given the inevitable appeal of most agency decisions, the impact of the federal court system is equally significant. With the rising national concern for the environment, federal regulation of water quality and, in cases like the dispute over the Platte, water quantity is likely to increase over the next few decades, a factor that Nebraskans must consider when making decisions on how to best use the resource.

Another important change from the 1980s that will carry over into the 21st Century is the establishment of the natural resources districts as the local political body charged with most water-related policy. In most cases, NRDs have largely been conservation-oriented groups, working on flood-control, soil conservation and recreation projects and handling localized problems. But as those districts that proposed major projects in the 1970s and 1980s discovered, NRDs can become political hotbeds. And given the post-Reagan propensity of government to push problems to the lowest level—with the federal government giving more responsibility to the states and the states more to local subdivisions—the NRDs will likely become the flash point for most water disputes in the future. With talk of campaign contributions from organized groups already occurring, the NRDs are likely to become rapidly politicized and serve, much as large school districts, as the contentious local debating point on the impact of federal and state decisions about natural resources. That change is likely to be as gradual as it is significant, but it appears inevitable.

A final, not-to-be-overlooked change in the late 1980s, the impact of which will be felt for years to come, is the increase in environmental education, particularly at the elementary, junior high and high school levels. This increased awareness, also reflected in the media, will percolate through American society and is likely to have far-reaching effects on future public policy. In this state, the Nebraska Groundwater Foundation is expanding the more general, recycling-is-good kind of environmental awareness to include the importance of water and the vulnerability of the water supply. The inevitable result of this heightened education is a broader based discussion of water issues, aiding in their destined rise to near the top of the public agenda.

Each year, a laundry list of water measures is presented to the Nebraska State Legislature. Congress and federal agencies also deal with issues having a direct impact on Nebraska's water, as do court and administrative decisions at both the federal and state level. That combination makes looking at the future guesswork. But some issues and actions can safely be identified as destined to be primary.

In surface-water development, the major issue to draw political heat is likely to be the assignment of instream flow water rights in Nebraska's waterways. As the state's urban population grows, the pressure for use of rivers and streams for recreation and wildlife habitat is likely to increase, as is the requirement for adequate streamflows for city drinking water supplies. If a rural/urban split on water issues is to occur, instream flows are likely to trigger it. Already Grand Island and Platte Valley farmers have had a set-to over the use of Platte river water, with threats of a boycott used for intimidation. To think that will be the only such conflict and that all will be settled amicably is to take a overly optimistic view toward the hard realities of a struggle over a limited, shared resource.

Efforts will continue to build water projects on the Platte River. But economics and environmental concerns appear to have made most such expensive and time-consuming proposals exercises in futility. Similarly, disputes over exactly how much water is needed in the Platte to maintain habitat for endangered species will likely continue for years. Essentially however, those arguments represent tinkering with an established fact—that the flows needed for habitat will be permanently required.

Increased federal attention toward the preservation of wetlands will have some impact in south-central Nebraska and the Sand Hills, as will continuing efforts on surface-water quality. Together, the changes in surface water will likely limit future irrigation development from rivers and streams.

Biological Systems Engineering, UNL

A meeting of the groundwater committee, Middle Republican Natural Resources District, June 1973

Nebraska's decades-old combination of conflict tempered by cooperation with neighboring states will continue as well. The issues here are clear. Nebraska wants to keep as much water as possible flowing into the state, and Colorado and Wyoming want to keep as much water as they can within their borders. The reverse is true in the state's relationship with Kansas. In both cases the ultimate arbitrator will be the federal government, ensuring, if nothing else, that any interstate dispute will take years, if not decades, to resolve.

Groundwater is likely to be the subject of more direct political action in the future than surface water, in large part because many groundwater issues are just emerging. Foremost among them must be addressing the future of the water beneath the Sand Hills, the state's most valuable "chunk" of a natural resource.

Local-control advocates would argue that the decision about what to do with Sand Hills groundwater should be left to those who live there and those who own the land under which the water lies. That, however, ignores political reality.

First and foremost, the Sand Hills aquifer serves as the chief water supply for the state's two major cities and for much of heavier populated eastern Nebraska. Sand Hills groundwater creates the Loup River system, and this system supplies most of the water in the Platte east of Columbus. Both Lincoln and the Omaha metropolitan area get drinking water from wells along the Platte. That makes Sand Hills policy a statewide concern, as urban state senators will act to protect the cities' interests in the legislature, and municipalities will do whatever is needed to continue to have a source of clean drinking water.

Before the Sand Hills water issue reaches the debate stage, much more must be understood about the aquifer and the ecosystem it supports. For example, could large-scale exploitation of the underground water have a devastating impact on the wet meadows, wetlands and lakes that make the region more than just another semi-arid grassland? Until that and similar questions are answered, at least as well as scientifically possible, political debate between parties inside

Ducks on a wetland in the Valentine National Wildlife Refuge

Charles Flowerday, CSD

and outside the state is premature at best. But the demands for that water will not recede with time, so such studies appear to be imperative regardless of one's position on further development. In like manner, the potential ways for dealing with efforts to purchase Nebraska water should be studied before a grab for the resource takes place. But such studies obviously require funding, and efforts to obtain that money will initially bring the Sand Hills issue to the forefront.

More immediately, though, groundwater quality concerns are likely to capture the public attention. In the 1980s, chemigation (the application of chemicals through irrigation systems) became an issue, and limitations on the use of ag chemicals became a reality. Those concerns are likely to increase in the environmentally aware decades to come, resulting in greater regulation on the part of both the federal and state governments and requiring work on the part of the irrigation industry to develop techniques to continue high levels of crop production under the new rules.

In some regions of the state, the central groundwater issue will remain not how to use an abundant resource but how to split up a dwindling amount of a valuable supply. Several of these painful battles have already been fought, most notably in the southwestern corner of the state. But they are likely to continue, increasing in number as groundwater is depleted and causing great local economic difficulty.

A legal conflict over groundwater seems inescapable as efforts to consider groundwater and surface water as part of the same system gain strength. Conjunctive-use bills have already been introduced into the legislature, and it is only a matter of time before it takes some kind of action. The ramifications of such a change will likely keep water lawyers employed for years, but unquestionably will allow for greater public involvement in the use of groundwater, formerly seen virtually as private property.

Any discussion of the future politics of water cannot ignore societal factors that will have an impact on how the resource is seen and used. Some of these factors, like the continuing growth of Denver, Lincoln and Omaha, are easy to observe. Others are not as obvious but will have great ramifications for the future. For example, the national trend away from beef consumption in favor of chicken is likely to eventually reduce the amount of water used to produce feed grains. If fewer cattle are being raised, requiring less feed, eventually the crops now being grown to support the cattle business will no longer be raised, particularly if federal farm subsidies are further reduced, a near certainty. Those kind of changes should but may not always be rationally considered by future policy makers because the desire to base political decisions on a romanticized past ideal sometimes can be irresistible.

IANR

Applying anhydrous ammonia to add nitrogen to the soil

Forecasting the future players in the water arena is a far less dicey proposition than picking the issues 20 years ahead. It is clear the players that began working with the resource in the 1800s will be around in the 2000s—farmers, the irrigation industry, government officials and scientists who study the resource. And those who have come along in the last few decades—environmental groups, municipalities, water lawyers, social scientists and civic groups—are also in the game to stay.

But as water issues become more visible and any crisis becomes more apparent, the number of players will increase to include people from all walks of life concerned about their future and that of their children. While it has little practical impact on water policy, the controversy over the proposed low-level nuclear waste storage facility in Boyd County provides vivid illustration of the emotional and political upheaval environmental issues can create. In that context, imagine the uproar if the drinking water supply for Lincoln or Omaha was contaminated or threatened by development. Even without such dire developments, as the value of water becomes clear, the interest in the issue will increase, taking the politics and policy of water to a brightly lit arena in which it has rarely appeared.

The key question here, however, is not whether there will be participants, as concern for water will increase through the years. Rather, the question is, who will lead? In Nebraska's governmental structure, with its one-house, nonpartisan, lobbyist-dominated legislature, public-policy leadership is nearly a sole function of the governor.

Past Nebraska governors, recognizing water as the political tinderbox that it is, have paid lip service to the issue, conducting studies and personally confronting water problems only on a case-by-case basis, usually involving major projects. That approach, while effective at lengthening political careers, shifts the responsibility for broader water policy to the scattershot method of assimilating court decisions into policy and the occasional important water law to come out of the legislative process. As water gains pre-eminence in the next few decades, the old way will no longer work.

Some future governor will have to face the fact that political capital must be expended to produce a comprehensive state water policy that deals with issues of development, water quality and supply. The only question is whether that politician's decision is made on principle—that the time has come to begin the inevitably divisive discussion of water and its role in the state—or if the policy is developed in response to some crisis that has befallen the resource. The interests of all Nebraskans would be best served by the former, so the future of the state's most valuable resource can be considered and debated without the pressure of calamity hanging over the discussion and coloring the result.

Nebraska Farmer

Past governors have grappled with creating consensus on water issues, and no doubt future governors will have even greater challenges facing them regarding water disputes. Here, then Governor Bob Kerrey signs legislation stemming from the Water Independence Congress, probably LB 1106, which created the Water Management Board, later disbanded, and state instream-flow rights.

Commentary:

An End of Century Reflection on the New Western Historians and the Uses of History

by Bob Manley

As the essays in this volume demonstrate, the story of water has been central to the history of Nebraska and the Great Plains for two centuries. In the 19th Century, history related attempts to understand the semi-arid land that lay beyond the Missouri River; the 20th Century has told the story of the "conquest" of this region in part through irrigation and reclamation strategies.

What will be the central theme of the 21st Century?

It is fair to say that, for the most part, past historians have supported an optimistic view of the semi-arid region during the 19th and 20th centuries. They recorded victories, social and technological, and successful irrigation and reclamation activities. The country did have a promising future!

The historical view of water in the 21st Century is likely to be different. Criticism of irrigation schemes and water-use programs has increased. Past water programs are being scrutinized and judged to have been both somewhat wasteful, now that crop surpluses are nearly chronic, and, in some instances, somewhat futile, now that mineralization problems are more understood and the ecological integrity and fragility of arid regions are better appreciated. The federal government has come to be regarded as the purveyor of "pork-barrel" legislation of the worst sort, sometimes in water projects. These critics feel there have been few sensible efforts by private or public developers to protect water, the plains' most precious resources.

That view, of course, is not entirely accurate. But what people believe happened in the past is more important than what actually transpired. The point is: the "revisionists" in water and Western history seem more pessimistic than some of us would feel is warranted. There still needs to be a conviction that the semi-arid West has a future. These historians are creating an image of the 19th and 20th centuries that tempers, perhaps even eradicates, the optimism of those days and replaces it with a critical, more guarded view, not only of the past, but, by implication and by pronouncement, the future.

This historical "environment" may change. It might be ascribed simply to the "malaise" of the 1990s. But it is a real possibility that how historians of the latter 20th and 21st centuries view our water heritage will have a decisive effect on future policy makers and voters.

In addition, there are those who insist that we are in the process of creating another "myth." Earlier generations had theirs: "Rainfall follows the plow"; "Cultivate with a harrow"; "Sown in hope, cultivated in doubt, harvested in despair." The new myth might be: "Science can solve our water problems."

Without in any sense deprecating scientific investigations, one must understand that scientific studies are subject to popularizations and by definition, distortion. And it must be understood, also, that scientific data never provides the only basis for decision-making. In this context, how citizens view their past, and by implication therefore, how they think about the future, is crucial to the process of discovering and defining that very future.

Generators housed in the Columbus hydroelectric powerhouse. Total nameplate capacity is 39,900 kilowatts.

Department of Water Resources

Future Uses of Nebraska Water

by John Janovy, Jr.

Introduction

The Future Is Both Inherited and Built

The future can be thought of in two very different ways: either it already exists and people must find their place in it, or it does not, and so it must be made by those alive today. In truth, however, the future is a mixture of the discovered and planned. This feature is especially characteristic of our relationships with water. We know that Nebraska will not get massive new supplies of water in the years ahead. But we also know that decisions made today will determine, and thereby build, our water future.

Natural Constraints on Planning and Development

Regional economies are shaped, to a large degree, by geological history and global climate patterns. Events that occurred hundreds of millions of years ago gave Kansas and Wyoming, but not Nebraska, abundant fossil fuels. In Nebraska, past geological events left behind a vast source of an underground supply of water—the High Plains (Ogallala) Aquifer. But present geological features, for example the Rocky Mountains, and our latitude influence rainfall and snowfall in the mountains to our west and our growing season. Because of our position on the plains, use of water in the Platte River system above the mouth of the Loup and in the Missouri is primarily dependent on snow that falls on distant mountains. So, we need to plan for the 21st Century, but at the same time are locked into the list of options dealt to North America long before there was a political entity called "Nebraska."

This dilemma shapes the state's current efforts to supply its citizens with jobs, food, shelter and recreation. The problem of finding compromises, ways to use natural resources so that children are clothed, fed and educated, the elderly are cared for and roads and buildings are maintained, will not go away. So in a very real sense, Nebraska's water future can be read in the newspapers. When irrigators and environmentalists confront one another, for example, we get a glimpse not only of past conflicts, but also of the future debate over how society as a whole should use a limited, but vital, resource.

Human-made Constraints on Planning and Development

Existing water laws, well-drilling permits, dams, gates, canals and streamflow allocations also can limit future options. That is, they must be considered in any plans developed for the next century. These laws, allocations and infrastructures represent attempts by previous generations to use a limited resource for the benefit of all. Everyone who has a direct interest in water, however, understands that some disagreements must be settled in the courts and some will be resolved by the federal government.

One human-made certainty about our water future remains: the products that people already have made, from laws to equipment, act as limits on the actions we can take. In a very real sense we have already built our water future, or at least some of it.

The Legal and Political Environment of the 21st Century

The General Nature of Existing Law

Surface waters are presently regulated by a complex system of laws and agreements involving states and power and irrigation districts (independent political units governed by elected boards of directors), and, to a lesser degree, natural resources districts. In general, water rights are owned by the first individual (or group) to claim them. Early developers established claims that later developers have had to respect. Today, the network of dams, canals and hydroelec-

Many controversies revolve around competing interests for the state's water, but few have received the attention of the dispute over Platte River flows. In the future, along with irrigators and environmental groups, municipalities drawing water from aquifers near the Platte may also be involved in conflicts over its water. This is an aerial view of the Platte River in central Nebraska.

Biological Systems Engineering, UNL

Soil Conservation Service

Endangered species legislation, particularly that involving the whooping crane, piping plover, least tern and bald eagle, shown here, has increased competition for Platte River flows.

tric plants across the plains west of Omaha operate according to established water rights, use permits and bypass obligations. In recent years, however, endangered species legislation also has entered the legal picture, and as licenses expire, political entities such as irrigation, public power and natural resources districts will have to comply with that legislation.

In general, the legal climate includes a growing body of laws and obligations dating back to before the turn of the century, and its quality changes as completely new considerations, such as the plights of rare species, particularly of certain birds, are taken into account by government.

Planning in the 1990s—A Peek Into the Future

Platte River Streamflow—A Familiar Case

The controversy over the relicensing of Kingsley Dam and the interests of irrigators and rare species dependent on water behind that dam is a highly educational look into our water future. This controversy clearly shows that the social issues that enter into long-range resource planning are quite different today from those of the past. And for Nebraskans, the most significant addition to the surface-water planning process in the past 25 years, mainly because of its extensive land purchases, has been the Platte River Whooping Crane Critical Habitat Maintenance Trust.

The Trust was formed in 1979 as part of the legal settlement between environmentalists and Nebraska's irrigation water users. Several groups, including the state of Nebraska, the National Audubon Society, the Nebraska Wildlife Federation and the National Wildlife Federation, filed a lawsuit contending that the environmental impact statement for the proposed Grayrocks Dam, to be built on a tributary of the North Platte, the Laramie River in Wyoming, did not take into account the effects of this dam on endangered species that use downstream habitats. This suit was settled by a complicated agreement restricting the amount of water Grayrocks could impound and establishing a $7.5 million trust fund to maintain habitat for threatened and endangered migratory birds.

From a historical perspective, the Trust represents a new kind of interested party, namely, a permanently funded, scientifically based unit with an expressed vested interest in habitat maintenance, one that exists as a result of compromise between antagonists in a *water use dispute*. Instead of simply providing mitigation for anticipated changes in a water resource, the Wyoming power and irrigation interests were obligated, through legal action, to establish an *institution* whose mission was mitigation. Here we have a case in which a water conflict was resolved by providing the expertise needed to assess scientifically and continuously the basis for wildlife-management decisions and environment impact, as well as to actually carry out such decisions and habitat maintenance.

Today the Trust finds itself an active, not always invited, and rarely neutral, participant in water-use planning. It commissions and pays for scientific research on Platte River hydrology and biology, asks the courts for permission to intervene in lawsuits, seeks to educate citizens and buys land it often converts to habitat. Its first director, John VanDerwalker, was a noted wildlife biologist. Its second executive director, Tom Emerton, is an attorney and a banker. This difference in the backgrounds of the first two directors clearly indicates that wildlife needs have become an integral part of the water-planning process for Nebraska.

Among Emerton's first tasks in the 1990s will be to straddle the controversy resulting from a decision by Nebraska's attorney general to refuse legal assistance regarding flows for wildlife to the state Game and Parks Commission. The Game and Parks Commission sought an instream-flow right that would guarantee adequate habitat for wildlife. The attorney general's action quickly drew the criticism of figures such as the National Audubon Society's president, Peter Berle, a New York lawyer. It is very doubtful that major decisions ever again can be made regarding important state water resources without participation, invited or not, from the Platte River Trust, the U.S. Fish and Wildlife Service, environmental groups such as the National Audubon Society, the state attorney general's office and a host of other individuals, agencies and organizations. Nor can one expect that Nebraska's water controversies will all remain within Nebraska's boundaries.

Water as an Industrial Commodity

Elsewhere in this book the relationships between Nebraska's water and its agriculture, especially between irrigation and corn production, are discussed from historical, economic, political and personal perspectives. No future can be discovered or built without considering these relationships. But if the present gives us any hints of the future, that consideration will include many competing interests.

One of the most uncertain, yet potentially controversial, issues that future Nebraskans may face is the commercial exploitation of water for interstate sale, rather than for intrastate agricultural, recreational and urban use. Already legislation has been introduced into the state legislature to allow, and indeed to promote, such exploitation. The arguments about this issue are likely to be lengthy, complicated and at times heated, if the debate already conducted is any indicator.

The results of this future debate will have a profound effect on the future of Nebraska's water. As an exercise in predicting the future, one can consider the extreme alternatives: on one hand, legal and heavy exploitation of water as a commodity; on the other, legal restriction of, or attempts to restrict, such exploitation. Possibly, extensive interstate sale of groundwater rights to thirsty western cities could deprive Nebraskans of access to groundwater for agriculture. Alternatively, one can easily envision a protracted court battle over the right to exploit aquifers in the same way that oil, gas and coal deposits have been mined.

The linkage between water and other "mineral resources" has always been obvious to planners. But today, ideas for expanding that linkage seem almost like science fiction. For example, about 70 percent of the cost of fuel for coal-burning electrical power plants comes from transporting that coal. A proposed coal slurry pipeline, carrying Wyoming coal to power plants to the south using Nebraska water, encountered significant opposition from agricultural and environmental groups alike.

But in 1992, engineers were quoted in Nebraska newspapers as claiming that coal transportation costs can be cut in half by compressing coal into logs and carrying them through pipelines by water. Estimates by the transportation and energy industries suggest that such water use would cost Nebraska hundreds of railroad jobs and idle millions of dollars worth of equipment. On the other hand, experts feel that much of the required water could come from the coal itself, thereby easing the burden on surface water and groundwater supplies. But, argue some economists, removal of water from coal would lighten it (and make it burn more efficiently), so rail transportation costs would fall! Although coal logs whipping through pipelines like money containers through a drive-in bank's pneumatic tube seem far removed from a field of healthy corn, both require water. And their mutual dependency on a limited resource graphically illustrates the complex interaction between the various activities requiring this resource.

Journal-Star Printing Company

John VanDerwalker, formerly director of the Platte River Whooping Crane Trust, in front of a crane-viewing blind

A Menagerie of Participants in Water Planning

An obvious lesson provided by the case studies above is that in the future, participants in water-related decision-making will be more varied, have a greater variety of interests and will represent a broader spectrum of citizens than in the past. If there is a future need, it will be for a citizenry that is better educated on water issues and more ready to participate in the debates than at present.

In order to participate actively in the decision-making, however, a person probably needs to understand existing law, the legal and political nature of resource-management districts, the flow of decision-making in goverment from the federal to the local level, as well as agricultural economics, hydrology, ecology and natural history! And there is no guarantee that any of these areas of knowledge will remain stable; indeed, young people of today should expect that the information needed to plan wisely will continue to grow in amount and complexity. For outright difficulty and persistence, few challenges match that of planning correctly for future water use!

Non-agricultural Dynamics in Future Water Use

Wildlife

"Wildlife" has traditionally been thought of as those (relatively) large species for which we hunt or fish. Fur-bearing species also have generally been included in the term. The value of such wildlife has been measured indirectly, for example, in terms of hunting and fishing permit sales, park permits and sales of equipment, clothing and other supplies unique to hunting and fishing. A significant fraction of species traditionally considered wildlife depend directly on water, especially surface water. This list includes ducks and geese, snapping turtles, bullfrogs and all game fish.

As a result of federal endangered species legislation, "wildlife" now technically includes non-game species such as piping plovers, least terns and rare fishes, for example, rainbow darters. And whereas in the past, wildlife meant human access and economic benefits, today, and presumably tomorrow, wildlife means in part denial of human access and subsequent benefits that are not easy to quantify. We have little basis for establishing the social "worth" of a wildlife sanctuary or defending the value of piping plovers. In the future, however, if conflicts over water use are to be resolved, someone will have to calculate the social value of preserving endangered species. Today that task falls to the philosophers and environmental activists; tomorrow it may fall on the shoulders of attorneys, economists and engineers.

Present and Future User Groups

Traditionally, hunters and anglers have had an intense interest in water resources and have supported that interest through their donations, political influence, organizations, license fees and educational activities. One would expect that as long as there are places to hunt and fish, there will be hunters and anglers. One virtually guaranteed element of the planning environment beyond the mid-1990s is the population of those who hunt and fish for sport. Their activities are an accepted part of our culture; a broad spectrum of society, across all socio-economic levels, participates in these outdoor sports; and news media regularly carry stories that focus on hunting and fishing. So it is reasonable to think that water-resource planning beyond domestic, municipal and agricultural use might begin with these groups' interests. Almost by definition, places to hunt and fish are also places where a great diversity of wildlife exists.

Scientists are a water user group whose needs are also tied to planning and policy decisions. These individuals range from hydrologists and geologists to limnologists (lake scientists) and other types of biologists. We often think of these professionals as people whose work provides solutions to problems or gives us information of value to decision makers. However, to a scientist, the object of study, be it a molecule, a cell, a river or a water-bearing rock formation, is vitally important as a model, a resource that can yield principles, concepts, theories, data and insights. So, if a river becomes dry or its streamflow greatly reduced and the valley and its sandbars filled with woody vegetation, scientists no longer have a river to study for discovering how best to manage rivers! And, although water resources have not always been important to society as objects of scientific study, future legal climates will demand information, sometimes vast amounts of it, that can only be supplied by scientists. We confront a circular situation with respect to science and water: we need one in order to plan how to use the other, but if we use the other incorrectly, we will have no source of information with which to plan!

Tourists are another user group that future planners will undoubtedly have to consider seriously. The term "tourism" includes all activities in which travelers to or through a region spend money acquired in another region. Birdwatchers from Omaha spending the night in Grand Island during the annual sandhill crane migration are tourists. More important to Nebraska's economy, however, are birdwatchers from New York and California who might spend the night in Grand Island, bringing with them, and leaving behind, money originally earned on either coast.

At present, tourism that depends directly on water resources includes hunting and fishing, boating, camping and wildlife watching. Although it is not easy to place a dollar value on this combination of sources, tourism is generally considered the third largest industry in Nebraska. In a few cases, for example, canoeing on the Niobrara and crane watching along the Platte, the connection between tourists' dollars and water resources are obvious. But in many other cases, for example, visits to museums and zoos or to a dude ranch in the Sand Hills with subirrigated meadows, the connection between tourism and water is indirect because the tourist attraction is embedded in the total resources and infrastructure of the state.

Nebraska Farmer

An unidentified fisherman at a Sand Hills lake

The Nebraska tourist industry has enormous potential for economic development, especially so when political instability, environmental degradation and unfavorable exchange rates discourage foreign travel. This development, however, is heavily tied to water uses that may violate traditions. The most notable example, though certainly not the only one, is managing the Platte River with the benefit of the cranes in mind. What must be remembered is that while the river may benefit the cranes, the presence of the cranes benefits the economy throughout a large stretch of the central Platte Valley. But none of these benefits come without costs. Wherever tourism has developed as a thriving industry, problems have accompanied the benefits. Humans in large numbers not only spend money, but they also require food, lodging, transportation and healthcare, police and fire-control services. And they tend to leave their trash behind.

The "take home lesson" from the 1980s and 1990s is that the active participants in water-resource planning, whether they participate by virtue of political office or economic, intellectual or ideological interests, are far more varied and sophisticated than they were a generation ago. It would seem foolish to expect either their sophistication or diversity of interest to disappear in the years to come.

Education for the 21st Century

Today, water-use plans are made mostly by engineers, attorneys and elected officials (who may or may not have specific training relative to water use). These are the people who are typically in positions of responsibility in state agencies, who participate in legal actions and who oversee policies and laws pertaining to water. These professionals have access to an enormous amount of information—scientific studies, newspaper and magazine stories, engineering reports, budgets and position papers, all written by various people trying to influence decisions. In fact, this book is part of the information people in decision-making positions will likely use with respect to water use!

In the United States, citizens outside of government have always felt free to participate in government, as individuals or members of an organization, through a variety of means: lobbying, letter writing, supporting causes financially or with volunteer work, protesting and circulating petitions. Sometimes these efforts affect how government conducts its business. Throughout the 1980s and 1990s, for example, Nebraska has witnessed and will continue to witness an ongoing battle over the rights to water in the Platte River system. Today this conflict is carried on by a wide variety of people on both sides who probably never could have foreseen themselves as social activists.

But there is one theme common to all participants in water-use decision-making, namely, the need for information and technical education. It is not likely that this need will disappear. Instead, in the next few decades, those who plan water use, either directly, because of their positions in government, or indirectly, because of their political activities, will need more information and better education than they have at present. Housewives will become familiar with computer models of watersheds; farmers already find themselves studying water law, and there is no reason to expect they will suddenly be relieved of that responsibility. And anyone running for public office who cannot discuss water issues is vulnerable to a knowledgeable opponent who speaks intelligently of the options available.

The experience of the past century in the Great Plains has shown us that water is the critical element for survival. But events and discoveries of the last decades of the 20th Century have shown us that the Great Plains are not an isolated ecosystem. Global economic and ecological factors influence Nebraska residents quickly and significantly; Nebraskans are no less likely to buy imported electronics and automobiles than are urbanites on either coast, and farmers welcome international markets for their grain. The surprise of the late 20th Century, however, has been the demonstration that global ecological events also exert a powerful influence on Nebraskans' lives, although not always as quickly, nor as dramatically, as economic ones. Therefore, in education our major need is for lessons that teach the nature and extent of our ecological and economic interdependence.

However, many of the educational goals, philosophies and techniques likely to be needed in the next century are not a part of the present educational system. Local opposition to certain kinds of content can easily hamstring the very educational efforts needed to provide planning expertise for the next generation. Many issues, such as population control, multicultural experiences, foreign language instruction, urban problems, federal budget priorities, "environmentalism," gender bias and the role of the visual arts, can be highly controversial; parents can exert enormous pressure on school administrators to avoid potentially controversial issues. Unfortunately, these are the very issues that young citizens must learn to address because they will have to deal with them as adults. And all play an important, though not always recognized, part in water and natural-resource planning. The average Nebraskan of tomorrow must be

The encroachment of woody vegetation on sandbars in the central Platte River is seen in this aerial photo taken in 1969 about 8 miles southwest of Shelton.

Conservation and Survey

as sophisticated, articulate and well educated as a resident of any other state simply because the decisions that tomorrow's citizens must make will demand a very deep, sophisticated and broad education.

Global Trends and Their Effect on Local Water Issues

Population Growth

There is no evidence whatsoever that humans are collectively capable of, or interested in, solving their most serious problem, the increase in their own numbers. Globally, the human population continues to increase exponentially, and even if limits on childbearing were put in force today—a politically impossible task—the numbers of humans would still increase because of the massive number of young people in developing nations.

Human beings require water. The body is 70 percent water. Furthermore, humans must eat or starve to death. These are not theoretical claims, but simple facts. For an agricultural economy, especially one like Nebraska's that is so heavily dependent on water for irrigation, the increase in human population represents a growing potential market for agricultural products. But this market is also limited by its ability to pay and by the available natural resources in the crop-growing regions.

Statistically speaking, a child born today in Nebraska can expect to live until after the middle of the 21st Century. What will that child see? We have no way of knowing exactly, with one exception: that child will see a staggering increase in the number of people on earth. And most of those new humans will be ill-fed, poorly educated and living in poor housing. It is naive to think that the students in the state's elementary schools today will not have to interact with and consider, politically, economically, socially, these other people being born every minute. And at some point, this mass of humanity will require water, either to drink or because of their needs for grain and protein. If there is one overriding educational need in resource planning, it is comprehending the effects that the burgeoning population of humans will have on local economies and lifestyles.

Tropical Deforestation

At present, the tropical forests are being cleared at the rate of about 50 acres a minute. At that rate, these forests are likely to be gone within the lifetimes of young people alive today. Some of the effects of tropical deforestation are predictable: political instability and loss of short-term agricultural potential in developing nations, destruction of plant species potentially useful to agriculture and medicine and extinction of most of the world's species of animals. Other effects are not so easily predicted, but scientists try not to be naive in these matters. We do not know, for example, whether global weather patterns will be altered as a result of tropical deforestation and the loss of the truly staggering amount of vegetation represented by tropical forests.

Plants are directly involved in water, nitrogen, carbon, phosphorus and oxygen cycles. Altering the world's plant cover is predicted to alter these cycles and the weather patterns associated with them. Will Nebraska's rainfall be reduced, increased or continue to fluctuate within present limits? Will Nebraska's average temperature and growing season be reduced, increased or remain more or less the same as today? These questions cannot be answered with certainty. What can be said for sure is that planners for Nebraska's future must take into account the activities of people with chainsaws buzzing through old-growth forests in far-off places.

Fossil Fuel Supplies

Scientists have been predicting for some time the end of the world's petroleum supplies. Although the United States has large coal reserves, those reserves are present largely because the nation has used oil instead, and the same scientists who predict the exhaustion of petroleum supplies also can envision the depletion of these coal reserves. Why is fossil fuel a consideration in the area of future water use? This is because agriculture, the industry that supplies food for a rapidly expanding population, is also a heavy user of fossil-fuel energy. This non-renewable resource is used for everything from fuel for tractors and other farm equipment, to transportation of agricultural products, to chemical feedstock for fertilizers and pesticides. A society with ample water but no fuel is as crippled as one with plenty of fuel but no water. So today's elementary school students, who will occupy the future we build today, must expect that we examine global fossil-fuel supply and economies as part of our water-resource planning.

A Worst Case Scenario

Best and worst case scenarios may differ depending on one's point of view. However, a worst-case scenario is clearly one in which everybody loses; that is, agricultural, environmental and recreational interests all suffer badly as a result of water-use decisions made today. In the case of Nebraska's water resources, we can envision a circumstance in which all of the negative processes at work today continue unchecked for the next quarter century.

In the worst imaginable future, during a prolonged, 10-to-20-year drought, municipal water supplies and wells provid-

Except for the Niobrara, Elkhorn and Loup rivers, most of Nebraska's river water derives from runoff, much of which emanates from states to the west. But nearly all of the state's groundwater derives from precipitation falling in the state, such as this thunderstorm near Neligh.

Jim Swinehart, CSD

ing water for rural residents become more and more contaminated with nitrates and other chemicals; the water-based tourist industry is lost because of reduced streamflow and drying lakes; groundwater levels are lowered significantly, not only reducing water for irrigation but increasing the cost of that irrigation—due to higher pumping costs—at a time when petroleum resources are becoming scarce; a seemingly never-ending progression of lawsuits, hearings and negotiations beset every attempt to resolve a water conflict in favor of one user over another; public school budgets are reduced, causing local school districts to cut back on educational activities that lead to young people's understanding of water problems; larger municipalities fight over water supplies, carrying their battles into the courts. And in a worst imaginable world of the early 21st Century, the Platte and other rivers are nearly dry and their valleys grown up with trees; hundreds of towns cannot use their wells because of nitrate contamination; abandoned center pivots mark the sites of massive erosion in the Sand Hills; every community—from Omaha to Arthur—has to ration water; and most of the state's lakes are largely silted in. Obviously, this picture is not a very pretty one.

Adding to a worst-case scenario are the natural phenomena over which Nebraskans, like people everywhere, have no control, for example, periodic drought or the effects of major volcanic eruptions that shorten the growing season or reduce average annual temperature. To these events must be added the ones over which humans theoretically have control but to date exhibit little ability or will to control: population growth, with its attendant tropical deforestation and consequent species extinction, fossil-fuel consumption, ozone depletion and the buildup of greenhouse gases in the atmosphere.

If this discussion sounds like a doomsday report, remember, it is a *worst-case* scenario. The critical question for all of us, however, is not, *Is it true*? but rather, *How much of a future worst-case scenario is due to our failure to act correctly today*? The answer to that last question is, of course: some. And the challenge for today, then, becomes one of minimizing our own contributions to the worst possible conditions in which our children and grandchildren must live.

A Best Case Scenario

In a best case scenario, heavy snow falls in the Rocky Mountains and perfectly timed seasonal rains, not too heavy, not too light, water wheat and corn throughout Nebraska two years out of every three; aquifers are recharged; surface streamflow is adequate to supply agriculture's needs while boosting a growing tourist industry; sustainable agricultural techniques and other technological advances solve the groundwater contamination problem; perfectly targeted and designed pesticides perform their functions efficiently with no toxic side effects for humans or plant or animal life; automotive engineers solve the greenhouse gas problem; scientists solve the ozone problem; world population stabilizes at a sustainable level; a major breakthrough in solar-energy technology solves the fossil-fuel shortage; the developing world quits burning the tropical forests while the developed nations quit clear-cutting their old-growth forests; the United States pays off its national debt; the illegal narcotics industry is shut down; and grain prices soar with political stability and improved economic growth throughout the developing world and eastern Europe.

If this latter scenario sounds like an utopian dream, remember it is a *best-case scenario*. The critical question for all of us, however, is not, *Will it come true*? but rather, *How much of the future best-case scenario is possible if we act correctly today*? The answer to that last question is, of course: some. And the challenge for today, then, becomes making the maximum contributions to the best conditions possible in which our children and grandchildren will live in the 21st Century.

Commentary:

A Letter to the Federal Energy Regulatory Commission About Nebraska's Platte River

by Vincent H. Dreeszen and Ray Bentall[1]

May 1990

In its "wild" state (that is, prior to any diversions of river water or to construction of any dams to create water storage) the Platte's flow was highly erratic, a large part of its annual discharge being due to springtime melting of the snowpack in its mountainous headwater area. Because the snowpack varied in volume from one year to another and because melting occurred more rapidly in some years than in others, the peak discharges and the duration of the high-flow period differed from one year to the next. The Platte's flow in summer, fall, and winter generally was low, and periods of no flow between North Platte and Columbus were not uncommon. Occasionally heavy rainfall in some part of the drainage area caused temporary high flows and even flooding during the normally low-flow seasons. A particularly interesting feature of the Platte's flow regime is that flood flows in one reach may be lost to seepage out of the river within a downstream reach. For example, the largest recorded flood flow contributed to the Platte by the South Platte River could not be detected as an increase of the Platte's flow at Grand Island. The gaining and losing reaches of the Platte tend to be seasonal, being closely related to variations in the distribution and amount of precipitation on valley lands.

Building of dams on the North Platte River has "tamed" the Platte River markedly. By holding back high flows in spring and releasing the stored water for irrigation and power generation, the high flows of the Platte have been reduced and the low flows increased. The flow of tributaries that drain irrigated lands changed from intermittent to perennial, and a few now have become suitable for spawning. Since the filling of Lake McConaughy, flooding by the Platte is a rarity, and the river rarely is without some flow owing to return flows from irrigation and hydro-power plants and to increased seepage of groundwater into the river. In our view, and in the opinion of many others familiar with the Platte, the now "controlled" flow of the Platte has not only been a boon to Nebraska's economy but also has *enhanced*

[1]CSD hydrogeolgists emeriti; written as a private letter during emeritus (retired) status

the environment for wildlife. For example, the number of sandhill cranes staging in the valley each spring has increased significantly since Lake McConaughy was created. Not only has the population of cranes increased, but many other life forms not known formerly to frequent the Platte now are there. Incidentally, to our knowledge, no one has shown that the Platte is essential to the well being of the endangered whooping crane, nor that it ever was. In other words, the existing water-resource developments for human benefit are responsible for creating a more salubrious environment for wildlife.

Requiring release of water from Lake McConaughy for the benefit of wildlife that probably did not frequent the valley *when the Platte still was "wild"* seems ludicrous when such changes are detrimental to present day irrigators, producers of power, and fishing and other recreational use of Lake McConaughy. Far wiser, in our view, would be to "leave well enough alone." Furthermore, it seems quite likely that giving in to demands of wildlife interests may lead to further and more environmentally deleterious demands in the future.

The current order that a minimum flow of 400 cfs [cubic feet per second] be maintained during specified periods by releasing water stored for irrigation and power generation is a flow rate that was rarely provided under pre-development conditions. Furthermore, owing to the existing hydrologic conditions along the Platte, releases from Lake McConaughy would need to be greater than 400 cfs because seepage losses and evaporation between the release point would occur between the lake and Grand Island, where the 400 cfs flow rate is to be provided. In some drought years, only a small part, if any, of the released water would reach Grand Island even if diversions in the intervening reach were to be curtailed completely.

At the risk of seeming pedantic, we are adding the following summary of our observations and conclusions about the Platte River:

1. The quantity of precipitation on the area drained by the Platte varies widely from one to another period but over the long run now is no more nor no less than in the past. In effect, the annual average water supply available for use has not changed. However, use of the river water has increased to the point that the average flow of the river at Duncan (just below the inflow from the Loup River) has been reduced as much as 50 percent. It is necessary, however, to point out that this large use would not be possible without storage of the high flows that formerly passed through the state unused. In other words, only the high flows have been decreased (and beneficially so) whereas the return flows and groundwater seepage into the river have augmented the river's dry-weather flow. Historically, the Platte River in the Big Bend area was a mile or more wide expanse of rushing water during the height of the snowmelt period and then became a ribbon of dry gravel interspersed with narrow rivulets of water for the remainder of the year. During some summers, and particularly during the 1930s drought period, even wading pools disappeared from the river bed at Grand Island. In fact and according to a July 1980 report of the Nebraska Department of Water Resources (DWR)—"in the twelve years prior to construction of Kingsley Dam, the key measuring station on the Platte River at Overton was dry for a portion of each year."

2. Whereas periods of no surface flow were not uncommon before the storage was created, they now are rare. Seepage to the river from the large mound of groundwater that extends from Sutherland east to Minden and that was created by canal losses and irrigation seepage has tended to increase in the Big Bend area and may be even greater in the fu-

The south channel of the Platte River near Cozad, July 20, 1938. This photo shows evidence of the historical low flow periods that were characteristic before the building of Kingsley Dam. On the next page is a view upstream on the same channel near Cozad in May of 1940.

Department of Water Resources

ture. We estimate that the increase to groundwater storage is about 7 to 10 million acre-feet and that annual seepage returns to the Platte are as much as 200,000 acre-feet/year from that source alone. Return flows upstream from Lake McConaughy also help stabilize the flow of the Platte River. DWR, in its 1980 report, states, "—summer inflow to Lake McConaughy is often 80 percent return flow from upstream developments."

3. Few people understand that the Platte between Kearney and the mouth of the Loup River naturally tends to lose flow by seepage to groundwater on both sides of the river. Before Europeans arrived the wide expanse of the valley land on the north side supported a stand of tall grasses that probably used as much or more groundwater than now is used by pump-irrigated crops. And on the south side, seepage loss was, and still is, to the groundwater supply of the Big Blue River basin.

4. Some people perceive the Platte as flowing in a lined conduit. That is, that the river's flow at a given location should be no less than at an upstream location. Such a concept is unrealistic because the Platte is intimately related to groundwater in the alluvial deposits over which it flows and the river and floodplain are vulnerable to high evapotranspiration losses. One reach may be gaining from groundwater and another reach simultaneously losing to groundwater. There is no guarantee that an addition to flow at an upstream point will still be in the river at a downstream point. History provides some insight in this regard. The following excerpt is from the Tenth Biennial Report of the State Board of Irrigation, Sept. 1, 1912 to Sept. 1, 1914. Precipitation during the period covered by the report was considered to be about average. We quote, "The weather conditions are such a strong factor governing the amount of water which is necessary to have passing both Bridgeport and North Platte that even with the records of the past season it is not possible to figure accurately the exact amounts. For instance, on August 6, 1914, there was several hundred feet of water passing the headgates of the Kearney Canal in excess of what they could use. On August 9, the river at Kearney was absolutely dry. No water was available for use by the Kearney Canal, yet during this period there had been nine hundred fifty to twelve hundred cubic feet of water passing our gaging station at North Platte, and the intervening ditches between Kearney and North Platte were using not to exceed three hundred forty feet. This extreme decrease in the flow of the river is easily accounted for considering the wide expanse of the river bed and the four days of exceedingly hot south winds on August 6th, 7th, 8th and 9th." Transpiration by vegetation on islands and on wetlands bordering the river also can account for water loss from the river. Existing models do not adequately consider these losses in planning river regulation.

5. Despite all claims to the contrary, the current hydrologic equilibrium reached through regulation of river flow has provided the benefits of flood protection, environmentally sound production of hydropower, water for irrigation, flat-water recreation, and a significantly enhanced riverine habitat for migratory waterfowl and indigenous species of birds, fish and mammals.

6. The order for re-regulation of releases from Lake McConaughy is both hydrologically and economically unsound and unrealistic.

7. In our judgment, the Platte River and its tributary flows are very poorly understood by anyone who has not studied the rivers and their history. Misinformation appears to be the rule rather than the exception, and that misinformation is being disseminated widely. The flow, storage, diversion, and management of the Platte River are exceedingly complex. Indeed, many changes have occurred over the past 100 years or so, and these changes, too, are complex. The issue of re-regulation of water flows from those reservoirs licensed by FERC in the western states for environmental purposes hopefully will be decided on a factual scientific basis case by case. Hard questions should be raised now.

Department of Water Resources

Mary Bargman, Nebraska Farmer

The Tri-County Canal (above) and the Platte River somewhere in eastern Nebraska during a low-flow period (below)

Department of Water Resources

The Future of Nebraska Agriculture

by A. L. (Roy) Frederick

Both dynamic changes and more of the same are in the offing for Nebraska farmers and ranchers during the next 25 years. The greatest changes will be in production practices, particularly with respect to the type and quantity of inputs used. Markets for agricultural commodities also will continue to evolve. Nonfood uses in the United States will become relatively more important while exports of raw commodities will account for a smaller relative share.

Commodities produced will remain much the same as today. In 2020, the big four in the crop sector will be corn, wheat, sorghum and soybeans, while beef cattle and swine will dominate on the livestock side. Hay production also will continue to be important, as will traditional specialty crops in the Panhandle (although sugar beet production could be vulnerable). Farm numbers will continue to decline but at a slower pace than in recent decades.

As in the past, Nebraska producers will base production decisions on market opportunities, new technology and availability of natural resources, especially water. Farm price and income support programs also will be a factor, although less so over time. What really will be different, however, is that society will limit production practices to those that are environmentally sound. Laws, regulations and incentives relating to the environment will contribute more than any other factor to the future structure of agriculture in Nebraska.

Agriculture's Continued Strength

Though significant changes will be occurring in Nebraska agriculture over the next quarter-century, the bottom line is that production agriculture will remain important in Nebraska. In 1991, personal income from farming and ranching in Nebraska amounted to $1.7 billion. While this represented only 6 percent of all personal income in the state, as much as 25 percent of Nebraska's personal income (and employment) was generated from economic activities associated with agriculture. Even if the relative contributions from agriculture decline in the future, the cause is not likely to be an absolute decline in agriculture. Rather, it will reflect growth in other sectors of the Nebraska economy.

The case for continued strength in Nebraska agriculture begins with the simplest of acknowledgments: food production is essential to meet a basic human need, and Nebraska's agricultural sector is positioned to respond to that need, both in domestic and international markets. In addition—and perhaps most significantly—the budding market for non-food, agriculturally based products will come into full flower. Most of those products will come from commodities that are important in Nebraska.

Charles Flowerday, CSD

Nebraska's ranchers represent an occasionally overlooked contribution to the state's agricultural economy. Cow-calf operations convert solar energy to food where row-crop farming generally would be next to impossible.

While farming and ranching account for 6 percent of the personal income in the state, 25 percent of the state's personal income and employment come from activities associated with these two enterprises.

IANR

Domestic Food Demand

Demand for food is primarily a function of population changes, although personal income is a factor as well. The U.S. population is expected to continue to grow, although at a modest rate. In the early 1990s, it was increasing by about two million persons per year, a growth rate of slightly less than 1 percent annually.

Because our population is growing older and relatively fewer people will be in child-bearing age groups than in the 1970s and 1980s, there is little chance for a spurt in natural population growth.

However, births should continue to exceed deaths, thereby keeping natural population changes on the positive side of the ledger.

Migration patterns also will continue to point toward population gains in the United States. While more open international trade of products and services would tend to reduce differences in personal incomes among countries, the United States will continue to be seen as a land of economic opportunities and political stability. Therefore, the rate of immigration probably will not change much from the early 1990s.

Most Americans are relatively well fed, so increased income would change the quantity of food consumed only marginally. However, higher incomes could be expected to change the nature of foodstuffs consumed to more high-value products, such as fresh fruits and vegetables.

Traditionally, higher incomes also have been linked to increased consumption of red meat, especially beef products. However, this was not the case during the 1976-90 period, when beef consumption fell from 94 to 68 pounds per capita. The challenge for the beef cattle industry is to educate skeptical, health-conscious consumers about the desirable nutritional qualities of beef so that consumption stabilizes at, or increases from, present levels.

Much rides on the beef industry's successful response to the challenge before it. Monthly cattle-on-feed reports from the U.S. Department of Agriculture (USDA) frequently show more cattle in Nebraska feedlots than in any other state. In addition, other sectors of Nebraskan agriculture, such as production of feed grains, are tied directly to beef cattle production. In candor, it would be difficult to make the case for continued vitality in Nebraska's agricultural sector if per capita beef consumption continued to decline at the pace of the 1970s and 1980s. Such a scenario does not seem likely because the beef industry is responding to consumers' concerns, some of which have been more apparent than real.

This is not to suggest that consumers will step back from their current demands for foods that contribute to good health. Foodstuffs that come directly from crops generally have been praised for not having undesirable characteristics such as saturated fats and cholesterol. But what is less well understood is that food products from animal agriculture often supply needed protein and vitamins to human diets.

Mary Crawford, Nebraska Farmer

The Snake Creek Ranch feedlot. Nebraska often leads the nation in cattle on feed. While some dramatic declines in beef consumption occurred during the 1970s and 1980s, leaner beef appears to have helped stabilize demand.

Future research and development will be directed at having the best of both worlds. For example, certain strains of hybrid corn will feature high levels of protein or unsaturated oil. And research on animal products such as red meat, eggs and dairy products will continue to be directed at reducing saturated fat and cholesterol. One thing seems certain, however: flavor, texture and general palatability will never go out of style as foodstuffs change to meet society's standards for a healthy and wholesome food supply.

Closely associated with the health issue will be increased efforts to assure food safety. The public will not tolerate chemical or biological additives that are perceived to be harmful to human health. Thus, the scientific and educational community will continue to be pressured to deliver accurate and timely information on a range of food-safety issues. A part of the continuing challenge will be to better match perceptions of food safety with scientific evidence. Clearer risk assessment regarding foodstuffs with and without specific additives can be expected from both the food industry and governmental agencies. Legislation that further addresses residue levels of additives in foodstuffs also is probable.

Foreign Food Demand

Worldwide population growth of about 1.7 percent (90 million persons) per year is nearly twice the U.S. growth rate. This will leave a total population of about 7 billion in 2010, compared to an estimated 5.4 billion in 1992.

Up to 90 percent of this growth will be in the developing countries of Africa, southern Asia and Latin America. Not coincidentally, these areas of the world also tend to be where the gap between food production and needs is most acute. Indeed, in many cases, it's growing. Unfortunately, many of these countries, particularly those in Africa, do not have sufficient capital to purchase needed foodstuffs from food exporters such as the United States. The dilemma for U.S. policy makers will continue to be as follows: should assistance be provided to develop the agricultural sectors of these countries, which might eventually also help them earn foreign exchange, or should foodstuffs be provided directly at less than the market price? In part, the U.S. response has depended on (and likely will continue to depend on) the actions of competing exporters, such as the European Economic Community. Our inclination, however, has been to focus on the first alternative.

The U.S. position in the world marketplace for agricultural products is important to Nebraska producers because we are a significant factor in five of the six most important export subgroups: wheat, feed grains, oil seeds, red meat and livestock by-products (Nebraska is not now, nor is it expected to be, a major exporter of horticultural products). Though Nebraska is disadvantaged by not being near any major port, the state's central location facilitates access to virtually all major export points by rail and truck carriers. Combined with the state's continuing production potential for commodities most likely to be demanded by exporters, there is little reason to believe that Nebraska won't continue to receive its fair share of the U.S. export market in the future.

But what about the matter of competing exporters? Questions relating both to the adoption of new technology and trade policy must be considered.

As recently as 1975, most American observers smugly concluded that the United States would always have a technological edge in agricultural production and therefore could expect the size of the agricultural export market to grow ever larger. However, subsequent developments in an array of countries around the world may invalidate such assertions. For example, economic studies now indicate that France can produce wheat at a lower price than we can, at least up to some absolute production limit. The same thing can be said for corn in Argentina and soybeans in Brazil. Thailand, a relatively new competitor in world export markets, also is showing signs of being a low-cost producer of a number of commodities. In short, the United States cannot rest on past laurels relating to production efficiency. We may even lose ground if we do not make the investment in basic research and technology that other countries are making.

Becky Ohlde, Nebraska Farmer

Along with crops such as grain sorghum and most soybeans, wheat is a dryland crop that generates much income for Nebraska's farmers.

One sector of the export market where the United States can gain ground is in value-added products (processing raw materials into finished food products), ranging from processed meat to a plethora of branded retail items. As recently as 1985, value-added agricultural exports amounted to only $11.5 billion. But by 1991, a record $21 billion worth of such products were exported. For the first time, the value-added sector accounted for more than half of all U.S. agricultural exports. However, the U.S. still accounted for only 14 percent of the world market for value-added agricultural products, a full 10 percentage points below the European Economic Community's share.

The beauty of value-added products is that they mean more jobs and incomes here at home. In particular, Nebraska has a stake in exporting more processed beef. The market potential is primarily in the industrialized coutries, including those in the Pacific Rim and western Europe. But that potential can be fully realized only if current beef import barriers are reduced in key countries.

As this is written, a number of potentially significant policy matters affecting trade are being discussed by the international community. Among the most important are formal discussions relating to the General Agreement on Tariffs and Trade (GATT), the North American Free Trade Agreement (NAFTA), and the European Economic Community unification. More informally, the United States has an important stake (and equally large questions) about future trade with the Commonwealth of Independent States (the former Soviet Union) and many of the developing countries.

Generally speaking, trade with fewer impediments (quotas, tariffs and subsidies) should be of benefit to consumers and taxpayers around the world. However, more open trade would not necessarily be desirable for producers, especially in the short run, or if considered from the perspective of specific commodity sectors in the United States such as dairy and sugar.

The challenge with all trade "reforms" is to convince those directly affected that short-run pain will be followed by long-run gain. No magic prescription is offered for the future. Therefore, while U.S. and Nebraska agricultural exports of raw commodities and value-added products probably will continue to "ratchet" higher, a sustained 1970s-style market boom is highly unlikely.

Canola, seen here, represents an alternative crop with marketable potential, mostly for its oil.

Becky Ohlde, Nebraska Farmer

Domestic Non-food Demand

While expansion of foreign trade always will depend heavily on others (both importers and competing exporters), developing non-traditional uses for agricultural products within the United States would provide more assurance of market growth. Progress to date provides a solid foundation for the future. For example, while corn continues to be used primarily for livestock feed, domestic use for other purposes—seed, food and industrial products—increased from 522 million bushels in 1975 to 1.4 billion bushels in 1991. Moreover, by 2000, non-feed

Becky Ohlde, Nebraska Farmer

Milkweed down, shown here, is grown as an alternative crop in Nebraska for stuffing in coats, jackets, vests and comforters.

uses are likely to approach 2.5 billion bushels annually.

The push to use agricultural commodities in non-traditional ways has economic roots, but in the future it will be driven at least as much by environmental concerns as economics. Ethanol production is a case in point.

In the 1970s, fear of serious shortages of petroleum-based liquid fuels spurred efforts to produce ethanol from corn. But it consistently has been more costly to produce ethanol than petroleum-based fuels, even with significant increases in ethanol production efficiency. Thus, ethanol production in the future might be in jeopardy if it were not for environmental concerns generally and the federal Clean Air Act in particular. This law has the potential to spur demand for ethanol in pollution-plagued cities throughout the country when the act is fully implemented in 1995.

In fairness, one must acknowledge that ethanol is not viewed as an environmental savior by all observers. While reputable studies indicate that ethanol would help reduce carbon monoxide and other "greenhouse gases" in the atmosphere by as much as 30 percent, critics worry that ethanol will push up the level of volatile organic compounds in the atmosphere. When these compounds react with summer sunlight, ozone is produced. Because ozone is unhealthy to breathe, the ethanol industry has been challenged to answer tough questions about the overall benefits of the product. If responses ultimately are convincing, current and planned ethanol production capacity may not be large enough to meet demand in the years ahead.

Environmental concerns also are prodding product development in other ways. Research directed at the problem of solid-waste management indicates that both corn and soybeans may be used to create at least partially biodegradable plastics and adhesives. Corn starch may be used for road de-icers rather than traditional chemicals, thereby reducing damage both to road surfaces and roadsides. Ink from soybean oil, which decomposes more rapidly than petroleum-based inks, has been an unqualified success; more than half of the nation's newspapers have adopted it for color printing. Research also is underway that would allow a soybean-oil product to serve as a substitute for diesel fuel or as an alternative to chemical preservatives on fresh fruits and vegetables.

At least for the immediate future, industrial uses for agriculturally based products will represent a low-value market; that is, full economic justification for using agricultural commodities in industrial products will occur only at low commodity prices. However, the economic stature of agriculturally based industrial products will improve—perhaps dramatically—in response to tightly focused research on alternative uses. An Alternative Agricultural Research and Commercialization Center has been established within USDA as a highly visible example of the attention and resources that will be brought to industrial-product research and development. In short, significant progress can be expected during the next 25 years, perhaps including areas that are not imaginable today.

In addition to new products from existing crops, some progress is likely in developing and commercializing new crops for both food and non-food markets. Mung beans, crambe, canola and milkweed are among the possibilities. However, at the present time it appears that no new crop will have the enormous market potential of, say, soybeans, which were developed a half century ago. New crops are more likely to be better suited for niche markets. This means that they will provide a profit opportunity for some producers in selected areas of Nebraska, but not for rank-and-file producers across the state.

Developing new products and new crops simultaneously is extremely costly. Perhaps that explains why market development seems to be more rapid when focused on new products from traditional crops. Simply put, alternative-use research is challenging enough without giving equal or greater attention to crop-production problems.

The research community's commitment to alternative crops research, particularly in the land-grant universities and USDA, appears to be somewhat fragmented at the present time. Soybean development was enhanced by the strong demand for vegetable oils during World War II. It provided the incentive to move quickly. But the development of another new crop with a market impact comparable to soybeans would require huge amounts of scientists' time and of financial resources from the public and private sectors. At this time, no new crop appears to be a likely beneficiary of such a commitment.

Crop Production Practices

Whatever the mix of agricultural commodities in the future, one can be certain that production practices will be more environmentally sensitive than in the past. In no small part, individual producers, acting on their own, will be the driving force. A survey of attitudes of Nebraska producers in conjunction with the 1990 farm-bill debate indicated that maintenance of soil and water resources ranked as the top priority among producers, even exceeding concerns about price and income supports. Producers and their families do not want to drink contaminated water or leave soil in poor condition for the next generation. Beyond that, preservation of the natural-resource base increasingly will be thought of as a community goal; other producers will be intolerant of the environmentally damaging practices of their peers.

That is not to say that all levels of government—federal, state and local—will not be a part of protecting the natural-resource base as well. A combination of compensatory incentives on one hand and regulatory restrictions on the other will continue. In recent years, the federal government has provided incentives through an array of new programs, including the conservation reserve program, the water-quality incentive program and the integrated farm management program. However, one can't help but speculate that the emphasis will be more on regulation (for example, as with conservation compliance and wetlands protection), given the continuing federal budget deficit.

Another alternative crop grown mostly for flour is amaranth. This crop was grown on Oak Creek Farms of Edgar.

Becky Ohlde, Nebraska Farmer

State agencies such as the Natural Resources Commission (NRC), the Department of Water Resources (DWR) and the Department of Environmental Quality (DEQ), plus the local natural resource districts (NRDs) may experience a collective increase in their responsibilities, especially if federal spending on soil and water conservation cannot be sustained at present levels.

Special Focus on Groundwater

Groundwater will continue to receive special attention. NRDs may establish groundwater management areas for the purpose of regulating groundwater depletion or contamination. In addition, state government and the NRDs may work together to reduce depletion and contamination through groundwater control areas or special protection areas (SPAs) for groundwater quality. Some outsiders observe that these programs are among the most forward-looking in the nation. Many Nebraskans respond that it only makes sense to do prudent planning given the vital importance of groundwater.

While public sensitivity to both the quantity and quality of groundwater is high in Nebraska, state law provides that landowners have wide-ranging property rights with respect to the underlying groundwater. Thus, conflicts are possible, perhaps inevitable, as public entities consider expanding the scope of groundwater regulation. The courts ultimately will have much to say about future groundwater use for agricultural purposes.

Education also will play a key role. The linkage between groundwater and agricultural production will become more benign. Much of the necessary technology and production practices to accomplish this already are ''shelf-ready'' but need to be more widely adopted. Irrigation scheduling is a good example. An important contributing factor to future adoption of such technology is that most of it will be feasible on all sizes of farms. With improved awareness, farms large and small should find it advantageous to adopt such practices.

A ''Go Big Red'' comforter made from Nebraska milkweed down

Sustainability

Calls for ''sustainable agriculture'' will increase in intensity. While this term means different things to different people, long-term agricultural production that neither degrades the natural-resource base nor damages the environment is at its core. The concept applies to both crop and animal agriculture. In the short run, moving toward sustainability may not be accomplished without some loss in production efficiency. Longer term evaluations, however, should show that sustainability and maximum production efficiency are much more likely to be mutually attainable goals. But even if these goals are not mutually attainable, society may dictate that protecting the environment is more important than lowest-cost production.

If one accepts the premise that production efficiency will still be important, even through sustainable systems, what are the prospects for future production?

Crop Production Prospects

Yields of major crops produced in Nebraska will continue to increase, but, as before, by varying amounts. Wheat yields, for example, have grown only about .7 percent per year over the last 20 years. Meanwhile, corn yields have grown nearly 2 percent annually.

The most daunting challenge will be to maintain or increase yields with reduced amounts of traditional inputs, including water, nitrogen fertilizers, and pesticides. It probably would not be possible to keep yields growing at the same pace as in the recent past if inputs were cut across the board and nothing else was added to the production equation. Fortunately, the future looks promising, even though production practices and inputs will change.

On some farms, a partial response to the need to conserve inputs will involve the use of a sophisticated, computerized information system that can give precise direction to the type and amount of inputs needed. In most cases producers will hire others to provide this information. In fact, it could be an extension of the work that crop consultants now do. Highly specific information within a farm field will be linked with computers on center-pivot irrigation systems to deliver specific and varying amounts of fertilizer, pesticides and water depending on specific conditions in the field. The idea, of course, is to provide only as much input as is needed for a particular area. In so doing, the overall level of inputs should be reduced.

Timeliness of input application also will receive increased attention. Fertilizers and pesticides will feature slower release of nutrients and toxic materials for the greater ultimate benefit of the crop.

New varieties will be tested and introduced that have a variety of desirable traits: tolerance or greater resistance to major insect and other pest problems; greater resistance to drought and efficiency of water use; improved tolerance of problem soil environments, including nutrient deficiencies and toxic levels of some chemical elements in the soil; and improved nutrient-use efficiency, especially for nitrogen, phosphorus and potassium. An underlying theme will be that biology is replacing chemistry as the science of choice in agriculture. Biotechnology, intended here to mean direct manipulation of the genetic material in a plant cell, will contribute significantly, but traditional plant breeding will not be abandoned in the search for more benign, yet efficient, crop production systems.

It should be emphasized that success in developing desirable traits in crops is not achievable with a single timetable, even for a single species. For example, in the case of hybrid corn, varieties are generally available that are resistant to herbicides. Next—perhaps by the mid-1990s—it's anticipated that significant progress will be made in incorporating disease resistance into the corn plant. Sometime toward the end of the decade, a biological breakthough in addressing insect problems such as those associated with corn borers or root worms seems possible. Then, sometime after 2000, scientists may have significant success in altering the intrinsic value of the seed produced by the plant; that is, the starch, oil and protein content.

Progress will not come easily, even with biological applications generally reducing the need for chemicals in crop production. Society will demand that beneficial biotechnology for one plant species not have disastrous, or even undesirable effects on another. Maintaining an ecological balance of plant and animal species will be paramount. Careful, painstaking testing will be required.

Finally, another benefit of moving to more sustainable crop production deserves mention. To the extent that crop production in the future requires reduced amounts of chemical fertilizers and pesticides, agriculture will be less dependent on petroleum and natural gas, the raw materials for most fertilizers and pesticides. Higher real (inflation-adjusted) costs for fossil fuels seem likely.

Livestock Production Prospects

More than 60 percent of the gross revenue from agriculture in Nebraska comes from livestock. Of the livestock total, about 90 percent is from beef cattle. Nebraska's unique combination of water, forage and feed-grain production has allowed the beef industry to grow.

Continued growth of the beef sector would appear to rest almost entirely on adequately addressing consumer health concerns about red meat. Nebraska's cattle feeders are responding in a variety of ways, ranging from feeding animals that produce more lean meat and less fat to participating in formal quality assurance programs. The purpose of such programs is to certify that drugs have not been administered illegally to animals and feeds do not contain harmful residues.

Notwithstanding the health issue, Nebraska's beef sector should grow relative to other states. The state's water-forage-feed combination is unmatched anywhere. In addition, Nebraska's central location makes the state's beef output easily accessible to markets on the east and west coasts. The state's beef producers should also get their share of a growing export market.

Swine production should also continue to grow on both an absolute and relative basis. While it does not seem likely that Nebraska will challenge Iowa for the top position anytime soon, we may move up from our current fifth-place national ranking to second or third in the next 25 years. (Nebraska's production has gone from 4.5 percent to 7 percent of the national total in the last 30 years.) Again, our abundant feed-grain production and water make for an important competitive factor, as is the lack of pork processing facilities in states west of Nebraska.

As in the beef sector, pork producers have encountered some diet and health concerns. The industry has responded by reducing fat content in pork and coupling that accomplishment with a highly effective advertising and promotion campaign. Continued research and education will be necessary.

Two other points relative to future livestock production in Nebraska deserve brief mention.

First, perhaps more so than in the crop sector, some biotechnology products applicable to animal agriculture are not being especially well received by either producers or consumers. Somatotropin, an additive that enhances production in both dairy cattle and swine, is a notable example. It is being panned as a price-breaker by some producers and a harmful additive by some consumers. Final government approval for widespread usage has not been granted. Still, it seems unlikely that somatotropin can be hidden away, never to be heard from again, simply because some producers fear it will drive prices lower. The fact of the matter is that in a competitive market system other producers are equally determined to use any new technology that will enhance profits.

Second, the animal rights movement is growing. Questions are being raised, for example, about keeping animals penned up in lots for extended periods, especially under adverse lot conditions such as excessive mud. Almost without exception, those involved in production agriculture insist that they are at least as concerned about the welfare of their animals as are the advocacy groups. There is disagreement, however, about what "rights" an animal should have. Some observers believe that animal rights advocates will have as large an impact on the livestock sector as the environmental lobby in the next 25 years, particularly in the areas of total confinement and inhospitable conditions.

Summary

While agriculture in Nebraska will continue to evolve in response to a number of forces, an absolute decline in its importance to the state is not expected.

Production practices will be significantly influenced by environmental concerns, particularly those relating to groundwater. Product markets will be linked to the environment as well, resulting in more agricultural commodities being used for industrial uses. The proportion of the state's agricultural production going to foreign markets, especially as raw commodities, probably will decline. However, it may not be cause for great concern. The value-added export market, particularly for processed beef, is expected to grow.

Government price and income supprts will provide less of a safety net for farm prices and profits. Thus, producers will need to find new ways to manage risk. In such an environment, the reward for good management will grow and the penalty for poor management will be more severe.

Commentary:

A New Relationship Should Emerge Between Agriculture and the Rest of Society: The Future of Agriculture and Water Use in Nebraska

by Kristie Thorp

Agriculture, as the single largest consumer of water in Nebraska, needs to play a key role in protecting our water resources for future generations. We all have an obligation to leave adequate unpolluted water so that future generations have it to drink and can use it to produce food and fiber. We also have a responsibility to preserve ecosystems for healthy and diverse populations of wildlife and other animals and plants, as well as for simple human enjoyment. The problem of groundwater pollution by nitrates from agricultural sources is becoming more widespread, and the competition for water from nonagricultural users is forcing agriculture to critically evaluate its water consumption. Among the first steps towards protecting water in Nebraska is to recognize the role agriculture now plays in jeopardizing water quantity and quality and to understand what impacts our choices of agricultural technologies have on our natural heritage and on rural communities.

The second step is to put to use knowledge and technologies needed to protect it. Among the most important technological choices in Nebraska has been the drive to develop irrigated agriculture. Irrigation development has dramatically improved agricultural production, but it has also contributed to water-quality problems and has sharpened the competition for water between agricultural and other users.

In Nebraska groundwater pollution from agriculture comes largely from leaching of commercial fertilizer in intensively irrigated row-cropped areas. Efforts to decrease nitrate contamination of groundwater, the primary drinking water source for Nebraskans, by the natural resources districts (NRDs) have resulted in the designation of three groundwater control areas and two special protection areas covering 5,352 square miles. According to information from NRD staff, 14 percent of all irrigated acreage is within these intervention areas today. This will jump to 26 percent if a third special protection area covering 2,858 square miles is designated in the Upper Big Blue NRD. In a period of five years more than one-quarter of the irrigated cropland in Nebraska may come under a water-quality protection program. Despite this impressive growth in the potential acreage under protection, more needs to be done.

Nebraska's groundwater protection programs have only been implemented to respond to pollution problems in areas where nitrates are near the maximum drinking-water standard. Although these are obviously priority areas, the programs can and should be used to prevent contamination problems in areas where the potential for contamination is high but not yet near the drinking-water standard. In addition, these programs could be used to prevent groundwater contamination by pesticides. Because we do not know the health effects of pesticides when we are exposed to long-term low doses, or to synergistic effects from multiple chemicals, prevention of contamination is the logical strategy to choose. The bottom line is that we do not know if pesticides are contaminating groundwater in the same areas where nitrate problems are severe, or in other vulnerable areas, because testing for their presence has been limited. Preventing contamination is also more reasonable than treating water to remove contamination. For example, two small rural communities have built nitrate treatment plants for their drinking water supply. The cost of building one treatment plant in a community with a population of 200 was $260,000. It is fortunate for these communities that federal assistance was available for construction of these plants. Given the tight budgets at both the state and federal level, similar assistance may not be available for other communities. It is critical that NRDs and farmers work together to prevent the need for individuals and small communities to treat drinking water for nitrate and pesticide contamination in the future.

Groundwater-quality protection efforts in Nebraska are based on getting farmers to use management practices that more carefully manage fertilizers and irrigation water. These are the same practices that many farmers say they are already using. Yet at recent public hearings to consider water quality studies done in the Upper Big Blue NRD, the few farmers who made public comments argued against protection status. This could be short-sighted, not only with regard to the public interest but also with regard to the long-term interests of farmers.

It is in the farmers' best interest to actively support these efforts because their families, present and future, and neighbors, rural and urban, are drinking the water that needs protecting. The practices that reduce sources of agricultural chemical pollution also benefit the farmer when purchased inputs causing contamination are used more efficiently.

Water conservation is demonstrated by this low-pressure center-pivot in Wheeler County.

Farmers should welcome the chance to document the practices they are using to protect water quality, and they should insist that studies be done to evaluate the effectiveness of these practices in preventing contamination. Farmers, however, should be willing to do more if these practices are not doing the job. The federal government also has to do its part by reforming farm programs in ways that encourage farming systems that better protect the environment through decreased chemical and energy use and more diverse cropping systems.

Water quantity is quickly becoming as important as water quality in several regions of Nebraska. Competition for water is an age-old story, and the competition for water in Nebraska is stiff. Three NRDs have established groundwater management areas to address aquifer-depletion problems. In addition, agricultural water use is being challenged by cities that are trying to protect their water sources and by environmental interests that want to protect instream flows for wildlife needs. The vagaries of weather also limit water availability, and although Nebraska has not yet reached a crisis similar to California's recent extended drought, which resulted in irrigation releases being severely limited, there are warning signs. The expanding beach areas on Lake McConaughy and the low water levels at Harlan Reservoir should signal that we need to begin to plan for equitable alternatives.

The future may bring other players into the competition for Nebraska water. In Arizona, California and Colorado, rural agricultural water has been sold to water-short urban areas. These kinds of water transfers drain local agricultural economies to benefit developers and industry in another area. The individuals who sell their water rights realize a short-term economic gain, and some of that wealth may recirculate locally, but only if those people remain in the community. Agricultural land that loses its access to water becomes less valuable, the local tax base erodes, and opportunities for a new generation of farmers may be lost. In addition, losses in a community's tax base will affect secondary industries, government services and other community business. Though plans to sell Sand Hills groundwater to Denver and other thirsty areas are not now feasible, they could be if economic and technological conditions change. Then we too may be faced with choosing between short-term gains for individuals and long-term losses to agricultural and rural communities.

At the same time Nebraska's water resources have become contaminated and competition for water has turned more fierce, rural communities have been struggling economically and farm numbers declining. Our choices of agricultural technology emphasizing capital-intensive production and high-input systems have contributed to environmental problems but have not brought a widespread increase in economic opportunities. Farm ownership may become even more consolidated in the near future as the U.S. farming population ages. Almost one-half of the country's farms are currently held by farmers aged 55 or older, and the number of beginning farmers fell by 29 percent from 1978 to 1987. In Nebraska, nearly half of the farmland is being farmed by people over age 55. A large body of research suggests that a continuation of this trend will further degrade the social and economic viability of farm communities. In short, we need a new generation of technologies that serve both social and environmental objectives better than those currently in use.

The choices made today will determine how Nebraska's agriculture will be allowed to use water in the future. Nebraskans should think in terms of what future we want for our children and not merely in terms of the lowest cost, grocery-store benefits of a faceless food-production industry. Producers farming today in ways that conserve and improve the soil and protect water quality are farming with consideration for future generations who will want to farm and live in rural communities.

We should be critical in our choice of technologies to ensure that economic opportunities and environmental values are enhanced. Agriculture need not view competing water users as adversaries of agriculture, or worse yet, dismiss them as illegitimate participants in debates because they don't know what it means to be a farmer. Agriculture needs to listen to and respect the many new voices of people who have a stake in how we farm. They are our neighbors, consumers of our products, and in many cases, are taxpayers who support farm, conservation and rural-assistance programs. They are potential allies. In turn, they should be willing to share the cost of resource protection through funding of conservation programs. But they are not likely to assume these responsibilities if agriculture does not take their concerns and knowledge seriously. It is in the interest of agriculture and rural communities to work for a set of policies that manages water for all users, so neither agriculture nor rural communities are slighted.

Nebraska Water History Chronology

1803 — Louisiana Purchase

1804-
1806 — Lewis and Clark make an expedition through the eastern and northeastern parts of Nebraska.

1806 — Lieutenant Zebulon Pike sets out to explore the central plains. He reaches the Republican River Valley before turning south. He says the area was desert-like.

1820 — Major Steven Long begins his expedition into the "plains country." An expedition cartographer named the region "The Great American Desert."

1850s— A few irrigation ditches are dug in the state, one in Cheyenne County and one in Buffalo County, among others.

1854 — Nebraska Territory is opened to settlement.

1857 — Led by J. Sterling Morton, residents living south of the Platte River say they will secede from Nebraska Territory. They are angered by federal plans to construct a wagon road north of the Platte, which they feel will encourage settlement there, not in their part of the state.

1860 — John Burke and W.M. Hinman dig a ditch near Fort McPherson in Lincoln County. They irrigate a small garden and sell vegetables to travelers and soldiers at the fort.

1862 — President Abraham Lincoln signs the Homestead Act. It made a quarter-section of land (160 acres) available for $10 to anyone who was the head of a family, over 21, and a U.S. citizen. If not a citizen, the applicant had to state an intention to become one. The land became the property of the applicant after he/she had lived on and cultivated a crop from it for 5 years. The act brought more than 100,000 settlers to Nebraska Territory.

1863 — Daniel Freeman receives the first homestead granted by the U.S. government. The land was located near present-day Beatrice.

1867 — Nebraska becomes a state.

— The riparian rights rule of the Common Law of England is adopted regarding water rights.

1870 — A small group of North Platte residents form an irrigation ditch company. The ditch is abandoned after several years of adequate rainfall.

1873 — Nebraska Governor Robert Furnas attends an irrigation convention in Denver. During a tour he sees the progress of Colorado irrigators.

1877 — Provisions of the law relating to internal improvements are extended to canal companies, allowing them to issue bonds and condemn rights of way for canals.

1879 — William Royal Vasey settles in the Platte River valley near Cozad.

— Samuel Aughey, a natural science professor at the University of Nebraska, and C.W. Wilber, an Illinois geologist, publish a pamphlet in response to Department of Interior Geologist John Powell's claim that settlement west of the 100th meridian should be restricted. In the pamphlet the two men state that to restrict settlement is to alter the natural laws at work to increase rainfall. Wilber later coined the phrase "rain follows the plow."

1880 — A local drought hits western Nebraska.

— The state has 144 operating water-powered mills.

1886 — Drought occurs locally in western Nebraska from 1886 through 1890.

1888 — Work begins on the Akers and Ford Irrigation Ditch, the first Nebraska diversion from the North Platte River. Later it was called the Farmers' and then the Tri-State ditch.

— The first University of Nebraska extension bulletin is published. Written by Lewis Hicks, it is titled "Irrigation in Nebraska."

1889 — St. Rayner irrigation laws are passed in Nebraska. These allowed people to take water from streams by posting a notice by their diversion. The water had to be put to beneficial use.

— An estimated 9,026 acres are being irrigated in the state.

1890 — Water is first diverted into the 10-mile Farmers' Ditch.

— U.S. Census Bureau announces that the American frontier had vanished.

— First irrigation project in Dawson County begun.

1891 — *The Omaha Bee* publishes William E. Smythe's articles promoting irrigation. A severe drought in 1890 prompted the articles.

— McCook Farmers' and Business Men's Club meets and decides to hold an irrigation convention in their city. Western Nebraska counties and counties in Kansas and Colrado sent delegates. Other conventions followed in Sidney and Ogallala.

— The first State Irrigation Convention is held in Lincoln, at which irrigation-regulation legislation is proposed, drafted and sent to the legislature. No action was taken on the bill in the 1891 and 1893 sessions.

— Smythe begins to publish his journal, *The Irrigation Age*.

— The year proves to be a wet one, ends the drought of 1896-1890 and quiets the talk of irrigation promoters.

1892 — The first National Irrigation Congress is held in Salt Lake City. Smythe is instrumental in its organization.

— Erwin H. Barbour comes to Nebraska to become state geologist. He starts a state soil survey and rejects the idea that rainfall is increasing.

1893 — After two years of above-average precipitation, drought again strikes, renewing the irrigation movement of 1891. The drought lasts through 1895.

— The first meeting of the Nebraska Irrigation Association is held in North Platte. Ivan A. Fort is elected president.

— William Akers is elected to the state Senate.

— Smythe, a nationally recognized authority on irrigation, is elected president of the National Irrigation Association. He helps organize an international congress held in Los Angeles.

1894 — Fort travels more than 30,000 miles throughout the Midwest promoting irrigation development. He helps organize many local irrigation associations.

— Lexington businessmen and farmers form the Farmers and Merchants Irrigation Company. The main canal is completed in July of 1895.

1895 — Legislation passes creating a "first-in-time, first-in-right" system of appropriation and establishing the state Board of Irrigation, commonly called the "Akers Law," after William Akers. Another bill passes, allowing for the organization of local irrigation districts.

— R. H. Willis is made State Board of Irrigation Water Commissioner for Banner, Cheyenne and Scotts Bluff counties.

1896 — The first irrigation annual, the proceedings of the 1895 meeting in Sidney, are published by A. G. Wolfenbarger.

— Akers is made state engineer and secretary of the state Board of Irrigation, positions he held for only about a year.

O.V.P. Stout begins to study the state's streamflow and water supply.

— The Great Nebraska Irrigation Fair is held at North Platte.

— Barbour begins his investigation of homemade windmills in the Platte River valley.

1898?— William Vasey joines the Farmers and Merchants Irrigation Company and begins to irrigate his crops.

1899 — The State Board of Irrigation becomes the State Board of Irrigation, Highways and Drainage.

1900 — Smythe publishes *The Conquering of Arid America*.

1902 — U.S. Congress passes the Reclamation Act, establishing the U.S. Bureau of Reclamation to build dams for irrigation in the West.

— An estimated 211,890 acreas are being irrigated in the state.

1903 — Congress authorizes five reclamation projects, including the North Platte Project in Nebraska and Wyoming.

1909 — Pathfinder Dam is completed and begins to deliver water to the North Platte Project in Nebraska and Wyoming.

1910 — Willis is appointed superintendent of Water Division No. 1, which is responsible for the state's irrigable lands drained by the Platte rivers and their tributaries west of the mouth of the Loup River and includes the Republican and Blue river systems.

— The Nebraska State Irrigation Association meets in Bridgeport. Delegates call it the first annual meeting. It was probably the first meeting held after a six-year lapse.

1911 — U.S. Congress passes the Warren Act, authorizing the sale of surplus water from federal projects to non-federal lands.

1913 — C. W. McConaughy and George P. Kingsley unveil their irrigation scheme for the Holdrege-Hastings area.

1914 — Ivan D. Wood is hired as first full-time agricultural engineer by the University of Nebraska Extension Service.

1915 — Akers dies while visiting a brother in Chicago.

— Royal William Vasey, youngest of William Royal Vasey's three sons, digs one of the first irrigation wells in Dawson County.

1917 — The Nebraska State Legislature approves the formation of an irrigation school to be opened in Scotts Bluff County.

— E.E. Brackett and O.W. Sjogren write second University of Nebraska Experiment Station Bulletin, titled "Pump Irrigation in Nebraska."

1918 — George E. Johnson appoints Willis assistant state engineer in charge of all irrigation.

1919 — The State Board of Irrigation is placed within the Department of Public Works. Willis's title is changed to chief of the Bureau of Irrigation, Water Power and Drainage.

— Willis is elected secretary of the Nebraska Irrigation Association.

— The state legislature appropriates $60,000 to the irrigation school in Scotts Bluff County.

1920 — Royal William Vasey takes over the operations of the family farm and marries Hilda German, sister of Milo German.

— An estimated 442,690 acres are being irrigated in Nebraska.

1921 — The Nebraska State Irrigation School opens near Scottsbluff.

1922 — Smythe dies.

1923 — Nebraska signs the South Platte Compact, apportioning flows of the South Platte River in Colorado.

— The Nebraska State Irrigation School closes due to low enrollment.

1929 — Harry Strunk organizes a "Flood Congress" in McCook. Representatives from 14 states establish a framework for what was later called the Pick-Sloan Plan.

1930s— During the Dust Bowl years, many farmers moved to the west coast or to larger cities in the Midwest. The drought lasts from 1931-1940.

1930 — An estimated 680,000 acres are being irrigated in Nebraska.

1933 — The State Enabling Act (SF 310) passes, authorizing formation of public districts for power and irrigation.

— The Bureau of Irrigation, Water Power and Drainage is placed in the Department of Roads and Irrigation.

— The Public Works Administration approves the Sutherland project.

1934 — Milo German develops the plastic siphon tube.

1935 — Governor Cochran places Scotts Bluff County under martial law over a water dispute in the Mitchell Irrigation District, the so-called "Water War of 1935."

— The worst dust storms in the Dust Bowl hit southern and western Nebraska.

— The Republican River floods, leaving 110 dead and $8.7 million in damage, equivalent to about $100 million today.

— President Franklin D. Roosevelt approves the Tri-County project. Public Works Administration funds the project.

1936 — The Nebraska Supreme Court rules that water cannot be diverted from one watershed to another. Ruling forces Tri-County to eliminate half of its planned irrigation area, to drop two planned reservoirs and to increase electricity-generating capacity.

— Private power companies sue to prevent federal funding of public power companies and win a temporary injunction.

1937 — One out of every six farm families in Nebraska depends on public assistance.

1938 — U.S. Supreme Court rules that the federal government can help finance public power projects.

— The Sutherland project is completed.

1939 — The state approves the creation of the Consumers Public Power District. The new district, with no prior indebtedness, was able to issue revenue bonds, buy private power companies, purchase power from the three hydro districts and resell that power to its consumers.

— John Alfs, a well driller from Shickley, develops a reverse-rotary hydraulic drilling rig, which revolutionizes the well-drilling industry. Developments by Meder Smith in Alliance at about the same time also lead to a reverse rotary hydraulic drill.

1940 — The Public Works Administration refinances the Tri-County, Sutherland and Loup projects to make loan repayment easier. It also lengthened the repayment period from 30 to 64 years. As a condition, the federal government insisted that the districts work together to obtain maximum efficiency. The Loup District supplied most of the electricity in the winter while the other two stored water. In the summer, the other two supplied most of the electricity while the Loup refilled. The Nebraska Public Power system was created.

— An estimated 750,000 acres are being irrigated in Nebraska.

1942 — Milo German begins to market his siphon tube.

1944 — U.S. Congress passes a bill containing the "Pick-Sloan Plan" provisions for water development in the Missouri River Basin.

1945 — U.S. Supreme Court renders the North Platte Decree, apportioning the flow of the North Platte River between Nebraska and Wyoming. In short, it stated that Nebraska was entitled to 75 percent of the natural flow in the North Platte River and its tributaries. It also placed limits on the amount of land Wyoming and Colorado irrigators could water.

— German incorporates the siphon tube manufacturing business under the name Nebraska Plastics Incorporated, uses the trade-name "COZAD."

1947 — The state Reclamation District Law passes.

— Frank Zybach attends a sprinkler-irrigation demonstration in Colorado and tells a friend there has to be a better way.

1948 — Zybach builds the first center-pivot irrigation system.

1949 — Willis retires as secretary of the Nebraska Irrigation Association. He stays on as secretary emeritus. Dan S. Jones is elected secretary for 1950.

— Congress approves another refinancing of the Tri-County, Sutherland and Loup projects, reducing the interest rate on the original loan to 2.5 percent.

1950 — Nebraska's first development farm started in the Upper Republican River valley near Arapahoe, used to demonstrate new irrigation technology. Seven additional farms open by 1965.

An estimated 1,100,000 acres is being irrigated in the state.

1951 — Willis retires from the Bureau of Irrigation, Water Power and Drainage. He stays on as a consultant.

1952 — Zybach receives a patent for his center-pivot irrigator. He goes into business with A. E. Trowbridge.

1953 — The onset of a severe drought spurs a period of non-stop well drilling. Some 13,000 new wells are drilled before the drought ends in 1957.

1954 — Willis dies.

— Center-pivot manufacturing rights are sold to Valley Manufacturing Company of Valley (now Valmont, Inc.).

1955 — Bill Vasey, Royal's son, returns to the family farm after military service.

1956 — CNPPID constructs the Canaday steam plant to meet increasing state power demands.

— Deon Axthelm, previously an extension agent in Clay County, comes to the University of Nebraska to start an irrigation-pump efficiency and irrigation-well testing program for the extension service.

1957 — The legislature creates the state Department of Water Resources.

— The legislature makes well registration mandatory.

1958 — Stan A. Matzke becomes secretary of the Nebraska Irrigation Association.

1960 — An estimated 2,700,000 acres are being irrigated in Nebraska.

1964 — Don P. Long becomes secretary of the State Irrigation Association.

— The State Health Department funds a multi-agency investigation of nitrates in Nebraska's groundwater, called the "Nebraska Water Quality Survey." Irrigation wells in all 93 counties were sampled.

1969 — The legislature creates Nebraska's natural resources districts, effective as of July 1, 1972.

1971 — The Nebraska Soil and Water Conservation Commission completes the "Report on the Framework Study" for the State Water Plan.

— Governor Exon signs a bill creating the Nebraska Department of Environmental Control to regulate land, air and water quality.

1973 — Passage of federal Endangered Species Act.

1974 — Passage of bill creating a Nebraska Resources Development Fund to help finance water projects.

1975 — Passage of state Endangered Species Act.

— Passage of state bill allowing creation of local groundwater control areas.

— Onset of severe drought. More than 12,000 new wells are drilled before the drought ended in 1977. Center-pivots gain widespread acceptance.

1977 — Valmont Industries, Inc., introduces the first lateral-move irrigation system.

— The U.S. Bureau of Reclamation provides a $500,000 grant to the UNL Department of Agricultural Engineering and the Cooperative Extension Service to train extension agents about irrigation scheduling. The ''Irrigation Scheduling Handbook'' is printed to teach producers to use irrigation water more efficiently.

1978 — Nebraska and Wyoming agree to settle the Greyrocks Dam dispute out of court. The settlement provides funds for creating the Platte River Whooping Crane Habitat Maintenance Trust.

— Passage of state bills redirecting state water-planning efforts and directing that water-policy issues be addressed.

1980 — Nebraska Supreme Court rules that inter-basin transfers are legal.

— The U.S. Department of Energy and the Nebraska Energy Office initiate the Pump Unit Management Program (PUMP) to help producers run their irrigation pumping plants more efficiently.

— Zybach dies.

1982 — Passage of state bill allowing creation of local Groundwater Management Areas. The emphasis was on dealing directly with grounwater quality problems rather than on doing so through a management area for quantity.

1983 — Governor Bob Kerrey convenes the Nebraska Water Congress to make water-policy recommendations.

1984 — Passage of bill allowing the state Game and Parks Commission and local natural resources districts to apply for instream-flow rights and requiring natural resources districts to prepare groundwater-management plans.

— Burlington-Northern Foundation donates $1 million to a 5-year study of chemigation, the application of fertilizers and pesticides through irrigation water, by the UNL Institute of Agriculture and Natural Resources.

— Federal farm bill of 1985 imposes a variety of restrictions to discourage irrigation development on erodible soils.

1986 — Chemigation legislation passes, requiring back-flow check valves and applicator testing and certification.

— Passage of bill allowing creation of special protection areas for water quality by state Department of Environmental Control if it is so requested by any political subdivision and determined to be appropriate by DEC.

— Passage of state wellhead-protection legislation for municipalities.

— The federal Tax Reform Act of 1986 drops the investment tax credit, eliminates agriculture as a tax shelter and restructures the treatment of capital gains. It also eliminates provisions that encourage conversion of wetlands to farmland during times of agricultural surplus.

1989 — The ''Federal Manual for Identifying and Delineating Jurisdictional Wetlands'' is published jointly by the U.S. Army Corps of Engineers, the U.S. Environmental Protection Agency, the Soil Conservation Service and the U.S. Fish and Wildlife Service.

1990 — Final report for the Burlington-Northern Foundation chemigation study is published. Finds that when good management practices are followed, chemigation poses no special threat to groundwater quality.

1991 — U.S. Congress passes legislation designating a 70-mile stretch of the Niobrara River as a Wild and Scenic River.

— The National Wetlands Inventory conducted by SCS is suspended by the Bush administration because of landowner protests stating that the criteria for determining wetlands is overly inclusive. The federal manual developed in 1989 is being revised.

Epilogue

by Susan S. Seacrest
President,
Nebraska Groundwater Foundation

A single, present moment contains the infinity of the past. Water exemplifies this truth. Our philosophy and beliefs about the proper management of our precious water resources rest on a constant process of learning from both failure and success.

A final note or epilogue to a history of water is impossible to write. More appropriately, we can pause and reflect on the myriad of experiences that have created Nebraska's bountiful groundwater and surface waters as they are today. And like a farm pond stirring in a Nebraska breeze, the reflective moment creates strong yet ever changing impressions. . . .

Admiration for the courage of the pioneers who settled and tamed this land and these waters mixed with regret for the price exacted from Nebraska's Native Americans.

Awe for the visionaries who built the life-saving and life-giving dam and irrigation systems of our state mixed with a renewed appreciation for the fragile yet stunning beauty of our prairie ecosystems.

Appreciation for the economic bounty the nation's greatest supply of groundwater provides mixed with an understanding that even the most infinite water supplies must be carefully managed for future generations.

Awareness that each of us can make a difference simply by becoming better informed about our water and its importance mixed with the realization that caring for our water is an enormous task and will be very complex.

Amazement that we have come so far only to find we have so far to go. And finally:

Assurance that we will become the stewards we need to be. The water itself connects and binds us, not only to a state we love, but most importantly, to each other. This poem by Walter McDonald expresses that empowering idea:

Living on Open Plains

When it doesn't rain, we water,
pumping the purest water three
hundred feet
straight up from nothing we've ever
seen,
raising bumper crops by faith alone.
. . .
Even in snow we sit in firelight and
watch our fields
fill up, stranded these dark evenings
between towns
and miles from mountains, but
knowing wherever others
on the road a mile away are going, we
are there.

And so we pause, reflect, and go on, together.

Biographies of the Authors and Project Management Team

Ray Bentall is a co-author of the commentary piece (a letter to the Federal Energy Regulatory Commission) for the "Future Uses" chapter. He came to Nebraska to work on the Missouri Basin Development Project in 1945 for the U.S. Geological Survey (USGS). He spent 30 years with the USGS as a publications reviewer and processor and as a hydrogeologist. He then went to work for the Conservation and Survey Division of the University of Nebraska-Lincoln as an editor and researcher. He retired from there in 1982 but continues to do scholarship and research, including a comprehensive look at Nebraska's rivers to be published soon by the division. He has a bachelor's degree in biology from the University of Michigan in Ann Arbor and a master's degree in geology from the University of Tennessee in Knoxville.

Ann Salomon Bleed, author of "Climate and Hydrology," since 1988 has been the state hydrologist with the Nebraska Department of Water Resources and an adjunct professor with the University of Nebraska-Lincoln (UNL) colleges of Agriculture and Engineering. She has a bachelor's degree in biology from Earlham College in Richmond, Indiana, a master's in zoology from Pennsylvania State University in State College, Pennsylvania, a Ph.D. in ecology from the University of Wisconsin-Madison and a second master's in industrial and management systems engineering from UNL.

Bleed also has worked as an assistant professor with the UNL Department of Forestry, Fisheries and Wildlife, Conservation and Survey Division, Water Resources Center and Department of Industrial Relations.

Vince Dreeszen is a member of the project management team, co-author of certain key sections of the "Climate and Hydrology" chapter, author of an endpiece to the "Industry" chapter on water availability and use and another endpiece on the history of early test-drilling technology for the "Technology" chapter. He is also co-author of the commentary to the "Future Uses" chapter. He is a professor emeritus with and formerly director of the Conservation and Survey Division of the University of Nebraska-Lincoln. He has a bachelor's degree from Peru (Nebraska) State College and a master's degree in geology from the University of Nebraska. He joined Conservation and Survey as a hydrogeologist in 1949 and became its director in 1967, serving in that capacity until 1987.

Dreeszen also serves on the board of trustees of the Nebraska Groundwater Foundation and the Nebraska State Irrigation Association and for 20 years was a member of the Nebraska Water Conference Council, which plans and coordinates the annual Nebraska Water Conference. In addition, he has been involved in planning for the 1993 water conference, which will honor the 100th anniversary of the founding of the State Irrigation Association.

Cindy Evert, author of the "Industry" chapter, currently is employed by American Gramaphone of Omaha. She earned a bachelor's degree in journalism with concentrations in news-editorial and sociology from the University of Nebraska-Lincoln and is pursuing a master's degree in adult education.

Evert has been writing poetry, newsletter and magazine articles, brochures and print and broadcast advertising for more than 20 years. She has served as director of communication with the League of Nebraska Municipalities and deputy director of communications for Governor Kay Orr, developing and coordinating the "Capitol for a Day" and "GREAT (Governor's Rural Economic Action Team) Town Hall Meeting" programs. Before that, she directed public information for the Nebraska Department of Economic Development for five years. In addition, she has worked for two newspapers and two state associations.

Charles Flowerday is the editor of *Flat Water: A History of Nebraska and Its Water*. He is also author of an endpiece to the pre-1900 chapter that is on Native American use of water, senior author of "Conservation Management and Environmental Protection," junior author of "Environmental Quality and Changes in Ecosystems" and co-photography editor. He is the editor for the Conservation and Survey Division, Institute of Agriculture and Natural Resources (IANR), University of Nebraska-Lincoln. He has a bachelor's degree in English from Princeton University in Princeton, New Jersey, a master's in education and theology from Union Theological Seminary in New York City and a second bachelor's in journalism from UNL with a minor in general agriculture.

He has been the editor of *Human Rights Perspectives*, a quarterly newsletter of the Human Rights Office of the National Council of Churches and an educator with the Amalgamated Clothing and Textile Workers Union, educating religious leaders and lay people about labor relations in the southern textile industry. He also has been an editorial assistant with the press section of IANR Communications Services at UNL, a public affairs assistant with the Nebraska state office of the USDA Soil Conservation Service, associate editor for the *Farm and Ranch Guide* of Bismarck, North Dakota, and a technical writer/editor and management consultant trainee with Team Development Group of Minneapolis, Minnesota. He is co-editor (with Ann Bleed) of *An Atlas of the Sand Hills*, a comprehensive look at the natural history of the Nebraska Sand Hills, as well as editor of *Oil in Nebraska* and *The Groundwater Atlas of Nebraska*, among others.

Roy Frederick, author of the "Future of Agriculture" chapter, is a professor and extension economist in the Department of Agricultural Economics at the University of Nebraska-Lincoln. He specializes in public policy, with particular emphasis on farm programs, taxation, the general economy and international trade.

Born and raised in Kearney, Frederick holds bachelor's and master's degrees from the University of Nebraska and a doctorate from Purdue University.

Frederick returned to UNL in 1989, having served as director of the Nebraska Department of Agriculture for 2-1/2 years. He had prior appointments at Kansas State University in Manhattan and with the Committee on Agriculture and Forestry of the U.S. Senate.

Steve Gaul, author of the "Politics and Policy" chapter, has been coordinator of planning and review processes at the Nebraska Natural Resources Commission since 1979. As such, he coordinates state water-planning and review-process activities, coordinates the commission's efforts in environmental education, participates in various commission studies and programs and provides staff assistance on the national "Take Pride in America" awards program. From 1979 to 1986, he coordinated a series of water-policy issue studies for the commission.

Gaul holds a bachelor's degree in geography from Kansas State University in Manhattan and master's in geography from the University of Nebraska-Lincoln. He also has worked as a planner-economist for the Nebraska State Office of Planning and Programming and, before that, as a planner for the North Central Regional Planning Commission in Devils Lake, N.D.

Sally J. Herrin, senior author of "Environmental Quality and Changes in Ecosystems, junior author of "Conservation Management and Environmental Protection" and co-photography editor, is a free-lance writer and editor. She is also working as a mentor in the Lincoln Public Schools mentoring program. In February 1993, she will begin a residency for the "Artists in Residency" program of the Nebraska Arts Council. She is also a reviewer for the Nebraska Arts Council. She holds a bachelor's degree in English from Indiana University in Bloomington, a master's in child development from Louisiana State University in Baton Rouge and a master's and doctorate in English from the University of Nebraska-Lincoln.

Herrin has published more than 20 poems and four short stories, including one anthologized in *Best of New Writers* in 1989 by Writer's Digest publishers, as well as many guest editorials in the *Lincoln Star* and *Journal-Star*. She has also been an assistant professor of English at Moorhead (Minnesota) State University, an instructor and lecturer in English at UNL, an editorial assistant and reader for Prairie Schooner, the UNL literary magazine, and a curriculum coordinator, designer and writer for Southeast Community College (Lincoln, Nebraska) in the areas of science and mental health.

John Janovy, author of the "Future Uses" chapter, has also authored *Keith County Journal, Back in Keith County, Yellowlegs*, a novel, *On Becoming a Biologist* and *Vermilion Sea*, reflections on Baja California. He is Varner Distinguished Professor of Biological Sciences at the University of Nebraska-Lincoln and holds bachelor's, master's and doctoral degrees in biology from the University of Oklahoma in Norman. He has authored more than 50 scientific articles for professional journals.

J. Michael Jess, a member of the project management team, was appointed director of the Nebraska Department of Water Resources in 1981, having served as deputy director since 1975. Jess was also appointed Nebraska representative and treasurer of the Missouri Basin States Association and chairman of the Nebraska Boundary Commission in 1981. He is a registered professional engineer in Nebraska and Illinois and holds bachelor's and master's degrees in civil engineering from the University of Nebraska-Lincoln.

Jess represents Nebraska regarding issues related to the U.S. Supreme Court Decree on the North Platte River and is responsible for the annual operation of the North Platte River system. In addition, he serves as Nebraska commissioner for the Big Blue River, Republican River, South Platte River and Upper Niobrara River interstate compacts.

He is a member of the Nebraska Water Conference Council, chairman of the state Water Data Coordination Council, ex-officio member of the board of directors of the Nebraska Water Resources Association, an advisor to the Nebraska State Irrigation Association and Nebraska coordinator for federal groundwater recharge demonstration projects. He has worked with the Illinois State Water Survey, the U.S. Army Corps of Engineers and the UNL Conservation and Survey Division.

Robert D. Kuzelka, project manager of *Flat Water: A History of Nebraska and Its Water* (the Nebraska Water History Project), is assistant director of the Water Center at the University of Nebraska-Lincoln, an associate professor in the UNL Forestry, Fisheries and Wildlife Department and formerly water-resources planner and assistant to the director for the UNL Conservation and Survey Division. He holds a bachelor's in architecture from UNL and a master's in community and regional planning from the University of Texas at Austin.

Kuzelka has also worked as the comprehensive planning coordinator with the Nebraska State Office of Planning and Programming, director of physical planning with the Model City Agency in Tulsa, Oklahoma, and as a tutor of architectural design at the University of Sydney, Australia.

Devorah Lanner, editorial associate and vignette author for *Flat Water*, is a paralegal/communications specialist with the Nebraska Center for Legal Services and an editorial assistant with the National Soil Survey of the USDA Soil Conservation Service. She is also completing a journalism degree at the University of Nebraska-Lincoln.

Lanner served as public affairs director and conservation policy analyst at the Center for Rural Affairs in Walthill, Nebraska, from 1985 to 1989. She has also been a correspondent for the Center for Investigative Reporting in Washington, D.C., and an editorial assistant with Communications Services in the UNL Institute of Agriculture and Natural Resources, where she reported on the Rural Revitalization program of the Cooperative Extension Service. She has also authored or co-authored numerous articles for national magazines, professional journals and newspapers. Lanner has recently worked as an editorial staff specialist with the Conservation and Survey Division.

Robert N. Manley, consulting historian and contributing editor for *Flat Water*, is a noted Nebraska historian. He currently works as a community ambassador and historian for the Nebraska Department of Economic Development, teaching community leaders to understand their past and to use that knowledge to help prepare for the future. He holds master's and doctoral degrees in history from the University of Nebraska-Lincoln and has authored Vol. I of *A Centennial History of the University of Nebraska* and several other history books.

Previously, Manley has worked as senior historian at the Stuhr Museum in Grand Island, as a private consultant to a publisher of educational materials, as a professor at Hiram Scott College in Scottsbluff, as a professor in the UNL history department and as a high school teacher in McCook, Osceola and Seward.

Don McCabe, author of the "Farm Family" chapter, is managing editor of the *Nebraska Farmer* magazine, for which he coordinates story assignments and layouts and covers Nebraska natural-resource issues. He holds a bachelor's degree in journalism from the University of Nebraska-Lincoln.

McCabe also serves as Publicity Committee Chair for the Nebraska Ag Relations Council and as a member of the Communications Committee of the Mid-Nebraska Water Quality Project.

Before becoming the managing editor at Nebraska Farmer in 1988, McCabe worked as an associate editor with the magazine for more than 10 years. He also worked as a farm and general-assignment reporter for the *York* (Nebraska) *News-Times*.

Brad Rundquist, editorial associate for the *Flat Water* project, is presently weekend and copy editor with the *Ft. Dodge (Iowa) Messenger.* He holds a bachelor's degree in journalism from the University of Nebraska-Lincoln. For *Flat Water*, he researched and wrote many vignettes and profiles, did other research for selected authors and coordinated chapter progress, deadlines and reviews.

Rundquist has worked as an editorial assistant with the UNL Conservation and Survey Division.

Susan Seacrest, a member of the project management team and author of the epilogue to Flat Water, is founder and president of the Nebraska Groundwater Foundation. As such, she oversees the educational programs of the foundation, which are designed to teach the general public about the conservation and management of groundwater. She holds a bachelor's degree from Saint Olaf College in Northfield, Minnesota, and a master's degree in education from the University of Rochester in Rochester, New York.

Seacrest also serves on the National Drinking Water Advisory Council, the U.S. Environmental Protection Agency Management Advisory Group, the board of trustees of the National Arbor Day Foundation, the advisory board of the University of Nebraska-Lincoln Conservation and Survey Division, the steering committee of the American Clean Water Foundation and as chair of the National Groundwater Education Consortium.

Steve Schafer, author of the "Economics and Finance" chapter, is senior budget management analyst for the Nebraska Department of Administrative Services. He is responsible for monitoring expenditures of and conducting budget reviews on the state departments of Agriculture, Water Resources, Environmental Quality and Economic Development, as well as the state Natural Resources Commission, the Game and Parks Commission, the Indian Commission and the State Board of Agriculture. Schafer is also responsible for monitoring legislation and analyzing the fiscal impact on assigned agencies. He holds a bachelor's degree in political science from the University of Nebraska-Lincoln and a master's of public affairs from the Woodrow Wilson School of Public and International Affairs at Princeton University in Princeton, New Jersey.

Les Sheffield, author of the "Technology" chapter and a member of the project management team, is a professor emeritus of agricultural economics at the University of Nebraska-Lincoln. He holds a bachelor's in agronomy and a master's and doctorate in agricultural economics from the University of Nebraska-Lincoln. He worked as an irrigation economics specialist with the UNL Cooperative Extension Service from 1971-1992.

Sheffield, a native Nebraskan, wrote the first graduate thesis on center-pivot irrigation in 1971 and organized the first annual Nebraska Irrigation Tour in 1974. He has been involved with planning those tours every year since. He also has worked as a specialist with the UNL Water Center, vice president of the University of Nebraska Foundation, assistant to the vice chancellor of the UNL Institute of Agriculture and Natural Resources, superintendent of the university's West-Central Research and Extension Center in North Platte, assistant to the dean of the College of Agriculture, executive secretary of Great Plains Wheat, Inc., in Garden City, Kansas, chief of the Nebraska Wheat Commission, executive secretary of the Nebraska Grain Improvement Association and as an extension agent in Dawson County. He currently serves as secretary of the Nebraska Water Conference Council, which has sponsored the annual Nebraska Water Conference every year since 1972.

Kathleen Sullivan, author of the "Extension and Education" chapter, is a free-lance communications specialist and writer working on issues related to agriculture, families and community development in Nebraska. She received her bachelor's degree from the University of Nebraska-Lincoln in home economics journalism in 1971 and holds a master's degree in home economics from UNL.

Sullivan is a native Nebraskan who grew up on a farm in Valley County. For a number of years, she worked for the UNL Cooperative Extension Service, first as a home economics communications specialist in the Department of Agricultural Communications (now IANR Communications Services) and later as associate state leader for extension home economics. In 1978, she and her husband moved to "greater Nebraska," where they became involved in rural banking, first in Mason City and then in Cedar Rapids.

Kristie Thorp, author of the commentary to the "Future of Agriculture" chapter, has been the conservation and environmental project leader for the Center for Rural Affairs since 1988. The center is a nonprofit research and advocacy organization for family farms, rural communities and the environment. She has co-authored several publications, including "Unsettled Ground: An Assessment of the 1985 Soil Conservation Provisions and Future Policy Recommendations" and the "Sustainable Options Guide to Farm and Conservation Programs in the 1990 Farm Bill." She also has developed an adult participatory workshop curriculum, "Groundwater and Agricultural Chemicals." She has a master's degree in resource management administration from Antioch College of Keene, New Hampshire.

Perry Wigley, a member of the project management team, is the director of the University of Nebraska-Lincoln Conservation and Survey Division and state geologist. He holds a bachelor's degree from Birmingham-Southern in Alabama and a master's and doctorate from Virginia Polytechnic Institute in Richmond.

Wigley has been exploration manager of onshore North America for Texas Eastern Exploration Co., a petroleum producing company in Houston, Texas, and program manager and assistant director of the Georgia Geological Survey in Atlanta, where his projects included the environmental geology of barrier islands, hazardous waste disposal, surface-mine reclamation and groundwater protection. He also has been a professor of geology at Eastern Kentucky University in Richmond, a visiting professor of geology at the University of Waterloo in Waterloo, Ontario, and has headed up a consulting company that provided expertise in engineering, coal and petroleum geology.

L. Kent Wolgamott, author of the "Future Control" chapter, is formerly the energy and natural resources writer for the *Lincoln Journal*. He is currently its entertainment writer, doing film and concert reviews, as well as entertainment news and features. He has also been the transportation writer and has done general assignment and sports reporting for that paper. He holds a bachelor's degree in journalism from the University of Nebraska-Lincoln.

Wolgamott also has been a legislative aide for the late Senator Edward Zorinsky, with responsibility for energy and environmental legislation, and has covered politics for the *Grand Island* (Nebraska) *Independent*.

Index

NOTES